AF869585

Smart Innovation, Systems and Technologies

Volume 239

Series Editors

Robert J. Howlett, Bournemouth University and KES International, Shoreham-by-sea, UK

Lakhmi C. Jain, KES International, Shoreham-by-Sea, UK

The Smart Innovation, Systems and Technologies book series encompasses the topics of knowledge, intelligence, innovation and sustainability. The aim of the series is to make available a platform for the publication of books on all aspects of single and multi-disciplinary research on these themes in order to make the latest results available in a readily-accessible form. Volumes on interdisciplinary research combining two or more of these areas is particularly sought.

The series covers systems and paradigms that employ knowledge and intelligence in a broad sense. Its scope is systems having embedded knowledge and intelligence, which may be applied to the solution of world problems in industry, the environment and the community. It also focusses on the knowledge-transfer methodologies and innovation strategies employed to make this happen effectively. The combination of intelligent systems tools and a broad range of applications introduces a need for a synergy of disciplines from science, technology, business and the humanities. The series will include conference proceedings, edited collections, monographs, handbooks, reference books, and other relevant types of book in areas of science and technology where smart systems and technologies can offer innovative solutions.

High quality content is an essential feature for all book proposals accepted for the series. It is expected that editors of all accepted volumes will ensure that contributions are subjected to an appropriate level of reviewing process and adhere to KES quality principles.

Indexed by SCOPUS, EI Compendex, INSPEC, WTI Frankfurt eG, zbMATH, Japanese Science and Technology Agency (JST), SCImago, DBLP.

All books published in the series are submitted for consideration in Web of Science.

More information about this series at http://www.springer.com/series/8767

Veeredhi Vasudeva Rao · Adepu Kumaraswamy ·
Sahil Kalra · Ambuj Saxena
Editors

Computational and Experimental Methods in Mechanical Engineering

Proceedings of ICCEMME 2021

Springer

Editors
Veeredhi Vasudeva Rao
Department of Mechanical & Industrial Engineering
College of Science, Engineering and Technology University of South Africa
Pretoria, South Africa

Sahil Kalra
Department of Mechanical Engineering
Indian Institute of Technology Jammu
Jammu, India

Adepu Kumaraswamy
Department of Mechanical Engineering
Defence Institute of Advanced Technology
Pune, India

Ambuj Saxena
Department of Mechanical Engineering
G. L. Bajaj Institute of Technology & Management
Greater Noida, India

ISSN 2190-3018 ISSN 2190-3026 (electronic)
Smart Innovation, Systems and Technologies
ISBN 978-981-16-2856-6 ISBN 978-981-16-2857-3 (eBook)
https://doi.org/10.1007/978-981-16-2857-3

© The Editor(s) (if applicable) and The Author(s), under exclusive license to Springer Nature Singapore Pte Ltd. 2022
This work is subject to copyright. All rights are solely and exclusively licensed by the Publisher, whether the whole or part of the material is concerned, specifically the rights of translation, reprinting, reuse of illustrations, recitation, broadcasting, reproduction on microfilms or in any other physical way, and transmission or information storage and retrieval, electronic adaptation, computer software, or by similar or dissimilar methodology now known or hereafter developed.
The use of general descriptive names, registered names, trademarks, service marks, etc. in this publication does not imply, even in the absence of a specific statement, that such names are exempt from the relevant protective laws and regulations and therefore free for general use.
The publisher, the authors and the editors are safe to assume that the advice and information in this book are believed to be true and accurate at the date of publication. Neither the publisher nor the authors or the editors give a warranty, expressed or implied, with respect to the material contained herein or for any errors or omissions that may have been made. The publisher remains neutral with regard to jurisdictional claims in published maps and institutional affiliations.

This Springer imprint is published by the registered company Springer Nature Singapore Pte Ltd.
The registered company address is: 152 Beach Road, #21-01/04 Gateway East, Singapore 189721, Singapore

Preface

International Conference on Computational and Experimental Methods in Mechanical Engineering (**ICCEMME-2021**) has been the third conference of its series organized by the Department of Mechanical Engineering of G. L. Bajaj Institute of Technology and Management, Greater Noida, Uttar Pradesh, India. The institute is located in the vicinity of industrial hub. Therefore, it was decided to provide a forum to bring together scientists, speakers from industries, university professors, graduate students and mechanical engineers, presenting new research in science, technology, and engineering.

The motive of this conference was to provide an opportunity to share their innovative ideas in the form of a paper presentation. The conference attracted many participants working in various fields of engineering. Research articles were based on Design, Mechanics, Materials, Energy, Industrial & Production Engineering areas such as Sustainable Manufacturing Systems, Rapid Prototyping, Manufacturing Process Optimization, Machining, and Machine Tools, Casting, Welding, Forming, Machining, Machine Tools, Computer-Aided Engineering, Manufacturing, Management, Automation and Metrology, Industrial, Management and Marketing, etc. During the conference, about ten delegates were joined from various contraries and delivered keynote lecture on the theme of the conference. All papers were critically reviewed by two reviewers from National/International authors.

More than 500 hundred manuscripts were submitted to the conference, topics ranging from the latest research in the field of aerodynamics and fluid mechanics, artificial intelligence, IOT, blockchain rapid manufacturing technologies, and prototyping, remanufacturing, refrigeration and air conditioning, renewable energies technology, IC engines, turbo machinery metrology, and computer-aided design and inspection.

Furthermore, we would like to extend our appreciation to all the authors for contributing their valuable research to the conference. The committee is also grateful to all the reviewers who spared their time to carefully review all the assigned research articles and to all the committee members for their great effort in making their conference into grant success.

We are thankful to all sponsored agencies who gave us their cooperation and funding support.

We are thankful to our management and the director of G L Bajaj Institute of Technology and Management, Greater Noida, Uttar Pradesh, India, for their continuous source of inspiration and valuable support. We are thankful to all the members of the organizing committee for their contribution in organizing the conference. Last but not least, we thank Springer for its professional assistance, particularly, Mr. Aninda Bose who supported this publication.

Pretoria, South Africa — Veeredhi Vasudeva Rao
Pune, India — Adepu Kumaraswamy
Jammu, India — Sahil Kalra
Greater Noida, India — Ambuj Saxena

Contents

Editors and Contributors

About the Editors

Veeredhi Vasudeva Rao holds Bachelor's Degree in Mechanical Engineering and Master's Degree with specialization in Heat Transfer from Andhra University, India. He holds a Doctoral Degree from the Indian Institute of Science (IISc) Bangalore, India, with specialization in Heat Transfer from faculty of engineering. During his studies at master's and doctoral level, he was a recipient of (GATE) National Scholarship and Council of Scientific and Industrial Research (CSIR) fellowships, respectively. He was Postdoctoral Fellow at Nanyang Technological University, Singapore. Formerly, he was Principal of SreeNidhi Institute of Science and Technology, Hyderabad, India. He was also Director of Technology Development and Test center (TDTC) recognized by Government of India. He has published more than 100 research papers in peer-reviewed National and International Journals of repute and peer-reviewed conference proceedings. He has guided 4 Doctoral Research Scholars and 25 master's students for successful completion. He has presented several research papers in international conference in India and the USA.

Dr. Adepu Kumaraswamy graduated in Mechanical Engineering from Kakatiya University, Warangal, Telangana, India, in 1992 and received Master's in Design and Production Engineering from National Institute of Technology (NIT), Warangal, in 1995. Subsequently, he has pursued doctoral research work in the area of static and dynamic indentation behavior of materials for defence applications at Osmania University, Hyderabad, in 2008. He has authored 36 publications in reputed journals and good number of publications in various conferences. He was Principal Investigator of three sponsored R&D projects in the area of tribological studies of hydraulic seals, impact mechanics and high strain rate deformation behavior of materials for defence applications. He has 13 years of post-M.Tech. and 12 years of post-Ph.D. teaching experience. He was Professor and Associate Dean (R&D) at SNIST (autonomous institution), Hyderabad, during 2008–2010. He is currently Professor in Mechanical Engineering, Head, Department of Technology Management, and Head, Materials Management Group, Defence Institute of Advanced Technology (DU),

Pune, funded by Department of Defence R&D, Ministry of Defence, Government of India. He has guided four Ph.Ds, one MS (R) and over fifty M.Tech. projects and currently guiding six Ph.D. students.

Sahil Kalra was born in Kaithal, a small city in Haryana. He completed his B.Tech. in Mechanical Engineering from Kurukshetra University in 2011. During his B.Tech. degree, he received a full tuition fee waiver from the university. In his first attempt, he qualified GATE exam with 96 percentile and joined M.Tech. in the Department of Mechanical Engineering, NIT Jalandhar. Based on the performance, he was given an opportunity to serve as Assistant Professor in the same institute. However, after serving for one semester, he moved to IIT Kanpur as a full-time Ph.D. student. There he worked on a research project sponsored by Space Application Centre, Ahmedabad, Indian Space Research and Organization (ISRO). He worked jointly with the ISRO scientists and developed a reconfigurable antenna for space applications. He defended his Ph.D. degree in May 2, 2019. From his Ph.D. research, he has published four reputed International Journal articles, two patents and six conferences, and symposiums. He is recipient of two international and two national awards at different conferences and symposiums. He has been offered a prestigious fellowship, namely Faculty in Science and Engineering, at the University of Groningen (QS ranking 65) in the Netherlands (Europe). Currently, he is serving as Assistant Professor in the Department of Mechanical Engineering, IIT Jammu. His research interests are system dynamics and control, mechatronics, robotics, vibration control, modal analysis, smart materials and smart structures. He is passionate about the algorithm development for different finite element method-based engineering problems and its solution using using Abaqus and MATLAB.

Dr. Ambuj Saxena completed his bachelor degree in Mechanical Engineering Department from Uttar Pradesh Technical University, Lucknow (Uttar Pradesh), India, in 2008. In his first attempt, he qualified GATE 2008 exam with 85.18 percentile and completed Master's degree in Mechanical Engineering from Dr. B R Ambedker National Institute of Technology (NIT), Jalandhar, in 2011. Further, he was completed his doctoral degree (Ph.D.) in the area of constitutive modelling and high strain rate deformation behavior of armor steel weldments for defence applications at Defence Institute of Advanced Technology (DIAT-DU), Pune, in 2019. He has completed his doctoral work research experimentation at Defence Metallurgical Research Laboratory, Hyderabad. He has authored 21 SCI/SCIE and 5 Scopus publications in reputed journals. He has published 3 research papers at the reputed international conference. He is the reviewer of several international journals. He is currently working as an Associate Professor in the Department of Mechanical Engineering, G. L. Bajaj Institute of Technology and Management, Greater Noida. He is a lifetime member of the Indian Structural Integrity Society (InSIS).

Contributors

Amardeep Department of Mechanical Engineering, GL Bajaj Institute of Technology and Management, Greater Noida, India

Aatif Ameer G.L. Bajaj Institute of Technology and Management, Greater Noida, UP, India

M. A. Ansari Department of Electrical Engineering, Gautam Buddha University, Greater Noida, India

N. A. Ansari Department of Mechanical Engineering, Delhi Technological University, Delhi, India

Jasmine Awatramani Amity University, Noida, Uttar Pradesh, India

Navneet Singh Baghel Department of Mechanical Engineering, G.L Bajaj Institute of Technology and Management Greater Noida, Greater Noida, India

Harshit Bahri Department of Mechanical Engineering, G. L. Bajaj Institute of Technology and Management, Greater Noida, UP, India

Ashish Bansal Department of Mechanical Engineering, G.L Bajaj Institute of Technology and Management Greater Noida, Greater Noida, India

Chetan Bansal Department of Mechanical Engineering, G L Bajaj Institute of Technology and Management, Greater Noida, India

Mohit Bansal G. L. Bajaj Institute of Technology & Management, Greater Noida, U.P., India

Aditya Kumar Bhati Department of Mechanical Engineering, G.L. Bajaj Institute of Technology & Management, Greater Noida, India

Siddhant Bhatia Department of Mechanical Engineering, GL Bajaj Institute of Technology and Management, Greater Noida, India

Rishabh Chaturvedi IET Department of Mechanical Engineering, GLA University, Mathura, Uttar Pradesh, India

Pankaj Kumar Chauhan Department of Mechanical Engineering, Jamia Millia Islamia, New Delhi, India

Sandeep Chauhan Department of Mechanical Engineering, G. L. Bajaj Institute of Technology & Management, Greater Noida, India

Shailendra Singh Chauhan Department of Mechanical Engineering, G.L. Bajaj Institute of Technology & Management, Greater Noida, India

Vinit Chauhan Department of Mechanical Engineering, The NorthCap University, Gurugram, Haryana, India

Yash Chauhan Department of Mechanical Engineering, G.L. Bajaj Institute of Technology & Management, Greater Noida, India

Ankit Chhonkar Department of Mechanical Engineering, G.L Bajaj Institute of Technology and Management Greater Noida, Greater Noida, India

Shilpa Choudhary Department of Electronics and Communication Engineering, GL Bajaj Institute of Technology and Management, Greater Noida, UP, India

Priyanka Datta G. L. Bajaj Institute of Technology & Management, Greater Noida, U.P., India

Deepti Dohare GL Bajaj Institute of Technology and Management, Greater Noida, India

Shama Farozan Aryabhatta Center for Nanoscience and Nanotechnology, Aryabhatta Knowledge University, Patna, Bihar, India

Sonal Gandhi G L Bajaj Institute of Technology and Management, Greater Noida, India

Vinamra Kumar Govil EE Department, I.E.T Lucknow, Lucknow, India

Anuj Gupta Department of Mechanical Engineering, Vishveshwarya Group of Institutions, Dadri, UP, India

Deepa Gupta Department of Management Studies, G.L. Bajaj Institute of Management & Research, A.P.J. Abdul Kalam Technical University, Gautam Buddha Nagar, India

Shorya Gupta Department of Mechanical Engineering, Amity University, Noida, Uttar Pradesh, India

Shradha Gupta Department of Applied Sciences, GL Bajaj Institute of Technology and Management, Greater Noida, Uttar Pradesh, India

Tarun Kumar Gupta Department of Mechanical Engineering, G. L. Bajaj Institute of Technology & Management, Greater Noida, India

Siddharth Singh Gurjar Department of Mechanical Engineering, G.L. Bajaj Institute of Technology & Management, Greater Noida, India

Mohammad Hanief Mechanical Engineering Department, National Institute of Technology, Srinagar, India

Nitasha Hasteer Amity University, Noida, Uttar Pradesh, India

Anas Islam IET Department of Mechanical Engineering, GLA University, Mathura, India

Shruti Jain Department of ECE, Jaypee University of Information Technology, Solan, HP, India

Mayank Shekhar Jha Centre de Recherche en Automatique de Nancy, Université de Lorraine, Nancy, France

Shivani Joshi G L Bajaj Institute of Technology and Management, Greater Noida, India

Sahil Kalra Indian Institute of Technology Jammu, Jammu & Kashmir, India

Shivendra Kaura G. L. Bajaj Institute of Technology & Management, Greater Noida, U.P., India

Sabah Khan Department of Mechanical Engineering, Jamia Millia Islamia, New Delhi, India

Aditya Kumar G.L. Bajaj Institute of Technology and Management, Greater Noida, India

Deepak Kumar Indian Institute of Technology Jammu, Jammu & Kashmir, India

Harshit Kumar Department of Mechanical Engineering, Dronacharya Group of Institutions, Greater Noida, India

Ishu Kumar Department of Mechanical Engineering, GL Bajaj Institute of Technology and Management, Greater Noida, India

Kuldeep Kumar GL Bajaj Institute of Technology and Management, Greater Noida, U.P., India

Manoj Kumar Greater Noida Institute of Technology, Noida, Uttar Pradesh, India

Nagendra Kumar G. L. Bajaj Institute of Technology & Management, Greater Noida, UP, India

Pradeep Kumar G. L. Bajaj Institute of Technology & Management, Greater Noida, UP, India

Rajiv Kumar G L Bajaj Institute of Technology and Management, Greater Noida, India

Rishabh Kumar Department of Mechanical Engineering, G. L. Bajaj Institute of Technology and Management, Greater Noida, UP, India

Singh Sonu Kumar Aryabhatta Center for Nanoscience and Nanotechnology, Aryabhatta Knowledge University, Patna, Bihar, India

Rohit Singh Lather Department of Mechanical Engineering, The NorthCap University, Gurugram, Haryana, India

Ajay Mahendru Department of Mechanical Engineering, Inderprastha Engineering College, Ghaziabad, India

Pranav Malhotra Department of Mechanical Engineering, Amity University, Noida, Uttar Pradesh, India

Aditya Mathur Department of Mechanical Engineering, G L Bajaj Institute of Technology and Management, Greater Noida, India

Akanksha Mathur Department of Mechanical Engineering, The NorthCap University, Gurugram, Haryana, India

Bajrangi Maurya G.L. Bajaj Institute of Technology and Management, Greater Noida, India

Kuwar Mausam IET Department of Mechanical Engineering, GLA University, Mathura, India

Pankaj Kumar Mavi Department of Mechanical Engineering, G.L. Bajaj Institute of Technology & Management, Greater Noida, India

Maninder Meenu Centre for Agricultural Research and Innovation, GNDU, Amritsar, India

Rajat Mehrotra Department of Electrical & Electronics Engineering, GL Bajaj Institute of Technology & Management, Greater Noida, India

Tushar Mehta Department of Mechanical Engineering, The NorthCap University, Gurugram, Haryana, India

Puneet Kumar Mishra ECE Department, G.L. Bajaj Institute of Technology and Management, Greater Noida, UP, India

Shivam Mishra G. L. Bajaj Institute of Technology & Management, Greater Noida, UP, India

Shivi Mittal Department of Management Studies, G.L. Bajaj Institute of Technology & Management, A.P.J. Abdul Kalam Technical University, Greater Noida, India

Khalid Nafees Department of Mechanical Engineering, Inderprastha Engineering College, Ghaziabad, India

Amit Pal Department of Mechanical Engineering, Delhi Technological University, Delhi, India

Nirbhay Singh Parmar Department of Mechanical Engineering, Indus Institute of Technology and Management, Kanpur, UP, India

Sachin Patel G.L. Bajaj Institute of Technology and Management, Greater Noida, India

Vansh Pokhriyal G.L. Bajaj Institute of Technology and Management, Greater Noida, UP, India

Ponam Bhagwant University Ajmer, Rajasthan, India

Brijesh Prasad G. L. Bajaj Institute of Technology & Management, Greater Noida, UP, India

Om Priya Aryabhatta Center for Nanoscience and Nanotechnology, Aryabhatta Knowledge University, Patna, Bihar, India

Amrita Rai ECE Department, G.L. Bajaj Institute of Technology and Management, Greater Noida, UP, India

Pooja Rani G L Bajaj Institute of Technology and Management, Greater Noida, UP, India

Ritwick Rawat G.L. Bajaj Institute of Technology and Management, Greater Noida, India

Rajesh Rohilla Delhi Technological University, Delhi, India

Manish Saraswat Department of Mechanical Engineering, ABES Engineering College, Ghaziabad, UP, India

Harendra Kumar Satyapal Aryabhatta Center for Nanoscience and Nanotechnology, Aryabhatta Knowledge University, Patna, Bihar, India

Alok Saxena G.L. Bajaj Institute of Technology and Management, Greater Noida, UP, India

Pooja Saxena G. L. Bajaj Institute of Technology and Management, Greater Noida, UP, India

S. S. Saxena Air Commodore VSM, Auto Wings Training and Consultancy Centre, Greater Noida, India

Dhruv Sethi Department of Mechanical Engineering, GL Bajaj Institute of Technology and Management, Greater Noida, India

Shantanu G.L. Bajaj Institute of Technology and Management, Greater Noida, India

Abhishek Sharma Department of Mechanical Engineering, GL Bajaj Institute of Technology and Management, Greater Noida, UP, India

Aman Sharma IET Department of Mechanical Engineering, GLA University, Mathura, India

Anil Kumar Sharma Department of Mathematics, SPC Degree College, Baghpat, Uttar Pradesh, India

Ankit Sharma Department of Mechanical Engineering, GL Bajaj Institute of Technology and Management, Greater Noida, India

Bhupendra Prakash Sharma Department of Mechanical Engineering, Amity University, Noida, Uttar Pradesh, India

Devesh Sharma Department of Mechanical Engineering, G.L Bajaj Institute of Technology and Management Greater Noida, Greater Noida, India

Himanshu Sharma Department of Mechanical Engineering, G.L Bajaj Institute of Technology and Management Greater Noida, Greater Noida, India

Kailash Sharma G. L. Bajaj Institute of Technology & Management, Greater Noida, UP, India

Mayank Sharma GL Bajaj Institute of Technology and Management, Greater Noida, U.P., India

Rahul Sharma Department of Mechanical Engineering, G.L Bajaj Institute of Technology and Management Greater Noida, Greater Noida, India

Rahul Sharma Department of Mechanical Engineering, The NorthCap University, Gurugram, Haryana, India

Sanjeev Sharma Department of Physics, GL Bajaj Institute of Technology and Management, Greater Noida, Uttar Pradesh, India;
Department of Applied Sciences, GL Bajaj Institute of Technology and Management, Greater Noida, Uttar Pradesh, India

Saurabh Sharma Aryabhatta Center for Nanoscience and Nanotechnology, Aryabhatta Knowledge University, Patna, Bihar, India

Vikas Kumar Sharma IET Department of Mechanical Engineering, GLA University, Mathura, India

Prashant Shukla Amity Institute for Advanced Research and Studies (M&D), Amity University Uttar Pradesh, Sec 125, Noida, UP, India

Vidit Shukla Department of Electronics and Communication Engineering, GL Bajaj Institute of Technology and Management, Greater Noida, UP, India

C. Shylaja ECE Department, G.L. Bajaj Institute of Technology and Management, Greater Noida, UP, India

Rahul Sindhwani Department of Mechanical Engineering, Amity University, Noida, Uttar Pradesh, India

Anil P. Singh Department of Mechanical Engineering, Inderprastha Engineering College, Ghaziabad, India

Harkirat Singh Department of Mechanical Engineering, Amity University, Noida, Uttar Pradesh, India

Harvendra Singh Department of Mechanical Engineering, G. L. Bajaj Institute of Technology and Management, Greater Noida, UP, India

Jaydeep Singh Department of Mechanical Engineering, Dronacharya Group of Institutions, Greater Noida, India

Karanjot Singh Department of Mechanical Engineering, Dronacharya Group of Institutions, Greater Noida, India

Kaushalendra Kumar Singh Department of Mechanical Engineering, G. L. Bajaj Institute of Technology and Management, Greater Noida, UP, India

Parshuram Singh Bhagwant University Ajmer, Rajasthan, India

Pradeep Kumar Singh IET Department of Mechanical Engineering, GLA University, Mathura, Uttar Pradesh, India

Prince Kumar Singh GL Bajaj Institute of Technology and Management, Greater Noida, U.P., India

Priyanka Singh Department of Mechanical Engineering, Amity University, Noida, Uttar Pradesh, India

Ranjeet Kumar Singh G.L. Bajaj Institute of Technology and Management, Greater Noida, UP, India

Ravindra Pratap Singh IET Department of Mechanical Engineering, GLA University, Mathura, India

Shailesh Kumar Singh Department of Mechanical Engineering, Dronacharya Group of Institutions, Greater Noida, India

Vimal Singh G L Bajaj Institute of Technology and Management, Greater Noida, India

Harsh Sohal Department of ECE, Jaypee University of Information Technology, Solan, HP, India

Ashish Kumar Srivastava GL Bajaj Institute of Technology and Management, Greater Noida, U.P., India

Prabhat Srivastava Shri Venkateshwara University, Gajraula, Uttar Pradesh, India

Priyank Srivastava Department of Mechanical Engineering, Amity University, Noida, Uttar Pradesh, India

Vishal Shankar Srivastava Department of Mechanical Engineering, G. L. Bajaj Institute of Technology & Management, Greater Noida, India

Mohd. Qamar Tanveer Department of Mechanical Engineering, Inderprastha Engineering College, Ghaziabad, India

Divakar Tiwari G.L. Bajaj Institute of Technology and Management, Greater Noida, UP, India

Rajesh Tiwari Department of Applied Science (Physics), ABES, Ghaziabad, Uttar Pradesh, India

Vikas Tiwari G.L. Bajaj Institute of Technology and Management, Greater Noida, UP, India

Mihir Tomar Department of Mechanical Engineering, G.L. Bajaj Institute of Technology & Management, Greater Noida, India

Ashish Tripathi Department of Information Technology, G. L. Bajaj Institute of Technology and Management, Greater Noida, Uttar Pradesh, India

Pragati Tripathi Department of Electrical Engineering, Gautam Buddha University, Greater Noida, India

Prem Chand Vashist Department of Information Technology, G. L. Bajaj Institute of Technology and Management, Greater Noida, Uttar Pradesh, India;
CST, Lucknow, Uttar Pradesh, India

Mudita Vats Department of Electronics and Communication Engineering, GL Bajaj Institute of Technology and Management, Greater Noida, UP, India

Aman Verma GL Bajaj Institute of Technology and Management, Greater Noida, U.P., India

Gaayan Verma Amity University, Noida, Uttar Pradesh, India

Vimal Department of Physics, GL Bajaj Institute of Technology and Management, Greater Noida, Uttar Pradesh, India

Himanshu Yadav GL Bajaj Institute of Technology and Management, Greater Noida, U.P., India

Jatin Yadav Department of Mechanical Engineering, GL Bajaj Institute of Technology and Management, Greater Noida, India

Omprakash Yadav Department of Mechanical Engineering, G L Bajaj Institute of Technology and Management, Greater Noida, India

Rajat Yadav IET Department of Mechanical Engineering, GLA University, Mathura, India

Ramashankar Yadav G L Bajaj Institute of Technology and Management, Greater Noida, India

Vinod Kumar Yadav Department of Mechanical Engineering, G. L. Bajaj Institute of Technology and Management, Greater Noida, UP, India

Chapter 1
Damage Detection in Eggshell Using Lamb Waves

Sahil Kalra, Maninder Meenu, and Deepak Kumar

Abstract The application of Lamb waves is gaining attention in the scientific community for diagnosing defects in the thin-walled structures. The lamb waves have also been extensively employed in the aerospace, civil, and mechanical industries for detection of damage especially in the thin-walled plate or cylindrical structures. In the same direction, we have explored the application of Lamb waves in crack detection of chicken's egg. Eggshell is primarily composed of calcium carbonate ($CaCO_3$) stabilized by a protein matrix. This protein matrix provides mechanical strength to the $CaCO_3$ layer. The eggshell damage while laying and transportation is a major problem faced by the poultry industry. Eggshell damage also results in major economic loss as the cracked eggs must be discarded while sorting and selling. The cracked eggshells have been identified manually which is a laborious and time-consuming process. Thus, for rapid and efficient detection of eggshell cracks various digital imaging-based algorithms, X-ray and gamma, optical transmission, nuclear magnetic resonance (NMR) and electrical properties-based techniques have been developed. However, the application of Lamb waves for the detection of eggshell crack has not been explored till now. This is the first study that reports the application of Lamb waves for detecting cracks in the eggshell. This technique is economical compared to other available techniques. We have carried out finite element (FE) modelling and simulations on the chicken eggshells. In order to capture the wave passing through the shell, we have calculated the minimum element size requirement for a given FE mesh and the time increments required to capture the wave in the shell. The simulation results show the accurate detection of eggshell cracks. This technique has a great potential to be employed at an industrial scale for sorting of the cracked and intact eggs.

S. Kalra (✉) · D. Kumar
Indian Institute of Technology Jammu, Jammu & Kashmir, India
e-mail: sahil.kalra@iitjammu.ac.in

M. Meenu
Centre for Agricultural Research and Innovation, GNDU, Amritsar, India

© The Author(s), under exclusive license to Springer Nature Singapore Pte Ltd. 2022
V. V. Rao et al. (eds.), *Computational and Experimental Methods in Mechanical Engineering*, Smart Innovation, Systems and Technologies 239,
https://doi.org/10.1007/978-981-16-2857-3_1

1.1 Introduction

Shell is an outer layer present in many agri-food products. The main function of the outer shell is to protect the inner food content from bacterial contamination. The food materials such as various fruits, dry fruits, seafood and eggs contain outer shells. The damage or crack in their shells while harvesting, laying eggs and transportation is a major problem that leads to bacterial contamination of these food materials which, in turn, rapidly degrade the food quality. Consumption of these poor-quality food items reported to impose several adverse effects on human health. Hence, there is an immediate demand for identifying and sorting these damaged and poor-quality food items. However, manual inspection is time consuming and laborious process. Thus, in literature, several state-of-the-art technologies and automation systems have been reported for detecting and sorting poor quality and damaged food items. Various nondestructive evaluation (NDE) techniques are used to estimate the defect in the shell structure [1–3]. The conventional NDE methods are based on optical or ultrasonic-based techniques. These techniques are the approximate methods. When the location of the defect is known these techniques may be applied directly over a region of interest. In this direction, the damage detection using Lamb wave for structural health monitoring is an emerging technology. It is an efficient technique for the damage detection for the thin structures.

In this paper, we have discussed the proof of concept for the application of Lamb wave to find the defect in the egg shells. Eggshell is a very delicate structure and faces many problems while laying, handling and transportation. The major problems associated with egg quality include shell defects and internal defects. These defects can be broadly categorized into three groups, namely, defects affecting yolk quality, defects affecting albumin quality and defects affecting the overall quality [4]. Egg quality defects are usually resolved easily; however, these quality and defect issues may result in economic losses if they are not addressed quickly. The shells are generally reported to have three types of cracks, namely, gross cracks, star cracks, and hairline cracks [5].

The NDE method using the Lamb wave is capable to investigate defects in a structure without prior information of the location of a defect which makes this technology superior compared to the other present state of the art. This technique also enables the rapid inspection of eggs without disassembly of the structure. Moreover, this method can also be applied to various types of defects by adjusting the frequency of the wave. The vibration approach may be used to detect such defects by comparing the dynamic behaviour of the cracked and uncracked shells. However, using the vibrational approach, it is unable to detect the defect if it is very small compared to the size of the structure. Thus, the wave propagation approach is a much more efficient method as it can detect a small and hidden defect in the structures.

1.2 Lamb Wave Analysis in Thin Shell Structures

Lamb waves were discovered by Horace Lamb and published in the book "Lamb wave theory" in 1917 [6]. The Lamb wave theory was inspired by the work of Lord Rayleigh on Rayleigh waves (1889) [7]. These waves are the guided elastic waves produced from the restoring forces which occur between particles of the material when they are displaced elastically. On excitation, the waves influence the characteristic properties of the material by changing the stress and velocity inside the material. These waves travel on the surface of the material or through the thickness of the material if the thickness is very less compared to the length and width of the material. They remain guided between two parallel surfaces. Lamb waves are plain strain waves occurring in free plate-like structures, ultimately removing the traction forces effect on the face of the structure. Lamb waves produce stress in the material on excitation, which is distributed throughout the material and over the entire thickness. The defect in the material disturbs the wave profile that, in turn, alters the stress distribution and affects the impulse signal, which is sensed by the sensor and can be visualized from the sensor response [8–12].

Considering the thickness d, Lamb waves follow the following characteristic equations:

$$\frac{\tanh\left({}^{\beta d}/_{2}\right)}{\tanh\left({}^{\alpha d}/_{2}\right)} = \frac{4\alpha\beta k^2}{\left(k^2+\beta^2\right)^2} \text{ for symmetric mode} \tag{1.1}$$

$$\frac{\tanh\left({}^{\beta d}/_{2}\right)}{\tanh\left({}^{\alpha d}/_{2}\right)} = \frac{\left(k^2+\beta^2\right)^2}{4\alpha\beta k^2} \text{ for asymmetric mode} \tag{1.2}$$

$$\text{where, } \alpha^2 = k^2 - \frac{\omega^2}{c_l^2} \text{ and } \beta^2 = k^2 - \frac{\omega^2}{c_t^2}$$

where ω is the angular frequency, k is the wave number, c_l is the longitudinal wave velocity and c_t is the transverse wave velocity.

1.3 Finite Element Analysis of Eggshell Using Lamb Waves

We have chicken's egg as a sample material, and the eggshell was a thin surface. The objective of this study was to detect any crack on the eggshell. We have modelled an eggshell for the crack detection, keeping its shell as the original target of application. As an important part of this study, a considerable amount of work has been done by performing an FE analysis of a sample egg by using appropriate dimensions mentioned in the literature. The work includes the establishment of a wave source and wave propagation from this actuation point to a sensor point both in case of an intact and cracked egg sample. The egg is assumed to have a symmetricity along its volumetric central point, and the shell is considered to be a curved plate.

The investigation is carried out using Abaqus modelling and simulation software [13]. Abaqus uses the finite element method to simulate the model in which materials' elastic modulus is one of the important parameters. In case of eggshell, the elastic modulus has been measured by several authors, both for static [14–18] as well as dynamic conditions [19, 20]. The prior knowledge of these crucial mechanical properties of biological materials would help to create a precise model and hence the improved quality of results. For the present study, the geometry of hen's egg was retrieved from the literature of Severa et al. [20] and the concept of applying mathematical modelling in food is inspired by Meenu et al. [21]. The finite element analysis simulation is done on an eggshell of identical dimensions, one without defect and the other with an elliptical hole as a defect.

1.4 Material Properties and Element Size

Dimensions of Oval: Shown in Fig. 1.1.

Material Properties: Elasticity modulus = 40 GPa.

Poisson's ratio = 0.29.

Density = 8027 kg/m^3

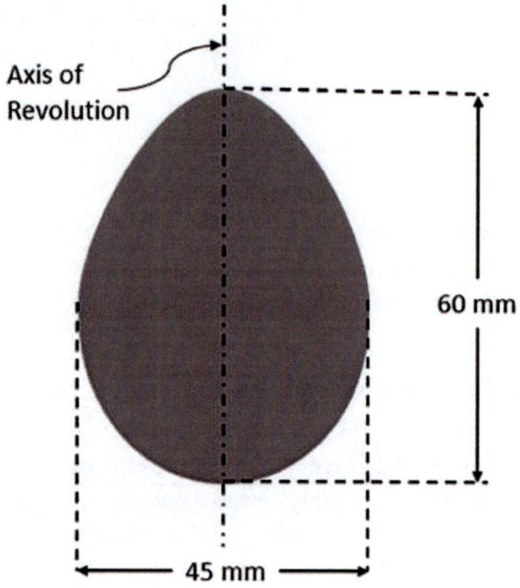

Fig. 1.1 Geometry of egg for modelling in Abaqus

$$\text{Speed of the wave, } c_L = \sqrt{E/\rho}$$

Loading: Impulse signal as a concentrated load on the exciter.
Boundary conditions: $U_1 = U_2 = U_3 = 0$ on edge surfaces of the plate
Elastic propagation speed: $c_L = 3162$ m/s.

Abaqus uses the finite element method to simulate the models in which the model is divided into finite elements, and the operation is performed on each element to get the required solution. While doing wave propagation, the element size and the increment time are important factors to be taken care of, as they influence the final solution.

Time increment size should be sufficient enough to capture the smallest natural period of interest, and element size should be taken such that it can capture the wavelength, but it should not be so small that the wave crosses one element in one increment. Element size can be determined based on the wavelength of impulse. Wavelength of impulse $\sim 2 \times c_L \times dt = 2 \times 3162 \times 10E-6 = 2 \times 6598 \times 2E-6 = 2.6E-2$ m

Hence, element size = 1E−3 m is appropriate

$$\text{Critical time increment} = \frac{\text{mesh element size}}{\text{theoritical elastic propagation speed}} = \frac{0.001}{6598} = 1.5E-7\ \text{s}$$

Hence, the maximum increment time step = 1E−7 s is found to be appropriate.

Figures 1.2 and 1.3 show different meshed models of intact and defected eggs with sensory nodes at different locations of the egg. Out of these nodes, one node is considered as actuator node which sends the input signal to the eggshell which is sensed at the sensory node as the output signal (Fig. 1.4).

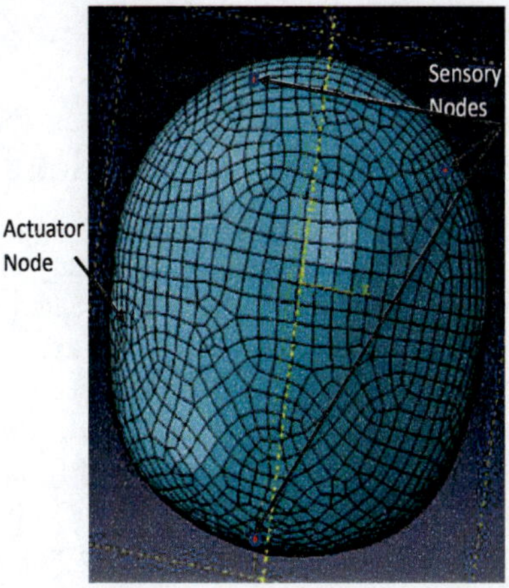

Fig. 1.2 Healthy egg with axis-symmetric problem and mesh

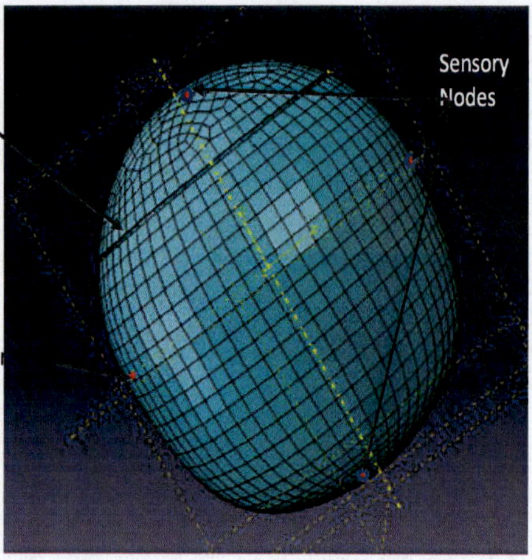

Fig. 1.3 Egg with hair line defect and axis-symmetric geometry

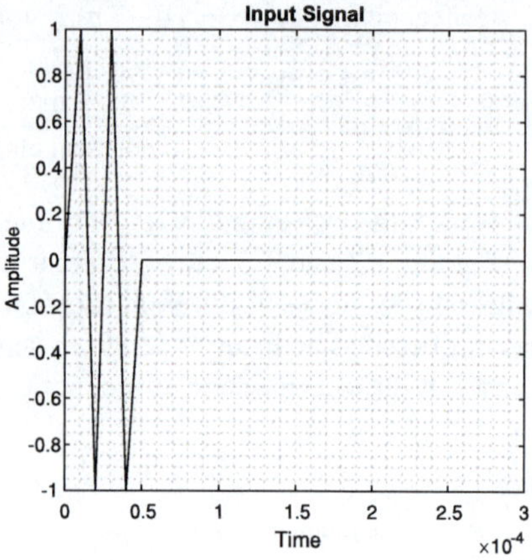

Fig. 1.4 Input signal sent to the eggshell

1.5 Post-processing of Finite Element Results

The input impulse signal, boundary conditions, along with required geometric and material parameters are implemented on the Finite Element model of the eggshell. The problem is considered as axis-symmetric case. To post-process the results, various frames of field output data are requested during simulation so that the propagation pattern can be observed. The displacement data on the top sensor node (Fig. 1.3) obtained are acquired from Abaqus and plotted in Matlab. The simulation results show that for the case of hairline crack, significant attenuation is clearly observed in the magnitude of the output signal, which is due to scattering around the defect. The difference in the propagation pattern is also clearly visible as shown in Fig. 1.5.

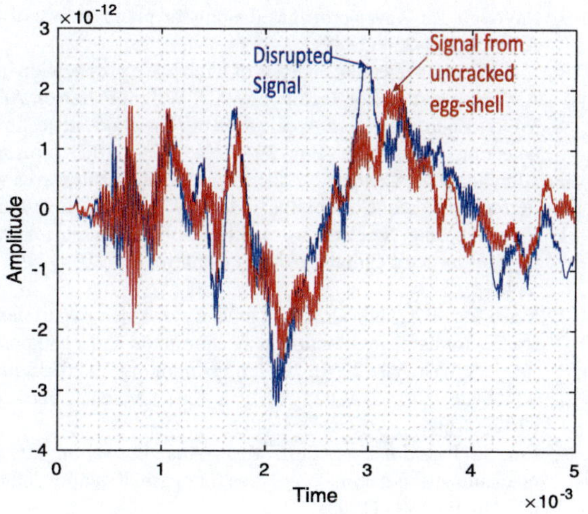

Fig. 1.5 Output signal received on one of the sensory nodes

1.6 Conclusion and Future Scope

The results from the Finite Element simulation show a significant difference in the output signal received from intact and hairline cracked egg. The simulations may be validated for different types of cracks. In future, this non-destructive technology may be validated experimentally. The results from Finite Element simulations may also be validated experimentally. In future, this technique may be implemented to sort and differentiate the defected and intact eggs on the industrial scale.

References

1. Meenu, M., Cai, Q., Xu, B.: A critical review on analytical techniques to detect adulteration of extra virgin olive oil. Trends Food Sci. Technol. **91**, 391–408 (2019). https://doi.org/10.1016/j.tifs.2019.07.045
2. Meenu, M., Kamboj, U., Sharma, A., Guha, P., Mishra, S.: Green method for determination of phenolic compounds in mung bean (*Vigna radiata* L.) based on near-infrared spectroscopy and chemometrics. Int. J. Food Sci. Technol. **51**(12), 2520–2527 (2016). doi:https://doi.org/10.1111/ijfs.13232
3. Meenu, M., Xu, B.: Application of vibrational spectroscopy for classification, authentication and quality analysis of mushroom: a concise review. Food Chem. **289**, 545–557 (2019). https://doi.org/10.1016/j.foodchem.2019.03.091
4. EPF and NZFS: Egg producers federation of New Zealand (Inc) code of practice (includes requirements for risk management programmes) (2002) http://www.nzfsa.govt.nz/consultation/egg-cop/discussion-paper/egg-cop-ch1.pdf
5. Elaroussi, M.A., et al.: Calcium homeostasis in the laying hen:1. age and dietary calcium effects. Poult. Sci. **73**(10), 1581–1589 (1994)
6. Lamb, H.: On waves in an elastic plate. Proc. R. Soc. A Math. Phys. Eng. Sci. **93**, 114–128 (1917)

7. Rayleigh, L.: Waves propagated along the plane surface of an elastic solid. Proc. Lond. Math. Soc. **20**, 225–234 (1889)
8. Han, S.: Finite element analysis of lamb wave propagation in a thin aluminium plate. MS thesis Air Force Institute of Technology AU, Wright Patterson AFB, Ohio (2007)
9. Yu, L., Zhongqing, Su.: Application of Kernel density estimation in lamb wave-based damage detection. Math. Probl. Eng. **2012**(20), 1–24 (2012). https://doi.org/10.1155/2012/406521
10. Alleyne, D.N., Cawley, P.: The interaction of lamb waves with defects. IEEE Tran. Ultrason. Ferroelectr. Freq. Control **39**(3), 381–397 (1992). https://doi.org/10.1109/58.143172
11. Wang, Q., Jing, X.: Lamb wave tomography technique for crack damage detection. In: Proceedings of the 33rd Chinese Control Conference, (2014) https://doi.org/10.1109/chicc.2014.6895446
12. Masurkar, F.A., Nitesh, P.Y. Lamb wave based experimental and finite element simulation studies for damage detection in an aluminium and a composite plate using geodesic algorithm. Int. J. Acoust. Vib. **22**(4), 2017. https://doi.org/10.20855/ijav.2017.22.4486
13. ABAQUS Inc.: ABAQUS Analysis User's Manual, Version 6.6. ABAQUS Inc., Providence, RI (2006)
14. Lin, J., Fajardo, T.A., Puri, V.M., Anantheswaran, R.C., Macneil, J.H.: Measurement of mechanical and thermal properties for eggshell quality determination. ASAE Paper, St Joseph, MI, No. 93-6503 (1993)
15. Dhanoa, P.S., Puri, V.M., Anantheswaran, R.C.: Thermal and mechanical properties of eggshell under different treatments. Trans. ASAE **39**(3), 999–1004 (1996)
16. Buchar, J., Simeonovova, J.: On the identification of the eggshell elastic properties under quasistatic compression. In: 19th CAD—FEM Users Meeting 2001, International Congress on FEM Technology, pp. 17–19. Potsdam, Berlin (2001)
17. De Ketelaere, B., Govaerts, T., Coucke, P., Dewil, E., Visscher, J., Decuypere, E., De Baerdemaker, J.: Measuring the eggshell strength of 6 different genetic strains of laying hens: techniques and comparisons. Br. Poult. Sci. **43**, 238–244 (2002)
18. Kemps, B., De Ketelaere, B., Bamelis, F., Govaerts, T., Mertens, K., Kamers, B., Tona, K., Decuypere, E., Bardemaeker, B.: Development of a methodology for the calculation of young's modulus of eggshell using vibration measurements. Biosys. Eng. **89**(2), 215–221 (2004)
19. Lin, H., Mertens, K., Kemps, B., Govaerts, T., De Ketelaere, B., De Baerdemaker, J., Decuypere, E., Buyse, J.: New approach of testing the effect of heat stress on eggshell quality: mechanical and material properties of eggshell and membrane. Br. Poult. Sci. **45**(4), 476–482 (2004)
20. Severa, L., Nedomová, Š, Buchar, J., Cupera, J.: Novel approaches in mathematical description of hen egg geometry. Int. J. Food Prop. **16**(7), 1472–1482 (2013)
21. Meenu, M., Guha, P., Mishra, S.: Coupled heat and moisture transfer of a single mung bean grain based on IR heating. Int. J. Model. Simul. Sci. Comput. **08**(02), 1740001 (2017). https://doi.org/10.1142/s1793962317400013

Chapter 2
Study of Steering System for an Electric Trike-Ackerman Steering

Sachin Patel, Ritwick Rawat, Shantanu, Aditya Kumar, and Amardeep

Abstract In small towns and cities, we need a lightweight vehicle to travel from one place to another. Every time we use a vehicle that has complex component assembly which uses a lot more amount of fuel because of the overall weight, and also because of complex steering systems, we have to put up some extra steering effort to give direction to our vehicle. Using a simple steering mechanism/system results in overall weight reduction, low manufacturing cost and low maintenance cost for the user. In this paper, the focus has been put on the simple design of the steering system based on Ackerman steering mechanism considering the factors that are related to the overall weight of the vehicle, driver's safety, easy maintenance and low cost.

2.1 Introduction

Reduction of vehicle weight is a recommended strategy to address growing concerns about emissions of greenhouse gases and fuel consumption. Generally, if we talk about a steering system, the things that come to our mind are rack and pinion arrangement/mechanism, and for steering control, we generally use universal joints and steering wheel which results in overall weight increment of vehicle. Using a simple steering mechanism/system can help us to cut off a lot of weight as compared to other complex steering systems used nowadays. Steering is needed to guide the vehicle in the desired direction. While turning, the wheel nearer to the centre of rotation are known as inner wheels, and the wheels farther from the centre are known as outer wheels. Ackerman steering is a basic concept that while turning around a corner, the inner tire must turn at a greater angle than the outside tire. But in racing, it gets more complex because not only do the wheels have to turn at a proper angle to propel the vehicle around the corner but also they must consider the proper slip angles required for the maximum tire lateral force. Without understanding these principles,

S. Patel · R. Rawat · Shantanu · A. Kumar · Amardeep (✉)
G.L. Bajaj Institute of Technology and Management, Greater Noida, India
e-mail: amardeep.college@gmail.com

© The Author(s), under exclusive license to Springer Nature Singapore Pte Ltd. 2022

V. V. Rao et al. (eds.), *Computational and Experimental Methods in Mechanical Engineering*, Smart Innovation, Systems and Technologies 239,
https://doi.org/10.1007/978-981-16-2857-3_3

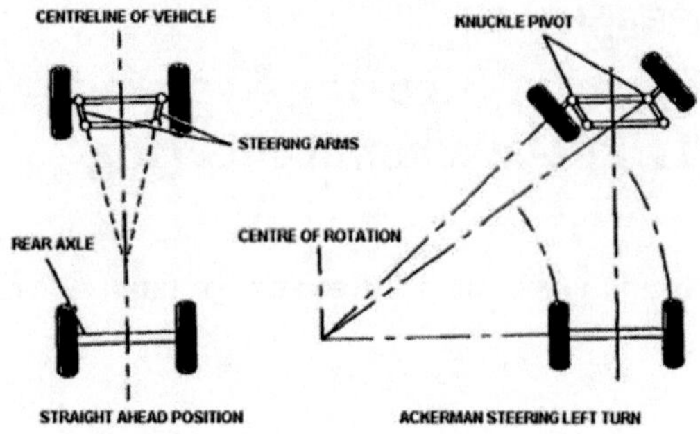

Fig. 2.1 The Ackerman turn centre [1]

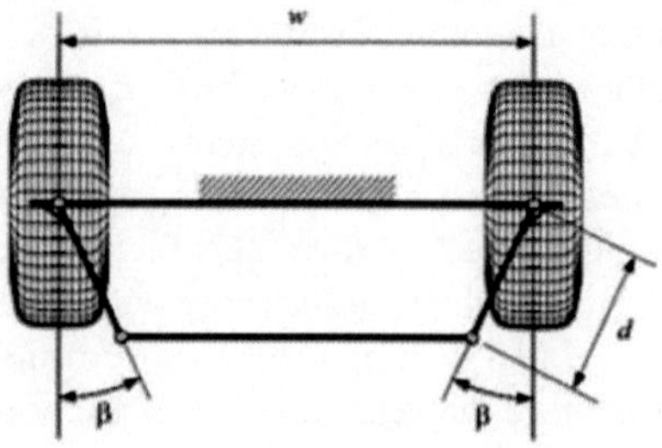

Fig. 2.2 Ackerman centre and centre of rotation [2]

the vehicle would not be able to maintain traction and speed while taking a turn on the corner which is shown in Fig. 2.1 [1] (Fig. 2.2).

2.2 Steering System and Mechanisms

The steering system starts with a steering wheel or a handle to give input. Driver's input from the steering wheel is passed through a shaft followed by a system of gear reduction, frequency rack and pinion, ball bearings or multiplication of ball bearings. It pulls the steering gear to the unruly wheels to produce movement in the direction of the direction. The steer of each wheels is controlled by the steering arm. The steering arm is attached to a wheelbarrow that controls the key, key lock, and harp. For some cars, it is an integral part of a one-piece harp and a driving punch. To achieve optimal mobility, a minimum operating length of 35 must be given on the front wheels of the passenger cars [3]. Two traditional mechanisms that are used till date are Rack and Pinion and Recirculating ball steering.

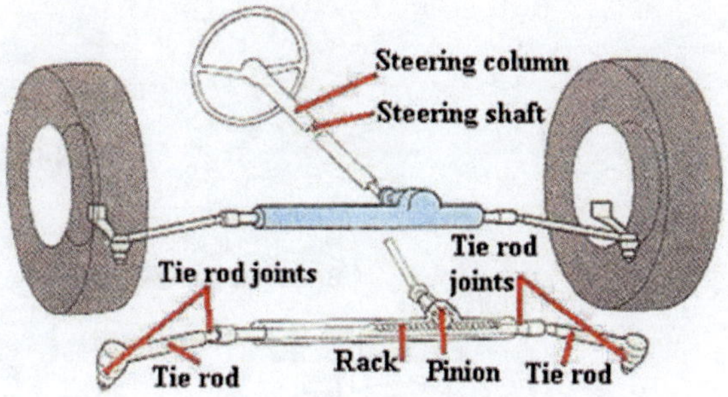

Fig. 2.3 Rack and pinion steering system [5]

2.3 Rack and Pinion

Rack and pinion steering as shown in Fig. 2.3 is the most common type of movement control systems in cars, small trucks and SUVs. The rack and pinion machine is when we open the steering wheel; the gear will change, moving rack. The tie rod connects to the steering arm, which is attached to the spinning rope. The purpose of Rack and Pinion gear is to rotate the steering wheel in a straight line. It allows gear reduction, which results in the easier wheel turning [4].

2.4 Variable Ratio Steering

Rack subtype and pinion gear steering is a variable ratio steering. This method of alignment has a different value for the height of the teeth in the middle than at the edges. This makes the steering wheel less sensitive when the steering wheel is close to the center. And when it turns on the building, the wheels are very sensitive to the circular motion of the steering wheel [4].

2.5 Re-circulating Ball/Steering Box

Re-circulating Ball Steering is shown in Fig. 2.4 and is the most commonly used steering system in heavy automobiles [6]. It runs on Parallelogram linkage, in which

- The Pitman and Idler arm remains parallel.
- The mechanism absorbs heavy shock loads and vibrations.

The mechanism of the Re-circulating ball is

(i) When the steering wheel is rotated, the rod turns.
(ii) When the wheel spins, the block moves.
(iii) The block moves another gear that in turn moves the Pitman arm.

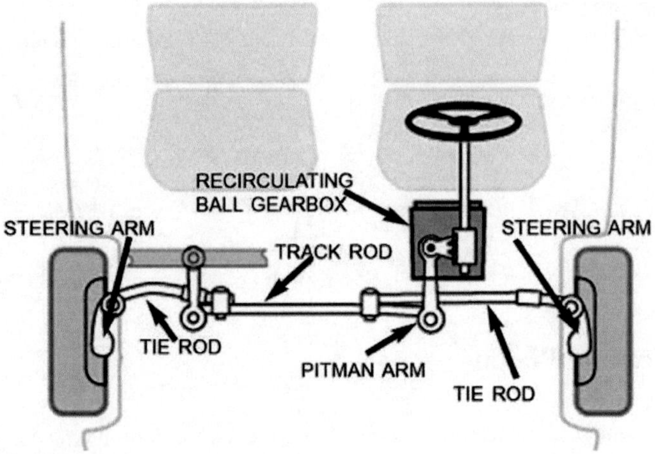

Fig. 2.4 Recirculating ball steering system [7]

(iv) The ball bearings in the threads re-circulate through the gear as it turns.

2.6 Sources of Ackerman

Ackerman distinguishes between direct and circular movements. The directional inputs move in sequence while the directional/spinning/wheel/wheel rotates around the steering axis, also called the kingpin axis. Directional guess is the line drawn from the kingpin axis using directional arm keeping the height constant, and if the steering wheel is the same as the center line of the vehicle, then both the directional angles will be the same and there will be 0% Ackerman. By carefully selecting the angles, we can use the emerging Ackermann [8]. If the steering rod is indefinitely longer or too long, this will be Ackerman's only source, but since the binding rods are limited, they represent the second source of circular motion, as the wheel rotates, the angle of the binding rod, as it is clearly visible, it should change as the point of the binding rod moves in length. The main movement goes later, but there is the second movement in length, and Ackerman is the result of a second order.

2.7 Ackerman Principle and Steering Geometry

For true rolling of the vehicle on a turn or a curved track, lines drawn with each wheel axis must pass through the instantaneous centre. The instantaneous centre position is constantly changing due to the change in the angular position of the front wheel to adjust the direction of the directed vehicle. Since both rear wheels

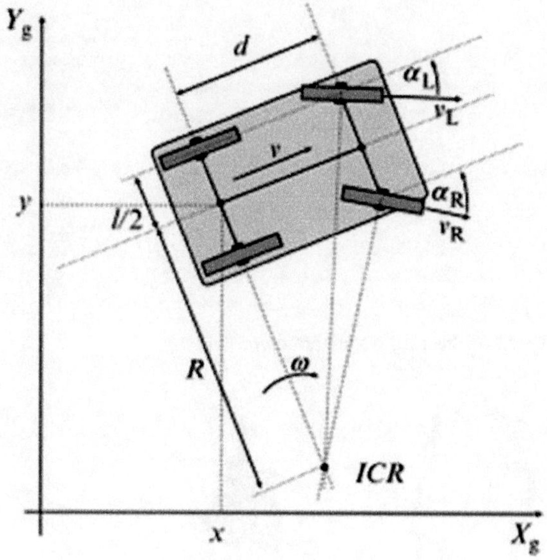

Fig. 2.5 Ackerman drive schematics [10]

are positioned in the same axis but the front axles are independent of each other, the current operating instantaneous center lies somewhere in a hypothetical line created by being pulled by the rear axles axis. Figure 2.5 shows the Ackerman drive schematics. Ackermann's principle is based on two front wheels with axles blown along the axle beam. Ackermann's original coupling has the same track-rod-arms so that both sensing wheels rotate at equal angles. As a result, the intersecting lines do not intersect at the same time. If both front wheels are free to follow their natural paths, they can come together and eventually cross over. As the car spins in the same direction, both wheels constantly collide and cause the tire to slip and slip. The subsequent modification connects using the track arms so that the inner wheels rotate around its kingpin slightly than the outer wheel. So, the lines drawn on the stub-axles meet in one place at a time by guessing the background axis [9]. The idea behind the Ackerman mechanism is that the outer wheel, i.e., the wheel closer to ICR must turn at a greater angle and thus moves with a slower speed while taking a turn [10].

2.8 Prototype

This prototype in Fig. 2.6 shows the basic four-bar linkage used in the Ackerman steering mechanism. This prototype was printed on a 3D printing machine. The prototype was verified by observing the linear displacement of link CD and link GH when the link FH is pulled to rotate mid-plate DEG [11] (Fig. 2.7).

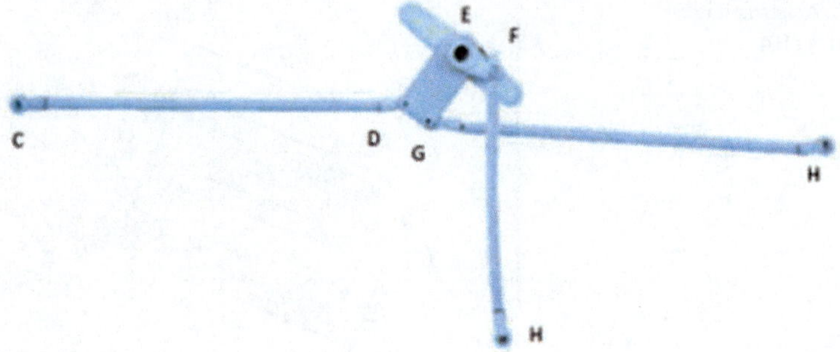

Fig. 2.6 Steering system prototype [11]

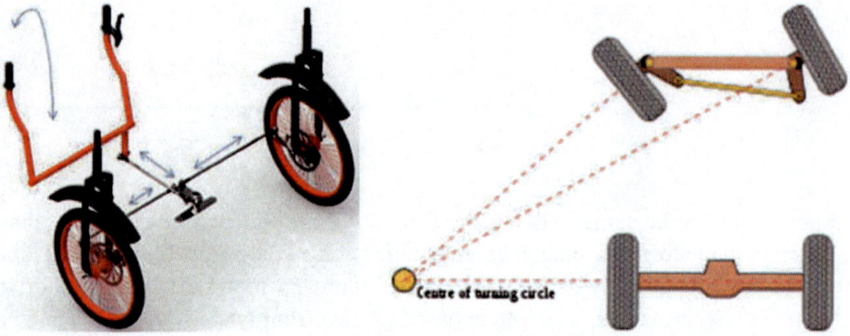

Fig. 2.7 Steering system overall geometry [11, 12]

2.9 Literature Survey

Wei et al. [13] have learned the powerful and dynamic aspects of the planetary connection process with permissions. Both movement accuracy and road stability have been improved. They also studied the effect of the system and the vibrations caused by the permission to connect the rod. After all these studies, they successfully detected the vibration of the rod caused by the permit. The numerical method is used to analyse the effect of the permit value on the powerful signals of the direction system, providing support for the theory of pinion formation and rack steering mechanism. Then they work on different system motion equations and find the amount of variable torque on the kingpin [14]. To increase the parameters of the objectives and to determine the size of the defined vehicle equipment, a genetic algorithm method was also used to summarize the size of the machine components, which were set as specific requirements, that were considered. They then work on the error operation that gives the business value at its fair value. After the completion of the error correction, they had performed a kinematic analysis of the correction method, with a well-executed magnitude, performed to confirm the results obtained in operation. The analysis was performed on the Multibody Simulation Software MSC-ADAMS, and the appropriate parameters were applied to the model to mimic the actual degrees of system freedom. Results of ADAMS were sent to MS Excel with appropriate graphs. The performance of a well-designed machine is compared to the construction of an existing train, and the result is a Steering and Tie Rod Displacement Error [15]. They suggested that the hand-operated joystick replace four-wheel drive models and show, for example, that this method works to improve mobility. In fact, they have shown that the differential gear mechanism works like a one-handed joystick. This made it easy to change locomotion mode easily. The rear-wheel drive has produced brake work without the use of special components such as discs, shoe pads or locking devices. Dipal Kumar Koladia studied mathematical models to design rack and pinion type of Ackerman steering geometry. He studied that by applying and solving three equations of mathematical model for any vehicle, rack and pinion type of Ackerman steering geometry can be designed for any vehicle [16]. Thirugnanam et al. [17] studied the different lifting mechanisms and concluded that, compared with other lifting mechanisms, no separate room is required for rack and pinion lift mechanisms. They are compact and save effort and time. Parvez et al. [18] conducted a case study in the existing column of the passenger car. That was a solid column in the event of a crash transferring power directly to the driver causing serious injury or damage. So, they fix the solid column with a collapsing steering column so that it is built securely. Wins et al. [19] take into account the fact that risks are increasing on a daily basis. In view of the strength of the damage to the steering wheel, they suggest that the construction of safety wheels be built. Pneumatic collapsible steering production is also less expensive and can achieve more comfort by increasing the gas or air pressure in it. They have made them more sensitive to the use of sensors and provided a more efficient airbag. Some more literature surveys are shown in Table 2.1.

2.10 Conclusion

- This type of steering system (Ackerman steering system) is easy to design and manufacture as compared to rack and pinion type steering system.
- The overall cost of this steering system is very less as compared with other steering systems.
- This review paper helps to understand the Ackerman principles, working and factors like minimizing slipping and tire wear.
- Directional geometry can be done using Ackerman's statistical model of the different arms of the inner wheels and selecting the geometry of what Ackermann percent and directional effort are right for.

Table 2.1 Exhaustive literature survey

Author/Year	Title	Mechanism	Efficiency	Applications	References
Akhilesh Kumar Verma, Sanjay Kumar Sharma, Sanjay Kumar Singh (2015)	All wheel steering system	Steering mechanism for four wheels	Higher as compared to two-wheel steering	Narrow roads, gentle curves, lane changes	[20]
Manzoor Hadi, A. Padma Rao (2015)	Chain drive steering mechanism	Davis steering mechanism, Ackerman steering mechanism	Low efficiency	Domestic vehicles, commercial vehicles	[21]
Lei Wang, Min Xu, Xiadong Wu (2013)	Differential speed steering control for four-wheeled independent driving electric vehicle	Differential steered mechanism	Efficiency is high	Electric vehicle motor driving	[22]
Pingxia Zhang, Li Gao, Yongqiang Zhu (2016)	Study on control schemes of flexible steering system of a multi-axle all-n wheel steering robot	Centre of rotation or steering angle of wheels	Efficiency is high	Situ rotation, multimode steering, lateral movement	[23]
P. L. Ramalingam, M. Prabhu, S. Subramaniyan, R. Vigithra (2016)	Implementation of variable turning radius by combining Ackerman and reverse Ackerman mechanism	Ackerman mechanism	Efficiency is high	High-speed vehicles, commercial and domestic vehicles	[24]
D. P. Miller, D. F. Flippo (2013)	Turning efficiency prediction for-skid steering via single wheel testing	Ackerman mechanism	Efficiency is higher	Commercial and domestic vehicles to avoid skidding	[25]
Er. Amitesh Kumar, Dr. Dinesh N. Kamble (2014)	Zero-turn four-wheel steering system	Linkage mechanism	Efficiency is low	Commercial vehicles	[26]
Yuan Shen, Biao Chu, Dongcai Liu, Chang Anzhu (2015)	Optimization of steering system of forklift vehicle for idle	Crank mechanism	Efficiency is high	Forklift vehicles	[27]

(continued)

Table 2.1 (continued)

Author/Year	Title	Mechanism	Efficiency	Applications	References
Yuhon Tee, Yongchai Tan, Boonyew Teoh, Engbeng Tan, Zhenyang Wong (2010)	A compact design of zero radius steering autonomous amphibious vehicle with direct differential drive-UTAR-AAV	Direct differential mechanism	Efficiency is high	Commercial vehicles	[28]

References

1. Steering system: types, parts, function, diagram & steering gears. www.theengineerspost.com. Accessed 03 Dec 2020
2. 19065_Team Synergy Racing_Design Report, 2019, Mechanical engineering. Department, G.L Bajaj Institute of Technology and Management, Greater Noida
3. Kadu, R., Gaikwad, P., Joshi, K., Gaikar, A., Mahjan, P., Patil, V.: Design and analysis of chassis, engine and steering system of an ATV. ISO 9001:2008 Certified J. **6**(4) (2019), e-ISSN: 2395-0056, p-ISSN: 2395-0072
4. http://blog.automotive-technology.com/types-of-automotive-steering-mechanism/. Accessed 20 Dec 2020
5. https://www.themotorombudsman.org/useful-information/motoring-advice/getting-to-know-rack-and-pinion-steering. Accessed 20 Dec 2020
6. https://blog.automotive-technology.com/types-of-automotive-steering-mechanism/. Accessed 20 Dec 2020.
7. https://www.larescorp.com/toolbox/skinned-knuckles-articles/recirculating-ball-nut-steering/. Accessed 20 Dec 2020
8. Mitchell, W.C., Staniforth, A., Scott, I.: Analysis of Ackermann steering geometry, series 2006-01-3638. https://vdocuments.mx/download/analysis-of-ackermann-steering-geometry
9. https://what-when-how.com/automobile/the-ackermann-principle-as-applied-to-steering-automobile/. Accessed 21 Dec 2020
10. Klančar, G., Zdešar, A., Blažič, S., Škrjanc, I.: Chapter 2—Motion modeling for mobile robots. In: Klančar, G., Zdešar, A., Blažič, S., Škrjanc, I. (eds.) Wheeled mobile robotics, pp. 13–59. Butterworth-Heinemann (2017). ISBN 9780128042045. https://doi.org/10.1016/B978-0-12-804204-5.00002-0
11. Kulkarni, S., Waddamwar, A., bhumkar, R.: Design of steering and suspension system for hybrid tadpole trike. e-ISSN: 2395-0056, p-ISSN: 2395-0072
12. Karthik, T.: Design and fabrication of multi mode steering system for cars. In: ETEDM—2019 Conference Proceedings, vol. 7, issue 06, ISSN: 2278-0181
13. Wei, D., Wang, Y., Jiang, T., Zheng, S., Zhao, W., Pan, Z.: Chaos vibration of pinion and rack steering trapezoidal mechanism containing two clearances. Mech. Syst. Signal Process. **92**, 146–155 (2017). ISSN 0888-3270. https://doi.org/10.1016/j.ymssp.2017.01.019
14. Kulkarni, A., Nanaware, R., Kulkarni, A., Garud, V.: Synthesis and optimization of steering mechanism of extra-long semi-trailer vehicle. Procedia Struct. Int. **14**, 158–167 (2019). ISSN 2452-3216. https://doi.org/10.1016/j.prostr.2019.05.021
15. Okada, T., Imura, T., Shimizu, T.: Smart one-hand operation mechanism for multi-purpose steering of a four-wheeled vehicle. ifac Proc. **40**(15), 343–348 (2007). ISSN 1474–6670. ISBN 9783902823656. https://doi.org/10.3182/20070903-3-FR-2921.00059
16. A review on design and manufacturing of steering system in an offroad vehicle. Int. J. Eng. Sci. Comput. IJESC **7**(5) (2017). D.Y Patil College of Engineering, Pune

17. Thirugnanam, A., Das, P., Rakesh, L.: Design and Fabrication of Rack and Pinion lift. Middle-East J. Sci. Res. **20**(6), 744–748 (2014). ISSN 1990-9233. https://doi.org/10.5829/idosi.mejsr.2014.20.06.11372
18. Parvez, M.S., Shaikh, I.J. Shakebuddin, M.: A literature review on collapsible steering column. IJIRST—Int. J. Innov. Res. Sci. Technol. **2**(5) (2015). ISSN (online): 2349-6010
19. Wins, R., Chatta, D., Nair, A.: Design of pneumatic collapsible steering. **2**(2) (2013). ISSN: 2319-3182
20. Singh, S.K., Sharma, S.K., Verma, A.K.: S-JPSET **7**(2). ISSN: 2229-7111(Print), ISSN: 2454-5767
21. Rao, A.P., Hadi, M.: Chain drive steering mechanism for a four-wheeler. **2**(2) (2015). ISSN (Online) 2393-8021 ISSN (Print) 2394-1588. https://doi.org/10.17148/IARJSET
22. Wu, X., Xu, M., Wang, L.: Int. J. Mater. Mech. Manufact. **1**(4) (2013). https://doi.org/10.7763/IJMMM.2013.V1.77
23. Zhang, P., Gao, L., Zhu, Y.: Study on control schemes of flexible steering system of a multi-axle all-wheel-steering robot. Adv. Mech. Eng. **8**(6), 1–13 (2016). The Author(s) 2016. https://doi.org/10.1177/1687814016651556
24. Ramalingam, P.L., Prabhu, M., Subramanian, S., Vigithra, R.: Implementation of variable turning radius by combining Ackermann and Reverse Ackermann mechanism. Middle-East J. Sci. Res. **24** (S1), 232–235 (2016). ISSN 1990-9233. https://doi.org/10.5829/idosi.mejsr.2016.24.S1.46
25. Flippo, D.F., Miller, D.P.: Turning efficiency prediction for skid steering via single wheel testing. J. Terramech. **52**, 23–29 (2014). ISSN 0022–4898. https://doi.org/10.1016/j.jterra.2014.02.001
26. Kumar, A., Kamble, D.N.: Zero turn four wheel steering system. Int. J. Sci. Eng. Res. **5**(12) (2014). ISSN 2229-5518
27. Shen, Y., Chu, B., Liu, D.C., Zhu, C.: Optimization of steering system of Forklift vehicle for idle performance. Math. Prob. Eng. **2015**, 9. https://doi.org/10.1155/2015/313182. Hindawi Publishing Corporation
28. Tee, Y.H., Tan, Y., Teoh, B.Y., Tan, E.B., Wong, Z.Y.: A compact design of zero-radius steering autonomous amphibious vehicle with direct differential directional drive—UTAR-AAV. In: 2010 IEEE Conference on Robotics, Automation and Mechatronics, Singapore, 2010, pp. 176–181. https://doi.org/10.1109/RAMECH.2010.5513192. https://ieeexplore.ieee.org/document/5513192

Chapter 3
Experimental Study of Parameter Affecting the Thermal Conductivity of Composite Materials and Alloy: A Review

Omprakash Yadav, Ankit Chhonkar, Rahul Sharma, Sandeep Chauhan, Ashish Bansal, Navneet Singh Baghel, Himanshu Sharma, and Devesh Sharma

Abstract In the new era, the development of composite materials and alloy is very demanding for the fulfilment of the present demand of industries and society. The thermal analysis of composite materials and alloy is requiring its thermal properties like thermal conductivity, specific heat, etc. before their uses. The main focus of this research work is to study the effect of different parameters on the thermal conductivity of composite material and alloy based upon the experimental and analytical data available in the literature. Furthermore, a different method of finding the thermal conductivity of composite materials and alloy like finite element method, finite volume method, statistical model, non-conventional laser flash technique, machine learning method, spatially varying relaxation parameter lattice Boltzman method using melting process, equivalent inclusion method, hot plate method cut bar method, etc. also has been presented in this work. The machine learning method is a very fast, accurate and efficient method for predicting the conductivity of composites and alloy. From the intensive studies, it was found that temperature, particle size, density of materials and composition are generally affecting the thermal conductivity of composite materials. Furthermore, it was found that conductivity of composites material increases with an increase in the composition of filler material and a decrease in particle size of the filler.

3.1 Introduction

For thousand of years, composite materials and alloy have been used in industries and societies. Nowadays, utilization and demand for composite material and alloy are increasing day by day. The composite materials are produced by combining two

O. Yadav (✉) · A. Chhonkar · R. Sharma · S. Chauhan · A. Bansal · N. S. Baghel · H. Sharma · D. Sharma
Department of Mechanical Engineering, G.L Bajaj Institute of Technology and Management Greater Noida, Greater Noida, India

© The Author(s), under exclusive license to Springer Nature Singapore Pte Ltd. 2022

V. V. Rao et al. (eds.), *Computational and Experimental Methods in Mechanical Engineering*, Smart Innovation, Systems and Technologies 239,
https://doi.org/10.1007/978-981-16-2857-3_4

or more than two materials with different physical, chemical and thermal properties. The material having higher concentration is matrix materials and material with lower concentration is filler or reinforced material. The properties of the composite materials and alloy are different compared to the property of the base element. Some of the commonly used composite materials are Reinforced concrete, plywood, composite wood, masonry brick, reinforced plastic, fibreglass, ceramic matrix composite, silica aerogel composite, carbon Fiber Reinforced Polymer composites, etc. and some of the advanced composite materials are polymer matrix composite and metal matrix composite. As per matrix phase is concerned, composite are classified as ceramic matrix composite, metal matrix composite and polymer matrix composite. The composite material finds its versatile use because of its inherent property like lightweight, lower cost, better toughness, hardness and strength.

A number of the experimental, computational and analytical models have been developed to estimate the effective thermal conductivity of composite material. Although a number of experimental works have been done by several authors to find the thermal conductivity of materials but only few researchers have developed analytical models and methods to find the conductivity of composites. The analytical model can save huge amount of energy, time and resources compared to the experimental model. Figure 3.1 represents some of the experimental and analytical models and methods which can be utilized to find the thermal conductivity of composite materials.

In the subsequent section, widely used experimental method viz Hot plate Method and Cut Bar Method has been discussed. Hot plate methods are generally employed

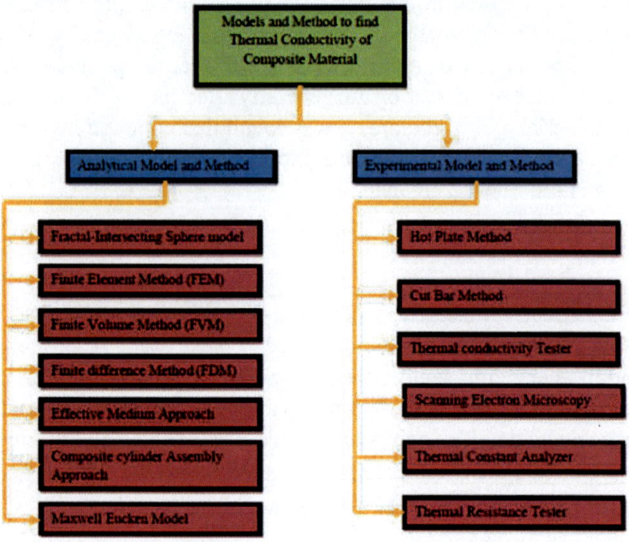

Fig. 3.1 Experimental and analytical method and model for finding thermal conductivity of composite material

to find the insulating material's thermal conductivity [5]. The main component of the hot plate apparatus are the main heater, hot plate, guard heater, cold plate, supply of cold fluid, thermocouple, sample holder, etc. The guard heater is used to maintain the temperature of the heater equals to the temperature of the hot plate to stop the flow of heat in the guard heater direction. The temperature of the cold plate was maintained very low by circulating cold water on the opposite side. Three thermocouples were inserted at the hot plate on the sample side, an average of three was taken as Th, and three more thermocouples were inserted on the cold plate on the sample side; average of three was taken as Tc. The following Fourier law of conduction can be utilized to find the thermal conductivity of sample material.

The systematic arrangement of components in the hot plate method is shown in Fig. 3.3.

Cut bar method is the steady state comparative technique in which the flow of heat is axial. In this method, the measurement of heat flow is not required which was the main cause of error in the absolute technique. In this method, two bars are used, one is the sample bar and the other is the standard bar. If the thermal conductivity of the standard bar is known, the conductivity of the sample bar can be found by the cut bar method. This method is most suitable when the conductivity of the sample bar is comparable to standard bar conductivity. A number of thermocouples are used to find the temperature at different sections. By knowing the temperature difference across the two bars, cross-sectional area and length of the bar, the thermal conductivity of the sample bar can be found by the following formula.

$K_1 = K_2 . \frac{A_2 l_1 \Delta T_2}{A_1 l_2 \Delta T_1}$,

where K_1 and K_2 are the thermal conductivity of the sample and standard bar, and A_1 and A_2 are the area of the sample and standard bar.

3.2 Literature Review

Zain-ul-Abdein et al. [1] computationally investigated the effective thermal conductivity of bakelite-graphite composites using Abaqus software. Furthermore, the author has studied the effect of particle size, composition and packing density on bakelite-graphite conductivity. The result reveals that the effective conductivity of 50% backelite-50% graphite with 70% packing density was 10.78 W/m-K. Furthermore, it was found that finer particle size results in better conductivity of Bakelite-graphite composites, and its variation was non-linear. Sang Hyun Lee et al. [2] investigated thermal expansion coefficient and thermal conductivity of tungsten-copper composite by varying the composition of copper between 5 and 30%. The spray conversion method was used to produce the powder, and their composite was developed by injection moulding. From this study, it was found that the thermal conductivity of 70% tungsten-30% copper composite by weight is 238 W/m-k at room temperature. Furthermore, its conductivity decreases with an increase in temperature. Tao Xie et al. [3] investigated the silica aerogel composite conductivity and effect of concentration, particle size and temperature by the computational tool heat conduction. The study reveals that thermal conductivity increases with doping concentration, reaches a maximum value and further decreases to a minimum value. As temperature increases, radiative conductivity increases but conductive conductivity is unaffected. Furthermore, with a decrease in particle size, radiative conductivity decreases. Gori et al. [4] computationally investigate the anisotropic thermal conductivity of silica matrix with asbestos, steel, copper composite by solving heat conduction equation. Assumption of parallel isothermal and heat flux lines was used to make the computational analysis simple. In this study, author concluded that the anisotropic efficiency of silica matrix with asbestos is the linear function potential anisotropy. Furthermore, copper has better conductivity compared to asbestos and steel. Cherki et al. [5] experimentally investigated the effect of size of cork on thermal properties of gypsum-cork composite by hot plate method, and the results have been compared with the analytical model. Due to light weight and better thermal property of cork-gypsum plaster, composite enhances the efficiency of the building. From this study, it was found that the effective thermal conductivity of cork-gypsum composite was 0.101–0.124 W/m-K while thermal conductivity of gypsum was 0.299 W/m-K. The author has reported a substantial decrease in conductivity of cork-gypsum composite compared to gypsum. Furthermore, it was reported that decrease in conductivity by factor 1/3 as a contrast with the gypsum without cork. The comparison of results of the different models has been presented in Fig. 3.3 by Cherki et al. [5]

Jong-Pil Won et al. [6] experimentally studied the thermal conductivity of composite which is based on cement matrix. For this, they placed composite on copper that has a protecting layer on a heating system, which is further used for snow melting. The temperature of the surface layer from outside as well as from inside is measured to find conductivity. During analysis, there is a separation between the plates of copper and concrete which are used as a variable. This gives the uniform distribution of temperature over the surface and is helpful to determine the thermal characteristics of cement-based composite. The results reveal that 1.8–2.05 kcal/m-hr-degree conductivity for this composite. Li et al. [7] investigated the thermal properties and surface structure of paraffin-expanded graphite composite by Scanning electron microscopy and Thermal constant analyzer. Expanded graphite is a composite powder that is taken under high compression pressure with the varieties of weight fraction. The author proposed a two-level model that tells us about the thermal conductivity of materials and for the particles' distribution in the material a mathematical function is used. From the results, it was found that thermal conductivity of 94% paraffin- 6% expanded graphite composite is 20 times more as compared to pure paraffin. S.I Kundalwal et al. [8] investigated the fuzzy fiber-reinforced composite(FFRC) conductivity by effective medium approach and composite cylinder assembly approach. From the results, it was concluded that thermal conductivity of FFRC was increased by 1040% and 400% for Carbon Nanotube (CNT) fraction of 6.88% and 4.27% than the composites without CNTs. Furthermore, with the increment in carbon fibre volume % and the temperature, the thermal conductivity of FFRC also increases. Gao et al. [9] experimentally and theoretically has found conductivity of silicon-rubber and alumina composite by thermal resistance tester. This composite is made up of silicon rubber and Al2O3 particles. From the results, it was found that effective thermal conductivity variation was non-linear with particle size and volume fraction. Comparison between experimental and theoretical method was made, and it was found that both was consistent with each other. Kovacik et al. [10] investigated the thermal conductivity of copper- graphite composite for 0–40% weight percentage of graphite using Cut bar method and condition of room temperature. There is a range of cu-graphite which has been prepared. The results reveal that the experimental value of thermal conductivity is less than the thermal conductivity of Cu due to the presence of copper oxide in it. Furthermore, it was found that thermal conductivity variation with the composition was non-linear at room temperature. Alok Agrawal et al. [11] have developed a mathematical model for the measurement of effective thermal conductivity of polymer filled with filler by using Unitherm model 2022 tester. From the results, it was found that effective conductivity increases non-linearly with the volume fraction of filler. The theoretical and experimental value of conductivity was approximately the same.

Feng et al. [12] investigated the thermal conductivity of the composites which are made of conductor and insulator having different geometries and structures. The experiment work to predict various effects on the thermal conductivity through these different geometries of second phase composites has been prepared to have spherical structure. Results reveal that the branch's fractal dimensions show lower value with the same filling elements. Hautcoeur et al. [13] experimentally studied the anisotropy thermal conductivity of ceramics-aluminium alloy composite and effect of aluminium infiltration in ceramic with a porous structure so that it can efficiently be used in thermal management fields. Ceramics-aluminium alloy composite was produced by freeze casting. From the result, it was found that along the freezing direction average conductivity was 80 W/m-K and along the direction normal to freezing direction was 13 W/m-K. Majid TabkiPaz et al. [14] investigated the thermal conductivity of carbon nanotubes and boron nitride with the use of the effective medium method and walk random logic. The results reveal that the thermal conductivity of nanotube composite increases with the alignment of carbon nanotubes which was studied by a random walk tool. The increase in conductivity is more in a parallel direction compared to the normal direction. Jinzao Xu et al. [15] have developed a statistical model to investigate the thermal conductivity of composite which was developed by polymer and metals with high thermal conductivity. In the electronics industries for cooling purposes, a material with high conductivity and light is required. The value of thermal conductivity was found by electric and thermal field analogies, which satisfy the Laplace equation. Furthermore, it was noted that with an increase in volume fraction of filler metals, the thermal conductivity of composite increases. Liu, Hua et al. [16] investigated the thermal conductivity silica aerogel composite, which is highly porous and has much potential for a number of insulation systems as insulating materials. A number of experiments were performed to predict the behaviour of heat transfer of silica aerogel (SAC) composite for a range of temperature between 290 and 1090 K and pressure range of 0.01–100 kPa. From this study, it was found that the conductivity of this composite increases linearly with a rise in temperature. Tingyu Wang et al. [17] experimentally and theoretically studied the behaviour of thermal conductivity of phase change composites which are microcapsules based. For the experimental work, microcapsule base composite mixed with carbon network has been prepared by a facile method. By the combination of the transient plane source method and effective medium theory, they are able to calculate the experimental and theoretical thermal conductivity, respectively. From the results, it was found that thermal conductivity of phase change composite increases with an increase in density of carbon network structure and it was nearly 24 times more for 24% expanded graphite. This increment on thermal conductivity helps in improving the efficiency of energy storage devices. Onitiri et al. [18] investigated the thermal conductivity and other thermal properties of iron Ore Tailings mixed with epoxy composite. Iron Ore is a waste material which is ejected from the iron ore's beneficiation. Furthermore, the effects of particle size and loading on thermal conductivity were investigated in this experiment. The results reveal that conductivity increases with an increase in the volumetric concentration of iron ore tailings filler (Fig. 3.4) and a decrease in particle size.

Wojciech Adamczyka et al. [19] investigated the thermal conductivity of the carbon fiber reinforced polymer composite by using the inverse technique. This technique uses a heat energy source and IR detector which gives the finite dimension of the sample. The multidimensional mathematical model was used, which describes the heat distribution on the entire body. The results reveal that the proposed methodology which was used to evaluate the thermal conductivity of fiber-reinforced polymer composite was in good agreement with other techniques. Han Wei et al. [20] studied thermal conductivity of composite material with porous media by the machine learning method. The machine learning method is a fast process that was newly sued to investigate the thermal conductivity of composite material and porous media. The machine learning method also can be used to investigate other physical properties of composite material and porous media. It is the best method for predicting the thermal conductivity. Gaurav Rajoriya et al. [21] computationally investigated the thermal conductivity of composite solid propellants by using the finite volume method. The solid loading of reinforced particles is very high in composite propellant which was used in the defence sector. This pack of propellant is changed in the form of cubical cells, and all cells have their properties based on their location. The results reveal that the thermal conductivity of these composite decreases with an increase in interface surface area. The simulation result was compared with experimental data and found that the value of simulation results is lower than the experimental value. Vasiliy Dokutovich et al. [22] investigated the effect of temperature and thickening agent on thermal conductivity of alkali carbonate which is filled with filler metal Al_2O_3 with the help of coaxial cylinder method under steady-state condition. From the result, it was found that variation of thermal conductivity was linear for low composition of Al2O3. For a high concentration of Al_2O_3, variation becomes non-linear. Jun-Ping Song et al. [23] investigated the thermal conductivity of different varieties of rubber materials which were filled by carbon black. The author has used microscopy of a transmission electron which tells about the connective structure of carbon black. From the results, it was found that found a larger number of carbon chains which help in increasing the connectivity structure and subsequently increases the thermal conductivity of the composite. Furthermore, the thermal conductivity of various types of carbon black-filled rubber composite was found high with parallel heat. Ke et al. [24] studied the thermal conductivity of composite material with the help of the Lattice Boltzmann method by varying the spatial relaxation parameter. It has high accuracy to measure the thermal conductivity of any material. The Lattice Boltzmann method result was compared with the experimental result, and it was found that both were consistent. The results reveal that thermal conductivity is more for the high volume fraction of filler and also for higher filler size. Liu et al. [25] experimentally investigated the thermal conductivity of plain woven composites mixed with carbon fiber. The effect of the composition of fiber and porosity was computationally investigated by the up-scaling method. By using the up-scaling approach, the thermal conductivity of Sic matrix, woven yard, and plain woven c/Sic composites were predicted. From the results, it was found that for the same porosity and carbon fiber volume fraction, the out-plane conductivity is smaller than the in-plane thermal conductivities. The decrease in thermal conductivities for in-yarn porosity was found as 3.1 and 5.6% for in-plane and out of the plane with respect to without in-yarn porosity.

Ashok Kumar et al. [26] experimentally investigated the thermal conductivity of elastospray composite for the purpose to evaluate their effect on the insulation of the wall and roof of the building by satisfying the requirements of ECBC, India. From the results, it was found that for wall insulation 50 mm thickness was sufficient while more thickness of conventional insulation layer was needed. Furthermore, it was found that lowering the U-value decreases the cooling load on the air conditioning. Chuan Li et al. [27] investigated the thermal conductivity of carbonate salt composite which is a phase change material, i.e., a material which contains a phase change material like in this case, it is carbonate salt. Two CPCMs are prepared, a mixture of NaLiCO3 and Na2CO3. To get the thermal conductivity of this material, a theoretical model is derived which consists of two MgO spheres with a mixture of PCM/TCEM and filled in interparticle voids; also a parallel-plate-based experiment was done to calculate the thermal conductivity of these composites and verify the theoretical model. After which the theoretical result agrees with data calculated experimentally. Furthermore, it was found that effective thermal conductivity increases with the volume fraction of salt. For the parallel plate model, its variation was found as linear. Gurushanth Vaggar et al. [28] experimentally investigated the effect of volume fraction of silicon cardide and glass fiber on the thermal conductivity of composite which was made by mixing silicon cardide, reinforced polymer and glass fiber by Lee experiments. This composite was made by a compression moulding machine for the experimental work. Mathematical models such as the Maxwell model with a difference of 10.40%, Rule of mixture with 16.12% and Hashin formula with 3.62 for 20% SiC content were found to be with very small differences with experimental results. The comparison showed that experimental results matched well with theoretical models for GFER10SI15 and GFIR10SI20. Ankush Sharma et al. [29] experimentally investigated the thermal conductivity of nonwoven jute epoxy composite with the marble dust as filler and without marble dust filler by hot disc transient plane source method. Furthermore, the Ansys workbench has been used to investigate theoretical thermal conductivity. From the experimental results, it was found that the thermal conductivity of jute epoxy composite increases with an increase in the volume fraction of marble dust filler. Furthermore, it was found that the computational result was in good agreement with the experimental results. Qu et al. [30] experimentally investigated the effects of expanded graphite-MWCNT and expanded graphite-CNF filler on the thermal conductivity of paraffin-HDPE SSPCP composite. The Maxwell Garnett model was used which found good agreement between the analytical model and experiment results. From the results, it was found that with the addition of only 5wt% of expanded graphite-CNF, the thermal conductivity increases to 21.2% of paraffin HDPE SSPCP. With the addition of 5wt% of expanded graphite-MWCNT, the thermal conductivity increases to 60% of paraffin HDPE SSPCP. Hence EG-MWCNT is more efficient than the other which is EG-CNF because the thermal resistance of EG-MWCNT is relatively smaller than the other.

3.3 Summary and discussion

From the extensive survey of previously reported experimental and computational research work, it has been found that a number of new composite materials have been developed to replace the conventional material because of its limitation to fulfill the demand of industries and society. Thermal property evaluation of these new composite materials is very essential for their analysis and design before their uses. In the present work, determination of thermal conductivity of composite materials, the effect of different parameter and different method used have been presented. A detailed description of different experimental and analytical methods and models has been presented in the introduction section of this paper in Fig. 3.2. It was found that generally used experimental methods are the hot plate method and cut bar method which has been described in the introduction section of this paper. Furthermore, generally used analytical methods are finite element method, finite volume method, Maxwell Eucken model, Effective medium model and Fractal intersecting sphere model. From the extensive study of past research work, some of the comparative results drawn by the author have been presented in Figs. 3.5, 3.6, 3.7.

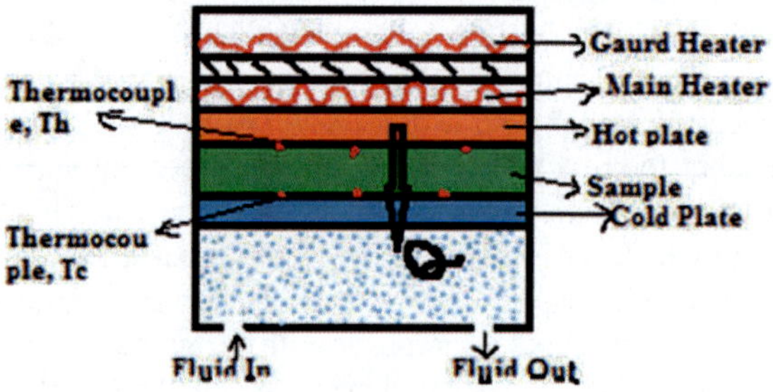

Fig. 3.2 Hot plate arrangement

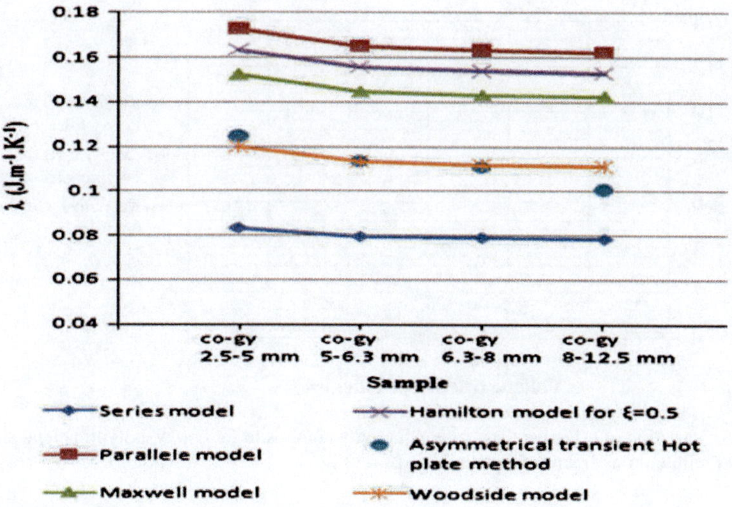

Fig. 3.3 Comparison of results of different model [5]

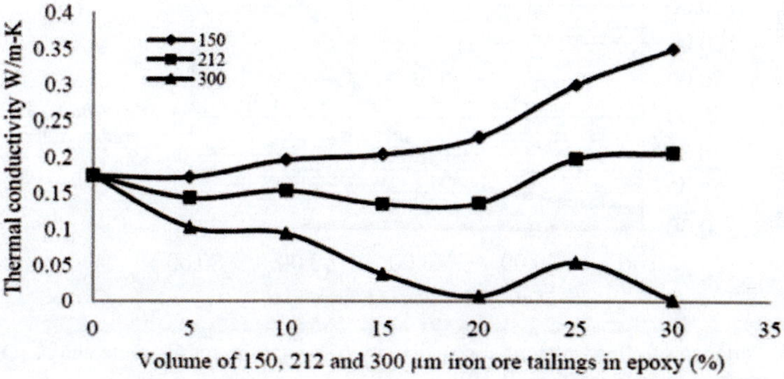

Fig. 3.4 Thermal conductivity of epoxy with varying volume % of iron ore [18]

From Figs. 3.5 and 3.6, it can be seen that the thermal conductivity of all the composite materials increases with an increase in percentage volume fraction of filler or reinforced material into the composite material. The rates of increase of thermal conductivity are small for the small amount of volume fraction of filler and further this rate increases at a faster rate with an increase in volume fraction of filler. From Fig. 3.7, also this can be seen that the effect of iron ore and rubber on the thermal conductivity of composite material are very small, while the effect of epoxy, polypropelene and graphit on thermal conductivity of composite is very significant.

From Fig. 3.7, it can be found that the thermal conductivity of all the composite material decreases with an increase in the particle size of filler or reinforced material into the composite material. The rates of decrease of thermal conductivity are fast

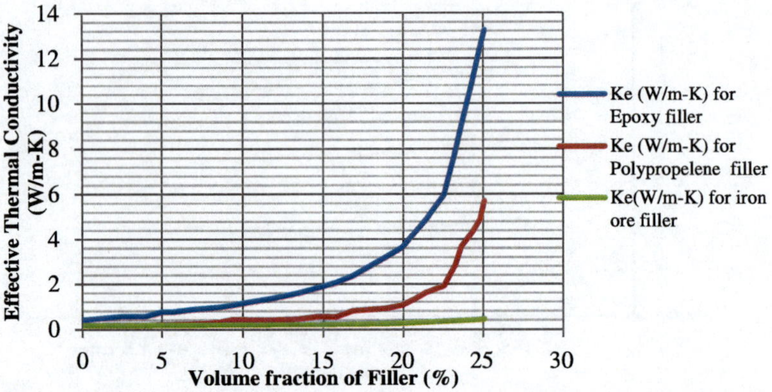

Fig. 3.5 Variation of effective thermal conductivity composite for epoxy, polypropelene and iron ore filler with volume fraction in percentage [3, 11]

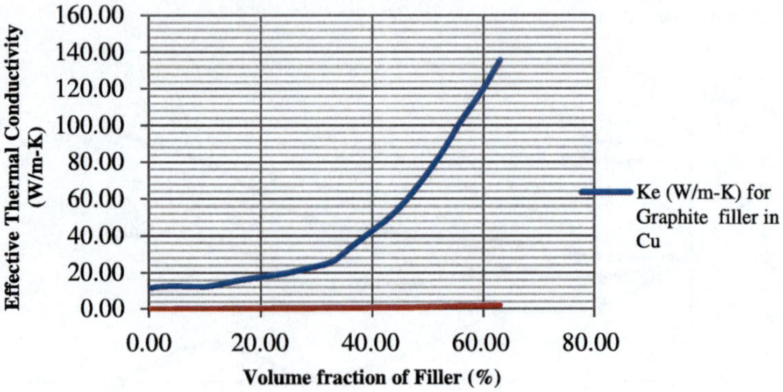

Fig. 3.6 Variation of effective thermal conductivity of composite for Graphite and Al_2O_3 filler with volume fraction in percentage [9, 10]

for small particle size of filler and further this rate decreases at a slower rate with an increase in particle size of filler. Furthermore, the conductivity of epoxy base, pp-based and graphite-based composite is higher compared to Al_2O_3 base and fibre based composite.

3.4 Conclusion

After an exhaustive literature survey of the past work done, the author has concluded the following observation.

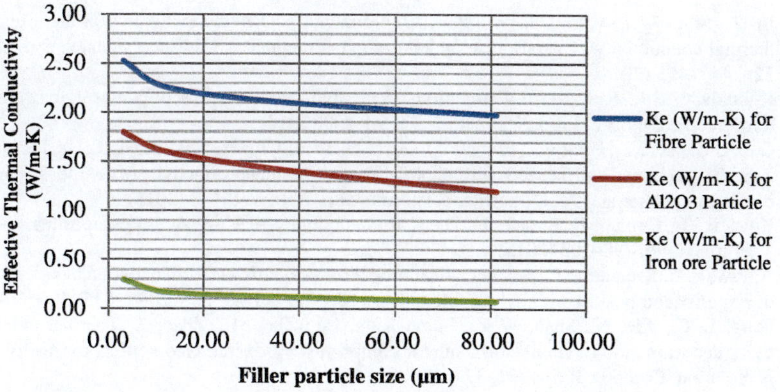

Fig. 3.7 Effect of particle size of fibre, Al_2O_3 and iron ore on thermal conductivity of composite material [3, 9, 18]

1. Thermal conductivity of composite materials was affected by the volume fraction of filler or reinforced material.
2. Thermal conductivity of composite material can be increase by increasing the volume fraction of reinforced materials. Graphite as a filler material increases the thermal conductivity substantially.
3. Thermal conductivity of composite materials was also affected by the particle size of filler or reinforced material.
4. Thermal conductivity of composite material can be increased by decreasing the particle size of reinforced or filler materials.

References

1. Zain-ul-Abdein, M., Azeem, S., Shah, S.M.:.Computational investigation of factors affecting thermal conductivity in a particulate filled composite using finite element method. Int. J. Eng. Sci. **56** (2012)
2. Lee, S.H., Kwon, S.Y., Ham, H.J.:Thermal conductivity of tungsten–copper composites. Thermochim. Acta **542**, 2–5 (2012)
3. Xie, T., He, Y.-L., Zi-Jun, Hu.: Theoretical study on thermal conductivities of silica aerogel composite insulating material. Int. J. Heat Mass Transf. **58**(1–2), 540–552 (2013)
4. Gori, F., Corasaniti, S.: Effective thermal conductivity of composites. Int. J. Heat Mass Transf. **77**, 653–661 (2014)
5. Cherki, A.-B., Remy, B., Khabbazi, A., Jannot, Y., Baillis, D.: Experimental thermal properties characterization of insulating cork–gypsum composite. Constr. Build. Mater. **54**, 202–209 (2014)
6. Won, J.-P., Kim, C.-K., Lee, S.-J., Lee, J.-H., Kim, R.-W.: Thermal characteristics of a conductive cement-based composite for a snow-melting heated pavement system. Compos. Struct. **118**, 106–111 (2014)

7. Li, Z., Sun, W.G., Wang, G., Wu, Z.G.: Experimental and numerical study on the effective thermal conductivity of paraffin/expanded graphite composite. Sol. Energy Mater. Sol. Cells **128**, 447–455 (2014)
8. Kundalwal, S.I., Ray, M.C.: Estimation of thermal conductivities of a novel fuzzy fiber reinforced composite. Int. J. Therm. Sci. **76**, 90–100 (2014)
9. Gao, B.Z., Xu, J.Z., Peng, J.J., Kang, F.Y., Du, H.D., Li, J., Chiang, S.W., Xu, C.J., Hu, N., Ning, X.S.: Experimental and theoretical studies of effective thermal conductivity of composites made of silicone rubber and Al_2O_3 particles. Thermochim. Acta **614**, 1–8 (2015)
10. Kováčik, J., Emmer, Š, Bielek, J.: Thermal conductivity of Cu-graphite composites. Int. J. Therm. Sci. **90**, 298–302 (2015)
11. Agrawal, A., Satapathy, A.: Mathematical model for evaluating effective thermal conductivity of polymer composites with hybrid fillers. Int. J. Therm. Sci. **89**, 203–209 (2015)
12. Feng, L.-C., Xie, N., Shao, W.-Z., Liang-Xing, Lu., Zhen, L., Zhao, J.: Thermal conductivity determination of conductor/insulator composites by fractal: Geometrical tortuosity and percolation. Compos. B Eng. **92**, 377–383 (2016)
13. Hautcoeur, D., Lorgouilloux, Y., Leriche, A., Gonon, M., Nait-Ali, B., Smith, D.S., Lardot, V., Cambier, F.:Thermal conductivity of ceramic/metal composites from preforms produced by freeze casting. Ceram. Int. **42**(12), 14077–14085 (2016)
14. TabkhPaz, M., Shajari, S., Mahmoodi, M., Park, D.-Y., Suresh, H., Park, S.S.: Thermal conductivity of carbon nanotube and hexagonal boron nitride polymer composites. Compos. B Eng. **100**, 19–30 (2016)
15. Xu, J., Gao, B., Hongda, Du., Kang, F.: A statistical model for effective thermal conductivity of composite materials. Int. J. Therm. Sci. **104**, 348–356 (2016)
16. Liu, H., Xia, X., Xie, X., Ai, Q., Li, D.: Experiment and identification of thermal conductivity and extinction coefficient of silica aerogel composite. Int. J. Therm. Sci. **121**, 192–203 (2017)
17. Wang, T., Wang, S., Wei, Wu.: Experimental study on effective thermal conductivity of microcapsules based phase change composites. Int. J. Heat Mass Transf. **109**, 930–937 (2017)
18. Onitiri, M.A., Akinlabi, E.T.: Thermal conductivity and impact properties of iron ore tailings filled epoxy composites. Mater. Today: Proc. **5**(1), 79–83 (2018)
19. Adamczyk, W.P., Pawlak, S., Ostrowski, Z.: Determination of thermal conductivity of CFRP composite materials using unconventional laser flash technique. Measurement **124**, 147–155 (2018)
20. Wei, H., Zhao, S., Rong, Q., Bao, H.:Predicting the effective thermal conductivities of composite materials and porous media by machine learning methods. Int. J. Heat Mass Transf. **127**, 908–916 (2018)
21. Rajoriya, G., Vijay, C., Ramakrishna, P.A.: Thermal conductivity estimation of high solid loading particulate composites: a numerical approach. Int. J. Therm. Sci. **127**, 252–265 (2018)
22. Cikutovich, V.N., Khokhlov, V.A., Zakir'yanova, I.D.: Thermal conductivity of composite materials: alkali carbonate-based melts filled with fine α-Al2O3. Int. J. Heat Mass Transf. **119**, 365–371 (2018)
23. Song, J.-P., Tian, K.-Y., Ma, L.-X., Li, W., Yao, S.-C.: The effect of carbon black morphology to the thermal conductivity of natural rubber composites. Int. J. Heat Mass Transf. **137**, 184–191 (2019)
24. Ke, X., Duan Y.: A spatially-varying relaxation parameter lattice boltzmann method (SVRP-LBM) for predicting the effective thermal conductivity of composite material. Comput. Mater. Sci. **169**, 109080 (2019)
25. Liu, Y., Qu, Z.G., Guo, J., Zhao, X.M.: Numerical study on effective thermal conductivities of plain woven C/SiC composites with considering pores in interlaced woven yarns. Int. J. Heat Mass Transf. **140**, 410–419 (2019)
26. Kumar, A., Suman, B.M.: Experimental evaluation of insulation materials for walls and roofs and their impact on indoor thermal comfort under composite climate. Build. Environ. **59**, 635–643 (2013)
27. Li, C., Li, Qi., Ding, Y.: Investigation on the effective thermal conductivity of carbonate salt based composite phase change materials for medium and high temperature thermal energy storage. Energy **176**, 728–741 (2019)

28. Vaggar, G.B., Kamate, S.C., Badyankal, P.V.: A study on thermal conductivity enhancement of silicon carbide filler glass fiber epoxy resin hybrid composites. Mater. Today: Proc. (2020)
29. Sharma, A., Choudhary, M., Agarwal, P., Patnaik, T.K., Biswas, S.K., Patnaik, A.: Experimental and numerical investigation of thermal conductivity of marble dust filled needle punched nonwoven jute-epoxy hybrid composite. Mater. Today: Proc. (2020)
30. Qu, Y., Wang, S., Zhou, D., Tian, Y.: Experimental study on thermal conductivity of paraffin-based shape-stabilized phase change material with hybrid carbon nano-additives. Renew. Energy **146**, 2637–2645 (2020)

Chapter 4
An Experimental and Mathematical Analysis for Improvement of Gas Stove Efficiency

Akanksha Mathur, Rohit Singh Lather, Vinit Chauhan, Rahul Sharma, and Tushar Mehta

Abstract There are various ways of enhancing the thermal efficiency of an LPG gas stove like variation of gas pressure, varying loading weight or loading height. This paper deals with the improvement of the thermal efficiency of LPG gas stoves by varying the loading height. Loading height is defined as the distance between the burner top and the bottom surface of the vessel when the vessel is kept on the stove. The experiment was performed as prescribed by the Bureau of Indian Standard (BIS). It was found that with an increase in the loading height there was an increase in thermal efficiency up to the optimum loading height of 17.5 mm at which maximum thermal efficiency was achieved. Further increase in the loading height resulted in a decrease in the thermal efficiency. A mathematical analysis was carried, and a correlation is developed which is compared with the experimental results. The maximum efficiency achieved was comparable to the value predicted from correlation. Uncertainty analysis was also computed for the efficiency obtained.

4.1 Introduction

Liquid petroleum gas (LPG) is a very common and essential fluid nowadays. It has many important properties which makes it very useful. Currently, 80% of the Indian households are using LPG as cooking gas, and three years ago, only 53% were using LPG, which is a revolutionary change. Earlier, fuel used for cooking mainly included coal and wood. These types of fuel have many properties which are harmful to both living beings and the environment. In terms of properties, LPG has benefits like it produces less pollution, easily transportable, high calorific value and many more [1, 2]. In India, the LPG market is mainly ruled by three major companies, they are Hindustan Petroleum Corporation Limited (HPCL), Indian Oil Corporation Limited (IOCL), and Bharat Petroleum Corporation Limited (BPCL). Nowadays

A. Mathur (✉) · R. S. Lather · V. Chauhan · R. Sharma · T. Mehta
Department of Mechanical Engineering, The NorthCap University, Gurugram 122017, Haryana, India
e-mail: akankshamathur@ncuindia.edu

© The Author(s), under exclusive license to Springer Nature Singapore Pte Ltd. 2022
V. V. Rao et al. (eds.), *Computational and Experimental Methods in Mechanical Engineering*, Smart Innovation, Systems and Technologies 239,
https://doi.org/10.1007/978-981-16-2857-3_5

since the fuel costs are rising sharply, improving the efficiency of combustion systems is the priority [3, 4]. Reducing the average gas consumption by achieving improved efficiency of gas stoves will save significant fuel costs for Indian consumers. Thus, it has become important to design gas burners and stoves to increase efficiency.

4.2 Literature Review

Various studies have been performed to enhance the LPG stove performances. The improvement through changes in the design of port diameters induces swirl in gas flow by proving inclined holes [5]. Some studies are there in the literature to minimize heat loss by including inserts and redesigning burner for energy saving with a comparison with numerical analysis [6, 7]. It was also seen that the thermal efficiency of the stove is significantly affected by the parameters like loading height and porosity of the burner [8, 9]. Hou and Ko [10] studied the effect of loading height on flame appearance, temperature field and efficiency of single impinging laminar jet flame. Single copper tube Bunsen burner was used which produced a double-flame fuel-rich flame. Thermal efficiency and temperature distribution around the flame were observed at different loading heights. When the loading height is highest, the inner premixed and outer diffusion flame was conical. When the loading height was decreased, the flames transited from conical to trumpet shape. The maximum thermal efficiency occurred at the loading height of 15 mm.

Ashman et al. [11] conducted experiments to study the effect of loading height and the emissions from natural gas-fired domestic cooktop burner using three different heights and efficiencies were observed. The set-up had a burner with three different thermal inputs and a sampling hood to collect the combustion products. It was noticed that with the increase in loading height of the stove, there was a decrease in thermal efficiency of the cooking stove with reducing heat transfer. Joshi et al. [12] did experiments for the development of efficient designs of the cooking system. Experimentation was done to find the optimum dimension of the vessel and efficiency was obtained according to BIS standard. The investigation was also done with real food items like rice, lentils and vegetables. They concluded that for any vessel of particular diameter, the optimum diameter of the flame is one-third of vessel diameter and the optimum distance between burner top and bottom of the vessel was 2–3 mm. The optimum heat flux which depended on heat supply rate and heat absorbed by the contents of the cooker was in the range of 20,000–25,000 kcal/h m^2.

In the work done by Agarwal et al. [13], the effect of thermal input on emission level was investigated. Maximum thermal efficiency of 61.66% was achieved with the present burner with 3 kg of LPG gas. It was found that the emission of pollutant gases increases with an increase in thermal input. Hydrocarbon has the maximum percentage, and it was present at 66.2 ppm at medium thermal input. A maximum thermal efficiency was observed with the largest diameter pot. With increasing loading height, thermal efficiency first increased, and then after attaining maximum at some specific loading height, it decreased.

The present work aims to analyze experimentally the effect of loading height of LPG burner on thermal efficiency and develop a correlation equation which predicts efficiency as a function of loading height.

4.3 Experimental Setup and Procedure

In this experiment, the thermal efficiency was determined by boiling water test according to the procedure mentioned in the Bureau of Indian Standard (BIS) IS 4246:2002 [14]. Different apparatus or equipment are shown in Fig. 4.1. In this study, an LPG cylinder weighing 7.5 kg when filled fully with gas and a weighing machine with the least count of 0.001 is used. Other essential equipment is pressure gauge, rotameter, gas pipe, aluminium vessel, aluminium stirrer, stove burner, and resistance temperature detector (RTD) with a display. Loading height arrangement was made by rack and pinion mechanism. Figure 4.2 shows the photograph of the setup for conducting thermal efficiency tests as per BIS standard. The experiment was performed at a uniform pressure of 2.492 KN/m. The aluminium vessel was loaded on the loading cell and weighed along with the stirrer and lid. The LPG cylinder was also weighed before starting the experiment. Now the vessel was filled with the distilled water amounting to 1 kg. The vessel was kept on the stove, when the loading height was at minimum, i.e., 14. 5 mm. Now the water is allowed to reach 31 ± 1 °C. As the vessel temperature is higher than that of water, the water finally reaches

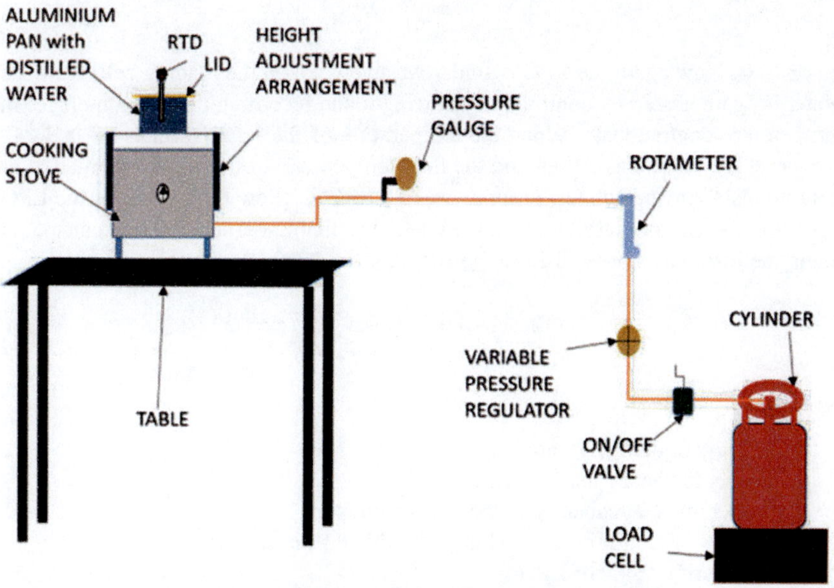

Fig. 4.1 Schematic of the test setup for thermal efficiency test (BIS standard)

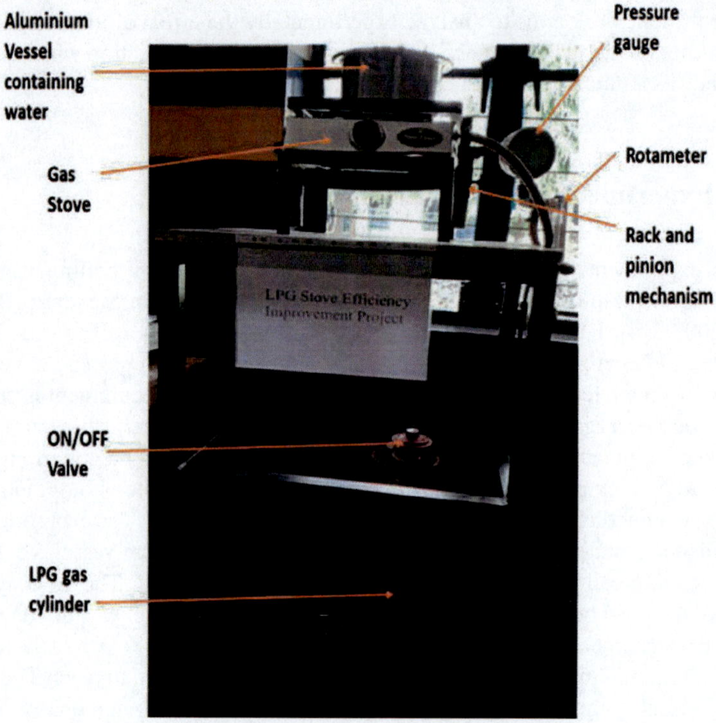

Fig. 4.2 A photograph showing thermal efficiency test setup with loading height arrangement

35 ± 1 °C. Now again the LPG cylinder was weighed on the loading cell. Now the water is again heated by continuously stirring it and recording the rising increase in temperature continuously. When the temperature of the water reaches 88 ± 1 °C, the gas stove valve was closed and the final temperature of the water recorded after continuously stirring for few seconds was 90 ± 1 °C. Now the weight of the LPG cylinder is again measured using a load cell. The thermal efficiency was calculated using the formula as prescribed in the BIS IS 4246:2002.

$$\eta = \frac{\left((\mathrm{m}_w \times C_w) + \left(\mathrm{m}_p \times C_p\right)\right)(T_2 - T_1)}{m_f \times CV} \tag{4.1}$$

where

m_w = mass of distilled water,
C_w = specific heat capacity of water,
C_p = specific heat capacity of the aluminium pan,
m_p = mass of aluminium pan along with lid and stirrer,
CV = calorific value of LPG,
m_f = mass of the fuel.

Table 4.1 Thermal efficiency corresponding to loading height

Loading height (mm)	Avg. fuel consumed (g)	Efficiency%
14.5	14.167	40.910
15.5	13.800	41.996
16.5	11.967	48.430
17	10.667	54.333
17.3	10.533	55.021
17.5	10.433	55.458
17.7	10.667	54.333
18	10.900	53.170
18.5	11.733	49.394
19.5	12.133	47.765
21.5	13.267	43.685

4.4 Calculation

The thermal efficiency was calculated using the formula prescribed by the BIS. Putting the following values in the formula mentioned in the Sect. 4.3: $m_w = 1000$ g, $C_w = 4.1826$ J/g k, $C_p = 0.896$ J/g k, $m_p = 689.9$ g, $m_f =$ mass of the fuel, CV = 45,636.12 J/g, $T_1 = 35$ °C and $T_2 = 90$ °C. The efficiency obtained is shown through a sample calculation.

$$\eta = \frac{((1000 \times 4.1826) + (689.9 \times 0.896))(90 - 35)}{m_f \times 45636.12} = \frac{5.786}{m_f}$$

At the loading height of 17.5 mm, $m_f = 10.433$. So, $\eta = \frac{5.786}{10.433} \times 100 = 55.458\%$. The thermal efficiency for each loading height was calculated. Table 4.1 shows the thermal efficiency corresponding to each loading height.

4.5 Uncertainty Analysis

An error analysis was done through uncertainty analysis on the calculated efficiency based on the experimental data. An uncertainty of ±1.142% in the derived efficiency from experiments is estimated which shows a low value of error. The formulation for uncertainty calculation used in the current work is described.

Suppose a function depends on $x_1, x_2, x_3 \ldots x_n$. If $y = f(x_1, x_2, x_3 \ldots x_n)$. Partially differentiating,

$$\partial_y = \left[\frac{dy}{dx1} \cdot \partial x1 + \frac{dy}{dx2} \cdot \partial x2 + \cdots \frac{dy}{dxn} \partial xn\right]$$

Squaring both sides,

$$\partial_{y^2} = \left(\frac{dy}{dx1}\right)^2 \cdot \partial x1^2 + \left(\frac{dy}{dx2}\right)^2 \cdot \partial x2^2 + \cdots \left(\frac{dy}{dxn}\right)^2 \cdot \partial xn^2$$

$$\partial y = \sum \left[\left(\frac{dy}{dxi}\right)^2 \cdot \partial xi^2\right]^{1/2}$$

The uncertainty formula was derived by using the gas stove efficiency formula (see Eq. 4.1). Since calorific value, mass of aluminium utensil, specific heat capacity of water and aluminium are fixed quantity; partial differentiating Eq. (4.1) gives,

$$\partial \eta^2 = \left(\frac{d\eta}{dm_w}\right)^2 \cdot \partial m_w^2 + \left(\frac{d\eta}{d(\Delta T)}\right)^2 \cdot \partial(\Delta T)^2 + \left(\frac{d\eta}{dm_f}\right)^2 \cdot dm_f^2$$

Partial differentiation w.r.t. mw,

$$\frac{d\eta}{dm_w} = \frac{c_{pw} \cdot \Delta T}{m_f * CV} \tag{4.2}$$

Partially differentiating w.r.t. ΔT,

$$\frac{d\eta}{d(\Delta T)} = \frac{\left[m_w \cdot c_{p_w} + m_a \cdot c_{p_a}\right]}{m_f * CV} \tag{4.3}$$

Partially differentiating w.r.t. m_f,

$$\frac{d\eta}{dm_f} = \frac{\left[m_w \cdot c_{p_w} + m_a \cdot c_{p_a}\right] \cdot \Delta T}{m_f^2 * CV} \tag{4.4}$$

Using values in formula based on physical experiments with $m_w = 1000$ g, $\Delta T = 55$ °C, $m_f = 10.433$ g, $\partial m_w = \pm 0.5$ g, $\partial(\Delta T) = \pm 1$°C, $\partial m_f = \pm 0.1$g.
Putting the above values in respective formulas,

$$\frac{d\eta}{dm_w} = \frac{(4.1826 * 55)}{(10.433 * 45636.12)} = 0.00048316$$

$$\frac{d\eta}{d(\Delta T)} = \frac{(1000 * 4.1826) + (698.9 * 0.896)}{(10.433 * 45636.12)} = .01009997$$

$$\frac{d\eta}{dm_f} = \frac{[(1000 * 4.1826) + (698.9 * 0.896)] * 55}{(10.433^2 * 45636.12)} = 0.053244355$$

Computing Uncertainty in efficiency,

$$\partial\eta^2 = (0.00048316)^2 \cdot (0.5)^2 + (0.01009997)^2 \cdot 1^2 + (0.053244355)^2 \cdot 0.1^2$$

$$\partial\eta^2 = 0.000130416. \text{ Therefore, } \partial\eta(\%) = \pm 1.142\%.$$

4.6 Result and Discussion

The variation of efficiency with increasing loading height is shown in Fig. 4.3. It can be seen from the plot that with an increase in loading height, the thermal efficiency first increases. There is an optimum loading height h_l at which the thermal efficiency reaches its peak value. Further increasing height decreases the efficiency.

For smaller values of h_l, the residence time of the gases in the combustion zone is less, and the flame is quenched by impinging on the relatively colder base of the vessel and the flame temperature does not rise to the maximum value. Beyond the optimum value of h_l, the flue gases cool before reaching the vessel bottom as residence time is greater. The trend of increasing efficiency with an increase in height is also related to the change in flame shape [10, 15]. At first, when the loading height is increased to some level, both of the flames' inner rich premixed and outer diffusion flame

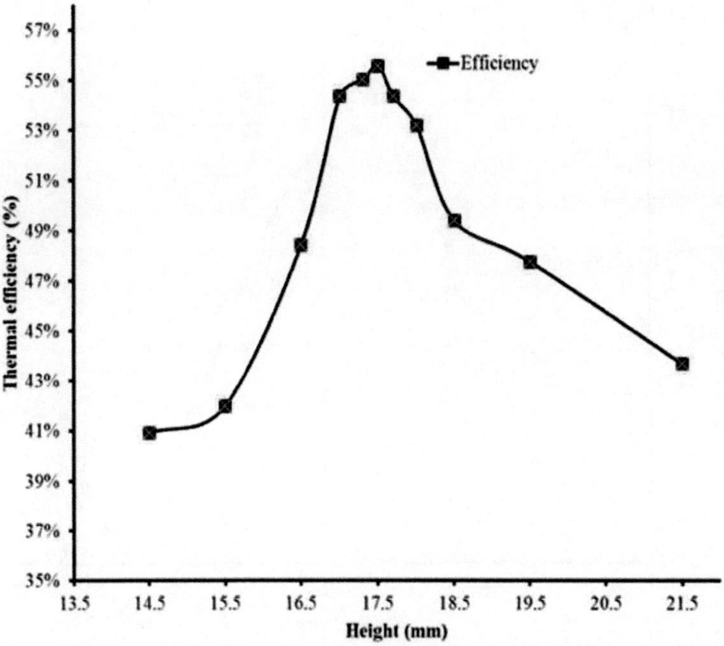

Fig. 4.3 A plot showing the variation of thermal efficiency with loading height

are trumpet-shaped, this type of flame produces a wide high-temperature region so flame covers most of the vessel heating surface. Hence heat transfer is higher and heat loss to the surrounding is less, as the distance between the burner and vessel surface is optimum, when the efficiency is highest. And as the loading height is further increased, the thermal efficiency starts decreasing. Flame shape changes as the loading height is increased. Usually rich premixed flame takes a converging shape with a decrease in impingement. This results in a reduction in the area covered by flame and less heat is transferred to the vessel. More heat is lost by convection and radiation to the surrounding air and hence lower efficiency is observed.

The values of efficiency were interpolated, and a fourth-degree polynomial equation was formulated by curve fitting. The correlation developed between thermal efficiency (ηth) with loading height (hl) is as follows:

$$\eta_{th} = 167.913 - 38.834h_i + 3.344h_i^2 - 0.126h_i^3 + 0.00178h_i^4$$

Using this equation and the calculated efficiency from the experimental data was plotted and compared on a single graph (see Fig. 4.4). The correlation closely follows the thermal efficiency obtained from experimental data with the r-square value of 0.92.

The co-relationship developed also shows that thermal efficiency increases to a maximum of 54.4% at an optimum value of h_l (17.7 mm). A small change in loading height changes the efficiency significantly as the latter is proportional to the fourth power of loading height.

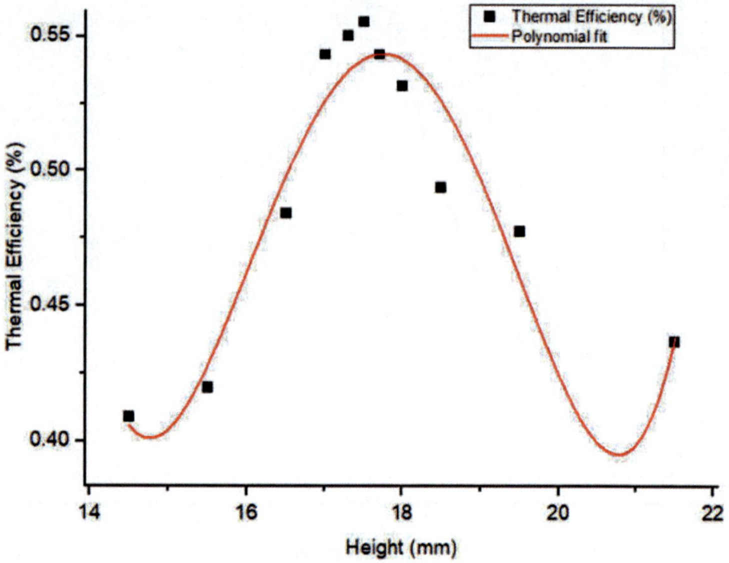

Fig. 4.4 Comparison of correlation developed and thermal efficiency from experiment data

4.7 Conclusions

In this study, the effect of loading height on the gas stove with one burner is experimentally determined. From the above discussion, it can be concluded that

- Thermal efficiency is sensitive to changes in loading height. The loading height has a significant effect on the thermal efficiency.
- When loading height is increased, thermal efficiency increases up to an optimum value of loading height. Further increase in loading height results in a decrease in thermal efficiency due to conical flame characteristics and increased heat losses to the environment.
- A relationship between loading height and thermal efficiency was developed in good agreement with experimental results.

Thermal efficiency can be enhanced by improving burner port design and swirl angle inclinations. Reduction in heat losses from flames by developing a radiation shield around the burner ring can be studied for enhancing the thermal performance of gas stoves. Also, the concave design of the base of vessels can be studied in which the impinging flame contact with the bottom and residence time of combustion products is increased.

Acknowledgements The authors would like to thank the management of The NorthCap University for providing the financial support to carry out the experimental work under the VCIF scheme. The author would like to thank Mr. Vijender Pal, laboratory technical assistant, for his contribution during experimental work.

References

1. Kandpal, J.B., Maheshwari, R.C., Kandpal, T.C.: Indoor air pollution from domesticcookstoves using coal, kerosene and lpg. Energy Convers. Manag. **36**(11), 1067–1072 (1995)
2. Behera, D., Balamugesh, T.: Indoor air pollution as a risk factor for lung cancer in women. J. Assoc. Phys. India **53**, 190–192 (2005)
3. Aisyaha, L., Ruliantoa, D., Wibowoa, C.S.: Analysis of the effect of preheating system to improve efficiency in LPG-fuelled small industrial burner. Energy Proc. **65**, 180–185 (2015)
4. Villacís, S., Martínez, J., Riofrío, A.J., Carrión, D.F., Orozco, M.A., Vaca, D.: Energy efficiency analysis of different materials for cookware commonly used in induction cookers. Energy Proc. **75**, 925–930 (2015)
5. Basu, D., Saha, R., Ganguly, R., Datta, A.: Performance improvement of LPG cook stoves through burner and nozzle modifications. J. Energy Inst. **81**(4), 218–225 (2008)
6. Boggavarapu, P., Ray, B., Ravikrishna, R.V.: Thermal efficiency of LPG and PNG-fired burners: experimental and numerical studies. Fuel **116**, 709–715 (2014)
7. Wichangarm, M., Matthujak, A., Sriveerakul, T., Sucharitpwatskul, S., Phongthanapanich, S.: Investigation on thermal efficiency of LPG cooking burner using computational fluid dynamics. Energy **203**, 117849 (2020)
8. Muthukumar, P., Shyamkumar, P.I.: Development of novel porous radiant burners for LPG cooking applications. Fuel **112**, 562–566 (2013)

9. Teotia, S., Yadav, V.K., Sharma, S., Yadav, J.P.: Effect of porosity and loading height on the performance of household LPG gas stoves. Proc. Inst. Mech. Eng. Part E J. Process Mech. Eng. 1–8 (2021). https://doi.org/10.1177/0954408920987024
10. Hou, S., Ko, Y.: Effects of heating height on flame appearance, temperature field and efficiency of an impinging laminar jet flame used in domestic gas stoves. Energy Convers. Manag. **45**, 1583–1595 (2004)
11. Ashman, P.J., Junus, R., Stubington, J.F., Sergeant, G.D.: The effects of load height on the emissions from a natural gas-fired domestic cooktop burner. Combust. Sci. Technol. **103**, 283–298 (1994)
12. Joshi, J.B., Pandit, A.B., Patel, S.B., Singhal, R.S., Bhide, G.K., Mariwala, K.V., Devidayal, B.A., Danao, S.P., Gudekar, A.S., Shinde, Y.H.: Development of efficient designs of cooking systems. I. Experimental. Ind. Eng. Chem. Res. **51**, 1878–1896 (2012)
13. Agarwal, P., Anand, A., Gupta, R.: Performance analysis of conventional LPG cooking stove. Int. J. Appl. Bioeng. **9**(1), 15–19 (2015)
14. Indian Standard, IS 4246: Domestic gas stoves for use with liquefied petroleum gases specification (fifth revision) (2002)
15. Bhatt, D.S., Mulla, I.A., Dhamangaonkar, P. R.: Design and development of burner ring to enhance thermal efficiency of domestic LPG stove and assessment of its performance. In: International Conference on Recent Innovations in Technology, R.I.T., Kottayam, Kerala, pp. 385–390 (2011)

Chapter 5
Theoretical Investigation of Physical, Mechanical and Thermal Properties of Al–Cu Functionally Graded Material Through Gradation Laws

Pankaj Kumar Chauhan and Sabah Khan

Abstract Functionally graded materials are tailor-made materials. In the present work, the theoretical study of Al–Cu functionally graded material through gradation laws is investigated. The properties like density, Young's modulus, shear modulus, specific heat and thermal conductivity are taken into consideration. The gradation laws such as power law, exponential law and sigmoid law gradation laws are applied to evaluate the behavior of these parameters at varying index numbers from one to three. It has been concluded that increasing the value of power law index (n), variation in all the effective material properties (mechanical and physical) of Al–Cu FGM, changing from linear into a polynomial way. Al (material at the bottom surface) properties are becoming more predominant up-to a larger range of z/h ratio with increasing value of power law index (n). At the value of $n = 1$, power law and Sigmoid law both are predicting the same results which are simply the rule of mixture.

5.1 Introduction

Engineering applications, such as aerospace, biomedical, space vehicles, electronics, and defense industries, have wide use of layered composite for decades due to their design flexibility to achieve required stiffness and strength, high strength to weight, low maintenance cost, lightweight and stiffness to weight properties. The problem of delamination is the prime concern in composites because of the extreme interlaminar stresses developed [1]. Incompetency of conventional composites for high-temperature uses is another grey area [2]. To encounter these problems, traditional composites are being replaced by functionally graded materials (FGMs) day by day. First time, concept of functionally graded material was introduced in 1984 in Japan by a team of researchers in the field of aerospace. FGMs are the new age materials, which are formed using two or more materials (typically metal and ceramics)

P. K. Chauhan (✉) · S. Khan
Department of Mechanical Engineering, Jamia Millia Islamia, New Delhi 11025, India

© The Author(s), under exclusive license to Springer Nature Singapore Pte Ltd. 2022

V. V. Rao et al. (eds.), *Computational and Experimental Methods in Mechanical Engineering*, Smart Innovation, Systems and Technologies 239,
https://doi.org/10.1007/978-981-16-2857-3_6

in such a way that constituents vary from one plane to another, along a direction (usually thickness) with continuous varying compositions [3]. FGMs may also be defined in another way as a group of materials showing continuously varying graded material properties [4]. FGMs show varying mechanical properties and microstructures in a continuous manner which also overcome the problems of composites of stress concentrations at internal boundaries [5]. Generally, metal and ceramics are used in FGMs as metals have greater fracture toughness and superior strength whereas ceramics have antioxidant property along with good resistance towards thermal loading [6]. In literature, a vast review towards advancement in analysis of FGMs, especially for beam-type structures, on the effect of system parameters, different deformation theories, boundary conditions and different laws for property distribution have been discussed [7]. Future aspects of the FGMs as a new class of material have been reported in the literature by various researchers in the form of literature reviews on analysis, modelling and optimization [6, 8–25]. Based on the literature, FGMs are mainly of two types for example materials with continuous gradation and materials with step-wise gradation. Usually, FGMs have two distinct material compositions having gradual continuous variation from one plane (bottom) to the other plane (top) along a direction as presented in Fig. 5.1a. In another type of FGMs, the material elements change with stepwise gradation illustrated in Fig. 5.1b.

In the upcoming time, FGMs have potential scope in different engineering applications. In this direction modelling of FGMs as per designer requirement using different gradation theories and different effective material properties have great importance. The present work is intended to fulfil such requirements of modelling and different effective material properties with an example of calculations.

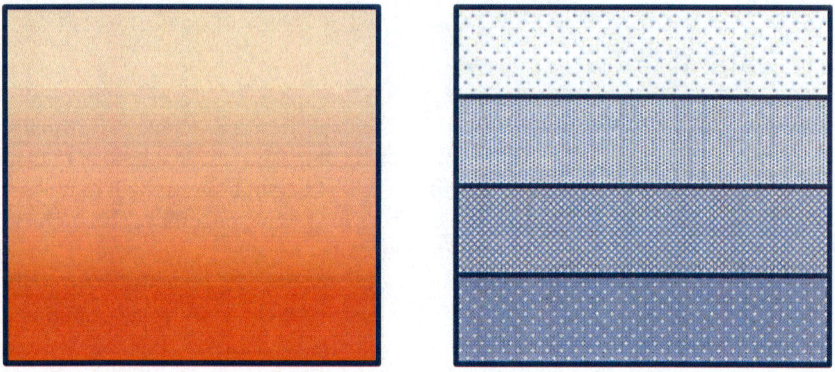

(a) (b)

Fig. 5.1 FGMs type **a** continuous and **b** discontinuous (step)

5.2 Materials and Methods

In the present study, we have taken the Al–Cu FGM for consideration. Figure 5.2 shows the gradation of Al-rich bottom with Cu gradation (Cu composition is varied as 90, 80, 70, 60, 50, 40, 30, 20, 10 wt%). Gradation of Cu is done with respect to the z-direction; x and y direction gradation is neglected. The rectangular cross-section of the sample is considered having a height h, made of a mixture of Cu and Al. Here the top surface is of pure Cu ($z = h/2$) and the bottom surface ($z = -h/2$) is of pure Al.

The FGM of Al–Cu was sintered by powder metallurgy methodology. The detailed discussion of the sintering process of Al–Cu FGM was discussed in our published literature [26]. There are several laws used for the modelling of functionally graded materials (Power law, exponential law and Sigmoid law) given below.

The power law is demonstrated by

$$V_{Cu}(z) = \left(\frac{h + 2z}{2h}\right)^n, \tag{5.1}$$

where n, h and z represent the power law index, the height of FGM layer in the z-direction and the distance from the reference plane where the property has to be measured, respectively. With respect to the reference plane at any point z, the effective material properties P_{eff} can be expressed as

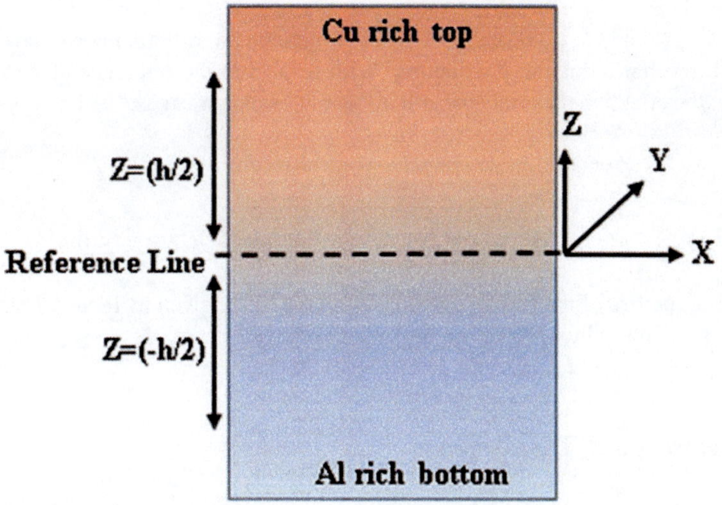

Fig. 5.2 Volume fraction of Cu and Al throughout the FGM layer

$$P_{\mathrm{eff}}(z) = (P_{\mathrm{Cu}} - P_{\mathrm{Al}})\left(\frac{h+2z}{2h}\right)^n + P_{\mathrm{Al}}, \tag{5.2}$$

where P_{Cu} and P_{Al} are the material property of Cu and Al, respectively. According to Exponential law for gradation, the material properties variation is expressed as per the following equation:

$$P_{\mathrm{eff}}(z) = P_{\mathrm{Cu}}\exp^{\left(-e\left(1-\frac{2z}{h}\right)\right)}, \tag{5.3}$$

where e may be given by $e = \frac{1}{2}\ln\left(\frac{P_{\mathrm{Cu}}}{P_{\mathrm{Al}}}\right)$,

where e is the factor used to describe the gradation along with the height of developed FGM in the z-direction. This law reveals the simple rule of mixtures. Sigmoid law for material properties distribution, using two power law functions, ensures even stress distribution at the layer interfaces. Expressions for volume fraction according to this law may be expressed in equations as.

$$V_{\mathrm{Cu1}}(z) = 1 - \frac{1}{2}\left(\frac{\frac{h}{2} - z}{\frac{h}{2}}\right)^n \quad \text{for } 0 \le z \le h/2 \tag{7.4a}$$

$$V_{\mathrm{Cu2}}(z) = \frac{1}{2}\left(\frac{\frac{h}{2} + z}{\frac{h}{2}}\right)^n \quad \text{for } -h/2 \le z \le 0 \tag{7.4b}$$

where $V_{\mathrm{Cu1}}(z)$ and $V_{\mathrm{Cu2}}(z)$ are the Sigmoid function (two different power law functions) for volume fraction distribution. With respect to the reference plane at any point z, the effective material properties P_{eff} can be expressed as.

$$P_{\mathrm{eff}}(z) = P_{\mathrm{Cu}}V_{\mathrm{Cu1}} + (1 - V_{\mathrm{Cu1}})P_{\mathrm{Al}} \quad \text{for } 0 \le z \le h/2 \tag{7.5a}$$

$$P_{\mathrm{eff}}(z) = P_{\mathrm{Cu}}V_{\mathrm{Cu2}} + (1 - V_{\mathrm{Cu2}})P_{\mathrm{Al}} \quad \text{for } -h/2 \le z \le 0 \tag{7.5b}$$

The properties of the matrix material (Al and Cu) are given in Table 5.1, and the same has been used for modelling of Al–Cu FGM.

5.3 Result and Discussion

The laws of gradation are applied for the Al–Cu FGM by using the mechanical and physical properties given in Table 5.1. The result comprises the study of physical and mechanical properties like Density (ρ), Young's Modulus (E), Shear modulus (G), Specific heat (C_p) and Thermal conductivity (k), along the thickness (z-direction)

Table 5.1 Different properties of constituent phases*

Sr. No.	Properties	Al	Cu	Unit
1	Density (ρ)	2700	8935	kg/m^3
2	Young's modulus (E)	75	130	GPa
3	Shear modulus (G)	29	46	GPa
4	Specific heat (C_p)	963	380	J/kg K
5	Thermal conductivity (k)	150	258	W/m K

*Constituent phases properties are taken as the average of min and max values

of the Al–Cu FGM. Figure 5.3a shows the variation of density (ρ) along the z-axis (according to Power and Exp. law). It has been observed that on increasing the value of power law index (n), variation in effective density of Al–Cu FGM, changing

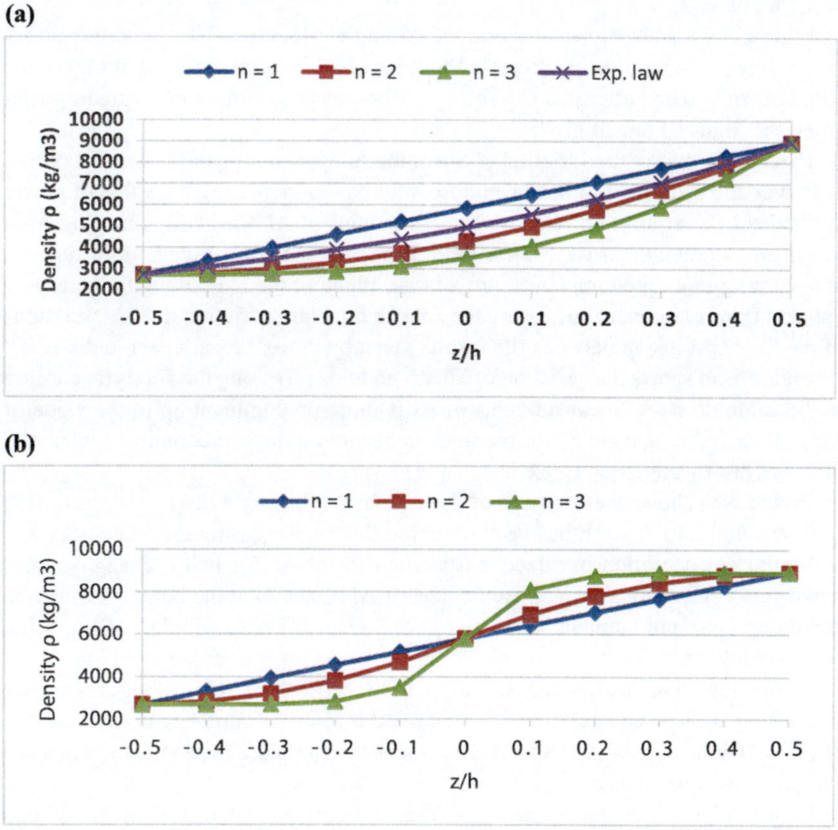

Fig. 5.3 **a** Variation of density (ρ) along the z-axis (according to Power and Exp. law). **b** Variation of density (ρ) along the z-axis (according to Sigmoid law)

from linear into a polynomial way. Density of Al (material at the bottom surface) is becoming more predominant up-to a larger range of z/h ratio with increasing value of power law index (n). Using Exponential law, variation in density of Al–Cu FGM lies in between the results given by power law at $n = 1$ and $n = 2$.

Figure 5.3b shows the variation of density (ρ) along the z-axis (according to Sigmoid law). The density of Al is more predominant up-to the value of $z/h = 0$ and after that Cu density is becoming more predominant when the Sigmoid law of gradation is used.

Figure 5.4a shows the variation of Young's Modulus (E) along the z-axis (according to Power and Exp. law). It has been observed that on increasing the value of power law index (n), variation in effective Young's Modulus of Al–Cu FGM, changing from linear into a polynomial way. Young's Modulus of Al (material at the bottom surface) is becoming more predominant up-to a larger range of z/h ratio with increasing value of power law index (n). Using Exponential law, variation in Young's Modulus of Al–Cu FGM lies in between the results given by power law at $n = 1$ and $n = 2$.

Figure 5.4b shows the variation of Young's Modulus (E) along the z-axis (according to Sigmoid law). Young's Modulus of Al is more predominant up-to the value of $z/h = 0$ and after that Cu Young's Modulus is becoming more predominant when the Sigmoid law of gradation is used.

Figure 5.5a shows the variation of Shear modulus (G) along the z-axis (according to Power and Exp. law). It has been observed that on increasing the value of power law index (n), variation in effective Shear modulus of Al–Cu FGM, changing from linear into a polynomial way. Shear modulus of Al (material at the bottom surface) is becoming more predominant up-to a larger range of the z/h ratio with increasing value of power law index (n). Using the Exponential law, variation in Shear modulus of Al–Cu FGM lies in between the results given by power law at $n = 1$ and $n = 2$.

Figure 5.5b shows the variation of Shear modulus (G) along the z-axis (according to the Sigmoid law). Shear modulus of Al is more predominant up-to the value of $z/h = 0$ and after that Cu Shear modulus is becoming more predominant when the Sigmoid law of gradation is used.

Figure 5.6a shows the variation of Specific heat (Cp) along the z-axis (according to Power and Exp. law). It has been observed that on increasing the value of power law index (n), variation in effective Specific heat of Al–Cu FGM, changing from linear into a polynomial way. Specific heat of Al (material at the bottom surface) is becoming more predominant up-to a larger range of z/h ratio with increasing value of power law index (n).

While Fig. 5.6b shows the variation of Specific heat (Cp) along the z-axis (according to Sigmoid law). Specific heat of Al is more predominant up-to the value of $z/h = 0$ and after that Cu Specific heat is becoming more predominant when the Sigmoid law of gradation is used.

Figure 5.7a shows the variation of Thermal conductivity (k) along the z-axis (according to Power and Exp. law). It has been observed that on increasing the value of power law index (n), variation in effective Thermal conductivity of Al–Cu FGM, changing from linear into a polynomial way. Thermal conductivity of Al (material at

(a)

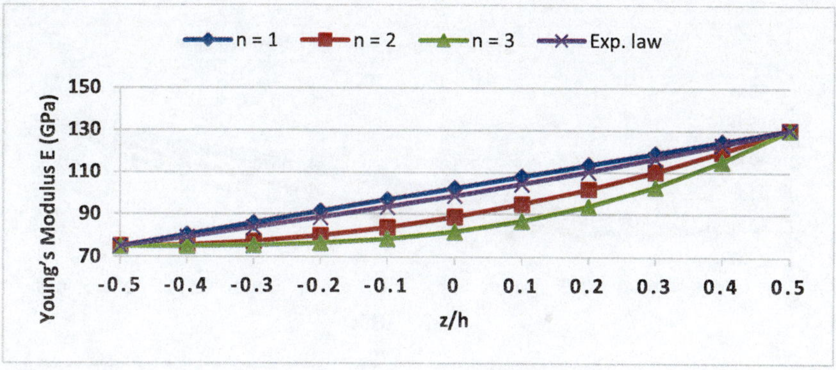

(b)

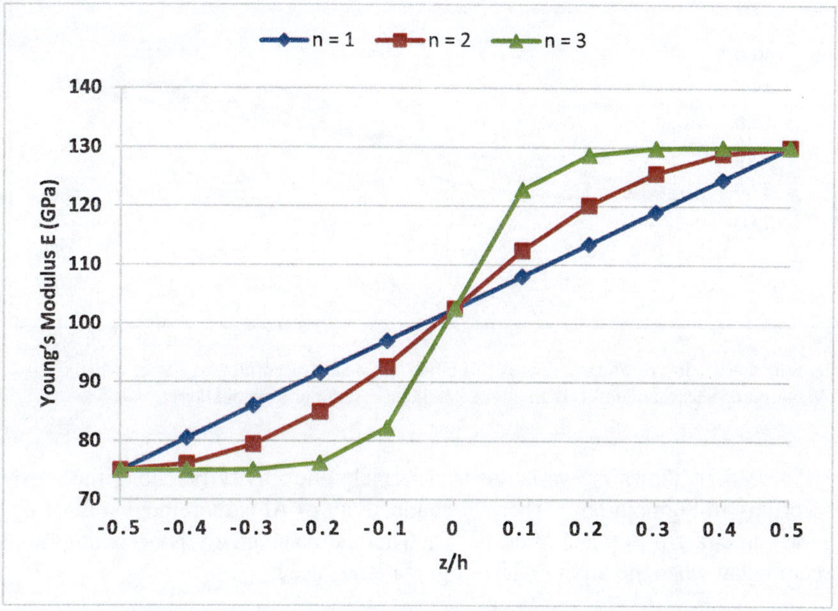

Fig. 5.4 **a** Variation of Young's modulus (E) along the z-axis (according to Power and Exp. law) **b** Variation of Young's modulus (E) along the z-axis (according to Sigmoid law)

the bottom surface) is becoming more predominant up-to a larger range of *z*/*h* ratio with increasing value of power law index (n). Using Exponential law, variation in Thermal conductivity of Al–Cu FGM lies in between the results given by power law at $n = 1$ and $n = 2$.

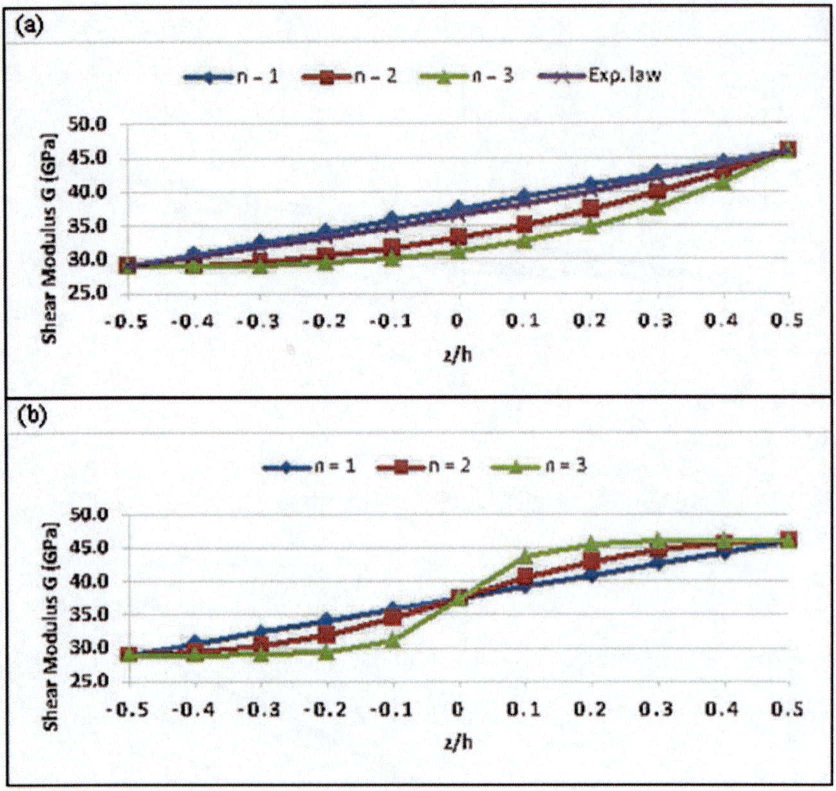

Fig. 5.5 **a** Variation of Shear modulus (G) along the z-axis (according to Power and Exp. law) **b** Variation of Shear modulus (G) along the z-axis (according to Sigmoid law)

Figure 5.7b shows the variation of Thermal conductivity (k) along the z-axis (according to Sigmoid law). Thermal conductivity of Al is more predominant up-to the value of z/h = 0 and after that Cu Thermal conductivity is becoming more predominant when the Sigmoid law of gradation is used.

5.4 Conclusion

The gradation laws are successfully applied. It has been concluded that the Al–Cu FGM comply with the gradation laws and the results observed are as follows:

1. It has been observed that on increasing the value of power law index (n), variation in all the effective material properties (mechanical and physical) of Al–Cu FGM, changing from linear into a polynomial way.

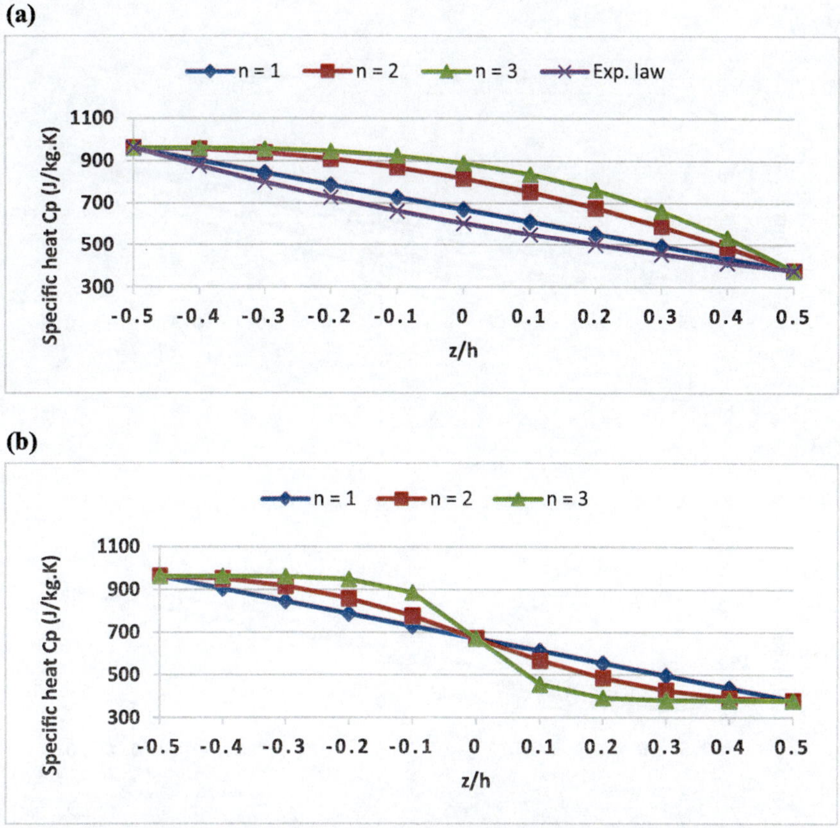

Fig. 5.6 **a** Variation of specific heat (Cp) along the z-axis (according to Power and Exp. law). **b** Variation of specific heat (Cp) along the z-axis (according to Sigmoid law)

2. Al (material at the bottom surface) properties are becoming more predominant up-to a larger range of the z/h ratio with increasing value of power law index (n).
3. At the value of $n = 1$, power law and Sigmoid law both are predicting the same results which are simply the rule of mixture.
4. It has been observed that following the Exponential law, variation in effective material properties of Al–Cu FGM lies in between the results given by power law at $n = 1$ and $n = 2$.
5. Power law predicts that effective material properties can be varied by varying the power law index, but using exponential law, it is not possible.
6. It has been observed that properties of Al are more predominant up-to the value of $z/h = 0$ and after that Cu properties are becoming more predominant when the Sigmoid law of gradation is used.

(a)

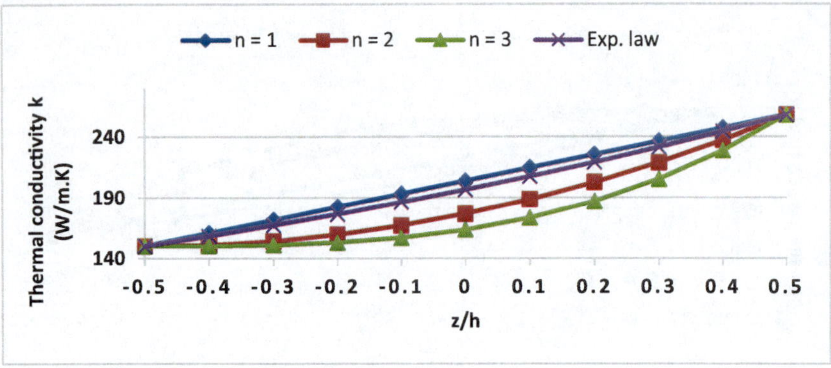

(b)

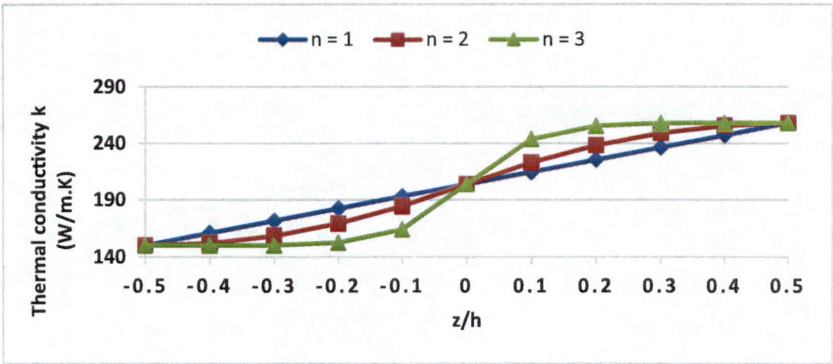

Fig. 5.7 **a** Variation of thermal conductivity (k) along the z-axis (according to Power and Exp. law) **b** Variation of Thermal conductivity (k) along the z-axis (according to Sigmoid law)

7. On the basis of results, it has been concluded that a wide range of spectrum of the Al–Cu FGM effective material properties can be estimated using the different laws of gradation. Indeed, experimental validation of these gradation laws to choose a suitable gradation law is an open-end for future scope of work.

References

1. Swaminathan, K., Naveenkumar, D.T., Zenkour, A.M., Carrera, E.: Stress, vibration and buckling analyses of FGM plates—a state-of-the-art review. Compos. Struct. **120**, 10–31 (2016)
2. Shen, H.S.: Functionally Graded Materials: Nonlinear Analysis of Plates and Shells. Taylor & Francis Group, CRC press, Boca Raton, FL, USA (2009)
3. Koizumi, M.F.G.M.: FGM activities in Japan. Compos. B Eng. **28**(1–2), 1–4 (1997)

4. Malekzadeh, P., Heydarpour, Y.: Free vibration analysis of rotating functionally graded cylindrical shells in thermal environment. Compos. Struct. **94**(9), 2971–2981 (2012)
5. Nie, G., Zhong, Z.: Dynamic analysis of multi-directional functionally graded annular plates. Appl. Math. Model. **34**(3), 608–616 (2010)
6. Jha, D.K., Kant, T., Singh, R.K.: A critical review of recent research on functionally graded plates. Compos. Struct. **96**, 833–849 (2013)
7. Chauhan, P.K., Khan, I.A.: Review on analysis of functionally graded material beam type structure. Int. J. Adv. Mech. Eng. **4**(3), 299–306 (2014)
8. Birman, V., Byrd, L.W.: Modeling and analysis of functionally graded materials and structures. Appl. Mech. Rev. **60**(5), 195–216 (2007)
9. Gibson, R.F.: A review of recent research on mechanics of multifunctional composite materials and structures. Compos. Struct. **92**(12), 2793–2810 (2010)
10. Liew, K.M., Zhao, X., Ferreira, A.J.: A review of meshless methods for laminated and functionally graded plates and shells. Compos. Struct. **93**(8), 2031–2041 (2011)
11. Mahamood, R.M., Akinlabi, E.T., Shukla, M., Pityana, S.: Functionally graded material: an overview. In: Proceedings of the World Congress on Engineering, vol. III, pp. 2–6 (2012)
12. Udupa, G., Rao, S.S., Gangadharan, K.V.: Functionally graded composite materials: an overview. Procedia Mater. Sci. **5**, 1291–1299 (2014)
13. Thai, H.T., Kim, S.E.: A review of theories for the modeling and analysis of functionally graded plates and shells. Compos. Struct. **128**, 70–86 (2015)
14. Gupta, A., Talha, M.: Recent development in modeling and analysis of functionally graded materials and structures. Prog. Aerosp. Sci. **79**, 1–14 (2015)
15. Dai, H.L., Rao, Y.N., Dai, T.: A review of recent researches on FGM cylindrical structures under coupled physical interactions, 2000–2015. Compos. Struct. **152**, 199–225 (2016)
16. Wu, C.P., Liu, Y.C.: A review of semi-analytical numerical methods for laminated composite and multilayered functionally graded elastic/piezoelectric plates and shells. Compos. Struct. **147**, 1–15 (2016)
17. Naebe, M., Shirvanimoghaddam, K.: Functionally graded materials: a review of fabrication and properties. Appl. Mater. Today **5**, 223–245 (2016)
18. Swaminathan, K., Sangeetha, D.M.: Thermal analysis of FGM plates—a critical review of various modeling techniques and solution methods. Compos. Struct. **160**, 43–60 (2017)
19. Kulkarni, P., Dhoble, A., Padole, P.: A review of research and recent trends in analysis of composite plates. Sādhanā **43**(6), 96 (2018)
20. Sayyad, A.S., Ghugal, Y.M.: Modeling and analysis of functionally graded sandwich beams: a review. Mech. Adv. Mater. Struct. 1–20 (2018)
21. Xu, F., Zhang, X., Zhang, H.: A review on functionally graded structures and materials for energy absorption. Eng. Struct. **171**, 309–325 (2018)
22. Loh, G.H., Pei, E., Harrison, D., Monzon, M.D.: An overview of functionally graded additive manufacturing. Addit. Manuf. **23**, 34–44 (2018)
23. Punera, D., Kant, T.: A critical review of stress and vibration analyses of functionally graded shell structures. Compos. Struct. **210**, 787–809 (2019)
24. Gayen, D., Tiwari, R., Chakraborty, D.: Static and dynamic analyses of cracked on functionally graded structural components: a review. Compos. Part B Eng. **217**, 106982 (2019)
25. Nikbakht, S., Kamarian, S., Shakeri, M.: A review on optimization of composite structures Part II: functionally graded materials. Compos. Struct. **214**, 83–102 (2019)
26. Chauhan, P.K., Khan, S.: Microstructural examination of aluminium-copper functionally graded material developed by powder metallurgy route. In: Materials Today: Proceedings, vol. 25, Part 4, pp. 833–837 (2020)

Chapter 6
Thermodynamic Analysis of N_2O Transcritical Cycle Using Dedicated Mechanical Subcooling

Pradeep Kumar and Shivam Mishra

Abstract The thermodynamic analysis to improve the energetic performance of N_2O transcritical cycle using dedicated mechanical subcooling is done for three different evaporator temperatures 5, −5 and −30 °C. The analysis is done for the environment temperature ranging from 30 to 40 °C. Propane is used as a refrigerant in mechanical subcooling cycle. It has been found that if combined cycle is used, the maximum increments in COP and specific cooling capacity (SCC) are 22% and 30%, respectively. It has also been found that using mechanical subcooling cycle, the reduction in optimum heat rejection pressure is higher at lower evaporator temperature, i.e. −30 °C and power consumption ratio is higher at higher evaporator temperature, i.e. 5 °C. Further, COP increment using different refrigerants in mechanical subcooling cycle has been presented, where no significant differences have been found. The increment indicates that this cycle is more efficient if the environment temperature is more than 30 °C.

6.1 Introduction

The world is facing energy crisis and environmental challenges due to increasing demand of power on account of rapid growth in population and industrial development. Refrigeration and air conditioning sectors are also adding to this problem due to high power consumption and environmental issues related to the usage of different refrigerants. In an article regarding future refrigerants [1], it has been argued how to consider all environmental issues or regulatory requirements together. From these perspectives, natural refrigerants do appear more attractive than others. Several natural refrigerants are regaining their importance because of their zero ODP and low GWP and are on a revival path [2]. Nitrous oxide (N_2O) may also be used as a future refrigerant [3]. While CO_2-based systems have already been developed and utilized for simultaneous cooling and heating applications, N_2O still needs to be explored.

P. Kumar · S. Mishra (✉)
G. L. Bajaj Institute of Technology & Management, Greater Noida, UP, India

© The Author(s), under exclusive license to Springer Nature Singapore Pte Ltd. 2022
V. V. Rao et al. (eds.), *Computational and Experimental Methods in Mechanical Engineering*, Smart Innovation, Systems and Technologies 239,
https://doi.org/10.1007/978-981-16-2857-3_7

However, there is a constraint with CO_2 that it cannot be used below an evaporation temperature of –53 °C as its triple point is nearly –56.56 °C. Since the triple point temperature of N_2O is –90.82 °C and the boiling point temperature (BPT) is –88.47 °C, hence N_2O can be preferred over CO_2 for lower evaporator temperature. Furthermore, as mentioned in article by [4], critical temperature, critical pressure and molecular weight of N_2O and CO_2 causes similar behaviour, and N_2O has lesser toxicity compared to CO_2, also its GWP is more than CO_2, but N_2O still comes under the consideration of low GWP.

6.2 Thermodynamic Analysis of the System

Thermodynamic analysis of the N_2O transcritical refrigeration cycle using dedicated mechanical subcooling (DMS) is presented. Engineering equation slover (EES) academic commercial V9.699 f-chart software has been used in the present analysis. Figure 6.1 represents the schematic for the N_2O transcritical refrigeration cycle which utilizes a dedicated mechanical subcooling process.

In dedicated mechanical subcooling, the main cycle and the subcooling cycle uses it's seperate condenser. Thornton [5] carried out a thermodynamic performance to study the performance of a dedicated mechanical subcooling system for different degrees of subcooling using a thermodynamic model. This system utilizes a single-stage compression system for medium- and high-evaporating temperatures and a two-stage compression system with intercooling for low-evaporating temperature

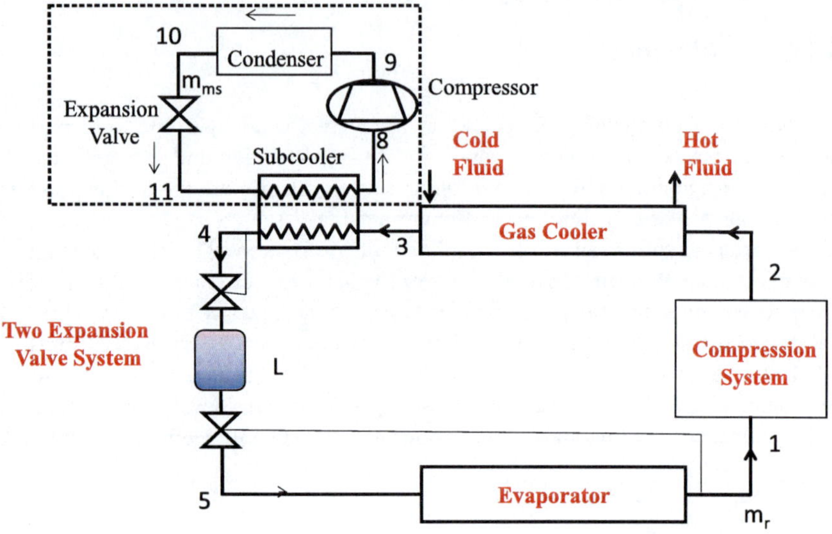

Fig. 6.1 Schematic representation of transcritical refrigeration cycle using mechanical subcooling

(LET) for the evaluation of N_2O transcritical cycle [6]. Both cycles reject the heat at the environment temperature. An optimum heat rejection pressure that is obtained from the iteration process is utilized while carrying out the calculation process. The assumptions for the analysis are mentioned below:

- The system is assumed to be in a steady-state condition and all the processes are considered to be steady flow processes.
- The compression process has been assumed to be non-isentropic for all the compressors.
- Heat losses have been neglected.
- Pressure losses in the pipelines are neglected for both cycles.
- The exit gas cooler temperature is 5 °C more than the environment temperature.
- The degree of superheat in the evaporator is the same, i.e. 10 °C for both cycles.
- The expansion process is an isenthalpic process for both cycles.

By applying the energy balance equations, to the subcooler section, we can calculate the performance of combined cycle depicted in Fig. 6.1. This further gives the relation between the mass flow rates of refrigerants in both cycles [7]. By using that, the power consumption ratio of both cycles is obtained by Eq. (6.1), and the overall COP of the complete cycle by Eq. (6.2).

$$\frac{P_{c.ms}}{P_{c,r}} = \frac{m_{ms} * w_{c,ms}}{m_r * w_{c,r}} = \frac{\Delta h_{sub} * w_{c,ms}}{q_{0,ms} * w_{c,r}} \tag{6.1}$$

$$COP = \frac{m_r * q_{0,r}}{m_r * w_{c,r} + m_{ms} * w_{c,ms}} = \frac{q_{0,r}}{w_{c,r} + \frac{\Delta h_{sub} * w_{c,ms}}{q_{0,ms}}} \tag{6.2}$$

Suffix *ms* denotes mechanical subcooling cycle while suffix *r* denotes simple refrigeration cycle.

6.3 Results and Discussion

Three different evaporator temperatures have been taken for the analysis that is mentioned as 5, −5 and −30 °C with different degrees of subcooling, and the environment temperatures range from 30 to 40 °C. Propane is taken as the refrigerant in the subcooling cycle, and EES [8] software has been used for the analysis of the cycle.

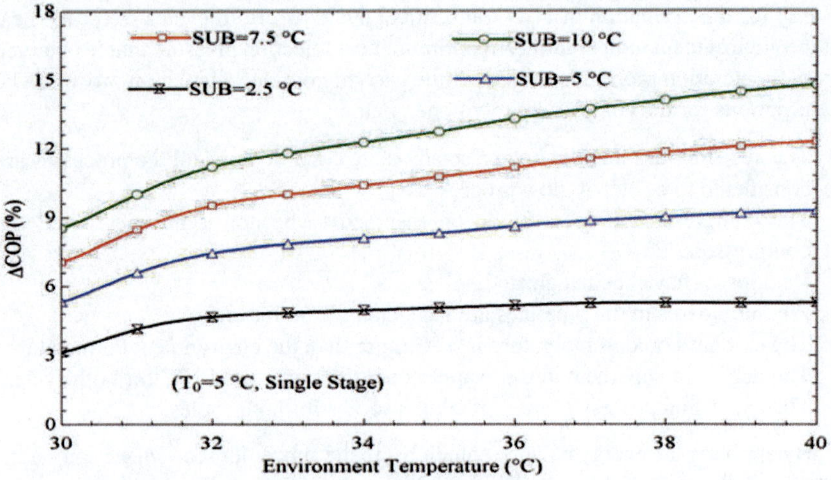

Fig. 6.2 Variation in COP increments with environment temperature at various degrees of subcooling ($T_0 = 5$ °C, single-stage)

6.3.1 *Effect of Environment Temperature on Percentage Improvement in COP at Different Degrees of Subcooling*

Variation of percentage increment in COP with environment temperature for different degrees of subcooling at different environment temperatures has been shown in Figs. 6.2, 6.3 and 6.4. It has been observed that the improvement in COP will be higher for all evaporator temperatures if the cycle is operated at an environment temperature above 30 °C. Because at high environment temperature, the optimum (HRP) of transcritical cycle is reduced. It is noticed that the improvement in COP depends on evaporator temperature also, and it is higher for evaporator temperatures of −5 and −30 over 5 °C. For all evaporator temperatures, the percentage improvement in COP increases as the degree of subcooling increases.

6.3.2 *Effect of Gas Cooler Pressure (GCP) on Specific Cooling Capacity (SCC) for Different Degrees of Subcooling*

Figure 6.5 shows a change in specific cooling capacity (SCC) with GCP (SCC) increase as the degree of subcooling increases, and also, the compression ratio is reduced by using mechanical subcooling. The effect has been studied for an evaporator temperature of −30 °C and environment temperature of 35 °C. It has been

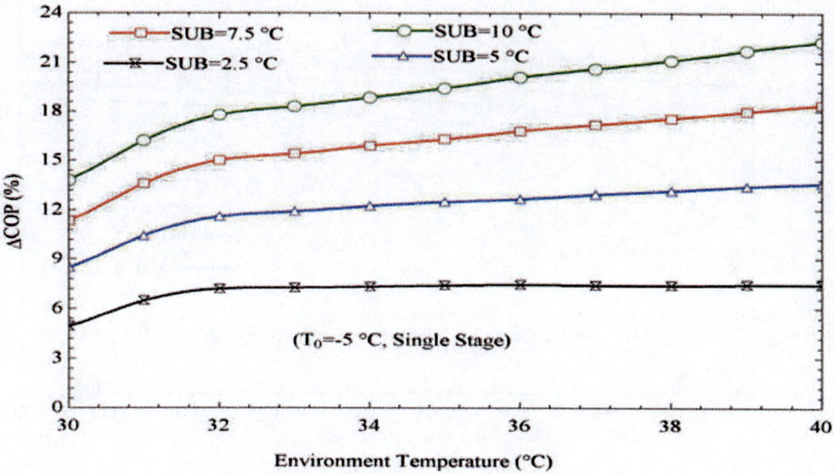

Fig. 6.3 Variation in COP increments with environment temperature at various degrees of subcooling ($T_0 = -5$ °C, single-stage)

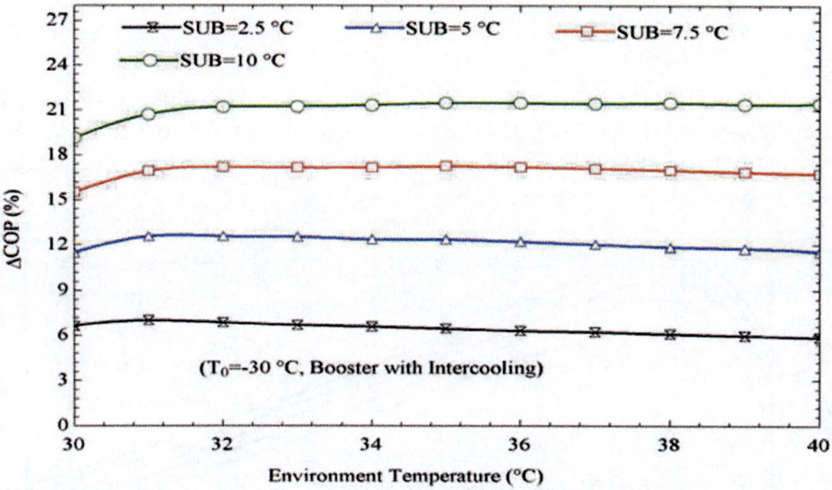

Fig. 6.4 Change in COP with environment temperature at various degrees of subcooling ($T_0 = -30$ °C, booster with intercooling)

observed that the (SCC) increases as gas-cooler exit pressure increases. Further, it can be concluded that COP of the cycle increases as the (SCC) increases.

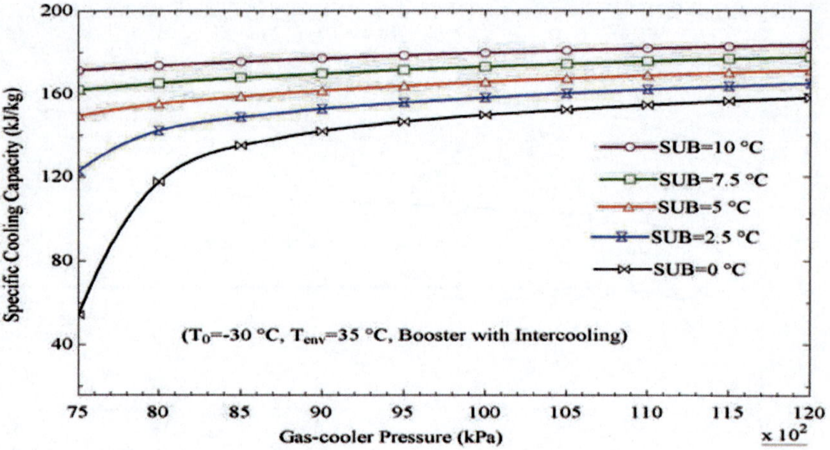

Fig. 6.5 Variation in (SCC) (SCC) with GCP at various degrees of subcooling ($T_0 = -30$ °C, $T_{env} = 35$ °C, booster with intercooling)

6.3.3 *Effect of GCP on COP at Different Degrees of Subcooling*

Variations in GCP for different degrees of subcooling and three different evaporator temperatures 5, −5 and −30 °C have been shown in Figs. 6.6, 6.7 and 6.8. The environment temperature is 35 °C for all the evaporator temperatures. It has been

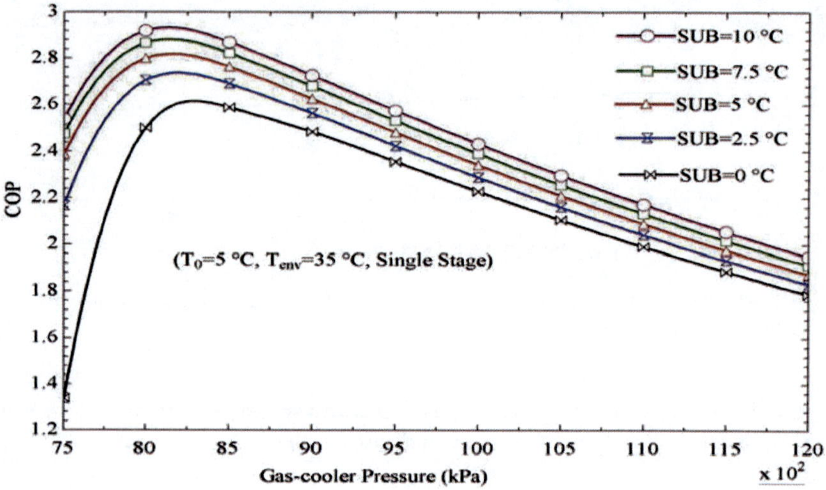

Fig. 6.6 Variation in COP with GCP at various degrees of subcooling ($T_0 = 5$ °C, $T_{env} = 35$ °C, single-stage)

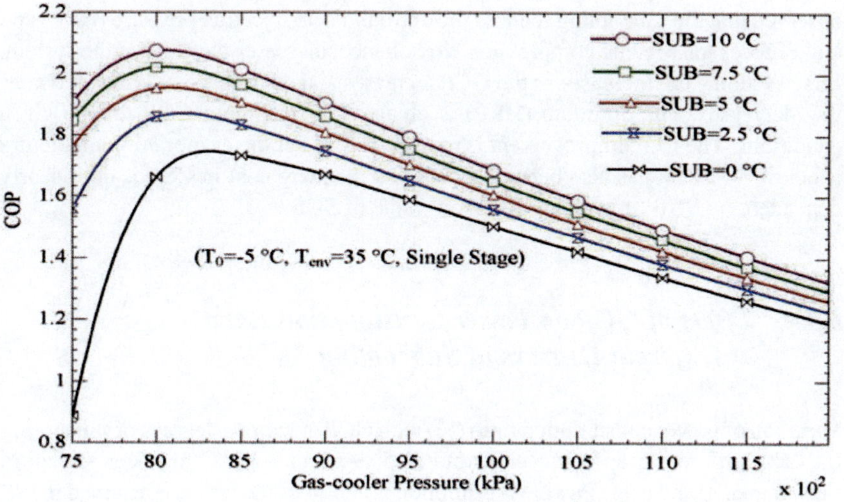

Fig. 6.7 Variation in COP with GCP at various degrees of subcooling ($T_0 = -5$ °C, $T_{env} = 35$ °C, single-stage)

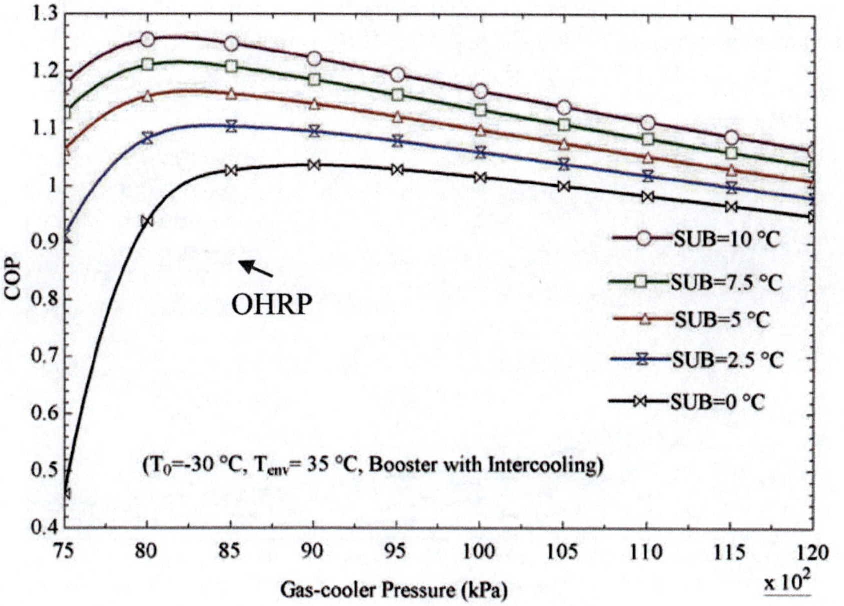

Fig. 6.8 Variation in COP with GCP at various degrees of subcooling ($T_0 = -30$ °C, $T_{env} = 35$ °C, booster with intercooling)

observed that the subcooling reduces the optimum heat rejection pressure (HRP) and also reduces the specific compression work, hence, increases the COP of the system. First, system COP increases as the GCP increases and reaches a maximum and after that decreases. The optimum (HRP) is obtained at maximum COP for the given conditions. The COP improvement is more with DMS if the evaporator temperature is below −5 °C. It has also been observed that the increment in COP is not linearly dependent on SUB, it is higher for a low value of SUB.

6.3.4 *Effect of GCP on Power Consumption Ratio at Different Degrees of Subcooling*

Variation of power consumption ratio (%) with GCP at various degrees of subcooling for three different evaporator temperatures 5, −5 and −30 °C has been shown in Figs. 6.9, 6.10 and 6.11. Power consumption is a ratio of the power consumed in MS cycle to the power consumed in N_2O transcritical cycle. The environment temperature is taken as 35 °C. It has also been observed that the power consumed in MS cycle is less than 15% of the power consumed in main cycle at the optimum HRP, but it increases for the heat rejection pressures below the optimum value. Hence, for the minimum power consumption, the optimum (HRP) is required.

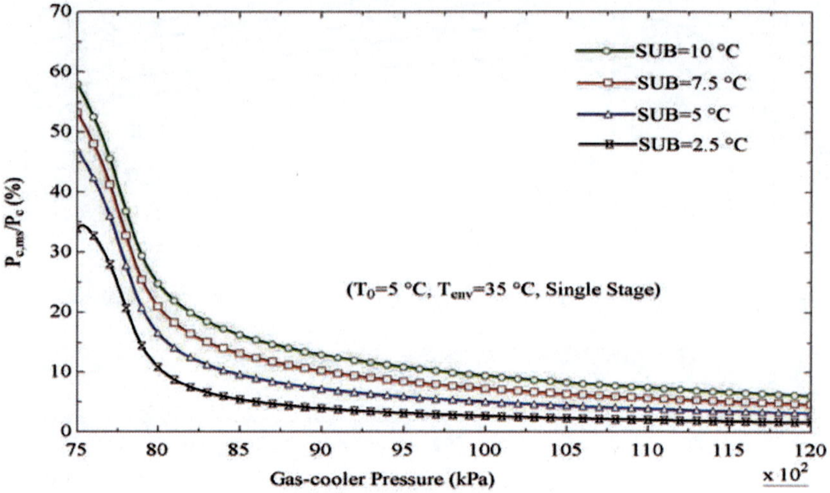

Fig. 6.9 Variation in power consumption ratio with GCP at various degrees of subcooling (T_0 = 5 °C, T_{env} = 35 °C, single-stage)

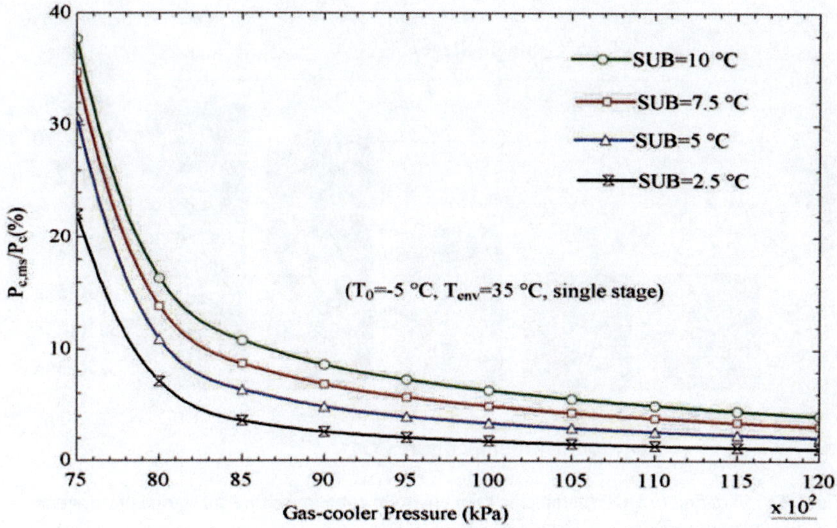

Fig. 6.10 Variation in power consumption ratio with GCP at various degrees of subcooling (T_0 = −5 °C, T_{env} = 35 °C, single-stage)

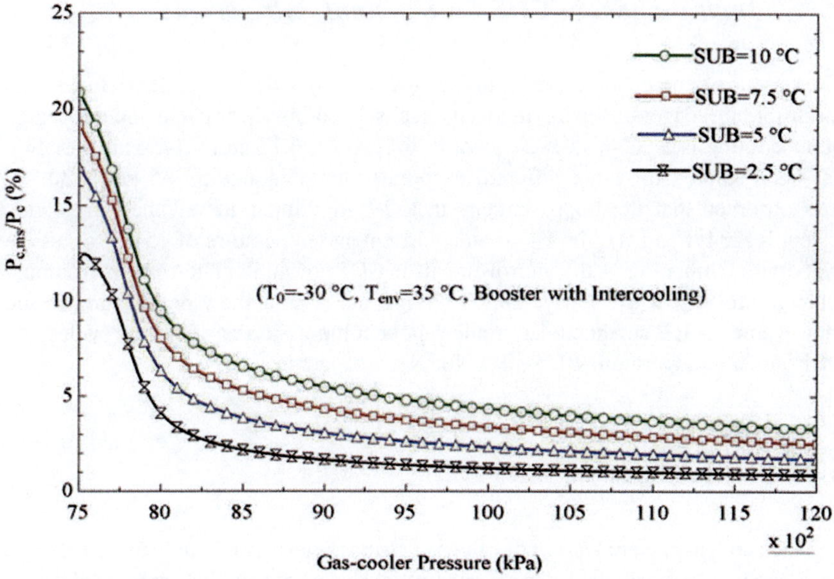

Fig. 6.11 Variation in power consumption ratio with GCP at various degrees of subcooling (T_0 = −30 °C, T_{env} = 35 °C, booster with intercooling)

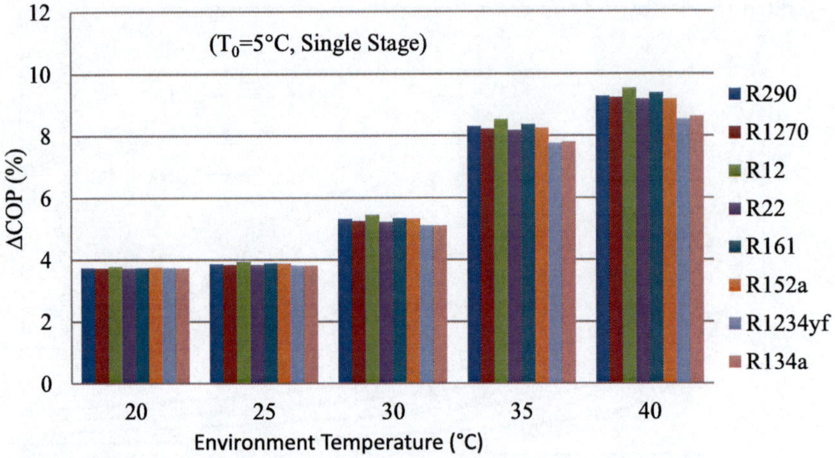

Fig. 6.12 Variation in COP increments with environment temperature for various refrigerants at 5 °C of subcooling (T_0 = 5 °C, single-stage)

6.3.5 Effect of Environment Temperature on Percentage Improvement in COP for Different Refrigerants

The variation of percentage change in COP (ΔCOP %) with environment temperature for different refrigerants using in mechanical subccoling (MS) cycle and the degree of subcooling is 5 °C has been shown in Figs. 6.12, 6.13 and 6.14. This variation has been studied for three different evaporator temperatures 5, −5 and −30 °C. It is observed that the improvements in COP are similar for all the refrigerants except R1234yf and R134a. Up to an environment temperature of 25 °C for all the evaporator temperatures, the improvements in COP are similar but from environment temperature 30 °C, the refrigerants R1234yf and R134a show degradation in the improvements. It is considered a simple vapour compression refrigeration cycle from environment temperature 20 to 30 °C for N_2O refrigerant.

6.4 Conclusions

The thermodynamic analysis of the given cycle has been carried out in this work. The performance improvements of transcritical cycle have been studied for three different evaporator temperatures and different degrees of subcooling. Finally, percentage improvement in COP has been determined and various refrigerants are used in the MS cycle. For this analysis, single-stage compression system has been used for the evaporator temperatures of 5 and −5 °C, and double-stage compression system has been used for the evaporator temperature of −30 °C. For this study, thermodynamic

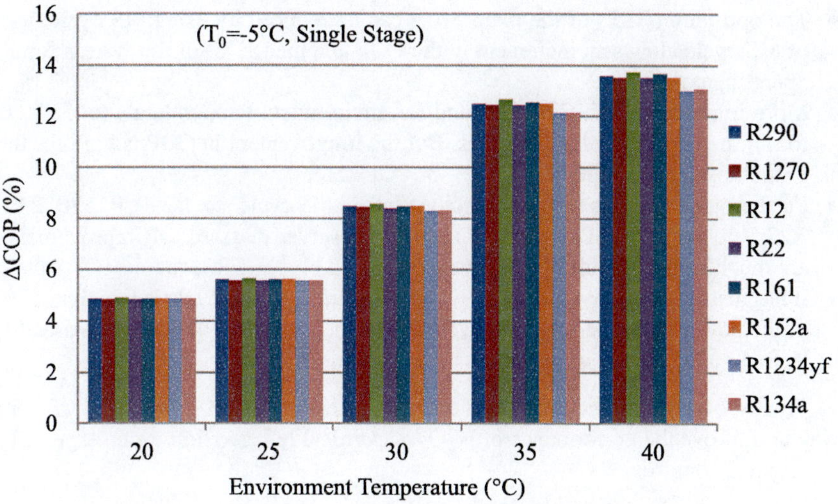

Fig. 6.13 Variation in COP increments with environment temperature for various refrigerants at 5 °C of subcooling ($T_0 = -5$ °C, single-stage)

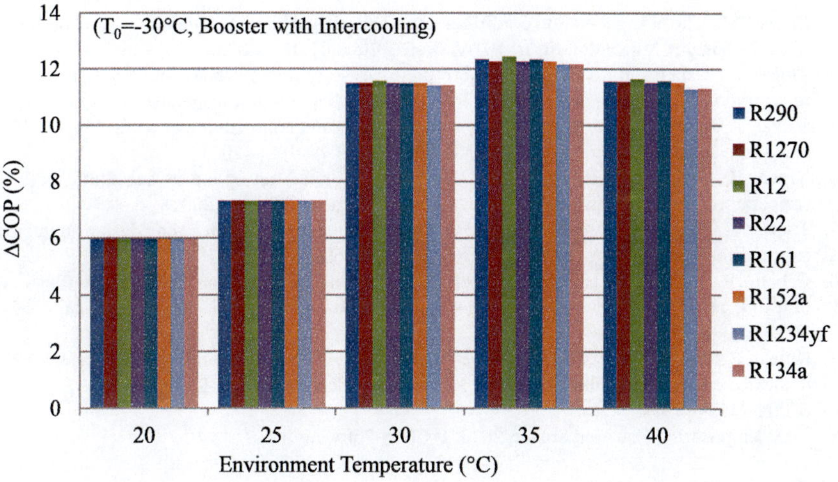

Fig. 6.14 Variation in COP increments with environment temperature for various refrigerants at 5 °C of subcooling ($T_0 = -30$ °C, booster with intercooling)

models have been developed in Engineering Equation Solver software. The following conclusions are pointed out from the present study:

- The optimum (HRP) of the main cycle can be reduced by using MS cycle. Use of MS cycle allows an increment in the COP and the (SCC) of the system for all operating conditions.
- More improvement in COP is found for environment temperature above 32 °C for all the evaporator temperatures. But the improvement in COP is high for the evaporator temperature of −5 and −30 °C.
- Furthermore, if we use different refrigerants in MS cycle, i.e. R290, R1270, R12, R22, R161, R152a, R1234yf and R134a, it is observed that the COP improvements are much similar for all refrigerants except R1234yf and R134a.
- It has also been observed that the power consumed in MS cycle is less than 15% of the power consumed in main cycle at the optimum (HRP), but it increases for the heat rejection pressures below the optimum value.

Finally, based on the conclusion of the results, the use of DMS cycle is an efficient way to improve the performance of N_2O transcritical refrigeration cycle, especially for tropical countries.

References

1. Calm, J.M.: The next generation of refrigerants—historical review, considerations, and outlook. Int. J. Refrigerat. https://doi.org/10.1016/j.ijrefrig.2008.01.013 (2008)
2. Llopis, R., Cabello, R., Sánchez, D., Torrella, E.: Energy improvements of CO_2 transcritical refrigeration cycles using dedicated mechanical subcooling. Int. J. Refrigerat., JIJR 3004 (2015)
3. Sarkar, J., Bhattacharyya, S.: Thermodynamic analysis and optimization of a novel N_2O–CO_2 cascade system for refrigeration and heating. IJR **32**, 1077–1084 (2008)
4. Kruse, H., Russmann, H.: The natural fluid nitrous oxide—an option as substitute for low temperature synthetic refrigerants. Int. J. Refrigerat. **29**(5), 799–806 (2006)
5. Thornton, J.W., Klein, S.A., Mitchell, J.W.: Dedicated mechanical-subcooling design strategies for supermarket applications. Int. J. Refrigerat. **17**(8), 508–515 (1994)
6. Cabello, R., Sánchez, D., Llopis, R., Torrella, E.: Experimental evaluation of the energy efficiency of a CO_2 refrigerating plant working in transcritical conditions. Appl. Therm. Eng. **28**, 1596–1604 (2008)
7. Kumar, P., Mishra, M.: Performance investigation of N_2O transcritical refrigeration cycle using dedicated mechanical subcooling. Int. J. Adv. Res. Sci. Eng. (IJARSE) **6**(2), 85–92 (2017) (ISSN-2319–8354)
8. EES: Engineering Equation Solver, 2014. F-Chart Software Inc.

Chapter 7
Effect of PEO Concentration on Electrochemical and Mechanical Properties of PVDF, PEO and LATP Blended Solid Polymer Electrolyte

Ponam and Parshuram Singh

Abstract The usage of solid polymer electrolytes is gaining popularity because of their safe and good performance in various batteries, especially in lithium-ion batteries. However, all these electrolytes have some limitations such as lower ionic conductivity at room temperature and deprived mechanical properties. In the present research, the electrolytes are prepared with PVDF, PEO and LATP blends to address all the mentioned issues. During the preparation of electrolytes, the contents of PEO are varied from 5 to 20% with PVDF. To improve the electrochemical properties of electrolyte, 10% (to the weight of PVDF and PEO mixture) of LATP is added to the electrolyte. The effect of variation in the proportion of PVDF and PEO contents on the electrical and mechanical properties of electrolytes is analysed in the present research. The addition of PEO in the electrolyte improves the ionic conductivity of the solid electrolyte. However, the yield strength of the electrolyte is decreased with an increase in PEO proportion. The detailed results are provided in this research article.

7.1 Introduction

The larger part of the energy requirement is fulfilled by crude oil and demand is continuously increasing. As a result, the available natural resources of crude oils are depleting rapidly [1, 2]. Therefore, the research is now shifting towards clean and renewable energy sources such as solar energy, wind energy, etc. The biggest hurdle in non-conventional energy sources is the storage of energy generated or produced. Among all energy storage devices, batteries are the most popular because of their durability, ease to connect and usage [3]. The batteries are categorized into primary battery and secondary battery. Primary batteries are not rechargeable and hence are very rarely utilized for renewable energy storage.

Ponam (✉) · P. Singh
Bhagwant University Ajmer, Rajasthan, India

© The Author(s), under exclusive license to Springer Nature Singapore Pte Ltd. 2022
V. V. Rao et al. (eds.), *Computational and Experimental Methods in Mechanical Engineering*, Smart Innovation, Systems and Technologies 239,
https://doi.org/10.1007/978-981-16-2857-3_8

The secondary batteries are rechargeable; therefore, these batteries are mostly used for the storage of renewable and non-renewable energies [4]. Further, among secondary batteries, lithium batteries are the first choice because of their multiple advantages such as higher energy density, low weight to energy ratio, durability, a lower rate of self-discharge and most important, higher working voltage. These batteries gain popularity from the date of invention, and this is the reason for their speedy improvements in the last three decades [5]. Therefore, its demand is continuously increasing in most applications and sectors like electric vehicles, electric toys, electronic gadgets, etc. This results in a continuous increase in the production of lithium batteries. According to a report, the production of lithium-based batteries surpasses the 300 Giga watt-hours annually [6]. This technique won the noble prize in 2019 as a revolutionary technique.

Highly conductive lithium metal anode enhances the performance of batteries compared to graphite anode which is generally used in primary batteries. The use of liquid electrolytes in secondary lithium-based batteries makes its usage dangerous even after many advantages. The lithium dendrites are formed during the electrochemical reaction because of the non-uniform stripping of the lithium anode. These dendrites can pierce the strainer of the battery which might cause short-circuit [7]. Most of the organic electrolytes can catch fire easily. Therefore short-circuit with combustible organic electrolyte and reactive lithium anode might lead to fire and in some cases explosion in the batteries [8]. The probability of such incidents is even higher in large appliances where large-size batteries are used. Much research is going on in parallel to overcome the safety concern of secondary lithium-based batteries. Solid electrolytes are the most optimum and best solution to the problem. The solid electrolytes not only overcome the safety concern but also provide the extra strength to the battery with an increase in energy density [9]. The solid electrolytes are not only inflammable but also reduce the formation of dendrites of lithium. Apart from this, solid electrolytes have good mechanical, thermal and electrical properties.

The solid electrolytes consist mainly of two broad categories:

- Solid electrolytes made up of organic polymer
- Solid electrolytes made up of inorganic polymers.

The solid organic polymer electrolytes have very good compatibility with the electrolyte because of a stronger adhesive bond among them. Therefore, this advantage of solid organic polymer electrolytes makes them suitable for any type of electrode. These electrolytes are easy to process and design which provides an additional benefit to the battery manufacturers and designers. Also, large-size batteries can be produced. These electrolytes are a good choice for lithium-based batteries [10]. However, organic electrolytes have a low-temperature range of operation and lower ionic conductivity. On the other hand, inorganic electrolytes have a wide range of operating temperatures and very good ionic conductivity [11].

The mechanical properties of inorganic electrolytes are better than organic electrolytes. Also, these inorganic electrolytes are more chemically stable which adds extra advantages for their usage. However, the inorganic electrolytes are brittle which makes them very difficult to process [12]. Therefore, it becomes difficult to redesign the battery size and shape while using inorganic electrolytes. These electrolytes are also having poor interface interaction and destitute processing recital, and it leads to great impedance. Some of the solid polymer electrolytes have poor ionic conductivity that limited their use in lithium-ion batteries. To overcome this problem, some organic and inorganic compounds can be used which enhance the ionic conductivity of the electrolytes [4].

In the current work, the electrochemical characteristics and mechanical properties of ion-conducting polymer electrolytes of poly (vinylidene fluoride) PVDF and poly (ethylene oxide) PEO are measured and analysed. The homogeneous mixtures of the above-mentioned polymers are prepared in different proportions to formulate the binary matrices. To enhance the properties such as electrochemical and mechanical, submicron lithium aluminium titanium phosphate (LATP) is added to the solution for preparation of solid polymer electrolytes. The prepared electrolytes with different compositions are analysed during the research, and the optimum proportion of organic–inorganic composites is found for further research. Various properties are measured for all the prepared electrolytes to find out the effect of these composites on the properties. Further analysis and results are discussed in detail in the result and discussion section.

7.2 Experimental

7.2.1 Material

The PEO (molecular weight 900,000 g/mol), PVDF (molecular weight 534,000 g/mol) and LATP are obtained from Sigma-Aldrich, USA. Whereas, dimethylformamide (DMF) is purchased from Pon Pure Chemicals Group, Chennai, India. All the polymers are vacuum-dried before use. The chemicals such as aluminium nitrate nonahydrate, lithium nitrate, titanium (IV) isopropoxide, phosphoric acid and absolute ethanol are purchased from local institute vendors.

7.2.2 Synthesis of Polymer Electrolytes

The ionic conductivity of electrolytes prepared from PVDF and PEO is not so impressive as compared to other polymer electrolytes [10, 13, 14]. Hence, to overcome this problem, LATP is added in a fixed quantity into the solid polymer electrolyte solution. First of all, the power of LATP is prepared, which is further used to enhance the ion conduction of the electrolytes. The sol–gel method is opted to prepare the LATP compound. First of all, 0.8 g of aluminium nitrate nonahydrate and 0.55 g lithium nitrate are dissolved in 80 ml of absolute ethanol. The magnetic stirrer is used to prepare the homogeneous solution and continued till a clear solution is attained. On the other hand, 2.7 mL titanium (IV) isopropoxide and 1.5 mL of phosphoric acid are dissolved in another 60 Ml of absolute ethanol in the separate mixture through magnetic stirring.

Once the homogeneous mixture of titanium (IV) isopropoxide, phosphoric acid and absolute ethanol is obtained, the first mixture is added to the solution. The solution is kept for vigorous stirring until a colourless solution is not formed. A small amount of phosphoric acid can also be added to the solution if required. The stirring will not be stopped until the solution is completely converted into a transparent gel. The completion of the process will take around 18–24 h depending on the reaction temperature and the catalyst used. Finally, the obtained mixture is kept in a water bath at 80 °C to convert into a gel. The obtained gel is dried at 80–100 °C. During this process, the colourless gel is converted into a yellowish gel. The gel is then dried at a temperature of 350 °C for 7 h. During the drying process, most reacting gases such as water vapours, NH_3, NO_2 and Cl_2 are eliminated from the gel. The dried gel is ground into powder and then calcined at 600 °C for 6 h. After that power will bring to room temperature and grind again into a very fine powder. To get the final LATP, the powder is calcined again at 900 °C for another 6 h in the air. This process completely removes any moisture left in the powder.

7.2.3 Preparation of Solid Polymer Electrolytes

To check the effect of the change of concentration of PVDF and PEO on the properties of electrolytes, five electrolyte firms are prepared. The concentration of PVDF is varied from 50 to 90% whereas the concentration of PEO is taken in the range of 05–20%. The mixture of PVDF and PEO are prepared with different weight ratios in the given range, i.e. PVDF: PEO (95:05, 90:10, 85:15 and 80:20). For each sample, firstly, the mixture of PVDF and PEO is taken into a conical flask with an airtight stopper. The mixture is dissolved into DMF at 70 °C for 10 h on a magnetic stirrer with a heating plate at a constant speed. Further, LATP of 10% by weight to the weight of PVDF + PEO mixture is added to the solution and let it react for another 10 h at the same reaction temperature.

Once the homogeneous slurry is prepared, spread it onto a glass plate and put it for drying for 10 h in a vacuum oven at a temperature of 50 °C. Repeat the process if the moisture is observed in the solid electrolyte. Cut the solid electrolytes into circular pellets of diameter 20 mm with the help of punch. Again, vacuum dry the pellets for another 2–4 h in a vacuum oven at 50 °C. Store the pellets into an airtight container for further analysis. The approach followed for the preparation of solid polymer electrolytes is very similar to the one followed by Wang et al. [15]. The schematic diagram for the solid polymer electrolyte preparation process is shown in Fig. 7.1. The nomenclature of the electrolytes sample prepared is given in Table 7.1.

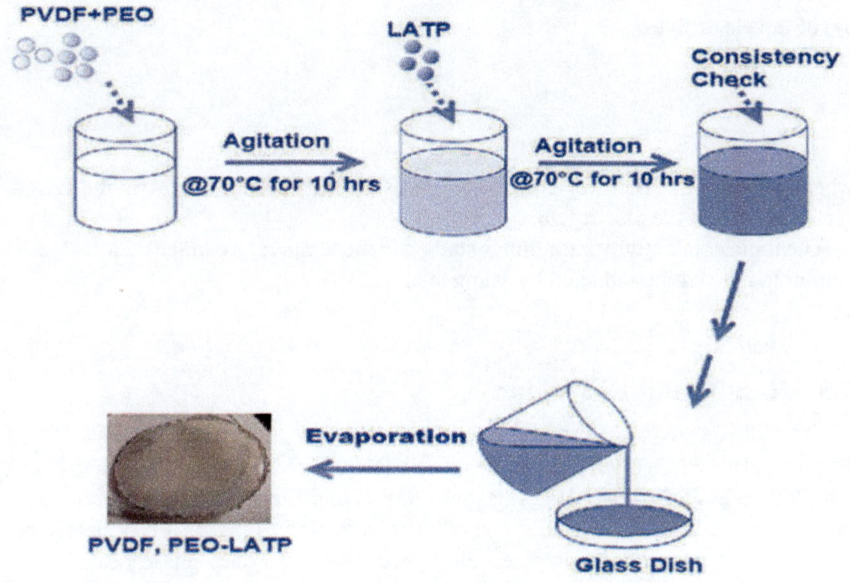

Fig. 7.1 The schematic diagram for the preparation of solid polymer electrolyte

Table 7.1 Nomenclature of the electrolytes

S. No	Name	Nomenclature
1	95PVDF5PEO-10LATP	(95% PVDF and 5% PEO) + 10% w/w LATP
2	90PVDF10PEO-10LATP	(90% PVDF and 10% PEO) + 10% w/w LATP
3	85PVDF15PEO-10LATP	(85% PVDF and 15% PEO) + 10% w/w LATP
4	80PVDF20PEO-10LATP	(80% PVDF and 20% PEO) + 10% w/w LATP

7.2.4 Characterization of Solid Polymer Electrolytes

The prepared solid polymer electrolytes are tested to analyse the various characteristics such as phase structure, electrochemical properties, thermal properties, etc. The phase structures are characterized using an X-ray diffractometer (XRD) with a scanning range of 10°–60°. Whereas, thermogravimetric analyser and Autolab are used to measure the thermal stability and ionic conductivity of solid polymer electrolytes. The thermal stability of electrolytes is measured in a nitrogen atmosphere with a heating rate of 10 °C per minute. The operating range for the process is from atmospheric temperature to 700 °C. The ionic conductivity of the electrolyte is also measured under a controlled nitrogen environment. To measure the ionic conductivity of the electrolyte, the electrolyte is kept in between two electrodes made up of stainless steel. The frequency range of current varies during the experiment from 10 Hz to 1 MHz. The following formula is used to measure the ionic conductivity

(γ) of the electrolyte:

$$\gamma = \frac{t}{R_b A}$$

where, t is the thickness of electrolyte, R_b is bulk resistance and A is the cross-sectional area of the electrolyte.

The ionic conductivity variation is studied in the temperature range of 300–353 K similar to the study conducted by Wang et al. [15].

7.3 Results and Discussion

The crystal structure of all the samples (80PVDF20PEO-10LATP, 85PVDF15PEO-10LATP, 90PVDF10PEO-10LATP and 95PVDF5PEO-10LATP) are analysed through an XRD analyser, and the results are shown in Fig. 7.2. It can clearly be observed that none of the prepared electrolytes has any sharp phase peak which shows the homogeneity among all the materials. The possession of OH group and fluoride present in PEO skeleton by PVDF molecules is one of the reasons for no sharp peak in the crystallinity structure. However, with an increase in the content of PEO in the electrolyte increases the crystallinity in the electrolyte. Because healthy H_2 bonding interaction among PEO and PVDF molecules in the electrolyte might result in a three-dimensional cross-linked matrix with lower crystallinity. The result obtained is very much similar to Wang et al. [15]. Also, no significant effect of the presence of LATP molecules is observed in the crystallinity of the solid polymer electrolyte.

However, the presence of PEO molecules and LATP molecules change the mechanical properties of electrolytes significantly, and the same can be observed in Fig. 7.3. The copious and asymmetrical bond in the structure matrix of the electrolyte leads to higher mechanical strength to the electrolyte bonding, and this results in improved Li^+ conduction by the electrolyte. However, the young's modulus of the electrolyte is not symmetrically varied with a change in PEO content in the electrolyte. With the increase in the composition of PEO molecules in the electrolyte from 5 to 10%, the young's modulus of the electrolyte decreases, and a further increase in the content of PEO from 10 to 15% increases the young's modulus to a large extent. For all the samples, the highest young's modulus is observed for the 85PVDF15PEO-10LATP sample.

The effect of change in PEO contents on the ionic conductivity of the electrolyte to operating temperature is shown in Fig. 7.4. It can be clearly observed from Fig. 7.3 that the increase in the PEO contents in the electrolyte increases the ionic conductivity of the electrolyte at ambient temperature which is taken 300 K in the present research. At room temperature, the highest ionic conductivity of 1.05×10^{-2} S-m is observed for the 80PVDF20PEO-10LATP electrolyte followed by 1.8×10^{-3} S-m, 6.8×10^{-4}

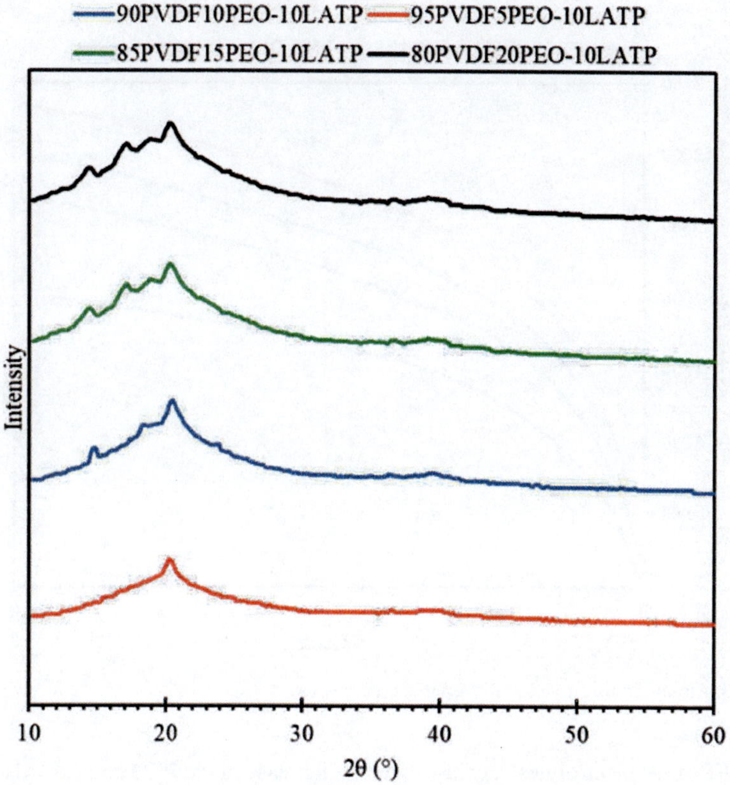

Fig. 7.2 XRD analysis of all prepared electrolytes

S-m and 3.7×10^{-4} S-m for 85PVDF15PEO-10LATP, 90PVDF10PEO-10LATP and 95PVDF5PEO-10LATP, respectively. Further, the ionic conductivity of all samples increases with the increase in operating temperature up to a melting point. However, the difference is to start decreasing as the temperature increases. At the melting point, the ionic conductivity of all sample become nearly equals. Therefore, it can be concluded that the number of energetic molecules of PEO increases with an increase in PEO content below the melting point that increases the ionic conductivity of the electrolyte [3].

The glass transition temperature (GTT) plays a very important role in the ionic conductivity of the electrolyte. The GTT for electrolyte is the temperature below which the electrolytes similarly changes their physical properties those of glass and crystal. The GTT of all samples of electrolyte is measured through differential scanning calorimetry (DSC). The result obtained for the GTT of all the samples is represented in Fig. 7.4. The GTT for 80PVDF20PEO-10LATP, 85PVDF15PEO-10LATP, 90PVDF10PEO-10LATP and 95PVDF5PEO-10LATP is −80.2 °C, −77.6 °C, −75.3 °C and −71.8 °C, respectively. From the above data, it can be concluded that

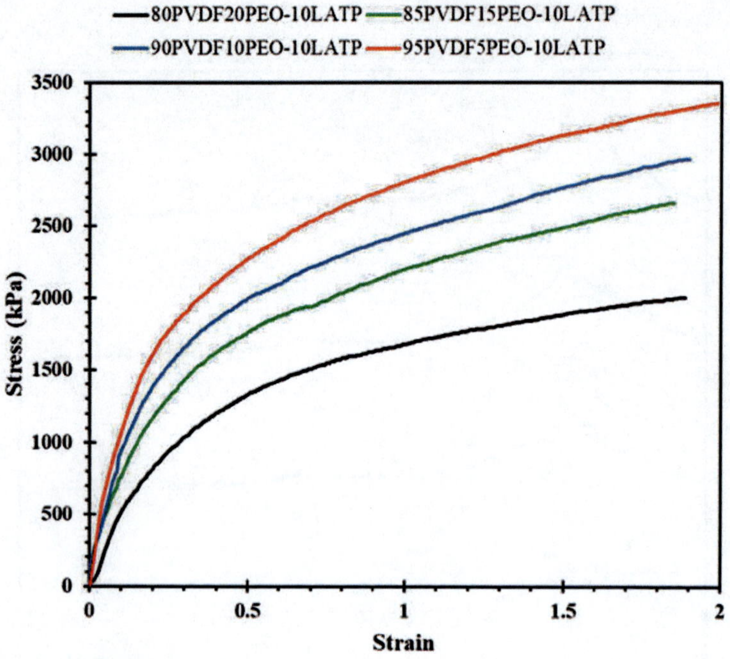

Fig. 7.3 Stress–Strain curve for all prepared electrolytes

the GTT of the electrolytes decreases with an increase in the PEO content. Also, the presence of LATP molecules in the electrolyte has decreased the GTT of the electrolytes. The ionic conductivity of the electrolyte generally is inversely proportional to the GTT, and this can be analysed from the results.

7.4 Conclusion

The effects of variation in the composition of PEO molecules on the mechanical and electrochemical properties of solid polymer electrolytes are analysed in the present research. The solid polymer electrolytes are prepared with a mixture of PVDF and PEO, also a small quantity of LATP is added to enhance the ionic conductivity of the electrolytes. Some major findings observed during the present study is concluded as:

- The homogeneity can be observed in all the prepared electrolytes due to the absorption of the OH bond of PEO by PVDF molecules. However, the presence of PEO molecules can be observed in the electrolyte with an increase in the PEO contents.
- The mechanical strength of the electrolyte decreases with an increase in the content of PEO in the electrolyte.

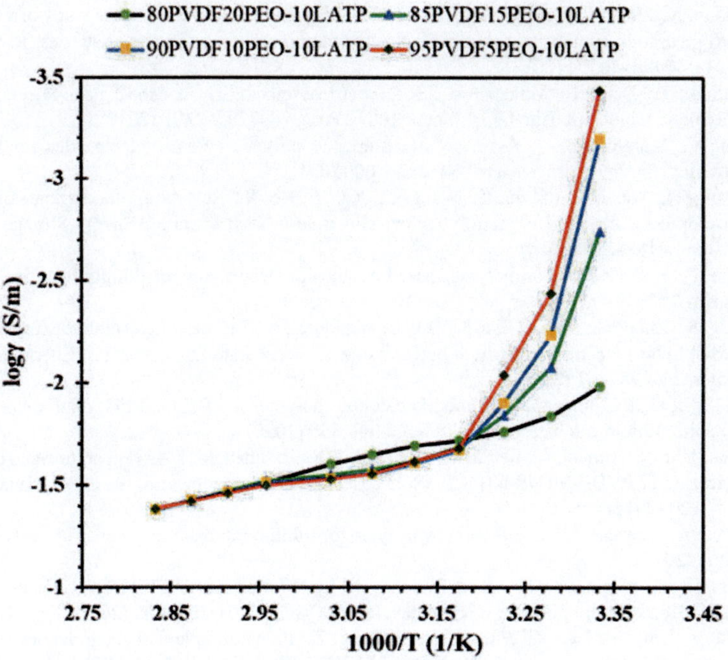

Fig. 7.4 Ionic conductivity versus temperature

- The ionic conductivity of the electrolyte increases with the increase in PEO contents in the electrolyte. The ionic conductivity also increases with an increase in operating temperature. However, near the melting point, the ionic conductivity of all electrolytes is found to be very similar to each other.
- The glass transition temperature of PEO is lower than PVDF, hence the increase in the PEO contents in electrolyte also decreases the GTT of the electrolyte which leads to higher ionic conductivity.

References

1. Kumar, P., Kumar, N.: Study of ignition delay period of n-Butanol blends with JOME and diesel under static loading conditions. Energy Sources. Part A Recover. Util. Environ. Eff. **40**(14), 1729–1736 (2018)
2. Kumar, P., Kumar, N.: Environmental effects process optimization for production of biodiesel from orange peel oil using response surface methodology. Energy Sources, Part A Recover. Util. Environ. Eff. **0**(0), 1–11 (2019)
3. Sengwa, R.J., Dhatarwal, P.: Predominantly chain segmental relaxation dependent ionic conductivity of multiphase semicrystalline PVDF/PEO/LiClO4 solid polymer electrolytes. Electrochim. Acta **338**, 135890 (2020)

4. Liang, X., Han, D., Wang, Y., Lan, L., Mao, J.: Preparation and performance study of a PVDF-LATP ceramic composite polymer electrolyte membrane for solid-state batteries. RSC Adv. **8**(71), 40498–40504 (2018)
5. Pathak, P., Deb, R., Mohapatra, S.R.: Electrical bistability in MoS2 nano-sheets doped polymeric nanocomposite films. Mater. Today Proc. **24**, 2295–2301 (2019)
6. Yu, X., Manthiram, A.: A review of composite polymer-ceramic electrolytes for lithium batteries. Energy Storage Mater. **34**, 282–300 (2021)
7. Wang, H., Sheng, L., Yasin, G., Wang, L., Xu, H., He, X.: Reviewing the current status and development of polymer electrolytes for solid-state lithium batteries. Energy Storage Mater. **33**(June), 188–215 (2020)
8. Yao, P., et al.: Review on polymer-based composite electrolytes for lithium batteries. Front. Chem. **7** (2019)
9. Bag, S., Zhou, C., Kim, P.J., Pol, V.G., Thangadurai, V.: LiF modified stable flexible PVDF-garnet hybrid electrolyte for high performance all-solid-state Li–S batteries. Energy Storage Mater. **24**, 198–207 (2020)
10. Yi, S., et al.: Fast ion conductor modified double-polymer (PVDF and PEO) matrix electrolyte for solid lithium-ion batteries. Solid State Ion. **355** (2020)
11. Ataollahi, N., Ahmad, A., Lee, T.K., Abdullah, A.R., Rahman, M.Y.A.: Preparation and characterization of PVDF-MG49-NH4CF 3SO3 based solid polymer electrolyte. E-Polymers **14**(2), 115–120 (2014)
12. Arya, A., Sharma, A.L.: Polymer electrolytes for lithium ion batteries: a critical study. Ionics **23**(3) (2017)
13. Ibrahim, S., Yassin, M.M., Ahmad, R., Johan, M.R.: Effects of various LiPF6 salt concentrations on PEO-based solid polymer electrolytes. Ionics (Kiel) **17**(5), 399–405 (2011)
14. Wu, F., Feng, T., Bai, Y., Wu, C., Ye, L., Feng, Z.: Preparation and characterization of solid polymer electrolytes based on PHEMO and PVDF-HFP. Solid State Ionics **180**(9–10), 677–680 (2009)
15. Wang, H., et al.: Mechanical property-reinforced PEO/PVDF/LiClO4/SN blend all solid polymer electrolyte for lithium ion batteries. J. Electroanal. Chem. **869** (2020)

Chapter 8
Applicability of Banana Fiber as Reinforcement in Composites

Sandeep Chauhan, Tarun Kumar Gupta, and Vishal Shankar Srivastava

Abstract The present study aims to provide a single platform for various researches on the mechanical properties of polymer composites reinforced with the banana fiber. The study also aims to spread awareness among future researchers, academicians, and practitioners regarding the potential use of the banana fiber to prepare the reinforced hybrid composites as a suitable material for various engineering and industrial applications.

8.1 Introduction

In the present time, engineers need some new or alternative materials for developing new products. Products used in modern times are subjected to static and dynamic loading and environmental conditions (dry or wet, dusty), and thus require specific properties as per the applications. For this, composite provides a solution as a strong material, as the weaker material of matrix is made strong by embedding the reinforcement. The reinforcement imparts desired properties in the composite materials as per the requirement of applications along with strength and the rigidity to withstand the structural load. Properties of composite materials can be controlled by careful selection of suitable material as matrix or reinforcement. Owing to the widespread use of composite materials in modern times, the present era is known as the era of composite materials. Composite materials find applications in almost every walk of modern life. Out of various composites, polymer matrix composites are widely used to make products that find applications in modern life. Although polymers have inadequate strength and stiffness for many structural purposes, these shortcomings are overcome by the addition of reinforcement materials. Natural fiber-reinforced polymer matrix composites attract the attention of modern researchers as these materials affect the environment less. Due to it, many research studies were undertaken by various

S. Chauhan · T. K. Gupta (✉) · V. S. Srivastava
Department of Mechanical Engineering, G. L. Bajaj Institute of Technology & Management, Greater Noida, India

© The Author(s), under exclusive license to Springer Nature Singapore Pte Ltd. 2022
V. V. Rao et al. (eds.), *Computational and Experimental Methods in Mechanical Engineering*, Smart Innovation, Systems and Technologies 239,
https://doi.org/10.1007/978-981-16-2857-3_9

researchers for developing natural fiber-reinforced composite materials for applications where high strength is not needed. These natural fiber-based composite materials possess properties suitable for applications like domestic appliances, cabinets, door and window frames, bathroom accessories, wardrobe accessories, office accessories, personal accessories, decorative articles, racks, etc. The polymer composites having a natural fiber's reinforcement are preferred in these applications due to ease in production with the cost economy and superior mechanical properties. These composites also find application for packaging of electronic gadgets due to the addition of reinforcement by the fibers with high thermal conductivity in a low thermal conductivity resin matrix which helps in keeping the item cool by dissipating the heat to the environment.

Natural fibers are less expensive, biodegradable, and possess no health hazards. Although plenty of natural fibers like banana, sisal, jute, etc., are available in the Indian subcontinent and African countries, these are rarely used as alternative materials. Presently, natural fibers are utilized to make yarn ropes, mats, articles like wall hangings, handbags, etc. Fibers like cotton, banana, and pineapple find applications in clothes and paper. Natural fibers are cellulose material and comprise microfibrils in an amorphous matrix of lignin and hemicellulose. These fibers include many fibrils that run parallel to the length. The hydrogen bonds impart the necessary strength and stiffness to the fibers. Normally, natural fibers are extracted from various plant parts and classified accordingly as shown along with examples in Fig. 8.1. The normally used natural fibers are sisal, wood, hemp, kenaf, wheat straw, coconut, flax, jute, banana, bamboo, etc. Normally, natural fibers possess some desirable properties as low-density biodegradability, renewability along with adequate strength, and stiffness. This leads to the widespread development of natural fiber-reinforced composites. Although natural fiber imparts desirable properties to composites, it has less adhesion with matrix due to the amorphous structure. Many researchers addressed this problem through suitable chemical treatment that increases direct contact between fiber and matrix.

Banana fibers possess properties like high specific strength, low elongation, fire-resistant, high moisture absorption capacity, and biodegradability. Due to these favorable properties, banana fiber-reinforced composites are attracting engineers to apply them to manufacture newer products. This paper aims to provide a detailed summary

Fig. 8.1 Classification of natural fibers with examples

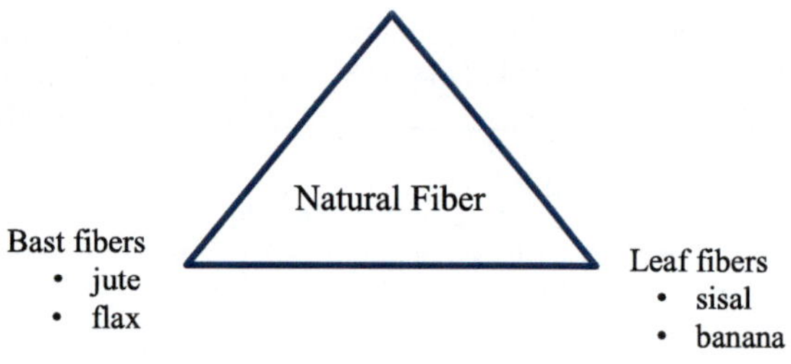

of researches carried out recently to analyze the reinforcement effects of banana fibers on the mechanical properties of developed composites. Section 8.2 describes the important chemical and mechanical properties of the banana fibers while Sect. 8.3 comprises a discussion regarding the studies conducted by different researchers to study the performance of banana fiber-reinforced composites. Section 8.4 presents the discussion on conclusions drawn from the present work.

8.2 Characteristics of Banana Fiber

This section presents a discussion on the important properties of the banana fibers. Chemically, banana fiber consists of cellulose, lignin, and hemicellulose. The usual ranges of components of banana fiber are presented in Table 8.1 along with other properties of the banana fibers. It is adequately strong with lightweight. It shows little elongation when subjected to tensile loading. Its shiny appearance depends upon the extraction process [1]. It easily absorbs moisture contents. It is recognized as an eco-friendly fiber due to its biodegradable properties.

8.3 Fiber Preparation

Banana fibers are extracted from the banana trunk in two phases. Firstly, the banana trunk is allowed to dry in sunlight for two weeks, and then it soaked into water for another two weeks and again dried in the sunlight to get the fibers [3, 4]. Afterward, the dried fibers are cut into the required length and passed through a rolling machine for thinning.

8.4 Methods of Composites Preparation

Two methods are frequently used by various researchers [5, 6] for producing the banana fibers composite—the hand layup molding technique and the compression molding technique. In the former technique, first, a de-bonding agent is applied to mold followed by a pigmented gel to obtain a good surface finish. When the gel

Table 8.1 Banana Fiber Properties (Kumar and Rajesh [2])

Particular	Cellulose %	Hemi cellulose %	Lignin %	Moisture content %	Density (kg/cm^3)	Lumen size (mm)	Tensile strength (MPa)	Young's modulus (GPa)
Value	63–64	6–19	5–10	10–11	1350	5	529–914	27–32

becomes tacky, the fibers and resins are manually applied in layers on the surface. When sufficient layers are added, then the mold was closed and left to cure for sufficient time at the required temperature under calculative load. The cured composite was ejected from the mold for further process.

In the later method, the known weight ratio of resin, hardener, and fiber was well mixed by a mixer and then poured into the mold cavity and closed. Sufficient pressure is applied for a known time for curing purposes. After that, the composite was ejected from the mold for further process.

It has been observed that the surface finish obtained by the former method is better than the later while the time taken to complete the process is more in the former method.

8.5 Banana Fiber-Reinforced Composites

In this section, some of the studies are discussed that took place to analyze the effects of banana fiber parameters on the behavior of composites manufactured. The summary of the studies carried out by different researchers in recent times to analyze the effects provided by the banana fibers used as reinforcement to the composites produced are presented in Table 8.2. Normally used matrix materials for developing banana fiber-reinforced composites are Epoxy resin, Polyester, Polypropylene, Phenol formaldehyde, and Vinyl ester. Although some research studies utilized cement and natural rubber as a matrix material to develop composite material for specific applications while some of these studies used hybrid matrix material as Vinyl ester and epoxy resin, Cashew nut shell resin liquid mixed with resin.

The length of fiber and fiber loading significantly affects the mechanical properties of composites. To understand the influence of the length of fiber on the mechanical behavior of the composites, Sumaila et al. [26] performed a study and introduced the concepts of the critical length of the fiber. While Ghosh et al. [8] undertook a study to analyze the effects of fiber percentages in a matrix on the mechanical behavior of the composite and observed an increasing trend in tensile strength with increment in the fiber fraction after an initial dip. This leads to the concept of critical volume of fibers.

Many research studies conducted by various researchers found improvement in the properties of banana fiber-reinforced composite if banana fibers are chemically treated before introducing into the polymer matrix. Some important studies conducted on this are discussed here. Among them, Joseph et al. [12] conducted a study to compare the performance of composites reinforced by banana fibers with various coatings and compared their performance with that of untreated fiber composite. They observed improvement in the mechanical behavior of composites with all coatings except latex coating. While Rajesh et al. [19] investigated the effects of pre-treatment of fibers with sodium hydroxide on strength and vibration behavior of composite. Short Banana and Sisal fibers with random orientations were used to develop the composite. They observed improvement in the behavior of composites

Table 8.2 Summary of research work carried out to analyze performance of banana fiber-reinforced composites

Sr. no.	Year	Author	Matrix material	Reinforced material	Curing time	Tests	Outcomes
1	2017	Amir et al. [7]	Polypropylene	Continuous banana fibers		Tensile, flexural, SEM	The work studied different reinforcement configurations of banana fiber, and results shows that the highest tensile and bending strength are obtained from banana yarn composites
2	2011	Ghosh et al. [8]	Vinyl ester resin	Banana fibers	24 h	Tensile	Volume fiber fraction influences the tensile strength greatly, and strength increases with increasing fiber volume fraction after an initial dip
3	2011	Hanifawati et al. [9]	Polyester	Banana fibers and glass fibers	24 h	Tensile, flexural	Improvement in mechanical properties observed by incorporation of small amount of glass fiber in banana fiber-reinforced composites
4	2005	Idicula et al. [10]	Polyester	Banana/sisal hybrid fiber	24 h	Tensile, dynamic, mechanical analysis, SEM	Developed hybrid composite by adding banana and sisal fibers together as reinforcement agent in polyester

(continued)

Table 8.2 (continued)

Sr. no.	Year	Author	Matrix material	Reinforced material	Curing time	Tests	Outcomes
5	2002	Joseph et al. [11]	Phenol formaldehyde	Banana fibers and glass fibers		Tensile, impact, SEM	The banana fiber-reinforced composites was developed which have mechanical properties to use as a structural material
6	2005	Joseph et al. [12]	Phenol formaldehyde	Banana fiber		SEM, tensile	On comparing performance of differently coated banana fiber-reinforced composites with untreated banana fiber-reinforced composite, they observed improvement in properties of the composites for all the modifications except latex coating
7	2018	Karthick et al. [13]	Epoxy resin	Banana fibers, Glass fiber		Tensile, flexural, impact, hardness	Developed and analyzed Banana–Glass fiber-reinforced hybrid epoxy composites. Results showed composite with 10 mm fiber length and 15% fiber loading possessed better mechanical properties

(continued)

Table 8.2 (continued)

Sr. no.	Year	Author	Matrix material	Reinforced material	Curing time	Tests	Outcomes
8	2016	Kumar and rajesh [2]	Natural rubber	Banana fibers		Abrasive wear, compressive strength, hardness	The studies and results strongly advocate that Banana fiber has immense potential for using as natural fiber to make reinforced rubber composites and find many applications
9	2012	Olusegun et al [14].	Epoxy resin	Ukam, hemp, banana, sisal coconut, and E-glass fiber	20 days	Tensile, bending, impact	Ukam and sisal fiber-reinforced composite materials have good mechanical properties, and these properties are enhanced by alkalization treatment. Thus, it can be considered as reinforcement for high performance polymer composites

(continued)

Table 8.2 (continued)

Sr. no.	Year	Author	Matrix material	Reinforced material	Curing time	Tests	Outcomes
10	2020	Ortega et al. [15]	Polypropylene	Banana, Banana with cotton, and linen fibers		Flame resistance, Impact, Tensile, Flexural	They developed and analyzed composites by utilizing magnesium hydroxide with matrix material and alkali treatment to the fiber material, and observed reduction in combustibility of the composites. They observed reduction in the flame propagation speed with percentage increment of magnesium hydroxide along with reduction in the mechanical properties. They established the relationship of fire properties with fiber types and number of layers
11	2015	Patel et al. [16]	Polyester resin	Banana fiber		Damage Test by machine vision technique	Developed and analyzed banana fiber-reinforced composite. They also determined optimum drilling parameters for the composites by utilizing ANOVA technique

(continued)

Table 8.2 (continued)

Sr. no.	Year	Author	Matrix material	Reinforced material	Curing time	Tests	Outcomes
12	2006	Pothan et al. [17]	Polyester resin	Banana fiber	24 h	Dynamic mechanical analysis, SEM	Developed composite by introducing banana fiber as reinforcement agent in polyester as matrix material for utilization as building material
13	2014	Prasad et al. [18]	hybrid polymer (Cashew Nut Shell Resin Liquid mixed with resin)	Jute and banana		Tensile, FEA, ANOVA	Developed jute and banana fiber-reinforced hybrid polymer matrix composite with variation in matrix composition and determined optimum composition through ANOVA technique
14	2016	Rajesh et al. [19]	Polyester resin	Banana and sisal fibers	5 h	Flexural, vibration	The work showed that stiffness of composite increases with (weight ratio) fiber up to 50% and then starts decreasing. It showed that maximum natural frequency was obtained for composites with 50% of fiber. The work also showed improvement in properties with chemical treatment of fibers with sodium hydroxide solution

(continued)

Table 8.2 (continued)

Sr. no.	Year	Author	Matrix material	Reinforced material	Curing time	Tests	Outcomes
15	2016	Ramachandran et al. [20]	Epoxy resin	Bamboo fiber, banana fiber and linen fiber	11–12 h	Impact, hardness, and fourier transform infra-red	Results showed that Bamboo–banana epoxy resin have highest strength along with better bonding between fibers and polymers. Results showed that bamboo–linen epoxy resin composite have highest hardness among all variants of composites
16	2013	Sakthivel and Ramesh [5]	Epoxy resin	banana, coir, sisal	24 h	Flexural test, impact test, hardness test, water absorption test, density test	Developed and analyzed epoxy resin-based composites with different natural fibers as the reinforcing agent. They found banana fiber-reinforced composite the best among the different composites
17	2014	Santhosh et al. [21]	Vinyl ester and epoxy resin	Banana fibers, coconut shell powder	24 h	Tensile test, flexural test, impact test	They observed improvement of properties in the composite when alkali treatment is provided to the fibers
18	2006	Sapuan et al. [3]	Epoxy resin	Woven banana fiber	24 h	Tensile and flexural, Anova	Developed woven banana fiber-reinforced composite and obtained maximum value of stress as 14.14 MN/m^2

Table 8.2 (continued)

Sr. no.	Year	Author	Matrix material	Reinforced material	Curing time	Tests	Outcomes
19	2007	Sapuan et al. [22]	Epoxy resin	Banana fiber	24 h		They showed suitability of banana fiber-reinforced composites for furniture-making
20	2000	Savastano et al. [23]	OPC cement, Adelaide Brighton brand Type GP	Sisal, banana, waste eucalyptus grandis pulp	7 days	Flexural, water absorption	Sisal and banana fibrous wastes were suitable for composite manufacture, whose physical properties are sufficient to use it in low-cost housing construction
21	2012	Shih et al. [24]	Epoxy resin	Banana fiber, pineapple leaf fiber, and disposable chopsticks fiber	25 min	TGA, tensile, SEM,	Developed epoxy resin-based composites with different fibers as reinforcement and observed better thermal properties in pineapple leaf fiber-reinforced composites while better mechanical behavior in recycled disposable chopstick fiber-reinforced composites
22	2014	Srinivasan et al. [25]	Epoxy resin	Banana and flax fibers	24 h	Tensile, flexural, impact, double shear, thermal analysis, SEM	Observed better strength in case of hybrid composite in comparison to single fiber composites

(continued)

Table 8.2 (continued)

Sr. no.	Year	Author	Matrix material	Reinforced material	Curing time	Tests	Outcomes
23	2013	Sumaila et al. [26]	Epoxy resin	Banana fibers	24 h.	Moisture absorption test, density, tensile, flexural, impact, hardness	The study showed increment in moisture absorption and the compressive strength with increment in fiber length along with a decrement in density. However, the tensile strength attain their highest values of 67.2 MPa at 15 mm fiber length
24	1994	Zhu et al. [6]	Portland cement	Banana fiber	21 days	Flexural strength, fracture energy, elastic modulus	Experimentation result shows that the kraft pulped banana fiber when loading between 8 and 16% by mass resulted in composites suitable to use as building material
25	1995	Zhu et al. [27]	Polyester	Banana fibers	36 h	Flexural, density	Developed banana fiber-reinforced composite suitable for industrial applications

due to the enhanced interfacial bond between matrix and fibers caused by chemical treatment.

Amir et al. [7] conducted a study to evaluate the influence of fiber orientation on the behavior of Banana Fiber-PP-MAPP composites and found that fiber configuration affects the tensile and flexural strength of the composites. They also proved that yarn fiber configuration provides the best mechanical properties in the composite.

Joseph et al. [11] developed banana fiber-reinforced composites that possess mechanical properties, comparable with synthetic glass fiber-reinforced composites, used as a structural material. While Zhu et al. [27] used kraft pulped banana fiber at a loading of 14% by mass with Portland cement to develop composites as commercially viable building materials. Zhu et al. [6] also developed banana fiber-reinforced polyester composites that exhibit the properties, which makes them suitable for industrial applications. Similarly, Olusegun [11] evaluated the mechanical behavior of banana, hemp, sisal, coconut, and E-glass fiber-reinforced laminates to ascertain the suitability of these materials for engineering applications, and Savastano et al. [23] used sisal and banana fibrous wastes with OPC cement for composite manufacture, whose physical properties are sufficient to use it in low-cost housing construction.

Kumar and Rajesh [2] produced composites by adding 30% loading of short fibers of banana with natural rubber matrix to prepare composites and performed various mechanical tests to understand the influence of the different treatments on the fiber surface. While Patel et al. [9] proposed a mathematical model using regression analysis and ANOVA to optimize drilling parameters for banana fiber-reinforced composites.

Some studies were initiated to develop hybrid composites by using two or more reinforcing elements with polymers. Idicula et al. [10] developed composites by adding short and thoroughly mixed banana and sisal fibers with random orientation in polyester to analyze the effects of relative volume fraction of the two fibers on the properties of the composites. While Ramachandran [20] developed and analyzed two-hybrid composites where the first one is developed by adding bamboo and banana fibers with epoxy resin while the second one is developed by adding bamboo and linen fibers with epoxy resin. They observed better Impact strength and compatibility of fibers with polymers in the case of bamboo–banana epoxy resin composite while better hardness properties were observed for bamboo–linen epoxy resin composite.

8.6 Conclusions

The present study explores the uses of the banana fiber as a reinforcement with the polymer resin and cement to develop composites with an emphasis on the chemical composition and their mechanical properties. The study also explores the effects of the chemical treatment of the banana fiber on the properties of composites and concludes that the chemical treatment of fibers augmented the reinforcement effect. The present study also demonstrates the potential of the banana fiber-reinforced composites in various areas like decoration and structural work due to less cost and lightweight with comparable tensile strength. This can be a great benefit to a country like India by producing useful components by using the banana fiber and its wastes in composites. Thus, the present work concludes that deterministic and sustained research in the future will augment the applications of banana fiber-reinforced composites.

References

1. Preethi, Palpandian, Balakrishna, G.: Physical and chemical properties of banana fibre extracted from commercial banana cultivars grown in Tamilnadu State. Agrotechnology. (2013). https://doi.org/10.4172/2168-9881.S11-008
2. Kumar, R.G., Rajesh, D.R.: A study on the abrasion resistance, compressive strength and hardness of banana–fibre reinforced natural rubber composites. Int. J. Adv. Res. Eng. Technol. **7**(3), 42–55 (2016)
3. Sapuan, S.M., Leenie, A., Harimi, M., Beng, Y.K.: Mechanical properties of woven banana fibre reinforced epoxy composites. Mater. Des. **27**(8), 689–693 (2006)
4. Cheung, Hoi-yan., Ho, Mei-po., Lau, Kin-tak., Cardona, Francisco, Hui, David: Natural fibre-reinforced composites for bioengineering and environmental engineering applications. Compos. B Eng. **40**, 655–663 (2009)
5. Sakthivel, M., Ramesh, S.: Mechanical properties of natural fibre (banana, coir, sisal) polymer composites. Sci. Park **1**(1), 1–6 (2013)
6. Zhu, W.H., Tobias, B.C., Coutts, R.S.P.: Banana fibre strands reinforced polyester composites. J. Mater. Sci. Lett. **14**(7), 508–510 (1995)
7. Amir, N., Abidin, K.A.Z., Shiri, F.B.M.: Effects of fibre configuration on mechanical properties of banana fibre/PP/MAPP natural fibre reinforced polymer composite. Procedia Eng. **184**, 573–580 (2017)
8. Ghosh, R., Reena, G., Krishna, A.R., Raju, B.H.L.: Effect of fibre volume fraction on the tensile strength of Banana fibre reinforced vinyl ester resin composites. Int J adv Eng sci technol **4**(1), 89–91 (2011)
9. Hanifawati, I.N., Hanim, A., Sapuan, S.M., Zainuddin, E.S.: Tensile and flexural behavior of hybrid banana Pseudostem/glass fibre reinforced polyester composites. In: Key Engineering Materials, vol. 471, pp. 686–691. Trans Tech Publications Ltd (2011)
10. Idicula, M., Malhotra, S.K., Joseph, K., Thomas, S.: Dynamic mechanical analysis of randomly oriented intimately mixed short banana/sisal hybrid fibre reinforced polyester composites. Compos. Sci. Technol. **65**(7–8), 1077–1087 (2005)
11. Joseph, S., Sreekala, M.S., Oommen, Z., Koshy, P., Thomas, S.: A comparison of the mechanical properties of phenol formaldehyde composites reinforced with banana fibres and glass fibres. Compos. Sci. Technol. **62**(14), 1857–1868 (2002)
12. Joseph, S., Koshy, P., Thomas, S.: The role of interfacial interactions on the mechanical properties of banana fibre reinforced phenol formaldehyde composites. Compos. Interfaces **12**(6), 581–600 (2005)
13. Karthick, R., Adithya, K., Hariharaprasath, C., Abhishek, V.: Evaluation of mechanical behavior of banana fibre reinforced hybrid epoxy composites. Mater. Today Proc. **5**(5), 12814–12820 (2018)
14. Olusegun, D.S., Stephen, A., Adekanye, T.A.: Assessing mechanical properties of natural fibre reinforced composites for engineering applications. J. Miner. Mater. Char. Eng. **11**(1), 780–784 (2012)
15. Ortega, R., Monzón, M.D., Ortega, Z.C., Cunningham, E.: Study and fire test of banana fibre reinforced composites with flame retardance properties. Open Chem. **18**(1), 275–286 (2020)

16. Patel, K., Chaudhary, V., Gohil, P.P., Patel, K.: Investigation on drilling of banana fibre reinforced composites. In: 2nd International Conference on Civil, Materials and Environmental Sciences. Atlantis Press (2015)
17. Pothan, L.A., Thomas, S., Groeninckx, G.: The role of fibre/matrix interactions on the dynamic mechanical properties of chemically modified banana fibre/polyester composites. Compos. A Appl. Sci. Manuf. **37**(9), 1260–1269 (2006)
18. Prasad, V., Joy, A., Venkatachalam, G., Narayanan, S., Rajakumar, S.: Finite Element analysis of jute and banana fibre reinforced hybrid polymer matrix composite and optimization of design parameters using ANOVA technique. Procedia Eng. **97**, 1116–1125 (2014)
19. Rajesh, M., Pitchaimani, J., Rajini, N.: Free vibration characteristics of banana/sisal natural fibers reinforced hybrid polymer composite beam. Procedia Eng. **144**, 1055–1059 (2016)
20. Ramachandran, M., Bansal, S., Raichurkar, P.: Experimental study of bamboo using banana and linen fibre reinforced polymeric composites. Perspect. Sci. **8**, 313–316 (2016)
21. Santhosh, J., Balanarasimman, N., Chandrasekar, R., Raja, S.: Study of properties of banana fiber reinforced composites. Int. J. Res. Eng. Technol. **3**(11), 144–150 (2014)
22. Sapuan, S.M., Harun, N., Abbas, K.A.: Design and fabrication of a multipurpose table using a composite of epoxy and banana pseudostem fibres. J. Trop. Agric. **45**(1), 66–68 (2008)
23. Savastano Jr., H., Warden, P.G., Coutts, R.S.P.: Brazilian waste fibres as reinforcement for cement-based composites. Cement Concr. Compos. **22**(5), 379–384 (2000)
24. Shih, Y.F., Cai, J.X., Kuan, C.S., Hsieh, C.F.: Plant fibers and wasted fiber/epoxy green composites. Compos. B Eng. **43**(7), 2817–2821 (2012)
25. Srinivasan, V.S., Boopathy, S.R., Sangeetha, D., Ramnath, B.V.: Evaluation of mechanical and thermal properties of banana–flax based natural fibre composite. Mater. Des. **60**, 620–627 (2014)
26. Sumaila, M., Amber, I., Bawa, M.: Effect of fiber length on the physical and mechanical properties of ramdom oriented, nonwoven short banana (musabalbisiana) fiber/epoxy composite. Cellulose **62**, 64 (2013)
27. Zhu, W.H., Tobias, B.C., Coutts, R.S.P., Langfors, G.: Air-cured banana-fibre-reinforced cement composites. Cement Concr. Compos. **16**(1), 3–8 (1994)

Chapter 9
State of Art on Microstructural and Mechanical Characterization of Wire and Arc Additive Manufacturing (WAAM)

Aman Verma, Himanshu Yadav, Kuldeep Kumar, Prince Kumar Singh, Mayank Sharma, Vishal Shankar Srivastava, and Ashish Kumar Srivastava

Abstract Wire and arc additive manufacturing (WAAM) technology is quite impressive and investigated in the last 30 years of development. It fascinates the scientists and manufacturers as it can produce densely deposited metal parts, and the objects produced are almost similar to the desired one. It is a process that includes the deposition of material layer-by-layer. It is gaining importance at a very fast pace as manufacturing industries find it cost- and time-efficient. Here, in this review paper, the mechanism of WAAM techniques, its type, and various components produced by WAAM have been discussed. It also includes the discussion on different surface deposit properties and microstructural properties of the products produced by WAAM. At last, the industrial application and future scope of this technique are also explored.

9.1 Introduction

Industry 4.0 has been known for digitalization of manufacturing industry [1]. It became the turning point in the field of automated manufacturing with innovations in additive manufacturing (AM). In the last two decades, the AM technique has shown great growth in industry because of directly able to produce objects from the CAD model [2]. Wire and arc additive manufacturing technique is one of the prominent and effective methods of AM that involves melting of metal parts and depositing the bead layer-by-layer [3]. Previously, subtractive manufacturing (SM) can be seen usually in every manufacturing industry. SM includes the raw material with a large size from which desired object is going to be made and then, unwanted material is machined out. Unlike SM, AM came as a revolution after SM, as it is highly cost-effective and uses the layer-by-layer method which saves lots of time as well as material [3]. AM also can obtain complex shapes better than SM does [4].

A. Verma · H. Yadav · K. Kumar · P. K. Singh · M. Sharma · V. S. Srivastava ·
A. K. Srivastava (✉)
GL Bajaj Institute of Technology and Management, Greater Noida, U.P., India

© The Author(s), under exclusive license to Springer Nature Singapore Pte Ltd. 2022
V. V. Rao et al. (eds.), *Computational and Experimental Methods in Mechanical Engineering*, Smart Innovation, Systems and Technologies 239,
https://doi.org/10.1007/978-981-16-2857-3_10

There are various AM techniques which are famous to develop the 3D components such as Binder jetting is an additive manufacturing process which includes a liquid binding agent that deposit on a layer of metal or sand powder particles to form a final product. However, it has certain limitations like low parts strength, not suitable for structural parts. Due to the use of binder material, it became less accurate than material jetting, as the part made requires post-processing which leads to significant time to the process [5].

Similarly, Powder bed fusion is an AM process where focused energy is used to melt a layer of powder. PBF is one of the AM techniques in which laser heat and electron bombardment are used as a heat source to melt and fuse the powder and material to produce a final component [5, 6]. Conventional manufacturing includes forging, casting, and machining. Components are usually manufactured by putting the molten metal into a casting mold of the required shape in the casting process [7, 8]. In forging, the required shape is produced by using compressive force with help of a hammer and die. Machining includes removal of wanted shape and material [9]. Wire and arc additive manufacturing (WAAM) includes great advantages in material and cost efficiency, good deposition rate, and have higher flexibility in design compared to traditional technique [10, 11]. WAAM generally includes robot manipulating devices and power sources used in other welding techniques like TIG and MIG [12]. In the last few years, WAAM technology has fascinated most of the manufacturing industries, as it can produce a complex part with ease and huge components with high deposition rate with almost 100% material usage [2, 5]. WAAM is a process of deposition of molten metal in a layer-by-layer form to achieve the final component [13]. Due to the increasing demand of novel materials in today's market leads to the rapid production of components with low material waste and low equipment cost [10]. WAAM for dissimilar alloys is one of the trending topics. WAAM components of dissimilar alloys are mostly used in special applications [14]. The high "investment-to-return" ratio, a ratio of initial required material cost to the material obtained as a finished product, is one of the biggest drawbacks of traditional and SM technologies, resulting in high cost and material waste [15, 16]. A study shows that 20:1 for conventional production of titanium aerospace brackets and 33:1 for titanium duct flanges for aero-engine components were recorded [17, 18]. WAAM has been recognized in recent years as a very efficient method of manufacturing large-scale structural parts of aluminum alloy that are large in shape and size, such as wing ribs, turbines blade, which are usually used in aircraft [13, 14]. The aerospace industry attracts toward WAAM as it has desired material strength, which is required for the components of WAAM aluminum alloy. Figure 9.1 shows the schematic representation of the WAAM process. WAAM has the main components like welding gun, shielding gas, and a special platform to perform its operation [19].

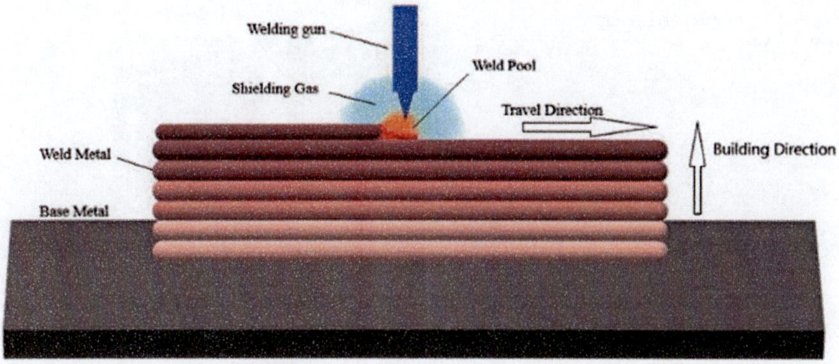

Fig. 9.1 Schematic representation of the WAAM process

9.2 Materials and Methods

Most of the study and research has found that there is not any noticeable difference in the WAAM component properties with the changes in heat source temperature [20, 21]. Previous experiments on WAAM aluminum alloys, such as 2000, 4000, and 5000 series, have shown that their mechanical properties are typically greatly altered with the chemical composition [22]. The aluminum alloy Al-Si_5 has shown lower ultimate tensile strength that was 142 MPa using GTAW, as observed by Wang et al. [23]. The tensile and yield strength of the alloy Al-Cu6 deposited by the pulse advanced CMT (CMT-PA) were 240 MPa and 114 MPa, respectively. These findings were identical to those of alloys with the same GTAW-manufactured composition and the electronic beam [24].

In comparison to other AM technique that is of particular interest, there are several advantages of WAAM [25]. WAAM component has a much more material volume, usually two to three times greater than other AM processes since it was not constrained with the viability and expense of a bed of powder or vacuum [26, 27]. The deposition rate of WAAM process is also high, especially in comparison with SLM [28]. The cost of capital is considerably lower with WAAM equipment costing about two to three times less [29]. Apart from these advantages, WAAM technique is associated with several issues related to application and assesories. WAAM is capable of producing larger components but complex and intricates shapes are still a challenging task [30–33].

There are several process parameters involved in WAAM process. Figure 9.2 shows the process parameters of WAAM. CMT (Cold Metal Transfer) technology has revolutionized the welding technique of similar or dissimilar metals [34]. It uses controlled metal deposition for producing good bead with low heat input for welding. Travel speed is the rate at which the welding torch moves with depositing the metal layer-by-layer [35–37]. Wire feed speed is responsible for the variation of current amperage and deposition temperature. Current plays an important role in

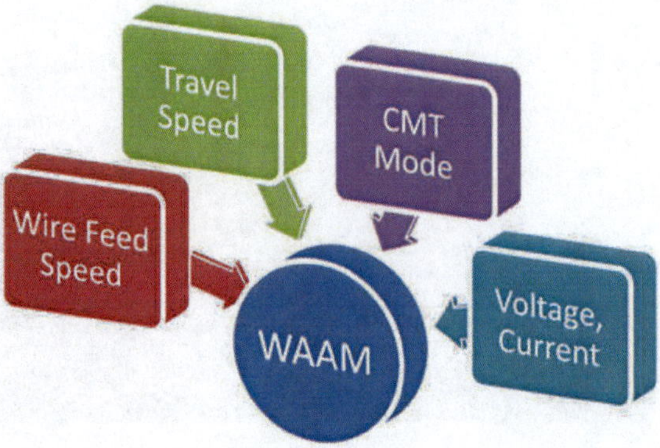

Fig. 9.2 Process parameters of WAAM

welding and melting the metal by providing high current value from 50 to 300 A and comparatively low voltage around 10–50 V [22]. Gordon et al. studied and found some optimum value regarding WAAM process base current as 50 A and voltage as 19 V, wire feed speed as 4.6 m/min and welding speed as 300 mm/min.

In the WAAM technique, these parameters identify the characteristics of the component [38]. The microscopic and structural characterizations by optical Microscopy, scanning electron microscopy, x-ray diffraction and x-ray photoelectron spectroscopy are reported by various authors working in this field [39–41]. The detailed description of characterization techniques of the WAAM-based component is shown in Fig. 9.3.

By using optical microscopy, the surface finish along with the cross-section and porosity of the component can be measured [42]. Scanning electron microscopy also

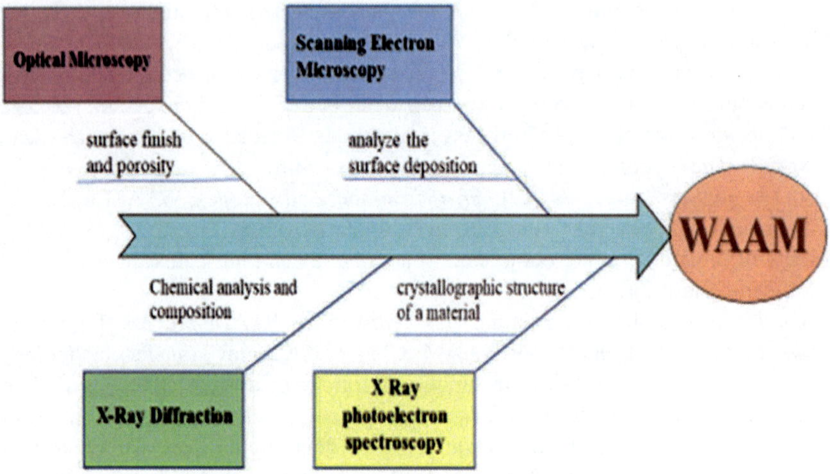

Fig. 9.3 Different characterization technique for WAAM

finds and analyzes the surface deposits, porosity, and the wire, as it used to provide the highly magnified image, with a great field of depth and structural formation [43–45]. Baei et al. [46] studied XPS (x-ray photoelectron spectroscopy) used for the chemical analysis and the chemical composition, particularly for analyzing the composition of wire and surface deposit. This technique is used to find the composition of raw materials. X-ray diffraction analysis (XRD) is a technique used to determine the microphysical structure of a material. It works by irradiating the component with falling X-rays on it and then measure the intensity and scattering angle of X-rays that leave the component [47].

9.3 Discussion

Analysis of development over the WAAM process focusing on the mechanical and microstructural properties of WAAM component has been done. In the WAAM process, the relation between the composition of material and microstructure of the material is responsible for the different properties of the component and quality of the component. Since the inter-pass or interlayer temperature varies along with the process, it becomes tough to control the microstructural evolution of the component deposited that is predominantly causing the variation of mechanical properties of the component produced [45]. Impurities generated in WAAM component nearly occurred due to the selected material properties and parameters of different processes used [46]. The improvement parameters of the WAAM process must be focused on the defect formation and its elimination. Seeing the need for the highly finished part, further analysis on it will result in a bunch of applications of WAAM process in upcoming years. Till now, many types of research have been done over various parameters such as different process planning like TIG, MIG, PMIG, CMT, GTAW, GMAW, etc. [43]. With the need for a variety of engineering materials and varying fabrication of material in the manufacturing industry, different WAAM process models will be developed [42].

In the context of the microstructural and mechanical properties, various authors have reported the improved tensile strength and hardness with homogeneous and isentropic microstructural features. Zhao et al. [47] studied the impact of various control techniques on the characteristic of delivered walls, evaluated by applying microstructural and tensile power along the horizontal line. Figure 9.4 shows the microstructural characterization of the WAAM component at different cooling conditions. Parts a-3 and b-3 depict the microstructure of the walls at various altitudes, and c-3 displays lower wall density and greater deposit over the first layer because of the greater heat exchanged to the base substance [47]. Moderately stretched surface grains can be observed around the heat impacted region. In between the median zones Fig. 9.4 a-2, b-2, and c-2, a powdery microstructure is retrieved, as the substance was burned many times throughout the process of deposition to the top layers and it increased the temperature in the walls. The part a-1, b-1, and c-1 in Fig. 9.4 depict the microstructure of the top deposit layer which show growth of large extended grains

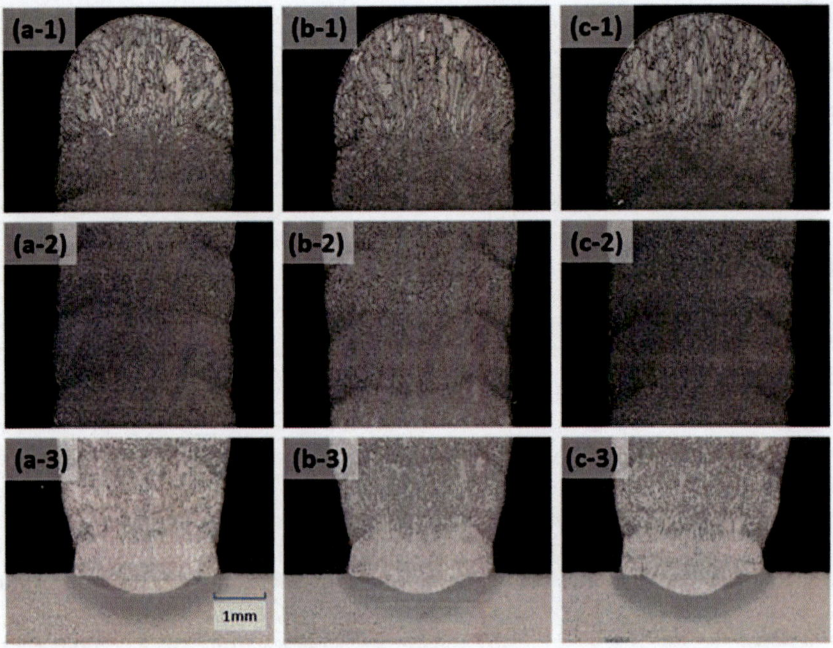

Fig. 9.4 Microstructure of the component using different cooling methods, **a** Forced air-cooled region, **b** Without forced air-cooled region, and **c** Natural cooled region [13]

toward the heat exchange point. All these three walls show identical microstructure all along with the wall height since the same amount of heat was used for depositions, and the inter-pass heat, element, and heat sink were alike. The static tensile test was useful for examining the mechanical properties of the walls [47]. As anticipated, no serious differences in the outcome and microstructures were set up. This can be concluded by the fact that inter-pass temperatures were not dissimilar for various samples developed by diverse control methods [26, 27]. The thermal range was quite high to affect the deposited shape, but not adequate to make a significant variation in the microstructural and mechanical characteristics of the wall produced.

Most important causes for suitable huge scale WAAM process is base plate distortion or control of distortion. The temperature variation while moving towards layers causes the distortion and results in residual tension, and to know about these changes, the study of AM layers is necessary [46]. While studying the thermal analysis for different deposition layers, it is found that temperature gradient always decreases with the increase of Number of layers as shown in Fig. 9.5.

A wire having a composition of AlCu-4.3% and Mg-1.5% was made and used to deposit the WAAM component. Microhardness test is conducted in the vertical direction of the deposited component. Value of microhardness of heat-treated components is increased by 52% from 106 to 165 HV [48]. There was a varying range 10HV

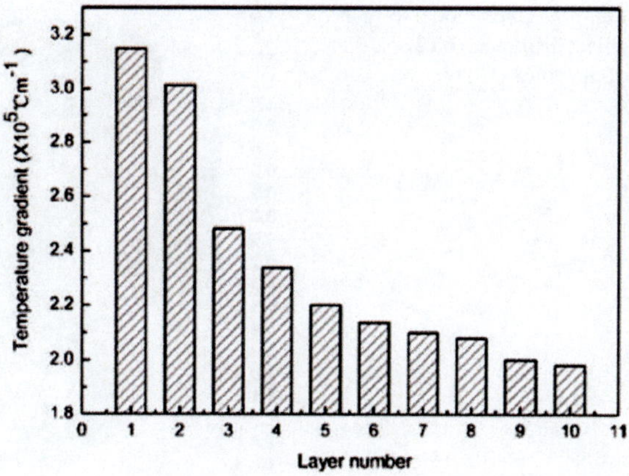

Fig. 9.5 Variation of a temperature gradient with layer number [25]

between 158 and 168 HV as shown in Fig. 9.6. The main reason for incrementation in microhardness is the heat treatment of the deposited component, as it evenly distributed the microhardness of the component. Because of the continuous scattering of grains structure, the variation of microhardness occurs, and significant numbers of defects and micro-pores might weaken the WAAM alloy component.

Nejc Kozamernik et al. [13] have done shape distortion analysis by analyzing the shape of a deposited component with the CAD software model. The component has a spiral cylindrical shape with a particular diameter. Here, inter-pass temperature control method at 140 °C and forced cooling is adopted. Figure 9.7 shows the scanned material color in reference to the difference between the printed component and 3D CAD design. This figure shows the variation of the net shape is mainly on the top side of the object that happened due to the cladding process.

Fend et al. [49] have used DWF-PAM and SWF-PAM with H00Cr21Ni10 stainless steel material focusing on Scanning Speed factor and found Ultimate Tensile Strength

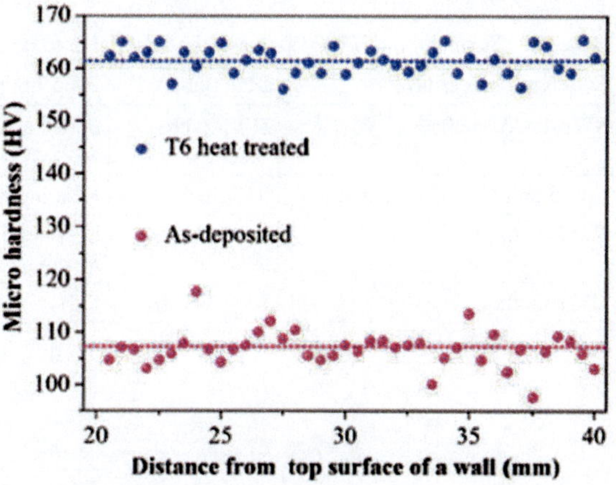

Fig. 9.6 Micro hardness at different points of the wall [48]

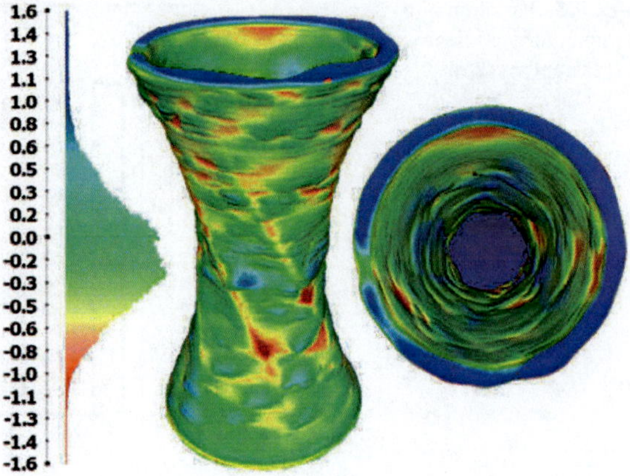

Fig. 9.7 CAD model to final object difference on a scale of the color [13]

similar in the vertical and horizontal directions. Wang et al. [23] have used Speed pulse WAAM with 315 stainless steel material focusing on Heat input and cooling rate factor, and found Ultimate Tensile Strength Horizontal 552 + 3. Caballero et al. [24] have used CMT with 17–4 PH stainless steel material focusing on Shielding gas 25% Co_2 + Ar factor and found Ultimate Tensile Strength Average 994.

Based on the above literature, it has been observed that the WAAM technique can provide the fruitful increment to the mechanical properties with fewer defects and wastage. A summary of the WAAM process, the material used, and the focusing process parameters are given in Table 9.1.

The improved mechanical properties like ultimate tensile strength (UTS), yield strength (YS), and elongation of two materials Al-Si5 and Al-Cu6.3, named A2 and B2, are analyzed. It is observed that elongation in the horizontal direction is much greater than the vertical direction and YS of material A2(Al-Si5) is greater than B2(Al-Cu6.3). Similarly, the UTS of A2 is greater than B2. Results of the comparative study are shown in Fig. 9.8.

Table 9.1 Summary on WAAM process, the material used, and focusing process parameters

Different process used	Material used	Focusing process factor	References
SWF-PAM Method	H0Cr21Ni10 Steel	Scanning speed	[49]
Speed pulse WAAM Technique	317 steel	Heat input, cooling rate	[23]
CMT method	17–3 PH steel	SG1 + post-deposition heat treatment	[24]
CMT method	17–3 PH steel	SG1 + solution treated	[24]

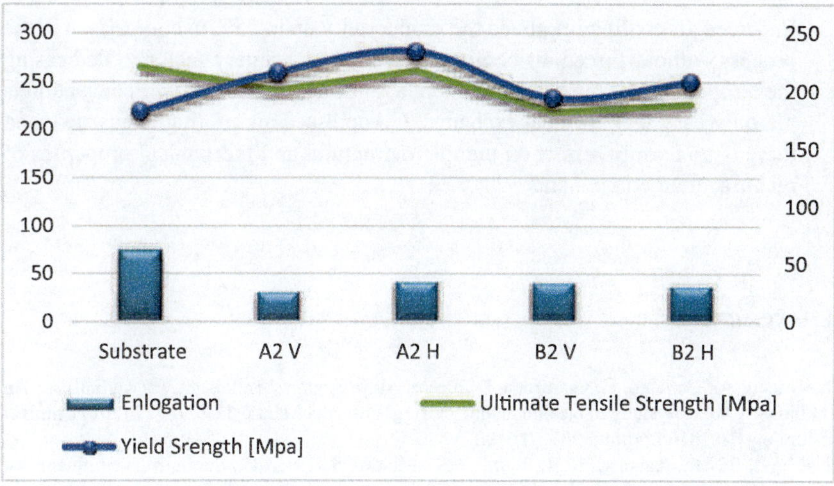

Fig. 9.8 Tensile Properties of the substrates, A2 and B2, in both vertical and horizontal direction

9.4 Conclusion and Future Scope

1. The history of WAAM module was solved in an elaborate form for a single set of specifications. The guidelines which defined the resulting microstructure were found to be the peak temperatures, heating and cooling rates, temperature gradients, strength, hardness, microhardness, and the surface parameter.
2. 386 °C/mm is liable for the emergence of the columnar microstructure during the process of solidification. For an equiaxed, the requirement would be 100 °C/mm.
3. Studies recommended inter-pass rolling as a methodology for issues concerning isotropic properties, while other analysts proposed inclusion of non-destructive testing in the process which could help find porosities in the early stage.
4. Further analysis should incorporate works on these objectives, besides improvement factors such as optimization, monitoring, process control, part plan, and thermal treatment, all of which if taken together can improvise understanding and execution of WAAM technology.
5. With advancements in these domains, WAAM can prove to be a better substitute for conventional production techniques such as casting, welding, and forging in specific applications. Present-day issues hinder the commercial usage and market approval of WAAM, whereas different analysis has already suggested lessening and eliminating them, so WAAM can be quite similar to the traditional system.

6. Enforced air cooling improved the productivity up to 25% in comparison to the process without forced air cooling. Cooling is a primary factor in decreasing the temperature of the base to the top layer of deposition, and air convection is the only easy way for heat exchange. Compressed air cooling is guarded, and there is no harmful effect on the microstructural and mechanical properties of manufactured components.

References

1. Grieco, A., Caricato, P., Gianfreda, D., Pesce, M., Rigon, V., Tregnaghi, L, Voglino, A.: An industry 4.0 case study in fashion manufacturing. Procedia Manuf. **11**:871–877 (2017). https://doi.org/10.1016/j.promfg.2017.07.190
2. Majid, S.N.A., Kalahari, M.R., Ramli, F.R., Maidin, S., Fai, T.C., Sudin, M.N.: Influence of integrated pressing during fused filament fabrication on tensile strength and porosity. J. Mech. Eng. **2**, 185–195 (2017)
3. Sharma, V., Singh, S.: Rapid prototyping: process advantage, comparison and application. Int. J. Comput. Intell. Res. **12**(1), 55–61 (2016)
4. Song, Y.-A., Park, S., Choi, D., Jee, H.: 3D welding and milling: Part I–a direct approach for freeform fabrication of metallic prototypes. Int. J. Mach. Tools Manuf. **45**, 1057–1062 (2005)
5. Srivastava, A.K., Kumar, N., Dixit, A.R.: Friction stir additive manufacturing—an innovative tool to enhance mechanical and microstructural properties. Mater. Sci. Eng. B **263**, 114832 (2021)
6. Nazan, M.A., Ramli, F.R., Alkahari, M.R., Sudin, M.N., Abdullah, M.A.: Optimization of warping deformation in open source 3D printer using response surface method. Proc. Mech. Eng. Res. Day 71–72 (2016)
7. Srivastava, A.K., Gupta, Y., Patel, S., Tiwari, S.K., Pandey, S.: Metal matrix composites-a review on synthesis and characterization. IOP Conf. Ser. Mater. Sci. Eng. **691**(1), 012077 (2019)
8. Song, Y.-A., Park, S., Chae, S.-W.: 3D welding and milling: Part II—optimization of the 3D welding process using an experimental design approach. Int. J. Mach. Tools Manuf. **45**, 1063–1069 (2005)
9. Katou, M., Oh, J., Miyamoto, Y., Matsuura, K., Kudoh, M.: Freeform fabrication of titanium metal and intermetallic alloys by three-dimensional micro welding. Mater. Des. **28**, 2093–2098 (2007)
10. Wong, K.V., Hernandez, A.: A review of additive manufacturing, international scholarly research notices 10. Article ID 208760 (2012). https://doi.org/10.5402/2012/208760
11. Jandric, Z., Labudovic, M., Kovacevic, R.: Effect of heat sink on microstructure of three-dimensional parts built by welding-based deposition. Int. J. Mach. Tools Manuf. **44**, 785–796 (2004)
12. Kwak, Y.-M., Doumanidis, C.: Geometry regulation of material deposition in near-net shape manufacturing by thermally scanned welding. J. Manuf. Process. **4**, 28–41 (2002)
13. Kozamernik, N., Bračun, D., Klobčar, D.: WAAM system with interpass temperature control and forced cooling for near-net-shape printing of small metal components. Int. J. Adv. Manuf. Technol. **110**, 1955–1968 (2020)
14. Nagamuthu, H., Sasahara, H., Mitsutake, Y., Hamamoto, T.: Development of a cooperative system for wire and arc additive manufacturing and machining. Addit. Manuf. **31**, 100896 (2020)
15. International Standard ISO Metallic materials—Vickers hardness test—Part 1: Test method (ISO 6507–1: 2018) Int. Stand. 1–49 (2018) ISO 6507-1:2018

16. Suryakumar, S., Karunakaran, K., Bernard, A., Chandrasekhar, U., Raghavender, N., Sharma, D.: Weld bead modeling and process optimization in hybrid layered manufacturing. Comput. Des. **43**, 331–344 (2011)
17. Gu, J., Gao, M., Yang, S., Bai, J., Zhai, Y., Ding, J.: Microstructure, defects, and mechanical properties of wire + arc additively manufactured AlCu4.3–Mg1.5 alloy. Mater. Des. **186**,108357 (2020)
18. Baufeld, B., Biest, O., Gault, R.: Additive manufacturing of Ti–6Al–4V components by shaped metal deposition: microstructure and mechanical properties. Mater. Des. **31**, 106–111 (2010)
19. Li, Y., Sun, Y., Han, Q., Zhang, G., Horvath, I.: Enhanced beads overlapping model for wire and arc additive manufacturing of multi-layer multi-bead metallic parts. J. Mater. Process. Technol. (2017). https://doi.org/10.1016/j.jmatprotec.2017.10.017
20. Ding, D., Pan, Z., Cuiuri, D., Li, H.: A tool-path generation strategy for wire and arc additive manufacturing. Int. J. Adv. Manuf. Technol. **73**, 173–183 (2014)
21. Ding, D., Pan, Z., Cuiuri, D., Li, H.: A practical path planning methodology for wire and arc additive manufacturing of thin-walled structures. Robot. Comput. Manuf. **34**, 8–19 (2015)
22. Ma, Y., Cuiuri, D., Li, H., Pan, Z., Shen, C.: The effect of postproduction heat treatment on γ-TiAl alloys produced by the GTAW-based additive manufacturing process. Mater. Sci. Eng. A **657**, 86–95 (2016)
23. Wang, L., Xue, J., Wang, Q.: Correlation between arc mode, microstructure, and mechanical properties during wire arc additive manufacturing of 316L stainless steel. Mater. Sci. Eng. A **751**, 183–190 (2019)
24. Caballero, A., Ding, J., Ganguly, S., Williams, S.: Wire + Arc Additive Manufacture of 17-4 PH stainless steel: effect of different processing conditions on microstructure, hardness, and tensile strength. J. Mater. Process. Technol. **268**, 54–62 (2019)
25. Suryakumar, S., Karunakaran, K.P., Chandrasekhar, U., Somashekara, M.A.: A study of the mechanical properties of objects built through weld-deposition. Proc. Inst. Mech. Eng. Part B: J. Eng. Manuf. **227**(8), 1138–1147 (2013)
26. Anzalone, G.C., Zhang, C, Wenjun, et al.: A low-cost opensource metal 3-D printer. https://doi.org/10.1109/ACCESS.2013.2293018 (2013)
27. Nilsiam, Y., Sanders, P.: Pearce, Slicer and process improvements for open-source GMAW-based metal 3-D printing. Addit. Manuf. **18**, 110–120 (2017)
28. Stützer, J., Totzauer, T., Wittig, B., Zinke, M., Jüttner, S.: GMAW cold wire technology for adjusting the ferrite-austenite ratio of wire and arc additive manufactured duplex stainless steel components. Metals **9**, 564 (2019)
29. Li, F., Chen, S., Wu, Z., et al.: Adaptive process control of wire and arc additive manufacturing for fabricating complex-shaped components. Int. J. Adv. Manuf. Technol. **96**, 871–879 (2018)
30. Manogharan, G., Wysk, R.A., Ola, L.A.: Harrysson, Additive manufacturing–integrated hybrid manufacturing and subtractive processes: economic model and analysis. Int. J. Comput. Integr. Manuf. **29**(5), 473–488 (2016)
31. Dwivedi, R.: Kovacevic R Automated torch path planning using polygon subdivision for solid freeform fabrication based on welding. J. Manuf. Syst. **23**(4), 278–291 (2004)
32. Ultimate Cora 3D Printing Slicing Software. https://ultimaker.com/software/ultimate-cora
33. Michel, F., Lockett, H., Ding, J., Martina, F., Marinelli, G., Williams, S.A.: Modular path planning solution for wire & arc additive manufacturing. Robot. Comput. Integer. Manuf. **60**, 1–11 (2019). https://doi.org/10.1016/j.rcim.2019.05.009
34. Gu, J., Gao, M., Yang, S., Bai, J., Zhai, Y., Ding, J.: Microstructure, defects, and mechanical properties of wire + arc additively manufactured AlCu4.3-Mg1.5 alloy. Mater. Des. **186**, 108357 (2020)
35. Martina, F., Colegrove, P.A., Williams, S.W., et al.: Microstructure of interpass rolled wire + Arc additive manufacturing Ti–6Al–4V components. Metall. Mater. Trans. A **46**, 6103–6118 (2015)
36. Chignon, P., Rochinich, C., Scoopingo, R.: Metro measuring error on simplified surfaces. Comput. Graph Forum **17**, 167–174 (1998). https://doi.org/10.1111/1467-8659.00236

37. Ding, D., Pan, Z., Coeur, D., Li, H.: A practical path planning methodology for wire and arc additive manufacturing of thin-walled structures. Robot. Comput. Integr. Manuf. **34**, 8–19 (2015)
38. Gu, J., Gao, M., Yang, S., Bai, J., Zhai, Y., Ding, J.: Microstructure, defects, and mechanical properties of wire + arc additively manufactured Al Cu4.3–Mg1.5 alloy. Mater. Des. **186**, 108357 (2020)
39. Oliveira, J.P., Santos, T.G., Miranda, R.M.: Revisiting fundamental welding concepts to improve additive manufacturing: from theory to practice. Prog. Mater. Sci. **107**, 100590 (2020)
40. Wager, V., Lui, J., Tang, S., Fantasy, J., Joel, L., Liu, C.: Reducing porosity and refining grains for arc additive manufacturing aluminum alloy by adjusting arc pulse frequency and current. Mater. (Basel) **11**(8), 1344 (2018)
41. Gunna, K., Cong, B.Q., Dinger, J., Willian, S.W., Share, C.: Wire & arc additive manufacturing of aluminium. Mater. Today: Proc. (2020). https://doi.org/10.1016/j.matpr.2020.09.153
42. Bai, J.Y., Yang, C.L., Lin, S.B., et al.: Mechanical properties of 2219-Al components produced by additive manufacturing with TIG. Int. J. Adv. Manuf. Technol. **86**, 479–485 (2016)
43. Wang, J.S., & Lee, P.D.: Simulating tortuous 3D morphology of microporosity formed during solidification of Al–Si–Cu alloys. Int. J. Cast Metals Res. **20**(3), 151–158 (2007)
44. Zhang, Y., Wu, L., Guo, X., et al.: Additive manufacturing of metallic materials. A Rev. J. Mater. Eng. Perform. **27**, 1–13 (2018)
45. Sue, C., Chena, Z., Gaoyal, C., Wan, Y.: Effect of heat input on microstructure and mechanical properties of Mg–Al alloys fabricated by WAAM. Appl. Surf. Sci. **486**, 431–440 (2019)
46. Baei, J., Ding, J., Williams, S., Zhao, Y.: Deformation microstructures and strength mechanisms for the wire & arc additively manufactured Al–Mg–Mn alloy with inter-layer rolling. Mater. Sci. Eng. A **712**, 292–301 (2018)
47. Zhao, H., Zhang, G., Yin, Z., Wu, L.: A 3d dynamic analysis of thermal behavior during single-pass multi-layer weld-based rapid prototyping. J. Mater. Process. Technol. **211**(3), 488–495 (2011)
48. Kramer, H.S., Starke, P., Klein, M., Eifler, D.: Cyclic hardness test Phybalcht—short-time procedure to evaluate fatigue properties of metallic materials. Int. J. Fatigue **63**, 78–84
49. Feng, Y., Zhan, B., He, J., Wang, K.: The double-wire feed and plasma arc additive manufacturing process for deposition in Cr–Ni stainless steel. J. Mater. Process. Technol. **259**, 206–215 (2018)

Chapter 10
Effect of La^{3+} Substitution on Structural, Magnetic, and Multiferroic Properties of Bismuth Ferrite ($Bi_{1-x}La_xFeO_3$) Nanoceramics

Shama Farozan, Harendra Kumar Satyapal, Om Priya, Saurabh Sharma, and Singh Sonu Kumar

Abstract Bi_{1-x} La_x FeO_3 ceramics with (x = 0.0, 0.025, 0.050, 0.075, 0.100) is prepared using citrate precursor-based sol–gel method. X-ray diffraction patterns reveal phase purity of all samples with rhombohedral crystal structure indexed to the *R3c* space group. FE-SEM micrographs confirm the average crystallite size to be ~57 nm. FTIR spectrum justifies structural distortion in $BiFeO_3$ crystal lattice with La^{3+} increasing composition. EDAX analysis confirms proper substitution of La^{3+} at bismuth site. Remanent magnetization (M_r) ranges from 5×10^{-3} to 9×10^{-3} emu/g, whereas Coercivity (H_c) ranges from 252 to 875 Oersted with La^{3+} increasing content. Moreover, doping of La^{3+} at the bismuth site causes an appreciable reduction in electrical leakages, resulting in clear observation of the ferroelectric polarization loop. Electrical polarization loops convert to a typical concave nature from an oval pattern with increasing La^{3+} ions, yielding remanence polarization (P_r) and coercive polarization (P_c) of the order 0.15 $\mu C/cm^2$ and 7.5 kV/cm, respectively.

10.1 Introduction

Materials showing multiferroic behavior have gained appreciable attention of researchers owing to their usefulness in electronics devices, as storage media, as magnetic sensors, etc. [1, 2]. The basis of ferromagnetism in any material is the availability of partially filled d-orbitals, on the contrary, the absence of electrons in d-orbitals is essential for ferroelectric nature [3]. As such, these two demands are difficult to get accomplished by any single material. However, few materials do exhibit ferromagnetic and ferroelectric properties simultaneously. One such material is bismuth ferrite (represented as $BiFeO_3$); with rhombohedral distorted structure indexed to the *R3c* space group. $BiFeO_3$ possesses appreciable multiferroic

S. Farozan (✉) · H. K. Satyapal · O. Priya · S. Sharma · S. S. Kumar
Aryabhatta Center for Nanoscience and Nanotechnology, Aryabhatta Knowledge University, Patna 800001, Bihar, India

© The Author(s), under exclusive license to Springer Nature Singapore Pte Ltd. 2022
V. V. Rao et al. (eds.), *Computational and Experimental Methods in Mechanical Engineering*, Smart Innovation, Systems and Technologies 239,
https://doi.org/10.1007/978-981-16-2857-3_11

properties, with simultaneous ferroelectric polarization (T_C ~1103 K) and anti-ferromagnetic ordering (T_N ~643 K) [4]. On one hand, the basis for ferrimagnetic polarization in $BiFeO_3$ crystals is the structural strain mediated by $Bi^{3+}(6s^2)$ lone pairs, while the basis of ferromagnetism is the super-exchange network of Fe^{3+}–O–Fe^{3+} causing G-type anti-ferromagnetic strains [5]. The emergence of extra phases and vacant oxygen sites amplifies leakage currents in $BiFeO_3$ at 300 K. The small-order ferromagnetism observed in $BiFeO_3$ is because of its spiral spin arrangements periodic with 620 Å along the $[110]_h$ axis [6]. But this phenomenon affects saturation magnetization and prohibits the generation of linear magnetoelectric behavior in bismuth ferrite. To combat this critical balance, nanoparticles of size < 600 Å (less than cycloid spin structure) are effective in dominating spiral spin patterns. Research works have been reported on enhancement in structural and magnetic properties of $BiFeO_3$ by substituting the Fe^{3+} and/or Bi sites with Gd^{3+}, Sm^{3+}, Eu^{3+}, Ca^{2+}, Pb^{2+}, La^{3+}, Cr^{3+}, etc. [7–13]. Recently, R. Pandey et al. have reported that La^{3+} doping at Bi^{3+} site easily dominates the cycloid structure of $BiFeO_3$ yielding refinement in the magnetic property. But an improvement in multiferroic property was dealt with only when bismuth ferrite crystal lattice was simultaneously substituted with Ti^{4+} at the iron site [14]. Therefore, in this paper, we report magnetic as well as multiferroic property enhancement in $BiFeO_3$ substituted with single cation La^{3+} at Bi^{3+} site. Doping with lanthanum, having a bigger ionic radius (1.16Å) as compared to Bi (1.06 Å), causes stabilization in the crystal symmetry and tuning of spiral spin arrangements within $BiFeO_3$ to yield appreciable improvement in the structural, magnetic, and multiferroic properties. Our efforts is to address the depleting research works available on physical property refinement in La^{3+}-substituted $BiFeO_3$, to the best of our knowledge.

10.2 Experimental

$Bi_{1-x}La_xFeO_3$ nanoceramics with (x = 0.025 to 0.100) are prepared using high purity grade Bi $(NO_3)_3$, Fe $(NO_3)_3$, La $(NO_3)_3$, and Citric acid as reactants. These reactant chemicals are first weighed in stoichiometric proportion and dissolved in individual beakers containing just suffice amount of de-ionized water. La $(NO_3)_3$ was preferably dissolved in HNO_3 for easy dilution. Next, all solutions were converted into a single beaker and kept on a magnetic stirrer set at 120 °C temperature for continuous stirring at 200 rpm, till a viscous jelly-like substance is obtained. Then, this jelly substance was kept in a hot air oven chamber at 200 °C for 30 min for auto combustion process. The resulting ash-like material was crushed and annealed in a microprocessor-based muffle furnace at 700 °C for 4 h to improve the crystallinity of $BiFeO_3$ material. Also, few amounts of each sample were pressed as a disk-like pellet (1 mm thick and 5 mm radius) and calcined at 900 °C for 6 h for multiferroic testings. The X-ray intensities were recorded at a wavelength (λ = 1.5418 Å) with X-ray Diffractometer (Bruker). The surface morphology of samples was scanned using Field Emission SEM (Carl Zeiss). The bond lengths and bond stretching in samples were traced with FTIR spectroscopy, Perkin Elmer. The Magnetic hysteresis loops at 300 K were measured using a Vibrating Sample Magnetometer (Lake Shore) of magnetic field up to 3.2 T capacities. The ferroelectric polarization patterns were obtained with Radiant Tech Precision 10 kV Multiferroic instrument.

10.3 Results and Discussions

10.3.1 XRD Analysis

The X-ray diffraction pattern of samples $Bi_{1-x}La_xFeO_3$ with (x = 0.025 to 0.100) is shown in Fig. 10.1, which reveals crystallinity and pure phase synthesis of the samples. All signature peaks related to bismuth ferrite are reflected in the XRD pattern. The samples were found to possess rhombohedral crystal symmetry belonging to the *R3c* space group, which is consistent with research groups [15]. All the prominent diffraction peaks of $BiFeO_3$, such as [012] and [110], diminishes in its intensity with La^{3+} doping. This suggests that substitution with La^{3+} prohibits the growth of other phases. Few secondary phases which are noticed in XRD pattern of samples $Bi_{1-x}La_xFeO_3$ with (x = 0.0, 0.025) is absent in samples with higher La^{3+} content. Merging of diffraction peaks into single peak near 2θ = 40° and 2θ = 57° in the X-ray diffraction patterns of the samples with La^{3+} content (x = 0 and 0.25) is evident in Fig. 10.1. The lattice constants are determined to be a = b = 5.569 Å and

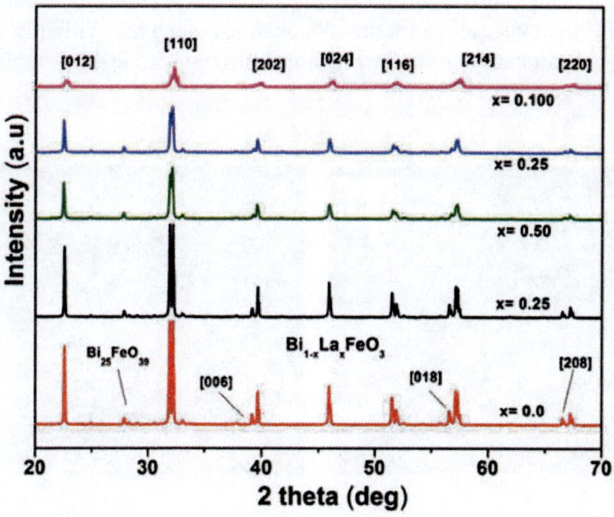

Fig. 10.1 XRD pattern of Bi_{1-x} La_x FeO_3 with (x = 0.025–0.10)

c = 13.801 Å for parent sample $BiFeO_3$. For finding lattice parameters of this rhombohedral crystal system, Eq. (10.1) is utilized. Crystal planes preferably selected to be replaced in Eq. (10.1) are [110], [012], [116], [024], [220], and [214].

$$\frac{1}{d^2} = \frac{4}{3}\left(\frac{h^2 + hk + k^2}{a^2}\right) + \frac{l^2}{c^2} \tag{10.1}$$

When we obtain lattice constants, we use Eq. (10.2) for calculating the lattice volume

$$V = \frac{\sqrt{3}}{2}\, a^2 c \tag{10.2}$$

Williamson-Hall plot helps find crystallite sizes [16]. The average crystal size is observed to be ~57 nm, which is further supported by FE-SEM morphological images.

10.3.2 FE-SEM Analysis

Figure 10.2(a, b) represents high resolution field emission SEM micrographs for sample $Bi_{1-x}La_xFeO_3$ with (x = 0.025) at different magnifications. Particles are seen as agglomerated owing to the annealing effect in the entire sample at 700 °C for 4 h. Proper incorporation of La^{3+} in $BiFeO_3$ crystal structure is visible in FE-SEM surface images describing grain boundaries clearly. Average crystal dimensions obtained using FE-SEM are consistent with the size obtained using the Williamson-Hall plot (W–H). ImageJ software is used for drawing histograms to idealize grain size with microscopy images Fig. 10.2(a, b).

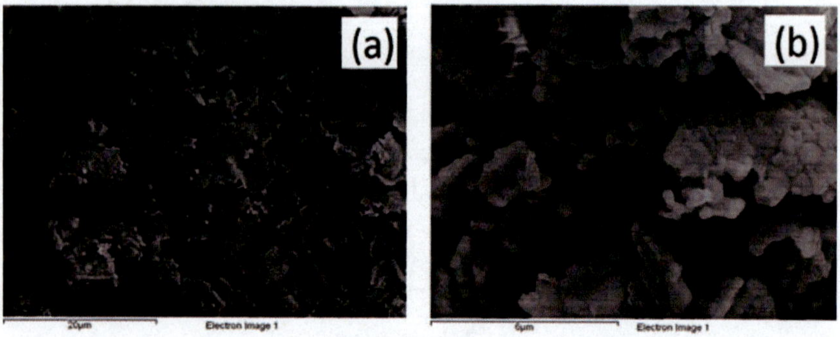

Fig. 10.2 **a, b** FE-SEM micrographs of Bi_{1-x} La_x FeO_3 with (x = 0.025) at different magnifications

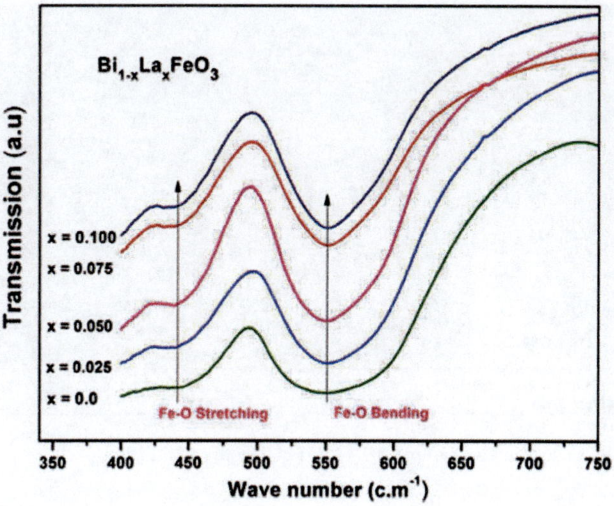

Fig. 10.3 FTIR spectra for Bi_{1-x} La_x FeO_3 with (x = 0.025 to 0.10)

10.3.3 FTIR Analysis

The FTIR spectrum of samples at 300 K is presented in Fig. 10.3 for the wavenumber 400–750 cm^{-1}. Generally, the absorption peak below 750 cm^{-1} belongs to the inorganic bonds. The absorption peak ~ wavenumber 450 cm^{-1} belongs to Fe–O bond stretching, and peak ~550 cm^{-1} signifies bending vibrations of Fe–O bonds in the crystal structure of $BiFeO_3$ [17, 18]. Such observations agree with characteristic FTIR bands of $BiFeO_3$ [8]. Moreover, the absorption peak ~450 and 550 cm^{-1} also signify the vibrational mode of Bi-O bonding of BiO_6 octahedral sites. The shift in these absorption peaks towards the right on the wavenumber axis can easily be visualized with respect to increasing La^{3+} content in $BiFeO_3$ crystal lattices.

10.3.4 EDAX Analysis

Chemical elemental analysis was done using an EDAX analyzer equipped in FE-SEM, as shown in Fig. 10.4. Atomic percentage and weigh percentage of constituent elements for sample Bi_{1-x} La_x FeO_3 with (x = 0.10) are mentioned in Table 10.1, which justifies proper substitution of La^{3+} in $BiFeO_3$. The atomic (%) composition obtained is consistent with the stoichiometric proportion used during the synthesis of sample $Bi_{0.9}La_{0.1}FeO_3$.

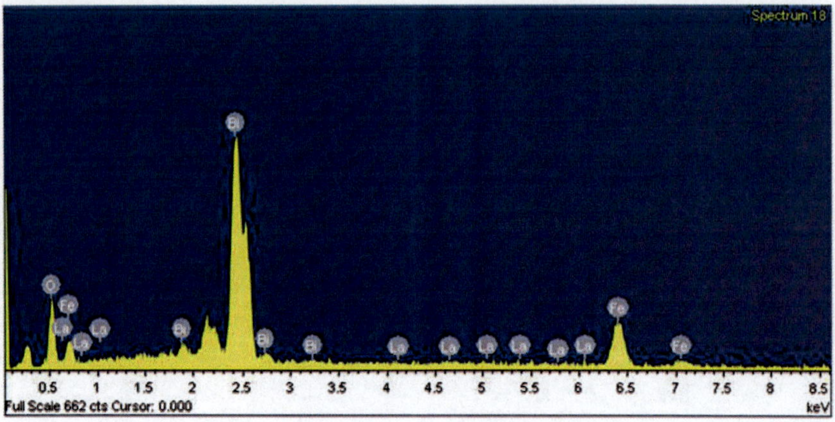

Fig. 10.4 EDAX spectrum for sample Bi_{1-x} La_x FeO_3 with (x = 0.10)

Table 10.1 Elemental composition for sample Bi_{1-x} La_x FeO_3 with (x = 0.10)

Element	Weight (%)	Atomic (%)
O	13.94	56.49
Fe	19.56	22.70
La	1.14	0.53
Bi	65.36	20.28

10.3.5 Magnetic Properties

Magnetic hysteresis (MH) loops in the range ±1.5 T were obtained using VSM (Lake Shore) for sample for Bi_{1-x} La_x FeO_3 with (x = 0.0, 0.10) as shown in Fig. 10.5, which suggests weak ferromagnetic nature of $BiFeO_3$. The MH loops appear unsaturated in the magnetic field of ±1.5 T. A magnetism of order ±5 T is generally reported to saturate such samples. Further, remanent magnetization (M_r) ranges from 5×10^{-3} to 9×10^{-3} emu/g, whereas coercivity (H_c) ranges from 252 to 875 Oersted with La^{3+} increasing composition. We can see in Fig. 10.5 that the virgin curve deviates from a linear relationship of the magnetic moment versus the applied magnetic field, indicating the hysteresis effect in La^{3+}-substituted bismuth ferrite. An improved MH loop is observed with La^{3+} (x = 0.10) with respect to pure phase $BiFeO_3$. The coercivity of the order 875 Oersted was noticed for $BiFeO_3$ with x = 0.10 composition.

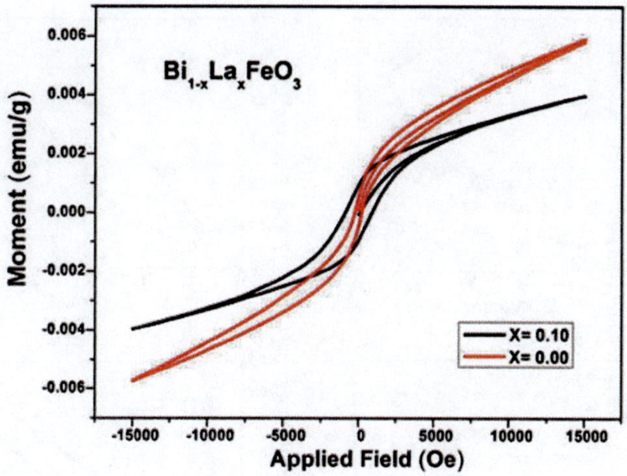

Fig. 10.5 Hysteresis loops for Bi_{1-x} La_x FeO_3 with (x = 0.0, 0.10)

10.3.6 *Multiferroic Analysis*

P-E loops of Bi_{1-x} La_x FeO_3 with (x = 0.025 to 0.100) were measured at 300 K, as shown in Fig. 10.6(a–e). The P-E loop of $BiFeO_3$ was obtained as full convex nature, indicating the samples have electrical leakage, and with increasing La^{3+} doping in $BiFeO_3$, the ferroelectric loop becomes more typical, suggesting reduction in electrical leakage. La^{3+} substitution in $Bi_{1-x}La_xFeO_3$ crystal geometry has considerably enhanced electrical polarization by significant inhibition of electrical leakage. It can be also visualized that incorporation of La^{3+} in $BiFeO_3$ lattice supports phase purity in samples and inhibits the generation of secondary phases.

10.4 Conclusion

Sol–gel technique has been utilized to successfully synthesize Bi_{1-x} La_x FeO_3 with (x = 0.025 to 0.100) with rhombohedral symmetry indexed to *R3c* space group. X-ray intensity pattern confirms pure phase synthesis of all samples. FTIR spectrum well supports the proper substitution of La^{3+} in $BiFeO_3$ crystal lattices with minor structural distortion. EDAX analysis confirms the proper incorporation of lanthanum in the $BiFeO_3$ lattice. Hysteresis loops suggest appreciable increment in magnetization with La^{3+} substitution. Moreover, La^{3+} doping in $BiFeO_3$ solid solutions has considerably enhanced electrical polarization by significant inhibition of electrical leakages. Hence, it can be inferred that the substitution of La^{3+} profitably tunes structural, magnetic, and multiferroic properties of bismuth ferrite for multifunctional applications.

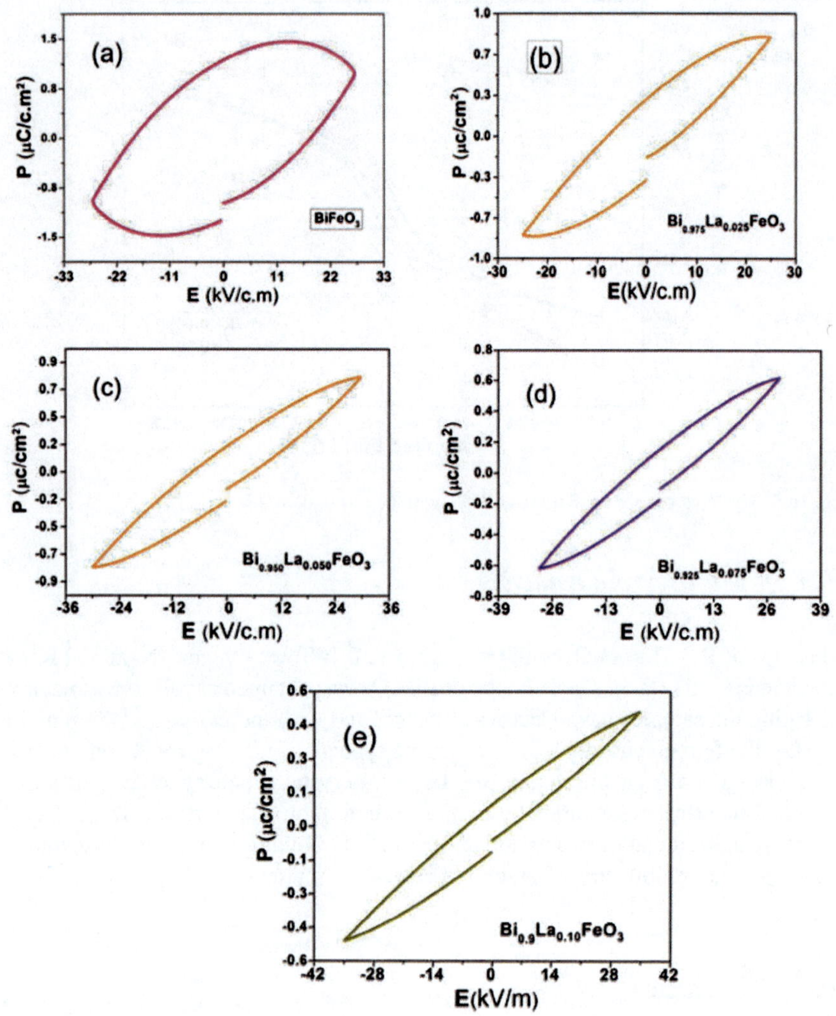

Fig. 10.6 **a–e** Ferroelectric loops for samples Bi_{1-x} La_x FeO_3 with (x = 0.025 to 0.100)

Acknowledgements The authors are thankful to Aryabhatta Knowledge University for providing all experimental facilities.

References

1. Eerenstein, W., Mathur, N.D., Scott, J.F.: Nature (London) **442**, 756–764 (2006). https://doi.org/10.1038/nature05023
2. Catalan, G., Scott, J.F.: Adv. Mater. **21**, 2463–2485 (2009). https://doi.org/10.1002/adma.200802849
3. Wang, D.H., Goh, W.C., Ning, M., Ong, C.K.: Appl. Phys. Lett. **88**, 212907 (2006). https://doi.org/10.1063/1.2208266
4. Pandey, R., Panda, C., Kumar, P., Kar, M.: J. Sol-Gel Sci. Technol. **85**, 166–177 (2018). https://doi.org/10.1007/s10971-017-4537-2
5. Ederer, C., Spaldin, N.A.: Phys. Rev. B **71**, 060401 (2005). https://doi.org/10.1103/PhysRevB.71.060401
6. Cheng, Z.X., Wang, X.L.: Phys. Rev. B **75**, 172406 (2007). https://doi.org/10.1103/PhysRevB.75.172406
7. Ablat, A., Wu, R., Mamat, M., Li, J., Muhemmed, E., Si, C., Wu, R., Wang, J., Qian, H., Ibrahim, K.: Int. **40**, 14083–14089 (2014). https://doi.org/10.1016/j.ceramint.2014.05.137
8. Ain, R.S.N., Halim, S.A., Hashim, M.: Adv. Mater. Res. **501**, 329–333 (2012). https://doi.org/10.4028/www.scientific.net/AMR.501.329
9. Huong, N.T., Thu, L.H.A., Long, N.N., Hong, N.H.: VNU J. Sci. Math. Phys. **33**, 35–40 (2017)
10. Chauhan, S., Kumar, M., Chhoker, S., Katyala, S.C., Singh, M.: RSC Adv. **6**, 43080–43090 (2016). https://doi.org/10.1039/C6RA02316A
11. Yuan, X., Shi, L., Zhao, J., Zhou, S., Li, Y., Xie, C., Guo, J.: J. Alloy. Comp. **708**, 93–98 (2017)
12. Sen, K., Singh, K., Gautam, A., Singh, M.: Ceram. Int. **38**, 243–249 (2012). https://doi.org/10.1016/j.ceramint.2011.06.059
13. Sinha, A.K., Bhushan, B., Sharma, R.K., Mishra, S.K., et al.: Results Phys. **13**, 102299 (2019). https://doi.org/10.1016/j.rinp.2019.102299
14. Pandey, R., Pradhan, L.K., Kumar, P., Kar, M.: J. Phys. Chem. Solids **119**, 107–113. https://doi.org/10.1016/j.jpcs.2017.12.001
15. Chen, X.K., Wu, Y.J., Zhang, J., Chen, X.J.: Sci. China Phys. Mech. Astron. **55**, 404–408 (2012). https://doi.org/10.1007/s11433-012-4635-7
16. Rafique, M., Herklotz, A., Dorr, K., Manzoor, S.: Appl. Phys. Lett. **110**, 202902 (2017). https://doi.org/10.1063/1.4983357
17. Sati, P.C., Arora, M., Chauhan, S., Chhoker, S., Kumar, M.: J. Appl. Phys. **112**, 094102 (2012). https://doi.org/10.1063/1.4761968
18. Garcia, F.G., Riccardi, C.S., Simões, A.Z.: J. Alloy. Comp. **501**, 25–29 (2010). https://doi.org/10.1016/j.jallcom.2010.04.049

Chapter 11
Experimental Analysis of Wear and Mechanical Characteristics of Aluminium Matrix Composite Fabricated Through Powder Metallurgy

Harvendra Singh, Harshit Bahri, and Kaushalendra Kumar Singh

Abstract Aluminium metal matrix composites have picked up significance in different industries due to their great mechanical properties. Aluminium metal matrix composites are favoured in the fields of aviation, military, car, marine and other homegrown applications. The various reinforcements with Aluminium metal matrix composites by powder metallurgy procedure brings about improved mechanical properties, for example, ultimate tensile strength, compressive strength, hardness, wear rate. Powder metallurgy is an ideal technique for manufacture for MMCs on account of the capacity to create close to net shapes and minimal material waste related with this cycle. In this paper, an endeavour has been made to unite a portion of the parts of mechanical properties of Aluminium MMCs manufactured by utilizing Powder Metallurgy Technique. This paper is also focused on the comparative analysis of different parameters on the AMCs.

11.1 Introduction

Traditional solid materials have restrictions in accomplishing great mix of solidarity, firmness, sturdiness, thickness and so forth. To defeat these weaknesses and to satisfy the regularly expanding need of advanced innovation, composites are the most encouraging materials of ongoing interest. Metal matrix composites (MMCs) have altogether improved properties including high specific strength, specific modulus, damping capacity and great wear resistance in contrast to composite compounds. Magnesium and its compounds did not contrast well with aluminum composites regarding outright quality, however; they are the lightest materials and have a great mix of low thickness and superb machinability in contrast with other basic materials. Elango et al. [1] investigated the behaviour of aluminium alloy LM25 with SiC and observed that when load and particle reinforcement increase, coefficient of friction decreases. Baradeswaran et al. [2] investigated the effect of graphite and Al_2O_3

H. Singh (✉) · H. Bahri · K. K. Singh
Department of Mechanical Engineering, G. L. Bajaj Institute of Technology and Management, Greater Noida, UP, India

© The Author(s), under exclusive license to Springer Nature Singapore Pte Ltd. 2022

V. V. Rao et al. (eds.), *Computational and Experimental Methods in Mechanical Engineering*, Smart Innovation, Systems and Technologies 239,
https://doi.org/10.1007/978-981-16-2857-3_12

on hybrid composite hardness increase and wear resistance. Sudha et al. [3] investigated the behaviour of aluminium matrix material and identified the main influencing parameter as sintering temperature, particle size, heat treatment condition and milling time. Chelladurai et al. [4] reviewed the mechanical properties and wear behaviour of aluminium-based matrix composites and concluded that stir casting is the cost-effective methodology for particulate and fibre-reinforced composites, and squeeze casting gives good surface finish and mechanical properties. Nayak et al. [5] worked on four types of powder metallurgy-based cylindrical sintered hybrid composites which were made through five different combinations of sintering temperature and time span and reported that the density and hardness of each specimen increased through the suggested powder metallurgy route. They also found that the powder metallurgy method efficiently reduces cluster formation in material when nano-size particle is used through milling process which results in densification of the material. Vasanth et al. [6] performed similar study and concluded that particle size, compaction pressure and sintering time potentially influenced the behaviour of proposed material. They also found optimum sintering time and compaction time in order to maximize the hardness of the material. They also reported that the material properties such as hardness and compressive strength can be improved by reinforcing aluminium matrix with the compounds like boron carbide, silicon carbide and aluminium oxide. The preparation of MMCs by conventional methods such as stir casting, squeeze casting and other techniques may lead to defects such as poor wetting, porosity and poor interface bonding between reinforcement and matrix material. The powder metallurgy method which includes ball milling gives homogenous distribution of the reinforcement particle in matrix alloy. Due to this, proper bonding between the matrix material and reinforcements will occur. It also improves mechanical and tribological properties.

11.2 Methodology and Experimental Setup

This section depicts the trial methodology in the current work. A point-by-point report is likewise given on the portrayal of crude materials utilized for creation of the MMC test examples. This part houses a depiction of the itemized venture savvy techniques embraced for manufacture of the test examples, the heat treatment conferred and the mechanical and electrical testing did. All the experimental work has been performed in Al-Falah School of Engineering and Technology, Dhauj, Faridabad, Haryana, India.

11.2.1 Rule of Mixtures

Rule of mixture is a strategy to deal with inexact assessment of composite material properties. This improves the reproducibility of the composite and helps to achieve identical properties. The mixture properties are as follows:

Density

$$d_c = d_m \times V_m + d_f \times V_f \tag{11.1}$$

where,

d_m, d_c, d_f—Matrix, composite and dispersed phase densities.
V_f and V_m—The volume fraction of the dispersed phase and matrix.

Thermal Expansion coefficient

Thermal Expansion coefficient (TEC) in linear direction

$$\alpha_{cl} = \frac{\alpha_m \times E_m \times V_m + \alpha_f \times E_f \times V_f}{E_m \times V_m + E_f \times V_f} \tag{11.2}$$

αcl, α_m, α_f—TEC of composite in linear direction, matrix and dispersed phase
E_m, E_f—modulus of elasticity of matrix and dispersed phase

Thermal Expansion coefficient (TEC) in lateral direction

$$\alpha_{ct} = (1 + P_m)\alpha_m \times V_m + \alpha_f \times V_f \tag{11.3}$$

where, P_m—Poisson ratio of matrix.

Modulus of Elasticity (Ecl)

$$E_{cl} = E_m \times V_m + E_f \times V_f \tag{11.4}$$

$$\frac{1}{E_{ct}} = \frac{V_m}{E_m} + V_f/E_f \tag{11.5}$$

Tensile Strength

Tensile strength of composite in linear direction.

$$\sigma_c = \sigma_m \times V_m + \sigma_f \tag{11.6}$$

where, σ_m, σ_c, σ_f—matrix, composite and dispersed phase tensile strength.

Selection of material: During my literature survey, I have noticed that in most high-precision applications of kinematics chain, the aluminium alloys from series 6000, namely Al 6063 and Al 6061, were used. These have good corrosion resistance

Table 11.1 For composite material density and mass of LM25 and mass of B_4C

Material	Density (g/cm^3)	Mass of LM25 in gms	Mass of B_4C in gms
LM25	2.68	42.09	0
LM25 + 5% B_4C	2.67	40.00	2
LM25 + 10% B_4C	2.66	38.00	4.2
LM + 25 + 15% B_4C	2.65	35.46	6.25

and extrusion properties with good strengths. However, they have high-temperature resistance and start losing their strength with increase in temperature. To overcome this limitation of Al 6000, I am considering LM 6, LM13 and LM25 from cast aluminium series at the initial stage because these materials can sustain high strength at high temperatures. After checking the compatibility of these matrix materials, I am going to reinforce fibrolite used as refractory to one or more aluminum alloy matrix materials to further enhance the heat resistance.

The material used in the current study is LM25 with different compositions of B_4C which has been listed in the Table underneath. The LM25 with B_4C has an outstanding mix of mechanical properties in the cast condition. This network was picked in light of the fact that it gives a superb blend of solidarity and harm resistance at raised and cryogenic temperatures. To do the investigation, unadulterated and three sorts of composites were readied. The composition of four samples are listed in Table 11.1.

Pure LM25

Al alloy LM25 95 wt% and 5 wt% of B_4C composite

Al alloy LM25 90 wt% and 10 wt% of B_4C composite

Al alloy LM25 85 wt% and 15 wt% of B_4C composite

11.2.2 Powder Metallurgy Method

The following steps are taken into consideration:

Mixing of powder

The MMC test specimens are produced by the powder metallurgy technique. 95%, 90% and 85% Aluminium alloy LM25 powder and 5%, 10% and 15% B_4C by weight are mixed for fabricating the composite. In total, four categories of mixture were prepared (pure LM25), 95% LM25 + 5% B_4C, 90% LM25 + 10% B_4C and 85% LM25 + 15% B_4C. Blending is performed in a ball planetary mill with the ball to power ratio of 15:1 and ball (chrome steel) size 10 mm. The mixing machine proceeds on 200–300 rpm to have a good homogeneous blend. Figure 11.1 shows the ball mill used for mixing.

Compaction of the powder

Compaction is done on cold uniaxial press in steps rather than pressing the whole powder mix in a single go. In each step, around 50 g powder mix is added and pressed to complete the specimen. The punch and die arrangement is shown in Fig. 11.2.

Cold uniaxial pressing

The powder test is squeezed in the cold uniaxial squeezing machine (UTM hydro-electromechanical) to deliver the green round tests of 20 mm external diameter across

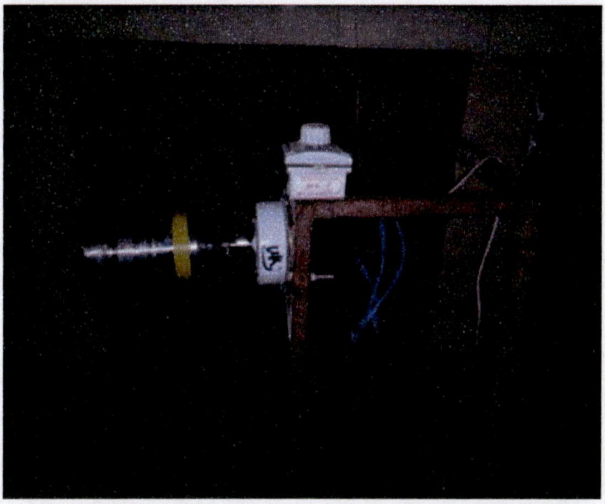

Fig. 11.1 Ball mill

Fig. 11.2 Punch and die assembly

Fig. 11.3 Universal testing machine

applying a load of 60 kN. A treated steel pass-on of 20 mm internal diameter was utilized for this reason. To allow the powder to stream uninhibitedly and to shield the model from holding fast onto the dividers, Graphite Grease was used as an oil that was applied to the dividers of the pass-on and punch. The squeezing machine is shown in Figs. 11.3.

Sintering of the green samples

The green sample is prepared at an elevated temperature in a controlled air climate below the melting point of matrix material for an adequate time. It is done in a muffle furnace as shown in the Fig. 11.4. A group of four samples was sintered at temperatures 580–600 °C individually. The time of holding was 2 h. By increasing boron caride content, the sintering temperature increases and high quality composite is obtained. At that point, heater is permitted to cool to room temperature for a range of 24 h. At that point, occupations are eliminated from the heater. The normal dia. and height of samples are 20 and 50 mm. Figure 11.5 shows the furnace used for sintering, and Fig. 11.6 shows the sample after the sintering.

11.2.3 Wear Test

Wear is a pattern with material removed from the two contact surfaces in solid state. Due to the cost factor and complexity in making wear resistance material, it is suitable to make the surface of material to be wear-resistant as wear is a phenomenon of surface removal.

Fig. 11.4 Specimen after green compaction

Fig. 11.5 Specimen in muffle furnace

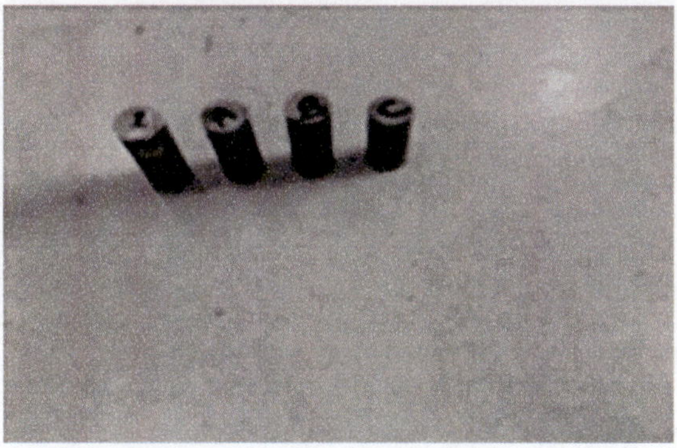

Fig. 11.6 Sample after sintering

Table 11.2 Value of parameters during sliding wear test

Material of pin	LM25, LM25/5B_4C, LM25/10B_4C, LM25/15B_4C
Material of disc	EN 31 Steel
Dimensions of pin	Cylinder with diameter 12 mm and height 30 mm
Sliding speed (m/s)	2,4
Applied load	20,40,60
Distance travelled (m)	2000

Experimental Procedure of Wear Test

The normal loads at which wear test was conducted were 20 N, 40 N, 60 N with a sliding velocity of 2 and 4 m/s. The total sliding distance for which the wear tests were carried out was 2000 m approx. The pin models were 50 mm in length with a diameter of 20 mm. Before the test, it was ensured that the surfaces are smooth for effective contact of flat surface with the steel disc. The process of cleaning the surfaces with C_3H_6O and weighing samples (up to an accuracy of 0.0001 gm using microbalance) is carried out after each test. The height loss technique is used to calculate the wear rate and expressed in terms of wear volume loss per unit sliding distance. The parameters involved in this experiment are applied load, speed of rotation and distance travelled. Values of these parameters assumed in the present study are shown in Table 11.2.

Pin-On-Disc Test

For tribological characterization, the Pin-on-Disc test (shown in Fig. 11.7) was performed with the following steps:

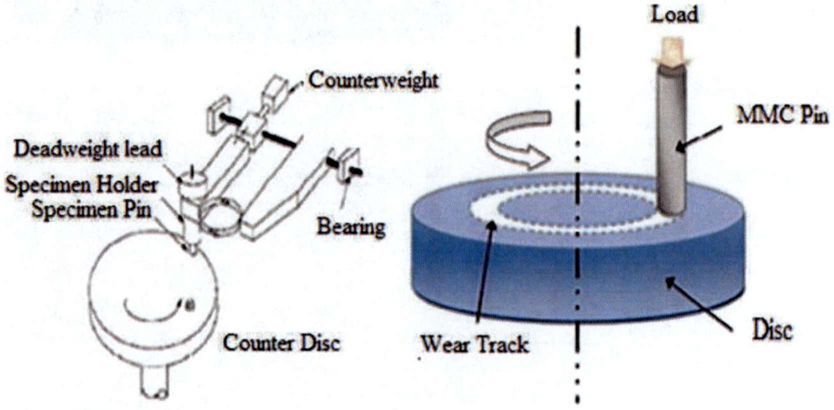

Fig. 11.7 Diagram of the pin on disc

- In the first stage, the pin surface was made level with the end goal that it will uphold the heap over its whole cross-area. It is gotten by utilizing emery paper (80 coarseness size) before testing
- In the second stage, a wear test was performed.
- In the final stage, both disc and pin were cleaned with ethanol. This stage is the dynamic competition between material transfer processes.

11.3 Results and Discussion

11.3.1 Weight Loss

The composite and alloy samples are cleaned with acetone. Each sample is weighed using a digital balance having an accuracy of ± 0.1 mg. After that, wear test is performed on tribometer. For all experiments, the sliding velocities are 2 m/s and 4 m/s, and loads are 20 N, 40 N and 60 N, respectively. Table 11.3 presents the results of wear test for 2 m/s and 4 m/s velocities and 20 N load. Similarly Tables 11.4 and 11.5 present the results of wear test for 40 N and 60 N loads at 2 m/s and 4 m/s velocities respectively.

Table 11.3 Data of wear loss of composite and alloy for 20 N

Weight loss of composite and alloy						
Specimen name	Sliding speed 2 m/s			Sliding speed 4 m/s		
	Initial weight (g)	Final weight (g)	Weight loss (g)	Initial weight (g)	Final weight (g)	Weight loss (g)
LM25	42.2712	42.246	0.0248	42.2712	8.2422	0.029
LM25/5B_4C	41.9724	41.962	0.0198	41.9724	7.985	0.0206
LM25/10B_4C	41.3557	41.344	0.0113	41.3557	8.3363	0.0194
LM25/15B_4C	41.2057	41.196	0.0101	41.2057	8.2047	0.0175

Table 11.4 Data of wear loss of alloy and composite for 40 N

Weight loss of composite and alloy						
Specimen Name	Sliding speed 2 m/s			Sliding speed 4 m/s		
	Initial weight (g)	Final weight (g)	Weight loss (g)	Initial weight (g)	Final weight (g)	Weight loss (g)
LM25	42.2712	42.2439	0.02732	42.2712	42.2371	0.03412
LM25/5B_4C	41.9724	41.95655	0.01585	41.9724	41.9469	0.02555
LM25/10B_4C	41.3557	41.34237	0.01335	41.3557	41.3343	0.02143
LM25/15B_4C	41.2057	41.1923	0.01342	41.2057	41.1862	0.01952

Table 11.5 Data of wear loss of alloy and composite for 60 N

Weight loss of composite and alloy						
Specimen Name	Sliding speed 2 m/s			Sliding speed 4 m/s		
	Initial weight (g)	Final weight (g)	Weight loss (g)	Initial weight (g)	Final weight (g)	Weight loss (g)
LM25	42.2712	42.2361	0. 03516	42.2712	42.2398	0.03142
LM25/5B $_4$C	41.9724	41.9459	0.02653	41.9724	41.9493	0.02312
LM25/10B$_4$C	41.3557	41.3337	0.02203	41.3557	41.3356	0.02013
LM25/15B$_4$C	41.2057	41.1862	0.01953	41.2057	41.1873	0.01842

After performing the wear tests of the aluminium alloy materials with reinforcement of different amounts of B_4C, following results were obtained.

Micro Hardness Measurement

Micro hardness testing is a procedure for assessing the hardness of a material. A micro hardness analyser MVH-1is used for the miniature hardness assessment. To check hardness, model with estimation (50 mm × 20 mm) is prepared. The required metallographic finish of the surface being pursued is obtained with the help of 100, 220, 400, 600 and 1000 coarseness size emery paper. Weight used on Vickers small-scale hardness analyser was 200 grams at 10X optical zoom with stay time 20 s for every model. The after-effect of Vickers micro hardness test for amalgam LM25 without support, and the wt% variety of reinforcement with B_4C in Al compound MMCs are shown in Table 11.6.

In the testing of Al-based MMCs reinforced with B_4C and micro hardness of Al alloy, it was seen that hardness of B_4C-strengthened composite is more than that of Al compound. Hardness of composite relies upon the hardness of the reinforcement and the lattice. The variation of Mean Micro Hardness No. with increasing B_4C is shown in Fig. 11.8.

It has been observed that the hardness of composite depends on the particle–matrix interface. Due to dislocations, hardening increases with increase in particle–matrix interface. As in the case of alumina reinforcement, if the composite is reinforced

Table 11.6 Data of micro hardness

Sample number	Sample name	Mean micro hardness no.
1	100%LM25 + 0%B_4C	90
2	95%LM25 + 5%B_4C	94
3	90%LM25 + 10%B_4C	97
4	85%LM25 + 15%B_4C	100

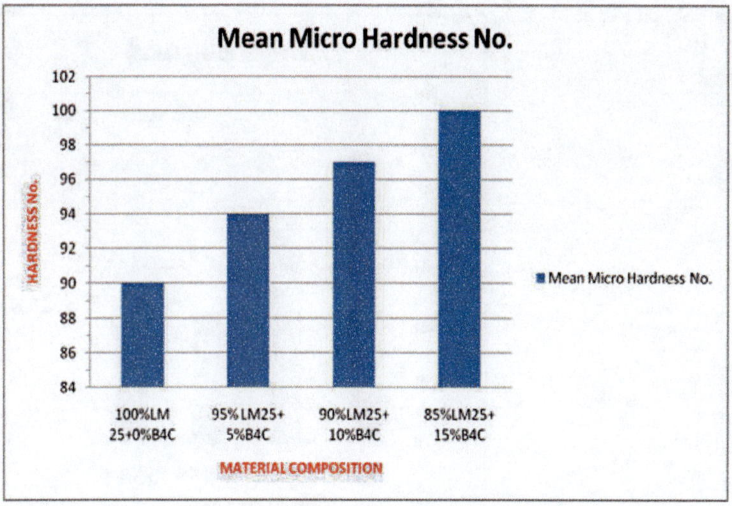

Fig. 11.8 Comparison of mean micro hardness of alloy and composite

with smaller ceramic particles, then it will have more particle–matrix interface as compared to reinforcement of large particles. Therefore, the hardness of composites can also be increased by decreasing the size of reinforced particles and increase the volume ratio of the reinforcement.

11.3.2 Wear Characteristics

The addition of boron carbide with the help of powder metallurgy results in cumulative weight loss. Further, the increase in boron carbide may lead to significant decrease in weight loss. Figure 11.9 shows the values of weight loss for different specimens for 20 N load at 2 m/s velocity. It can be seen that the weight loss is highest for LM25 and minimum for LM25/15B4C. Similarly the weight losses in different specimens for 40 N-2 m/s, 40 N-4 m/s, 60 N-2 m/s and 60 N-4 m/s; load-velocity combinations are shown in Figs. 11.9, 11.10, 11.11, 11.12, 11.13 and 11.14 respectively.

11.4 Conclusion

The following conclusions may be drawn from the work:

1. LM25 aluminium alloy has superior mechanical properties, high hardness, low weight and high wear resistance which can be improved with reinforcement of B_4C.

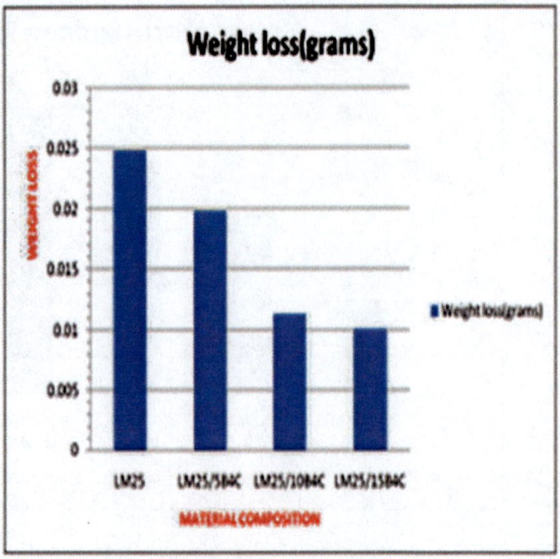

Fig. 11.9 Weight loss at 2 m/s for 20 N

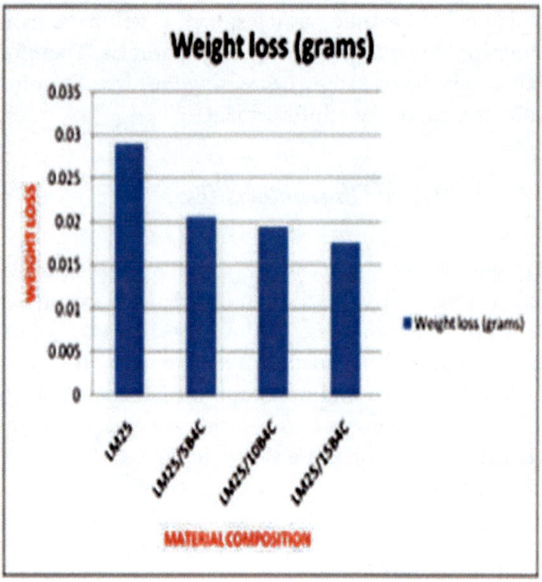

Fig. 11.10 Weight loss at 4 m/s for 20 N

2. B_4C has outstanding hardness, so as reinforcement material is used to improve the hardness of LM25. The hardness of composite is increased by decreasing the size of reinforced particles and increasing the volume ratio of the reinforcement.
3. Powder metallurgy gives the proper bonding between matrix material and reinforcement, and also, improves mechanical and tribological properties.

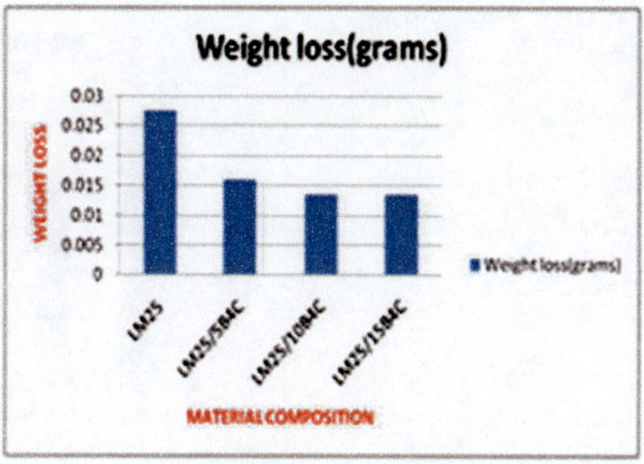

Fig. 11.11 Weight loss at 2 m/s for 40 N

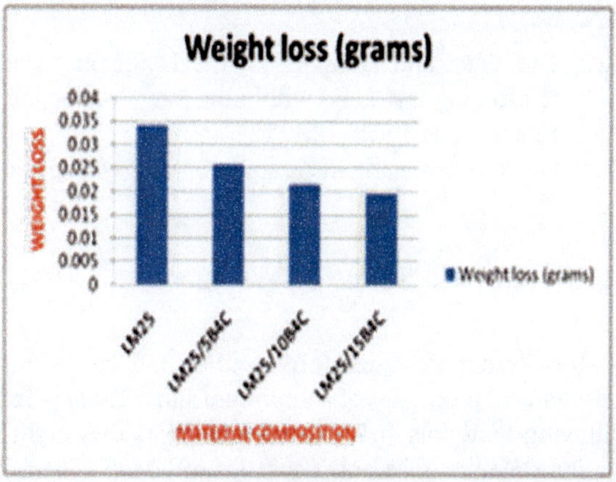

Fig. 11.12 Weight loss at 4 m/s for 40

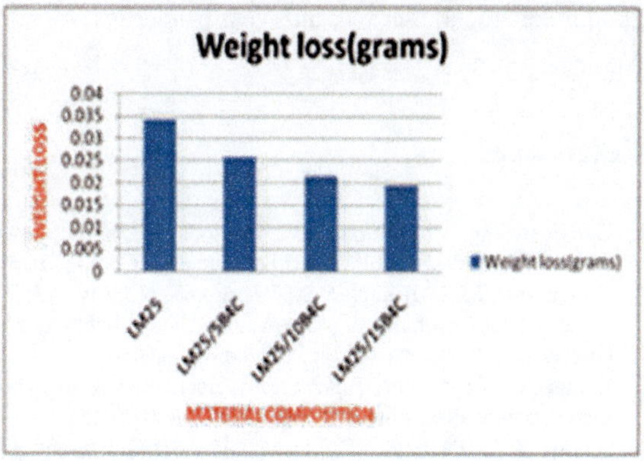

Fig. 11.13 Weight loss at 2 m/s for 60 N

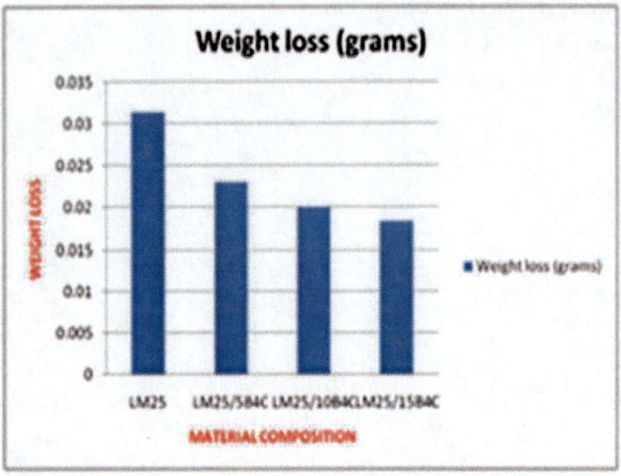

Fig. 11.14 Weight loss at 4 m/s for 60

4. The Wear and Hardness of the LM25 material increase very effectively by reinforcing B_4C, and within the range of our interest, best result is obtained at 15% B_4C and 85% LM25.
5. Further increasing the amount of B_4C will result in cost elevation.

11.5 Future Scope of Work

In the present work, reinforcement of B_4C has been done to improve the wear and mechanical properties of Aluminium alloy. There is an enormous scope of many more alloying materials with different amounts of weight percentage and compositions to improve the different properties of LM25. There is also a possibility of getting optimum results of reinforcements to obtain an Aluminium Matrix material having good mechanical, thermal and wear properties.

References

1. Baradeswaran, A., Perumal, A.E.: Composites: Part B study on mechanical and wear properties of Al7075/Al_2O_3/graphite hybrid composites. Compos Part B **56**, 464–471 (2014)
2. Celladurai, S.J.S., Kumar, S.S., Venugopal, N., Ray, A.P., Manjunath, T.C., Gnansekaran, S.: A review on mechanical properties and wear behaviour of aluminium based metal matrix composites. Materials Today: Proceedings (2020)
3. Elango, G., Raghunath, B.K.: Tribological behavior of hybrid (LM25Al + SiC + TiO_2) metal matrix composites. Procedia Eng **64**, 671–680 (2013)
4. Nayak, K.C., Pandey, A.K., Date, P.P.: Mechanical and physical characterization of powder metallurgy based aluminium metal matrix hybrid composite. Mater. Today: Proc. (2020)
5. Sudha, G.T., Ravichandran, M., Balasubramanian, M.: Mechanical properties, characterization and wear behavior of powder metallurgy composites: a review. Mater. Today Proc. **22**, 2582–2596 (2020)

6. Vasanth, K.M.J., Ram, P.S.L., Anand, V.P., Prabhu, M., Rahul, S.: Experimental investigation of mechanical and tribological properties of Aluminium metal matrix composites fabricated by powder metallurgy route: a review. Mater. Today Proc. (2020)

Chapter 12
Recent Advancements on Structural Health Monitoring Using Lamb Waves

Deepak Kumar, Sahil Kalra, and Mayank Shekhar Jha

Abstract Structural health monitoring (SHM) is an inspection study that deals with the diagnosis and prognosis of damages/faults in structures. SHM plays a crucial role in diagnosing the metallic and composite-based thin/shell structure. In this paper, critical analysis and discussion are provided on the SHM methods focusing on state of the art Lamb waves. The sensors and actuators, especially piezoelectric sensors, are explained from the viewpoint of health monitoring on thin structures. A tabular survey of the findings from the literature and existing lacunas are reported for the last 5 research years. It becomes evident from the research work that SHM methods with piezoelectric material-based sensors and actuators are more pronounced techniques in comparison to optic sensors, electrical resistance, electromagnetic techniques, and capacitive methods.

12.1 Introduction

Structural health monitoring (SHM) is the interdisciplinary field of carrying out diagnosis and prognosis processes in different domains, including aerospace, civil, mechanical, and naval structures. By definition, it is the mechanism of enforcing damage revelation techniques and characterization methods for monitoring defect-prone systems affected by external factors [1]. It utilizes the concept of failsafe design, which affirms that any structure can have defects as long as they do not lead to failure to the system. In other words, SHM intends to move toward the performance-based design philosophy [2]. Various state of the art technologies such

D. Kumar (✉) · S. Kalra
Indian Institute of Technology Jammu, Jammu & Kashmir, India
e-mail: 2019rme0031@iitjammu.ac.in

S. Kalra
e-mail: sahil.kalra@iitjammu.ac.in

M. S. Jha
Centre de Recherche en Automatique de Nancy, Université de Lorraine, Nancy, France
e-mail: mayank-shekhar.jha@univ-lorraine.fr

© The Author(s), under exclusive license to Springer Nature Singapore Pte Ltd. 2022
V. V. Rao et al. (eds.), *Computational and Experimental Methods in Mechanical Engineering*, Smart Innovation, Systems and Technologies 239,
https://doi.org/10.1007/978-981-16-2857-3_13

as acoustic emission, ultrasonic, thermal imaging, etc., are being utilized to monitor structural health. The diagnosis is generally carried out either actively or passively. In the passive system, the measurement of various parameters is carried out on the passive structure. However, in the active system, the measurement is done in real time. Hence, the methodologies focused on the active process are generally non-destructive based in nature.

The main objective of performing SHM is to decline the growth of maintenance needs and hence reduce the overall cost required for prolonged functioning of the system or structure. It also aims to enhance existing design performance and provide feedback that helps improve the future design based on the experience. The techniques for SHM vary from simple human sense like visual inspection to machining operations, which require skill like in the case of non-destructive evaluation (NDE). Based on the premise that damage will alter the characteristic properties of the structure, SHM is classified into two main categories [3]. The first category is the static-based SHM performed when the damage affects the static properties of structure like displacement and rotation. The second category is vibration-based SHM which is done when the damage affects the dynamic properties of the structure like frequency, modes and mode shapes, etc. In each of these categories, SHM is performed in four fundamental and essential steps [4]:

(i) Detection: determining whether there is any damage present in the structure,
(ii) Localization: localization of the damage (if detected),
(iii) Characterization: quantification of damage severity, and
(iv) Prognosis: prediction of remaining functioning life of the structure.

To detect any fault, the array of sensors is placed strategically on the structure that collects dynamic structural response either continuously or at regular intervals. By smartly analyzing these measured responses, one can identify damage occurrence. During data acquisition, environmental effects, such as temperature and humidity levels, and measurement noise create uncertainty which causes error in the signal processing. If non-stationary inputs like traffic, wind, and earthquake are avoided and if every point on the structure is observed, it is possible to detect any damage and its location. Though it is computationally expensive, it may even be used to assess the damage type and estimate the structure's remaining life if used under the supervision of high user expertise. Too many unknowns and modeling assumptions from physical to modal domain like boundary conditions, number of DOFs and material properties, etc., affect the analysis method. SHM can be used for monitoring metals, composites, laminates, and sandwich materials.

12.2 Structural Health Monitoring Using Lamb Waves

Lamb waves are unique ultrasonic waves that travel the controlled path between two free parallel surfaces, such as a thin plate or shell's upper and lower surfaces. For this reason, Lamb waves are often referred to as guided plate waves [5]. It was discovered by Horace Lamb in 1917, taking inspiration from the Rayleigh wave by Lord Rayleigh [6, 7]. The detailed theoretical development of the wave was set up by Mindlin in 1950 [8], supported by the experimental work carried out by Schoch in 1952 and Frederick in 1962 [9, 10]. After its discovery, most further improvements and applications were aimed mainly at the medical field during World War II [11–13]. Later in 1961, Worlton introduced Lamb waves as a damage detection means, which may be extended for smart structures [14, 15]. Subsequently, all these initial studies helped establish the basics of Lamb waves as an outstanding NDE technique.

Lamb waves comprise elastic-wave-based propagation, which causes specific scattering of wave and mode reconstruction depending upon structural damages,. A quantifiable assessment of faults and defects can be achieved by processing and analyzing the wave signals dispersed by damage.

Advantages of Lamb waves: some of the advantages are: cost-viability, quick and repeatable, a short-term inspection of huge structures, responsive to smaller defects, no need for transducer movement, using up little energy, capable of detecting surface and internal defects.

Limitations of Lamb waves: some of the limitations include: the need for practical and refined signal analysis techniques because of composite wave signal generation and output, collectively more than one wave modes accessible concurrently, wave propagation in complex structures difficult to simulate, heavy reliance on previous models or standard signals.

12.3 Sensors and Actuators in SHM

Following sensors are generally used for various signal detection in the devices [16]:

1. Fiber optic sensors,
2. Piezoelectric sensors,
3. Electrical resistance,
4. Electromagnetic techniques, and
5. Capacitive methods.

Out of these sensors, piezoelectric sensors are the extensively used active sensors for monitoring the health of various structures, thus have been discussed in detail in the present paper.

Piezoelectric wafer active sensors (PWAS) are the most used transducers for detecting acoustic signals. The specialty of these sensors is that they can be used for both actuation and sensing purposes. These transducers work on the piezoelectric concept and bring together the effects of the electrical and mechanical output. When working as an actuator, these piezoelectric devices convert electrical energy directly into mechanical energy, stimulating waves and vibrations in a structure. These can act as high-frequency vibrations generators in the supervised structure. Similarly, PWAS transducers can further be used for perceiving stress–strain because they directly convert the mechanical stress–strain energy into electrical energy. Since the output voltage and the strain rate are proportional, this type of measurement would be highly useful at high frequencies. In the case of Lamb waves, PWAS transducers act as both transmitting and receiving devices for the Lamb waves passing through any structure. When PWAS transmitters are excited with an electrical signal, it generates Lamb waves in the structure. These Lamb waves travel in the thin-line structure following a guided path and are reflected or diffracted when they encounter its boundaries, any discontinuities, and damages. These reflected or diffracted waves then reach the PWAS receiver, where they are all converted into electric signals. The receiving of reflected signals is performed using either of the two configurations—pitch-catch configuration, where one device transmits the signal, at the same time, another one captures it, or a pulse-echo configuration where the same device is used for transmitting as well as receiving the signal [3].

PWAS transducers are mostly used nowadays because they have few useful advantages over conventional ultrasonic transducers. By adhesive bonding, PWAS is strongly joined to the structure, whereas traditional ultrasonic transducers are joined by gel, water, or air which provides weaker bonding. These modern non-resonant devices can be tuned selectively into multiple guided-wave modes, whereas the conventional transducers are single-ultrasonic-resonance devices. The PWAS are small in size, light in weight, and economical to use, so more number of these transducers can be used simultaneously on structures, which was a limitation in conventional ultrasonic transducers because of being relatively expensive bigger in size. A detailed summary of the research work done in detecting and locating damages in various structures using Lamb waves is provided in a tabular form in Table 12.1.

Table 12.1 Detecting and locating damages in various structures using Lamb waves

Sl. no.	Author(s)	Material and structure	Method of detection/location	Result/Remarks
1	Zhang et al. [17]	Steel pipe	Used bidirectional piezoelectric transducer for CSH_0 and CSH_1 waves to monitor corrosion depth change	CSH_0 is more useful than CSH_1 in order to monitor the defect depth change
2	Majhi et al. [18]	Steel pipe	Evaluation of spread of corrosion by time–frequency analysis using guided waves, and tensile test	Destructive test
3	Zhao et al. [19]	Grounded rod	Detection of corrosion pit using acoustic pulse-echo signal, additionally for simulating the soil influence wet-clay wrapped rods are used	Principal longitudinal mode L(0,1) at low frequencies is best suitable for detecting corrosion damage
4	Livadiotis et al. [20]	Steel pipe (D = 12 in.)	Damage monitoring by algebraic reconstruction technique (ART) using helical guided ultrasonic wave (HGUW)	ART imaging algorithm precisely localizes defects on the circumference of the pipe
5	Zhang et al. [21]	Galvanized steel wire	Detection of the severity of depth increment by time-reversal method (TRM) while using longitudinal ultrasonic guided waves	Instead of mode L(0, 2), mode L(0, 3) is turned out to track the deep inside defects
6	Li et al. [22]	Aluminum pipe	Quantitative detection of the defect by wavenumber analysis	The changing wavenumber of the modes reveals the existence of a defect

(continued)

Table 12.1 (continued)

Sl. no.	Author(s)	Material and structure	Method of detection/location	Result/Remarks
7	Li et al. [23]	Composite	Use of non-linear Lamb wave and characterization by acoustic nonlinearity parameters (ANP)	ANPs increase with micro-crack density and with a propagating distance of wave
8	Chen et al. [24]	Stainless steel pipe	Detection of axial as well as circumferential cracks by using linearly polarized circular TE_{11} microwaves	Cracks successfully detected and located on the interior surface of the pipe
9	Rommeler et al. [25]	Polymer	Detection of crack and weld inspection by the method of air-coupled ultrasonic (ACUS) inspection	Contactless inspection
10	Du et al. [26]	Steel pipe	Damage identification using piezoceramic transducers for stress waves-based active sensing method	Multi-crack detection
11	Nirbhay et al. [27]	Brass	Detecting shallow to deep cracks by finite element simulation of Lamb wave response	Also reported spectrogram analysis, Fourier and Hilbert transforms
12	Ni et al. [28]	Steel	Detect and locate bend and cracks by finite element simulation of modes in different bend condition	Configuration type: Space-Z, plane-Z, and U type

(continued)

Table 12.1 (continued)

Sl. no.	Author(s)	Material and structure	Method of detection/location	Result/Remarks
13	Liu et al. [29]	Stainless steel	Circumferential crack identification using cross-sectional diagnostic imaging algorithm and an angular profile based frequency selection method	Accurate circumferential location of small cracks in pipe weld
14	Zhou et al. [30]	Aluminum	Damage detection using ultrasonic fiber optic sensors (UFOS) based on Mach–Zehnder interferometer	More sensitive to L mode guided waves
15	Li et al. [31]	Aluminum	Use of flexible polyvinylidene fluoride (PVDF) comb transducers for measuring fundamental and second harmonics of guided-wave propagation	Thermal fatigue damage assessment
16	Hu et al. [32]	Aluminum and Carbon Fiber Reinforced Plastic (CFRP)	Tomographic reconstruction of internal surface damage images by the two-stage inverse algorithm for Lamb wave tomography (LWT)	For the aluminum cylinder, S_0 mode is found to be more sensitive to the internal surface pit
17	Li et al. [33]	Metal laminated composite	Detect and locate interlaminar damage using finite element simulation of Stoneley wave	Better identification than Lamb wave
18	Baltazar et al. [34]	Aluminum	Reduced thickness detection by using macro fiber composite (MFC) sensors and Time–frequency representation (TFR) using wavelets	Short time wavelet entropy (STWE) analysis gives faster defect detection

(continued)

Table 12.1 (continued)

Sl. no.	Author(s)	Material and structure	Method of detection/location	Result/Remarks
19	Golato et al. [35]	Aluminum pipe	Using a sparse array of transducers to record signals scattered by defects	The superior performance of the proposed spiral path block reconstruction approach
20	Wang et al. [36]	Steel	Quantitative mono and multi-damage detection by a sparse sensor network, based on the principle of semi-decentralized and standardized sensing	Feasible and convenient for large cylindrical structures
21	Yaacoubi et al. [37]	Composite	Optimizing damage detection by finite element method	Hydrogen-powered aircrafts
22	Eybpoosh et al. [38]	Aluminum	Online damage detection by proposing an unsupervised feature-extraction method	Can only compensate 0.5–1 °C temperature variation
23	Gresil et al. [39]	Composite	Damage inspection by FE simulation techniques	Longitudinal guided modes L(0, 1) has high sensitivity
24	Munoz et al. [40]	Austenitic stainless steel	Detection, location, and size determination of cracks using novel signal processing approach for a short distance	Identification of edges or welds of the pipe
25	Rojas et al. [41]	Stainless steel	Damage detection by use of 1st order torsional mode and post-processing of signals with Fourier transform analysis	Signals detected by micro-fiber composite (MFC) sensor
26	Chandarana et al. [42]	Composite	Impact damage detection and monitoring by use of piezoelectric transducers for SHM	Location of damage by triangulation of AE signals

(continued)

Table 12.1 (continued)

Sl. no.	Author(s)	Material and structure	Method of detection/location	Result/Remarks
27	Song et al. [43]	Steel	Transversal defect detection by time–frequency analysis for guided waves	Analysis of vacant and liquid-filled pipes
28	Wang et al. [44]	Composite plate	Structural damage detection by use of Gaussian mixture model (GMM)	Use of principal component analysis (PCA) for reducing the dimension
29	Carrino et al. [45]	FRP Composite	Location of the artificial defect by use of piezoelectric wafer active sensors (PWAS)	Effective in locating the source of nonlinearity with reliance
30	Vogelaar et al. [46]	Steel	Locating internal damage by the reflection of torsional wave	Reflection co-efficient gradually decreases with increasing wall loss

12.4 Conclusion

The present paper has provided a brief survey of the SHM methods considering Lamb waves and piezoelectric material-based sensors and actuators. The thin structures comprising metallic and composite-based material were focused on revealing the efficacy of Lamb waves-based SHM monitoring for passive and active structures. Thus, the Lamb wave-based SHM technique is valid when the wavelength is five to ten times the element size. In the near future, the diagnostic and prognostic procedure for SHMs can be explored from the viewpoint of different domains, including mechanical, food, or aerospace industries.

References

1. Ciang, C.C., Lee, J.-R., Bang, H.-J.: Structural health monitoring for a wind turbine system: a review of damage detection methods. Measure. Sci. Technol. **19**(12), 122001 (2008)
2. Haugse, E.D., et al.: Design principles for aerospace structures utilizing SHM. Encyclopedia of Aerospace Eng. (2010)
3. Giurgiutiu, V., Santoni-Bottai, G.: Structural health monitoring of composite structures with piezoelectric-wafer active sensors. AIAA J. **49**(3), 565–581 (2011)
4. Saxena, M., Bannett, O.O., Sharma, V.: Bearing fault evaluation for structural health monitoring, fault detection, failure prevention and prognosis. Procedia Eng. **144**, 208–214 (2016)
5. Su, Z., Ye, L.: Identification of Damage using Lamb Waves: From Fundamentals to Applications, vol. 48. Springer Science & Business Media (2009)
6. Lamb, H.: On waves in an elastic plate. Proc. R. Soc. A Math. Phys. Eng. Sci. **93**, 114–128 (1917)
7. Rayleigh, L.: Waves propagated along the plane surface of an elastic solid. Proc. Lond. Math. Soc. **20**, 225–234 (1889)
8. Gazis, D.C.: Exact analysis of the plane-strain vibrations of thick-walled hollow cylinders. J. Acoust. Soc. Am. **30**, 786–794 (1958)
9. Firestone, F.A., Ling, D.S.: Propagation of waves in plates, technical report, Sperry Products, Danbury, CT, USA (1945)
10. Worlton, D.C.: Experimental confirmation of Lamb waves at megacycle frequencies. J. Appl. Phys. **32**, 967–971 (1961)
11. Wadhera, T., Kakkar, D.: Modeling risk perception using independent and social learning: application to individuals with autism spectrum disorder. J. Math. Sociol. 1–23 (2020)
12. Wadhera, T., Kakkar, D.: Conditional entropy approach to analyze cognitive dynamics in autism spectrum disorder. Neurol. Res. **42**(10), 869–878 (2020)
13. Wadhera, T., Kakkar, D.: Multiplex temporal measures reflecting neural underpinnings of brain functional connectivity under cognitive load in Autism spectrum disorder. Neurol. Res. **42**(4), 327–337 (2020)
14. Kalra, S., Bhattacharya, B., Munjal, B.S.: Design of shape memory alloy actuated intelligent parabolic antenna for space applications. Smart Mater. Struct. **26**(9), 095015 (2017)
15. Kalra, S., Bhattacharya, B., Munjal, B.S.: Development of shape memory alloy actuator integrated flexible poly-ether-ether-ketone antenna with simultaneous beam steering and shaping ability. J. Intell. Mater. Syst. Struct. **29**(18), 3634–3647 (2018)
16. Balageas, D., Fritzen, C.-P., Güemes, A. (eds.): (2010) Structural Health Monitoring, vol. 90. Wiley

17. Zhang, H., et al.: Circumferential SH wave piezoelectric transducer system for monitoring corrosion-like defect in large-diameter pipes. Sensors **20**(2), 460 (2020)
18. Majhi, S., et al.: Corrosion detection in steel bar: a time-frequency approach. NDT & E Int. **107**, 102150 (2019)
19. Zhao, J., et al.: Acoustic guided wave techniques for detecting corrosion damage of electrical grounding rods. Measurement **147**, 106858 (2019)
20. Livadiotis, S., Ebrahimkhanlou, A., Salamone, S.: An algebraic reconstruction imaging approach for corrosion damage monitoring of pipelines. Smart Mater. Struct. **28**(5), 055036 (2019)
21. Zhang, Y., Li, D., Zhou, Z.: Time reversal method for guided waves with multimode and multipath on corrosion defect detection in wire. Appl. Sci. **7**(4), 424 (2017)
22. Li, Z., et al.: Quantitative detection of lamination defect in thin-walled metallic pipe by using circumferential Lamb waves based on wavenumber analysis method. NDT & E Int. 102, 56–67 (2019)
23. Li, X., et al.: Propagation of non-linear Lamb waves in adhesive joint with micro-cracks distributing randomly. Appl Sci **10**(3), 741 (2020)
24. Chen, G., et al.: Detection of cracks with arbitrary orientations in a metal pipe using linearly-polarized circular TE11 mode microwaves. NDT & E Int. **107**, 102125 (2019)
25. Römmeler, A., et al.: Air coupled ultrasonic defect detection in polymer pipes. NDT & E Int. **102**, 244–253 (2019)
26. Du, G., et al.: Damage detection of pipeline multiple cracks using piezoceramic transducers. J. Vibroeng. **18**(5), 2828–2838 (2016)
27. Nirbhay, M., Dixit, A., Misra, R.K.: Finite element modelling of lamb waves propagation in 3D plates and brass tubes for damage detection. Russ. J. Nondest. Test. **53**(4), 308–329 (2017)
28. Ni, J., et al. Effect of pipe bend configuration on guided waves-based defects detection: an experimental study. J. Press. Vessel Technol. **138**(2) (2016)
29. Liu, K., et al.: Guided waves based diagnostic imaging of circumferential cracks in small-diameter pipe. Ultrasonics **65**, 34–42 (2016)
30. Zhou, W., et al.: Guided wave-based pipe damage inspection by ultrasonic fiber optic sensor
31. Li, W., Cho, Y.: Thermal fatigue damage assessment in an isotropic pipe using nonlinear ultrasonic guided waves. Exp. Mech. **54**(8), 1309–1318 (2014)
32. Hu, B., et al.: Tomographic reconstruction of damage images in hollow cylinders using Lamb waves. Ultrasonics **54**(7), 2015–2023
33. Li, B., et al.: A Stoneley wave method to detect interlaminar damage of metal layer composite pipe. Front. Mech. Eng. **10**(1), 89–94 (2015)
34. Baltazar, A., Rojas, E., Mijarez, R.: Structural health monitoring in cylindrical structures using helical guided wave propagation. Phys. Procedia **70**, 686–689 (2015)
35. Golato, A., et al.: Sparsity based defect imaging in pipes using guided waves. In: Compressive Sensing V: From Diverse Modalities to Big Data Analytics, vol. 9857. International Society for Optics and Photonics (2016)
36. Wang, Q., Hong, M., Su, Z.: A sparse sensor network topologized for cylindrical wave-based identification of damage in pipeline structures. Smart Mater. Struct. **25**(7), 075015 (2016)
37. Yaacoubi, S., et al.: Towards an ultrasonic guided wave procedure for health monitoring of composite vessels: application to hydrogen-powered aircraft. Materials **10**(9), 1097 (2017)
38. Eybpoosh, M., Berges, M., Noh, H.Y.: An energy-based sparse representation of ultrasonic guided-waves for online damage detection of pipelines under varying environmental and operational conditions. Mech. Syst. Signal Process. **82**, 260–278 (2017)
39. Gresil, M., Poohsai, A., Chandarana, N.: Guided wave propagation and damage detection in composite pipes using piezoelectric sensors. Procedia Eng. **188**, 148–155 (2017)
40. Muñoz, C.Q.G., et al.: New pipe notch detection and location method for short distances employing ultrasonic guided waves. Acta Acustica United with Acustica **103**(5), 772–781 (2017)
41. Rojas, E., Baltazar, A., Treesatayapun, C.: Investigation on damage identification in a pipe using torsional guided waves. AIP Conf. Proc. **1806**(1) (2017). AIP Publishing LLC

42. Chandarana, N., Gresil, M., Soutis, C.: Damage detection and monitoring in composite pipes using piezoelectric sensors
43. Song, Z., et al.: Experimental study of guided wave propagation and damage detection in large diameter pipe filled by different fluids. NDT & E Int. **93**, 78–85 (2018)
44. Wang, Q., Ma, S., Yue, D.: Identification of damage in composite structures using Gaussian mixture model-processed Lamb waves. Smart Mater. Struct. **27**(4), 045007 (2018)
45. Carrino, S., Maffezzoli, A., Scarselli, G.: Active SHM for composite pipes using piezoelectric sensors. Mater. Today Proc. (2020)
46. Vogelaar, B., Golombok, M.: Quantification and localization of internal pipe damage. Mech. Syst. Signal Process. **78**, 107–117 (2016)

Chapter 13
Global Overview of Large-Scale Photovoltaic System and Its Electrical Energy Storage Implementation

Bajrangi Maurya

Abstract Carbon-emission reduction is one of the great challenges emitted by fossil fuels during power generation. This can be mitigated at a large scale by using solar photovoltaic technology, which is one of the unmatched electrical energy generation techniques with the help of solar energy. Energy crises are increasing day by day as the demand for electrical energy is increasing, and hence to meet the global demand, we need to expand the capacity of PV generation at a large scale. The increased capacity of the solar photovoltaic system may vary from few megawatts (MWs) to few kilowatts (KWs) depending upon the types of level of generation. A traditional generating plant emits carbon and to eliminate this carbon emission, solar PV penetration in the power system can be done at a large scale. Since solar energy is a daily phenomenon and due to this uncertainty in PV power generation, electrical energy storage (EES) systems need to be installed to enhance system capacity and performance. Due to advanced technological development in the electrical storage system, such as Li-ion batteries, lead acid, and nickel–cadmium batteries, has given increasing energy sector participants by providing a cost-effective approach, quality improvement, and energy management. This paper focuses on the PV system penetration at a large scale to the existing grid system and a complete and comprehensive overview of the electrical storage system. Forecasting of new technologies and recent development in EES has been discussed. The current scenario and basic building block of the PV solar system have also been focused. Different gaps in recent work have been identified and indicated about the future development of EES for large-scale PV systems.

13.1 Introduction

Electrical energy is one of the essential things for human, and without electricity, it is difficult to complete daily routine either in home or in office. Due to industrialization and increased population, demand for electricity is increasing, and hence to fulfill

B. Maurya (✉)
G.L. Bajaj Institute of Technology and Management, Greater Noida, India

© The Author(s), under exclusive license to Springer Nature Singapore Pte Ltd. 2022
V. V. Rao et al. (eds.), *Computational and Experimental Methods in Mechanical Engineering*, Smart Innovation, Systems and Technologies 239,
https://doi.org/10.1007/978-981-16-2857-3_14

this requirement, more electricity capacity needs to be increased. For several years, conventional energy sources have been harnessed to generate electricity, but the emission of carbon dioxide gases from such generating plants has a significant impact on the environment. Due to carbon emission from burning fossils fuels, several other adverse effects on the environment have been created, such as global warming, environmental pollution, and greenhouse effect. Government from a different country are thinking about such issues to minimize carbon emission and country like India is giving entitlement their people to set up a solar PV system.

In recent years, it has been seen that there is a boom in renewable energy industries, and penetration of renewable energy with the grid system has been increased in both developing and developed countries. A developing country like India has great use of renewable energy sources, especially solar energy, to meet the increasing demand for electricity. Although, penetration of renewable energy with the grid creates technical and economic challenges. However, the modern power system uses energy storage systems, power electronics converter, intelligent communication networks, and advanced controllers to beat these challenges and revolutionary changes in renewable energy utilization. As we know that, solar energy is a neat and clean source of energy and sustainable in nature. Solar energy falls on earth, when there is clear sky irradiance. Nearly 50ZJ amount solar energy reaches on earth, which is the maximum annual potential on land having clear sky irradiance [1]. This amount of solar energy falling on earth before conversion into other forms is much larger than the total energy consumed annually worldwide [2].

The scope of capacity adjustment of power output regularly avoids fluctuations of dispatchable generating plants such as coal-fired plants or gas power plants. Solar energy can be utilized at a large scale by generating electricity with the help of photovoltaic (PV) solar panels, and this can be penetrated into the grid for mass consumption. Penetration of large PV-generated energy with grid may cause hindrance, and it is up to policymakers to increase system flexibility for proper functioning with an improved capacity [3]. Using electrical energy storage (EES) in connection with large-scale PV system penetration may provide energy management and quality improvement of electrical energy services. In the current scenario of the electricity market, the smart grid and EES play a key role in maintaining the quality and services of the electricity supply. In this paper, the installation of together with large-scale PV power plants has been reviewed [4].

Different storage system used to improve energy management generated by PV power plant has been discussed briefly. Due to the sporadic nature of solar energy, EES uses in power systems improve grid stability. Under irregular solar energy, EES can help in maintaining the stability and quality of the power system grid in an effective and economical way under irregular solar energy. The application and contribution of EES could be found in different stages of generation, transmission, and distribution in power systems. Details description of PV module power generation and its penetration with grid system at large-sale has been overviewed. Recent development in PV module system to increase solar energy generation capacity has been reviewed in Sect. 2. Technology plays a crucial role in fulfilling the target, and Sect. 3 describes different technologies used for PV cell. Due to the irregularity of solar irradiance, electricity generation through PV module is instable, and hence to minimize such effect, different issues and techniques have been discussed in Sect. 4. Integration of EES with PV system and different advantages of EES-PV system has been review in Sect. 5. Section 6 explains an overview of the technology used to improve energy management and power quality of system. The future scope of large-scale photovoltaic power plants interconnected with energy storage systems has been reviewed in conclusion. How to increase the potential of PV generation and make the use of EES economically, all feasible scenarios have been included in this research work.

13.2 Global Scenario of PV System

Different countries working toward the maximization of heir's solar energy department to harness solar energy and convert it into electrical energy. It is expected that by 2050, renewable energy contributes 100% to the power generation sector, and hence installation of renewable energy, especially solar energy, is increasing rapidly throughout the world. The impact of increased PV capacity in power systems has a significant effect and is termed as PV penetration [5, 6].

13.2.1 Germany

Germany has long been at the forefront of solar power producers and produced a total of 38.2 GW power out of 177 GW produced globally in 2014 [7, 8]. Germany has met over 50% of the nation's daily energy needs from solar power [9]. Its long-term shift to cleaner energy has made Germany's economy the world's largest to rely on renewable energy. Germany can achieve nearly 80% production of electricity with the help of renewable energy till 2050, and 52GW electrical energy may gain by 2020 by PV penetration [10]. Although Germany is far from a Sun-drenched nation, it is still aiming to expand the EES market to improve the performance and efficiency of the PV system.

13.2.2 China

Solar energy capacity in China is 44 GW, according to data released in 2020 [11]. It has ample solar energy, and it is harnessing this solar energy to increase its power sector capacity. China has nearly 60% share of coal to produce electrical energy as compared to 5% of solar energy of total its power generating capacity [12]. Keeping its reputation as the leading market, China made a big splash announcing the world's largest solar installation up to 2.2GW in the northwestern Chinese province of

Qinghai. With rapid development in new technology, China's solar power industry is developing fastly, producing more than 18% of world's total PV module [13]. Hundreds of companies in this country are involved in PV items manufacturing and exporting globally [14]. The top listed companies for PV cell manufacturing are Suntech, Canadian Solar, Jinco Solar, Trina Solar, Yingli Solar, etc. [15]. Technical research work and development put a great involvement for PV industries in China. It is expected that 85% contribution in the energy sector will be from renewable energy by 2050, in which solar energy plays a big role [16].

13.2.3 United States

USA had over 71.3 GW capacities by solar energy to generate electricity. The country is planning to achieve 70% energy demand with the help of PV at the end of 2050 [16, 17]. In terms of cumulative installed capacity, the USA is the second largest country after China to use solar energy. The country is working to increase PV installation capacity to eliminate carbon dioxide emissions. Many states of the country are working individually to raise the capacity of renewable energy usage with the inclusion of solar power [18].

13.2.4 Italy

Italy is the fourth largest country in terms of solar energy utilization. Italy produces 7% of total electricity energy by solar PV system during 2013–14 [18]. It has increased its capacity gradually up to 20GW till 2018. According to National Energy Report, Italy will install 31GW by 2020 [19]. The share of electricity consumption converted by PV power in Italy was 9.2% in 2018 [20].

13.3 Technologies Used in PV Cell

Photovoltaic cells are made of semiconductor materials that convert solar energy into electricity by the photovoltaic effect. In few decades, PV cell becomes the most powerful items to convert solar energy into electrical energy. Due to advancements in technologies, PV cell qualities improved a lot in terms of electricity generation [21]. Crystalline silicon is used for PV manufacturing in large amounts, but its cost is high comparatively. About 90% of solar panels use crystalline silicon [22]. Thin-film solar cells like cadmium telluride (CdTe) are the most effective and least expensive to produce electricity are amorphous silicon, copper indium-gallium selenide (CIGS), and gallim arsenide (GaAs). But cadmium is toxic in nature and creates problems in PV cell design. For residential and commercial use, CIGS solar cell play a prominent role with high efficiency and better economy [23]. A different emerging technology in PV cell has been reviewed as follows:

13.3.1 Multi-Junction Solar Cells

For utilizing the full potential of CPV (concentrator photovoltaic) technology, it is better to use multi-junction solar cell. It contains multiple p-n junctions connected in series/parallel combinations. These are made up of semiconductor materials. Traditional crystalline silicon (c-si) solar cells show 20–25% efficiency; on the other hand, up to 46% efficiency has been shown by multi-junction cells in full concentrated sunlight [24].

13.3.2 Organic Cells

Organic cells are manufactured with organic electronics, a field of electronics, linked with small organic molecules. It is also known as a conductive organic polymer. An example of an organic solar cell is polymer solar. Using organic cells in PV is due to high throughput for module creation possibility. It is obtained by coating or printing from continuous production solutions. The module cost may reduce to $1.1/watt with the help of a cheaper manufacturing process in addition to the use of a low amount of organic semiconductor operator [33]. It was confirmed that the tested product typically produced about 25 or 60% more than amorphous silicon solar cells [25].

13.3.3 Quantum Dot Solar Cell

A quantum dot solar cell (QDSC) is a solar cell designed to use quantum dots as an absorbing photovoltaic material. QDC is highly accepted for implementing solar cells due to the tunable band gap, which can be achieved by varying the size of the dots [26]. The reviewed stream on solar cells showed that penetration of PV cells globally in order to be affordable, highly efficient, and ergonomically friendly, important work needs to be done to accelerate PV panel [27]. The performances of the solar cell are prominently affected by external factors such as ambient temperature [28]. Quantum dots have intermediate properties between bulk semiconductors and discrete atoms or molecules [29]. Their optoelectronic properties change as a function of both size and shape. The power conversion efficiency of this solar cell technology is about 10– 11% [30].

13.4 Stability and Integration Issue for PV

There are different challenges faced by PV systems to generate electricity. The cloud shape and size speed of cloud cover and many other factors effect PV power generation highly [31]. Irregular irradiance of solar due to continuous passing of clouds can largely affect power quality, regular supply system, and VAR control and system operation [32]. Technical designing issues are also great challenges. Different research organizations such as IEEE and Underwriters Laboratories are helping in designing PV cell, defining the performance, and analyzing the capacity of PV systems.

13.4.1 Intermittent Issue for Solar Power

A policy of interconnection with grid systems, which is beneficial for both distributed generation owners and operators, is a big issue and needs to be addressed. Such issues include communication problems, control, and coordination, short-circuit protection, electrical isolation, and protection due to surge [33]. For the connection of PV panels, utility companies must follow the standards as EN61000-3-2, IEEE1547, IEC61850, National Electrical Code 690, and the potential international standard IEC61727 [34]. In the current scenario, many software packages are used for power system stability to deal with grounding, power quality, and detection of islanding operation issues [35].

13.4.2 Irradiance Forecasting of Solar Light

For the planning of PV penetration and EES systems and their operation, irradiance of solar energy plays a major effect. Irregular irradiance and forecasting could result in inappropriate designing of size and operation of PV cell components. The discontinuous and dynamic nature of solar resources creates unpredictable situations. Some accurate prediction methods need to be discovered to better estimate solar irradiance for PV systems. In [36, 37] a detailed and comprehensive forecasting methods for the solar resource have been reviewed. Different techniques need to be implemented for better forecasting for PV power systems applications, reviewed in detail [44]. Artificial intelligence (AI) approach, physical approach, and hybrid approach are some techniques used in irradiance forecasting are the statistical approach. Depending on the types of parameters such as long-term planning or short-term scheduling, forecasting techniques are used for the system for better operation [38]. Recent advancements in technological development and using satellite images are effective techniques for computation and physical approach toward forecasting. The alliance between forecasting time horizon and the applications to PV renewable power system has been shown in Fig. 13.1 [42].

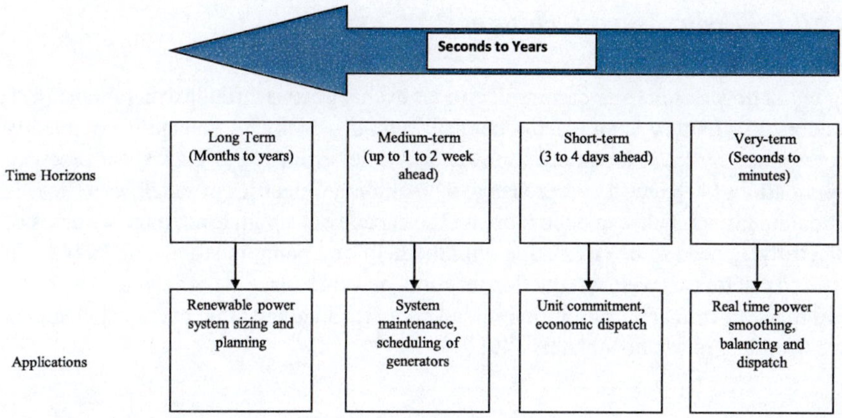

Fig. 13.1 PV system applications and forecasting horizons

13.5 Integration of EES with PV System

Solar energy is present during day, and due to this uncertainty in PV power generation, electrical energy storage (EES) systems need to be installed to enhance system capacity and performance. Using electrical energy storage (EES) in connection with large-scale PV system penetration may provide energy management and quality improvement of electrical energy services [39]. In the current scenario of the electricity market, the smart grid and EES play a key role in maintaining the quality and services of the electricity supply. EES have many applications and can be utilized at each stage of generation, transmission, and distribution of power system, whereas conventional generators are only generate electricity at a particular place. In [40] 100 KW distributed self-governing micro-grid in association with 50KW EES has been studied in terms of the impact of short-term frequency stability.

13.6 Various EES Technology

Rechargeable batteries are enormously used for storing surplus energy in PV systems. Grid-connected storage batteries play an important role as they are charged during the minimum demand of electricity, and energy has been supplied to customers through the grid at the time of high demand. In this section, various technologies used to improve the performance of storage systems have been discussed [46].

13.6.1 Redox Flow Technology

In redox flow technology, carbon electrodes exchange ions through a membrane [47]. Redox flow battery is an electrochemical type in which cell voltage is chemically determined by the 'Nernst equation' and it varies from 1.0 to 2.43 V for practical applications [41]. Due to ion exchange by membrane, electric current flows as liquids circulate in each half-cycle. To increase the capacity at a high level, huge volumes of electrolytes need to be circulating with the help of a pump in storage tanks [48]. In RFB, oxidation, as well as reduction reaction, may take place at both electrodes. Zinc-bromine, all-iron and iron-chromium, all cooper, and many other chemical elements are used in a redox flow battery [49].

13.6.2 Lead-Acid Battery

Due to high energy density, lead-acid batteries have been in use for a long time in terms of energy storage devices [51]. One of the main advantages is having low cost, and hence this makes the lead-acid battery a range of applications such as UPS, off-grid power systems and electric vehicles, etc [52]. The lead-acid battery has low self-discharge characteristics and having a large current capability [53]. This battery has a mature technology that works on the basis of recycling format. In addition of ultra-capacitors in a standard lead-acid battery provide good advantages and enhanced quality. However, it is not environment friendly and has a short lifespan and low depth of discharge issues [54].

13.6.3 Nickel–cadmium (Ni–Cd) Used Technology

Nickel–cadmium (Ni–Cd) batteries are rechargeable, and for a long time, these are used in portable electronic gadgets as well as in EES technology. This energy storage technology found a wide range of applications in communication systems and UPS for storing electrical energy for future used. Nickel–cadmium (Ni–Cd) batteries are having higher energy density (nearly 50–75 wh/kg) and longer life (2000–2500 cycles) as compared to lead-acid batteries. Hence, they are extensively used in generator-starting applications and uninterruptible power supply [45]. The 27 MW rated capacity battery used for spinning reserves and grid stabilization is a nickel–cadmium type battery. This is the second largest power saving battery set in 2003 in Golden Valley, Fairbanks, Alaska. Since metals (Cd and Ni) used in this batteries are more toxic, they create the environmental problem and have expensive EES technology compared to other EES technology [50].

13.7 Conclusion

This paper overviews the global scenario of large-scale photovoltaic system penetration with smart grid, PV generating system and focuses on its electrical energy storage implementation. Different electrical energy storage systems and technology have been discussed in this research work. This paper focuses on the PV system penetration at a large scale to the existing grid system and a complete and comprehensive overview of the electrical storage system. Forecasting of new technologies and recent development in EES has been discussed. The current scenario and basic building blocks of the PV solar system have also been focused. However, there are many challenges regarding the PV penetration issue with the grid system that need to be focused. Recent developments in the electrical storage system, such as Li-Ion batteries, lead acid, and nickel cadmium batteries, have given increasing energy sector participants by providing cost-effective approaches, quality improvement, and energy management. PV-EES integration and solar irradiance forecasting are the big issues in the successful operation of the PV system. There is s scope of research to meet PV adoption at a large scale for commercial purposes. Different cons in recent work have been identified and indicated for the future improvement of EES for large-scale PV systems. Using excellent prediction model with the help of better knowledge of the risks and uncertainty, solar irradiance forecasting can be improved. The future scope of large-scale photovoltaic power plants interconnected with energy storage systems has been reviewed in the section. How to increase the potential of PV generation and make the use of EES economically, all feasible scenarios have been included in this research work.

References

1. Photovoltaic research: National Centre for Photovoltaics. National Renewable Energy Laboratory. http://www.nrel.gov/ncpv/
2. Renewable 2016 global status report. Renewable Energy Policy Network for the 21st Century (2016). http://www.ren21.net/wp-content/uploads/2016/06/GSR_2016_Full_Report.pdf
3. Díaz-González, F., Sumper, A., Gomis-Bellmunt, O., Villafáfila-Robles, R.: A review of energy storage technologies for wind power applications. Renew. Sustain Energy Rev. **16**(4), 2154–71 (2012). https://doi.org/10.1016/j.rser.2012.01.029. http://www.sciencedirect.com/science/article/pii/S1364032112000305
4. Akinyele, D., Rayudu, R.: Review of energy storage technologies for sustainable power networks. Sustain. Energy Echnol. Assess. **8**, 74–91 (2014). https://doi.org/10.1016/j.seta.2014.07.004
5. Hosenuzzaman, M., Rahim, N., Selvaraj, J., Hasanuzzaman, M., Malek, A., Nahar, A.: Global prospects, progress, policies, and environmental impact of solar photovoltaic power generation. Renew. Sustain. Energy Rev. **41**, 284–297 (2015)
6. Delfanti M., Falabretti D., Merlo M.: Energy storage for pv power plant dispatching. Renewable Energy 2015; 80:61–72. https://doi.org/10.1016/j.renene.2015.01. 047. http://www.sciencedirect.com/science/article/pii/S0960148115000658.

7. Chung, S.W.: 100% renewable energy by 2050? why wait? Greenpeace.org (2015). http://www.greenpeace.org/international/en/news/Blogs/makingwaves/renewable-energy-revolution-2050-Shell/blog/54248/
8. Doukas, D.I., Bakas, P., Marinopoulos, A., Kim, H.: Energy storage integration into grid connected utility-scale photovoltaic (PV) systems. In: 2013 IEEE Grenoble Conference 2013, pp. 1–6. https://doi.org/10.1109/PTC.2013.6652294
9. Hosenuzzaman, M., Rahim, N., Several, J., Hasanuzzaman, M., Malek, A., Nahar, A.: Global prospects, progress, policies, and environmental impact of solar photovoltaic power generation. Renew. Sustain. Energy Rev. **41**, 284–297 (2015)
10. GTECH international project's references.: http://www.greenpower.es/en/corporation/references/. Last accessed 01 May 2020
11. IEEE guide for the interoperability of energy storage systems integrated with the electric power infrastructure, IEEE Std 2030.2TM-2015. In: IEEE Standards Association (2015). https://doi.org/10.1109/IEEESTD.2015.7140715
12. Department of energy global energy storage database, office of electricity delivery and energy reliability, Scandia national laboratories. https://www.energystorageexchange.org/projects/1955/. Last accessed 06 Sept 2019
13. Araiza, J., Hambrick, J., Moon, J., Starke, M., Vartanian, C.: Grid energy-storage projects: engineers building and using knowledge in emerging projects. IEEE Electrification Mag. **6**(3), 14–19 (2018). https://doi.org/10.1109/MELE.2018.849842
14. Eftekharnejad, S., Vittal, V., Heydt, G.T., Keel, B., Loehr, J.: Impact of increased penetration of photovoltaic generation on power systems. IEEE Trans. Power Syst. **28**, 893–901 (2013)
15. Energy and Environmental Economics. Inc. Investigating the economic value of flexible solar power plant operation (2018)
16. Renewables and electricity storage: a technology roadmap for Remap 2030. IRENA (2015). https://www.irena.org/DocumentDownloads/Publications/IRENA_REmap_Electricity_Storage_2015.pdf
17. Kroposki, B.: Reliable, low cost distributed generator/utility systems interconnect: 2001 annual report, NREL/SR-560-34634. In: National Renewable Energy Laboratory (2003). http://www.nrel.gov/docs/fy03osti/34634.Pdf
18. User guide for India's 2047 energy calculator, electrical energy storage (EES). In: National Instituition for Transforming India. http://www.indiaenergy.gov.in/iess/docs/Storage%20Documentation.pdf
19. Fthenakis, V.: Considering the total cost of electricity from sunlight and the alternatives [point of view]. Proc. IEEE **103**, 283–286 (2015)
20. Gong, K., Ma, X., Conforti, K.M., Kuttler, K.J., Grunewald, J.B., Yeager, K.L., et al.: A zinc–iron redox-flow battery under $100 per kW h of system capital cost. Energy Environ. Sci. **8**, 2941–2945 (2015)
21. Sunshot 2030.: Office of Energy Efficiency & Renewable Energy, Energy.gov. https://energy.gov/eere/sunshot/sunshot-2030
22. Battery storage for renewable: Market status and technology outlook. IRENA (2015). http://www.irena.org/documentdownloads/publications/irena_battery_storage_report_2015
23. Yang, X., Song, Y., Wang, G., Wang, W.: A comprehensive review on the development of sustainable energy strategy and implementation in China. IEEE Trans. Sustain. Energy **1**, 57–65 (2010)
24. Industry solar: Crossing the charm. Deutsche Bank Markets Research (2015). https://www.db.com/cr/en/docs/solar_report_full_length
25. Sun, H., Zhi, Q., Wang, Y., Yao, Q., Su, J.: China's solar photovoltaic industry development: the status quo, problems and approaches. Appl. Energy **118**, 221–230 (2014)
26. Zipp, K.: Energy storage prices expected to drop 70% by 2030. Solar PowerWorld. http://www.solarpowerworldonline.com/2016/01/ideal-power-partners-with-austin-energy-on-us-doe-funded-projectsto-
27. Song, D., Jiao, H., Te Fan, C.: Overview of the photovoltaic technology status and perspective in China. Renew. Sustain. Energy Rev. **48**, 848–856 (2015)

28. Kenning, T.: Residential storage costs will fall 84% globally by 2040—BNEF. Energy Storage News (2015). https://www.energy-storage.news/news/residential-storage-system-costs-to-fall-by-84-globally-by-2040-%20bnef. Last accessed 6 Jan 2017
29. Power development 13th Five-Year Plan (In Chinese).: National Development and Reform Commission of the People's Republic of China (2016). http://www.ndrc.gov.cn/zcfb/zcfbghwb/201612/P020161222570036010274.pdf
30. Yang, K., Walid, A.: Outage-storage tradeoff in frequency regulation for smart grid with renewables. IEEE Trans. Smart Grid. **4**, 245–252 (2013)
31. Manju, S., Sagar, N.: Progressing towards the development of sustainable energy: a critical review on the current status, applications, developmental barriers and prospects of solar photovoltaic systems in India. Renew. Sustain. Energy Rev. **70**, 298–313 (2017)
32. Serban, I., Teodorescu, R., Marinescu, C.: Energy storage systems impact on the shortterm frequency stability of distributed autonomous microgrids, an analysis using aggregate models. IET Renew. Power Gener. **7**, 531–539 (2013)
33. JNN solar mission: scheme/documents. Ministry of New and Renewable Energy, Government of India. http://www.mnre.gov.in/solarmission/jnnsm/introduction-2/
34. Robyns, B., Francois, B., Delille, G., Saudemont, C.: Energy storage in electric power grids. Wiley, Hoboken, New Jersey (2015)
35. Mountain, B., Szuster, P.: Solar, solar everywhere: opportunities and challenges for Australia's rooftop PV systems. IEEE Power Energy Mag. **13**, 53–60 (2015)
36. Díaz-González, F., Sumper, A., DÃaz-GonzÃ, F., Gomis-Bellmunt, O.: Energy storage in power systems. Wiley, Hoboken, New Jersey (2016)
37. Stetz, T., Von Appen, J., Niedermeyer, F., Scheibner, G., Sikora, R., Braun, M.: Twilight of the grids: the impact of distributed solar on Germany's energy transition. IEEE Power Energy Mag. **13**, 50–61 (2015)
38. Chai, S., Xu, Z., Wong, W.K.: Optimal granule-based PIs construction for solar irradiance forecast. IEEE Trans. Power Syst. **31**, 3332–3333 (2016)
39. Von Appen, J., Braun, M., Stetz, T., Diwold, K., Geibel, D.: Time in the sun: the challenge of high PV penetration in the German electric grid. IEEE Power Energy Mag. **11**, 55–64 (2013)
40. Fernandez-Jimenez, L.A., Muñoz-Jimenez, A., Falces, A., Mendoza-Villena, M., Garcia-, G.E., Lara-Santillan, P.M., et al.: Short-term power forecasting system for photovoltaic plants. Renew. Energy **44**, 311–317 (2012)
41. Ogimoto, K., Kaizuka, I., Ueda, Y., Oozeki, T.: A good fit: japan's solar power program and prospects for the new power system. IEEE Power Energy Mag. **11**, 65–74 (2013)
42. Geraldi, E., Romano, F., Ricciardelli, E.: An advanced model for the estimation of the surface solar irradiance under all atmospheric conditions using MSG/SEVIRI data. IEEE Trans. Geosci. Remote Sens. **50**, 2934–2953 (2012)
43. REmap 2030 Renewable energy prospect: Germany.: IRENA 2(015). http://www.irena.org/DocumentDownloads/Publications/IRENA_REmap_Germany_report_2015
44. Chen, S., Gooi, H., Wang, M.: Solar radiation forecast based on fuzzy logic and neural networks. Renew. Energy **60**, 195–201 (2013)
45. Lai L.L.: Global energy Internet and interconnection. IEEE Smart Grid Newsl. (2015). http://smartgrid.ieee.org/newsletters/october-2015/global-energy-internet-and-interconnection
46. Liu, J., Fang, W., Zhang, X., Yang, C.: An improved photovoltaic power forecasting model with the assistance of aerosol index data. IEEE Trans. Sustain. Energy **6**, 434–442 (2015)
47. Fact sheet: The solar star projects.: SunPower Corporation (2016) https://us.sunpower.com/sites/sunpower/files/media-library/factsheets/fs-solar-star-projects-factsheet
48. Mellit, A., Pavan, A.M.: A 24-h forecast of solar irradiance using artificial neural network: application for performance prediction of a grid-connected PV plant at Trieste. Italy. Sol. Energy **84**, 807–821 (2010)
49. Case study: solar PV-hydro hybrid system at Longyangxia, China.: International HydroAssociation (2015). http://www.hydropower.org/blog/case-study-solar-pv-hydro-hybrid-system-at-longyangxia-china

50. Jang, H.S., Bae, K.Y., Park, H., Sung, D.K.: Solar power prediction based on satellite images and support vector machine. IEEE Trans. Sustain. Energy **7**, 1255–1263 (2016)
51. Bacher, P., Madsen, H., Nielsen, H.A.: Online short-term solar power forecasting. Sol. Energy **83**, 1772–1783 (2009)
52. Real-time generation of solar plants in Gujarat.: Gujarat Energy Transmission Corporation Limited (2016). https://www.sldcguj.com/RealTimeData/GujSolar.asp. Last accessed 6 Jan 2017
53. Marquez, R., Coimbra, C.F.: Intra-hour DNI forecasting based on cloud tracking image analysis. Sol. Energy **91**, 327–336 (2013)
54. Neoen breaks ground on 300 MW French solar plant.: PV Mag. (2014). http://www.pv-magazine.com/news/details/beitrag/neoenbreaks-ground-on-300-mw-french-solar-plant_100017099/-axzz4BLrsOucx

Chapter 14
Consideration Analysis of Stress Distribution Using Automotive Chassis for Heavy Vehicle Transports

Manish Saraswat, Pradeep Kumar Singh, and Rajat Yadav

Abstract The automotive chassis work to support the body and various parts of the vehicle. Additionally, it needs to withstand the stun, contort, vibration, and different anxieties caused because of unexpected braking, speeding up, stunning street condition, radiating power while cornering, and powers initiated by its segments. The undercarriage goes about as the foundation of a substantial vehicle which conveys the most extreme burden for all planned working conditions. This paper depicts the plan and examination of weighty vehicle frames as the prime goal of any car business in the present quick evolving world. In the current paper, the relevant data of a current substantial vehicle suspension supplanting materials of high explicit load with lower thickness materials without decreasing inflexibility and solidness. Supplanting of steel with aluminum, magnesium, composites are taken for displaying. Planning utilizing PC helped plan programming Catia and investigation by exposed to the indistinguishable burden as that of a frame. The mathematical outcomes are approved with expository figuring thinking about the pressure dissemination and distribution utilizing the ANSYS programming.

14.1 Introduction

Otherwise called its frame, a case is the primary supporting structure motor vehicle to which each and every other part is associated skeleton's life.

Edge Basic Framework Automobile. It underpins all parts vehicle connected. It is made Drop Forged Steel. All parts for cars can just append car-related frameworks;

M. Saraswat
Department of Mechanical Engineering, Lloyd Institute of Engineering and technology, Greater Noida, UP, India

P. K. Singh · R. Yadav (✉)
IET Department of Mechanical Engineering, GLA University, Mathura, India
e-mail: rajat.yadav@gla.ac.in

P. K. Singh
e-mail: pradeep.kumar@gla.ac.in

© The Author(s), under exclusive license to Springer Nature Singapore Pte Ltd. 2022

V. V. Rao et al. (eds.), *Computational and Experimental Methods in Mechanical Engineering*, Smart Innovation, Systems and Technologies 239,
https://doi.org/10.1007/978-981-16-2857-3_15

for example, power plant, transmission, directing, suspension stopping mechanism are joined and bolstered it was [1–4]. The skeleton gives quality expected to help payload alongside various vehicle parts assists keep with auto mobiling unbending and tight. Thus, the suspension is a significant part by and large security framework. Besides, it guarantees low clamor levels, vibrations, and unpleasantness all through the vehicle [5–7]. The skeleton must adequately be unbending endure daze, curve, shaking diverse trouble. Notwithstanding excellence, one significant thing is suspension configuration must have enough adaptability and inflexibility for good taking care characteristics [8–10]. Along these lines, quality inflexibilitysignificantly measures undercarriage structure. Heap bearing structure is a frame, so undercarriage must be planned withstand stacks on it [11–13].

14.1.1 *Layout of Chassis and Its Main Components*

"Frame" is French word important full body autos, and incorporates all frameworks, for example, power plant, transmission, guiding, suspension, wheel tires, auto electric framework. Including body is called explicit configuration [14] (Fig. 14.1).

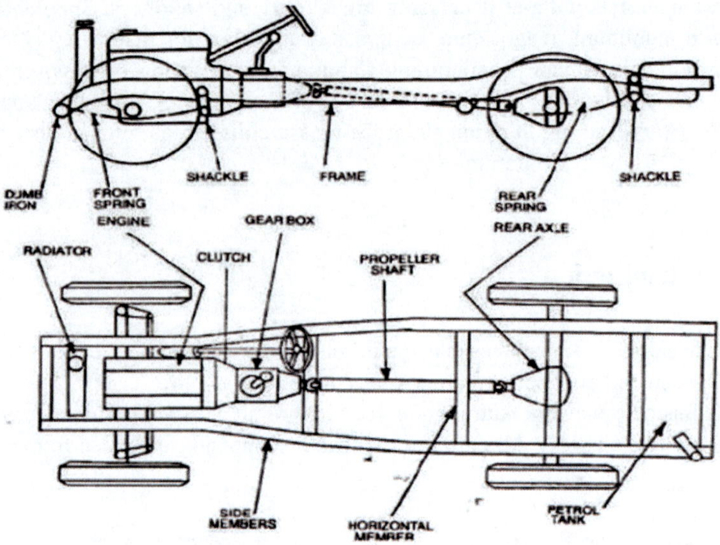

Fig. 14.1 Chassis and its main components

14.1.2 Materials

Distinctive suspension materials can lessen the weight vehicle, improving vehicle capacity to weight proportion. Material choice likewise profits by lessening part redirection, expanding outline quality, and deciding measure of support required [14].

14.1.3 ALUMINUM 6060

Unadulterated aluminum is also practical material sensibly moderate lightweight yet powerless and requires additional support to deliver an inflexible frame. Aluminum is hard to work within the light fact that it requires exceptionally talented welding entire delicate metal. There are normally few sorts of aluminum. For this venture, I chose to test with an aluminum amalgam 6060.

Aluminum composite 6060 is the most affordable and generally flexible of warmth treatable aluminum combinations. It has great properties aluminum. It offers great mechanical properties great consumption obstruction. This is encouraged by most ordinarily utilized strategies [15, 16]. It has great working capacity in the state. The unmistakable highlights aluminum composite 6060 are medium to high quality, great unbending nature, great surface completion, fantastic consumption protection from climate conditions, great usefulness, and broadly accessible. It tends to welded techniques heater can be brassed. It comes in a wearable structure ("Alclad") with a slight surface layer of high-immaculateness aluminum improve both appearance and erosion obstruction. This aluminum type is utilized for assortment items applications, from truck bodies edges to screw machine parts basic parts. 6060 is utilized acceptable consumption opposition with great quality. The highlights Al-6060 are appeared underneath in the table STEEL.

Composite steel iron and various parts, prevalently carbon is commonly used being developed various applications due to its high tenseness simplicity. Steel's base is metal, which can take two crystalline structures (allotropic structures), body-centered cubic (BCC) and face-centered cubic (FCC), dependent upon its temperature. Carbon in like way, steel composites contributes up to 2.1% of its weight. Changing the total alloying segments, either as dissolvable components quick strides in steel, lessens development those separations that make iron generally flexible frail, in this manner controlling its hardness, malleability, and rigidity. The outcome is steel. The quality steel diverged from unadulterated iron is simply possible at cost pliability iron, which iron is high.

The attributes of steel have appeared beneath the table (Table 14.1).

Table 14.1 The attributes of steel have appeared beneath the table

	Density	Young's modulus, GPa	Shear modulus, GPa	Bulk modulus, GPa	Poisson's ratio
Min	2.71	66.1	22	96.8	0.27
Max	2.71	70	26.1	153	0.31
Min	7.75	207	40	150	0.27
Max	8.1	210	45	172	0.31

Table 14.2 Ti-6Al-4V have appeared

	Density	Young's modulus, GPa	Shear modulus, GPa	Bulk modulus, GPa	Poisson's ratio
Min	4.429	104	40	96.8	0.31
Max	4.512	113	45	153	0.37

14.1.4 TITANIUM-6AL-4V

Ti-6Al-4 V, sometimes called Ti64, is an alpha–beta titanium combination with high proportion and astounding erosion obstruction. It is one most regularly utilized titanium composites and is applied in wide scope utilizations that require low thickness great erosion obstruction. Avionic business and biomechanical applications.

Ti-6Al-4V titanium amalgam is, for the most part, in alpha, with hcp valuable stone structure and beta, with bcc jewel structure. The mechanical properties may differ contingent upon capacity and attribute heat treatment state compound, and the general property goes very much prepared Ti-6Al-4V have appeared underneath the table (Table 14.2).

14.2 Literature Review

Most early exploration take a shot at undercarriage plan examination is restricted figuring pressure conveyance and exhaustion life in the case with few ump Minor (1945) depicted weariness misfortune during break commencement stage. Harm during the commencement stage may identify with disengagement, slip groups, and micro cracks. Since these marvels can estimate in exceptionally controlled research center conditions, most harm summation methods for the inception stage are observational in nature [17].

These methods can harm the life spent for little research facility tests. For this reason, life is characterized as crossing point design, which is equal to the development of little splits in huge part structure examined in this work is restricted to results acquired for K butt joints under hub stacking and transverse non-load bearing filet welds under the hub and bowing burdens. Notwithstanding, welded transverse filet welds under hub stacking are connected. The thickness goes considered in all cases did not expand more than 10–26 mm.

Contemplated pressure investigation truck skeleton with inflexible joints using FEM. The business restricted segment group ANSYS form 5.3 has been utilized for critical thinking. Because of plan enhancements, it is imperative to decide pressure truck suspension before production. Sidebar thickness, association plate thickness, and association plate length were differed to lessen sum pressure close to the bolted joint edge. The numerical outcomes show expanding sidebar thickness locally can lessen pressure on the side part. On the off chance that thickness change is unimaginable, expanding association plate length is an acceptable other option [18].

Portrayed static unique burdens vertical even and torsion powers following up on outline outlines. The torsion that conducts most business vehicle outlines is ruled by distorting torsions, since twisting is forestalled in joints where cross-individuals are appended to side-individuals. This paper presents half breed strategy examination that consolidates limited component admiration joint districts with diagnostically acquired bar components for cross-part and side part segments. The bar component incorporates distorting torsion power uprooting ties. The adaptability joints are combined with their uprooting similarity. The technique gives close concurrence with test results [19, 20].

Spatial elements of high portability track vehicle suspension frameworks. Utilizing the outcomes from the partner paper, conditions movement for suspension framework with discretionary number street wheels are methodically inferred. The track is a spoken mind-boggling interior vitality component that works between grounds, wheels body vehicle. Track strain is determined from Relaxed Caterers Relationship track crossing over impacts are displayed. Numerical outcomes for driver increasing speed retention, just as track strain, are introduced. A 90 calculate decrease register time is accomplished on multi-body model single-vehicle diagrams generally speaking way thinking producing knowledge from car guarantee information and gives instrument upheld by situational models, where investigation guarantee information has prompted extra understanding into hierarchical hazard appraisal and past activity. The underlying progress to handle solidness issue, evaluation guarantee hazard from weakened "quality breaks," has been utilized for quite while is additionally improved by considering double nature car guarantee chance in wording contingent time and use [21].

14.3 Methodology

Specification of Chassis frame

- Model: TATA ACE HT
- Length: 3800 mm Width (mm): 1500
- Height(mm): 1845 Wheelbase (mm): 2100

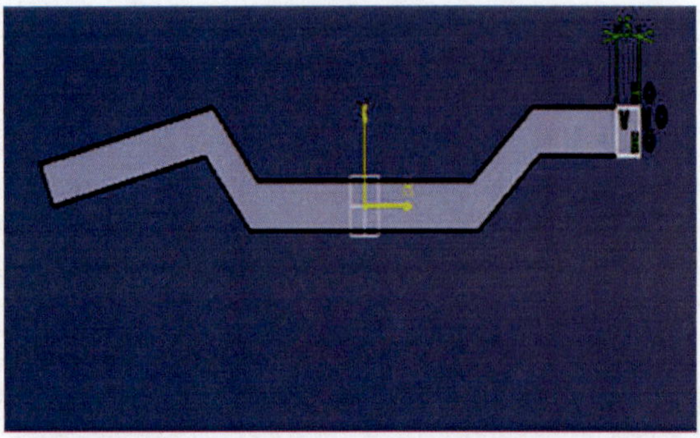

Fig. 14.2 Dimensions of chassis given in CATIA

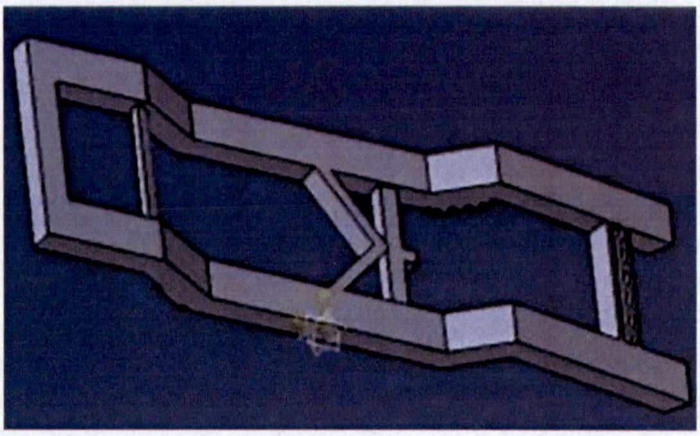

Fig. 14.3 Design of chasis in CATIA

- Loading Deck Length (mm): 2140 Loading Deck Width (mm): 1430 Height of Side Panels (mm): 300
- Min. Turning Circle Diameter (mm): 8600
- Gross vehicle weight (G.V.W) = 1200 kg
- Kerb weight = 800 kg.

There is a different component in CATIA using which different assignments can execute. The essential transom and component CATIA showed up in the outline (Figs. 14.2 and 14.3):

14.4 Results and Discussion

These grants decrease chance level cost trivial structures. The adaptable nature ANSYS offers to ensure customers can see influence structure on, for the most part, direct thing, be it electromagnetic, warm, mechanical, etc. (Figs. 14.4, 14.5, 14.6 and 14.7).

Now that the part exists define a library of the necessary materials that compose the object (or project) being modeled. This includes thermal and mechanical properties.

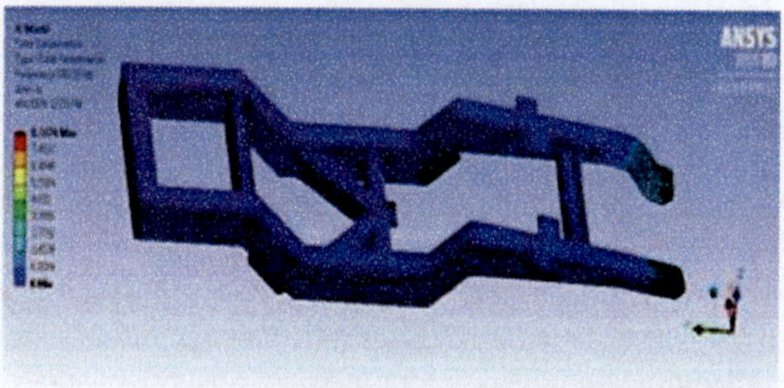

Fig. 14.4 Total deformation 1

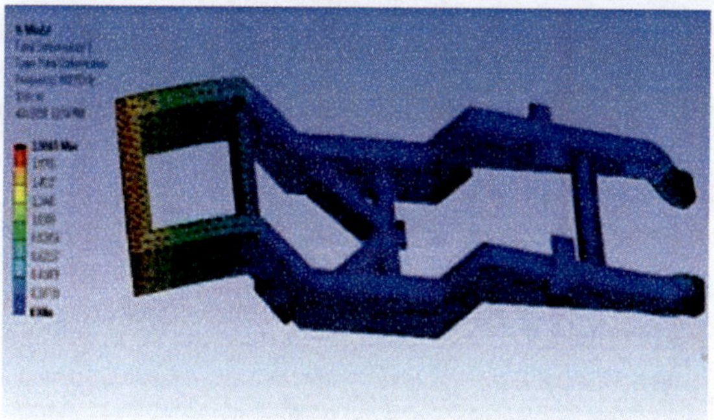

Fig. 14.5 Total deformation 2

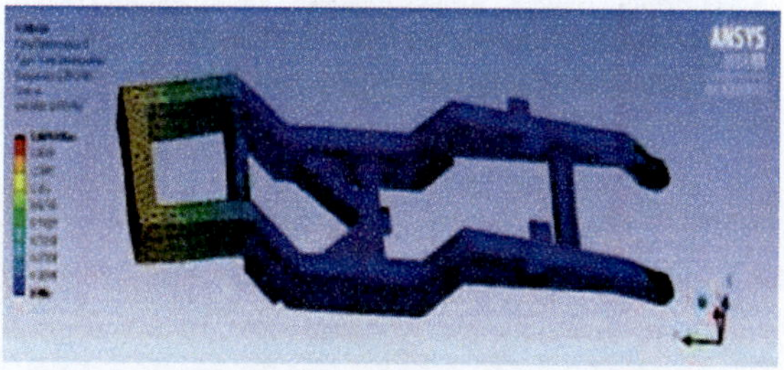

Fig. 14.6 Total deformation 3

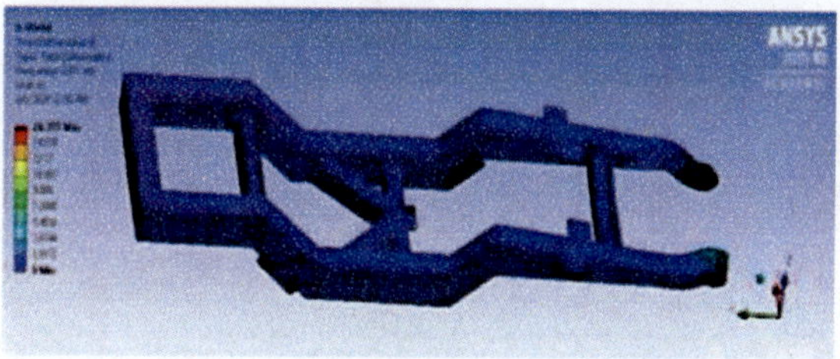

Fig. 14.7 Total deformation 4

14.5 Conclusion

In the current work, the stepping stool type case outline for the car truck was broken down utilizing ANSYS 2019 R3 programming. The part is protected in given stacking circumstances. To improve execution, geometry has been altered, permitting constrain levels to be diminished beneath the yield limit. The structure is alright for every three materials since produced van pressure, and most extreme shear pressure are not exactly suitable. Natural frequency mode shapes obtained by analysis model and temp Although previous tests have found that material is not solid due to chipping problem on rough streets, it is satisfactory. These specific looks are made especially for context research/Tata Ace/light vehicles. Can simulate loading condition without prototyping make necessary changes to design level if necessary, for proper operation original frame investigate frequency and mode shapes truck chassis, mounting areas components on truck chassis, and observe response truck chassis under static loading conditions.

References

1. Design Data Hand Book.: 3rd edn. In: Mahadevan K, Balaveera Reddy K (2010)
2. Shigley's Mechanical Engineering Design.: 8th edn. McGraw-Hill (2006)
3. The Finite Element Analysis and The Optimization Design of The Yj3128- type Dump Truck's Sub-Frames Based on ANSYS—Science Direct (2011) by Chen Yanhong, ZhuFeng
4. Rajappan, Dr. R., Vivekanandhan, M.: Static and modal analysis of chassis by using Fea. Int. J. Eng. Sci. (Ijes) **2**(2), 63–73 (2013)
5. Singh, P.K., Sharma, K.: Mechanical and viscoelastic properties of in-situ amine functionalized multiple layer grpahene/epoxy nanocomposites. Curr. Nanosci. **14**(3), 252–262
6. PATEL, R.L., Gawande, K.R., Morabiya, D.B.: Design and analysis of chassis frame of TATA 2516TC. Int. J. Res. Appl. Sci. Eng. Technol. (IJRASET) **2**(III) (2014)
7. Singh, P.K., et al.: Effect of sonication parameters on mechanical properties of in-situ amine functionalized multiple layer graphene/epoxy nanocomposites. J. Sci. Ind. Res. (79), 985–989 (2020)
8. Singh, A., Soni, V., Singh, A.: Structural analysis of ladder chassis for higherstrength. Int. J. Emerg. Technol. Adv. Eng. **4**(2) (2014)
9. Patel, V.V., Patel, R.I.: Structural analysis of a ladder chassis frame. World J. Sci. Technol. **2**(4), 05–08 (2012). ISSN: 2231-2587. Last accessed 21 April 2012
10. Agrawal, M.S.: Finite element analysis of truck chassis frame. Int. Res. J. Impact Factor (JCC): 3.6234 NAAS Rating: 2.02 Eng. Technol. (IRJET) **02**(03) (2015)
11. Yadav, A., Kumar, A., Singh, P.K., Sharma, K.: Glass transition temperature of functionalized graphene epoxy composites using molecular dynamics simulation. Integr. Ferroelectr. **186**(1), 106–114
12. Chaturvedi, R., Islam, A., Sharma, A., Sharma, K., Sharma, R.: Design and Analysis of Mechanical Gripper of Aristo-Robot for Welding (2020) ISSN: 0193-4120, Article Info vol. 83, pp. 23202–23209
13. Kumar, A., et al.: Evaluation of elastic moduli for different patterns of stone-thrower-wales defect in carbon nanotubes/epoxy composites. Mater. Today: Proc. **4**(9), 9423–9428 (2017)
14. Shukla, M.K., Sharma, K.: Improvement in mechanical and thermal properties of epoxy hybrid composites by functionalized graphene and carbon-nanotubes. Mater. Res. Express **6**(12), 125323 (2019)

15. Chaturvedi, R., Islam, A.: SOL-GEL synthesis of mullite coatings. Mater. Today: Proc. (2020)
16. Patil, H.B., Kachave, S.D., Deore, E.R.: Stress analysis of automotive chassis with various thicknesses. IOSR J. Mech. Civ. Eng. **6**(1), 44–49 (2013)
17. Singh, A., Soni, V., Singh, A.: Structural analysis of ladder chassis for higher strength. Int. J. Emerg. Technol. Adv. Eng. **4**(2), 253–259 (2014)
18. Gosavi, A., Shrivastava, A.K., Sinha, A.K.: Structural analysis of six axle trailer frame design and modification for weight reduction. Int. J. Emerg. Technol. Adv. Eng. **4**(1), 523–531 (2014)
19. Paul, I.D., Sarange, S.M., Bhole, G.P., Chaudhari, J.R.: Structural analysis of truck chassis using finite element method. Int. J. Multidiscip. Res. Adv. Eng. **4**(I), 85–98 (2012)
20. Patel, T.M., Bhatt, M.G., Patel, H.K.: Analysis and validation of Eicher 11.10 chassis frame using ansys. Int. J. Emerg. Trends Technol. Comput. Sci. **2**(2), 85–88 (2013)
21. Sanjay, S.G., Abhijeet, K., Pradeep, G.P., Baskar, P.: Finite element analysis of fire truck chassis for steel and carbon fiber materials. J. Eng. Res. Appl. **4**(7), 69–74 (2014)

Chapter 15
A Review on Gas Sensor Technology and Its Applications

Pooja Saxena and Prashant Shukla

Abstract Gas sensors have drawn the researchers' community's attention for a few decades due to their numerous applications in different areas of environmental monitoring, biomedical devices, and pharma industries, etc. Normally gas sensing research includes polymers, metal-oxide-based semiconducting materials, porous silicon, etc. The performance of gas sensors is characterized by considering various scientific parameters, such as its sensitivity, lowest detection value, response time, recovery time, selectivity, and working temperature. Polymers are known for their sensitive thermal, electrical, mechanical, and dielectric properties. Due to such characteristics, they have gained a wide interest in theoretical research as well as practical applications in sensor technology and devices. This paper reveals a wide research review about the gas sensor structure as a sensing device in the field of the sensor. In the end, the current status, future perspectives as well as advantages of particular polymer-based sensors are summarized.

15.1 Introduction

Gas sensors have extensive significant applications in the following areas which includes industrial production, such as detection of methane gas in mines [1–6], automobile sector in detections of toxic gases from vehicles [7–12], medical field applications [13–16], indoor air quality monitoring devices [17–20], and environmental monitoring, such as greenhouse gas monitoring, etc. [21–24].

In addition to this, many studies have been performed in the last 50 years to understand a detailed view of gas sensing technology and its various branches. The most important areas cover the analysis of different sensors, working principles, and fabrication techniques [25, 26]. This article comprehensively reviews the sensor

P. Saxena (✉)
G. L. Bajaj Institute of Technology and Management, Greater Noida, UP, India

P. Shukla
Amity Institute for Advanced Research and Studies (M&D), Amity University Uttar Pradesh, Sec 125, Noida, UP, India

© The Author(s), under exclusive license to Springer Nature Singapore Pte Ltd. 2022

V. V. Rao et al. (eds.), *Computational and Experimental Methods in Mechanical Engineering*, Smart Innovation, Systems and Technologies 239,
https://doi.org/10.1007/978-981-16-2857-3_16

technology and its classification based on polymer gas sensors [27, 28]. This presents the classification of gas sensing methods, introduces performance assessment of various gas sensing methods, technologies proposed to improvise the selectivity and sensitivity of gas sensors, and then finally concluded.

15.2 Gas Sensors: Physical Significance

Gas sensors have proved to be a significant device due to their wide application areas in the industrial sector [29], environmental monitoring [30, 31], biomedical devices [32], Pharma industry [33], Space exploration [34], food industry [35], biomedicines [36], etc. The performance of gas sensors depends on various technical parameters, such as sensitivity, selectivity, lowest detection value, response time, recovery time, and the most important factor is working temperature. The fabrication technology of gas sensors has been well established, but it still has several disadvantages and limitations, due to which it needs to be studied more broadly. Since there are many gas sensors available in the markets, metal-oxide-based sensors work at very high temperatures. It is very difficult to keep the sensor at this high temperature day and night continuously using Ac mains; otherwise, it is not sensitive to gaseous analytes. The high temperature is maintained by a microheater below the sensing material. Due to this heating, these sensors cannot be used on batteries. Apart from the heating problem, there is another issue with the currently available sensors is the proper mounting of the overall hot ceramic plate to maintain good thermal separation between the sensor element and housing. The highly sensitive sensors are also in demand for the exposure of explosive and poisonous gases [37], which is a challenging problem relates to human health and safety. Gas sensors also contribute to the environmental monitoring sector since it can detect the toxic gases released by the different kinds of pollutions and thus can help reduce global warming. Since the proper monitoring and detection can help us to find a way to control or reduce the pollutants level.

15.3 Criterion of Selection of an Efficient Gas Sensor

- It should have a simple fabrication technique.
- It should be a low-cost device.
- It should be highly sensitive and selective.
- It should have a fast response and recovery time.
- It should be preferably a compact battery-operated device.
- It should work at ambient temperature for minimum power consumption.
- User and Environment Friendly.
- It should have a low detection threshold value.
- It should be stable while repeatedly used.

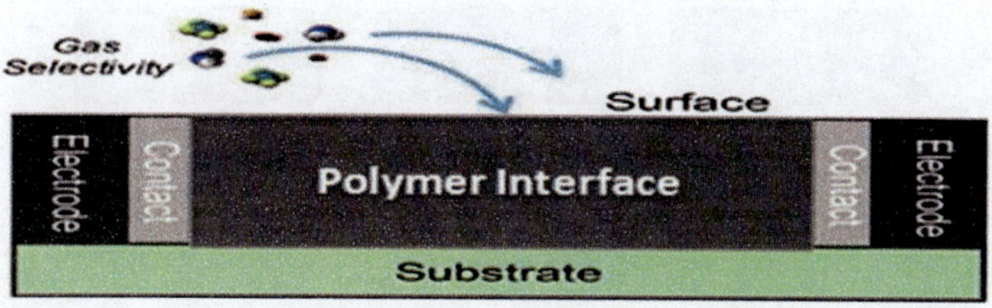

Fig. 15.1 Schematic of the contact between the target gas molecules and gas sensing material

To meet the above criterion for gas sensors, generally a polymer, semiconducting based metal oxides [38, 39], and porous silicon as sensing materials are used [40]. A typical metal-oxide-based gas sensor [41] consists of a substrate, sensing material, electrodes, and heater, as shown in Fig. 15.1. To enhance the sensitivity of the gas sensor, the contact between the target gas molecule and gas sensing material should be maximized.

Detection of air pollutants originating from sources such as vehicles, power plants, refineries, industries, and laboratories, include HCl, H_2S, and VOCs, such as benzene, toluene, xylene, etc., NH_3, CO, CO_2, NOx, etc. These pollutants may have deleterious effects on ecology, flora and fauna, and human health. Therefore, environmental monitoring is crucial to protect public health from toxic contaminants that are continuously released into breathable air. Consequently, there is a considerable need to find a user-friendly and eco-friendly sensing system for real-time monitoring of air pollutant emissions, which are very dangerous for the environment and human health. Recent studies in the literature have reported the harmful effects on human health internally as well as externally, such as irritation in the eyes, restless on being exposed to acetone for quite a long time, breathlessness in patients with respiratory problems, asthmatic patients, etc. Proper monitoring and early detection have thus become necessary for the well-being of human health and society [42, 43] (Fig. 15.2).

15.4 Performance Parameters for Gas Sensing

Gas sensors are utilized and widely used in gas detection. But due to some inherent limitations of gas sensors, many researchers are working to enhance the performance of gas sensing devices based on their technical parameters. For an effective gas sensor, we need to assess several characteristics parameters to check materials' performance and be selective [3]. Brief descriptions of few technical parameters are listed under in Table 15.1.

All above technical parameters discussed above are generally used to define the characteristics properties of a sensor device. For a perfect chemical sensor, it should have all these properties. Researchers have built up an approach to reach up to these characteristics, but cold get succeeded in touching the few only. Also, the designing

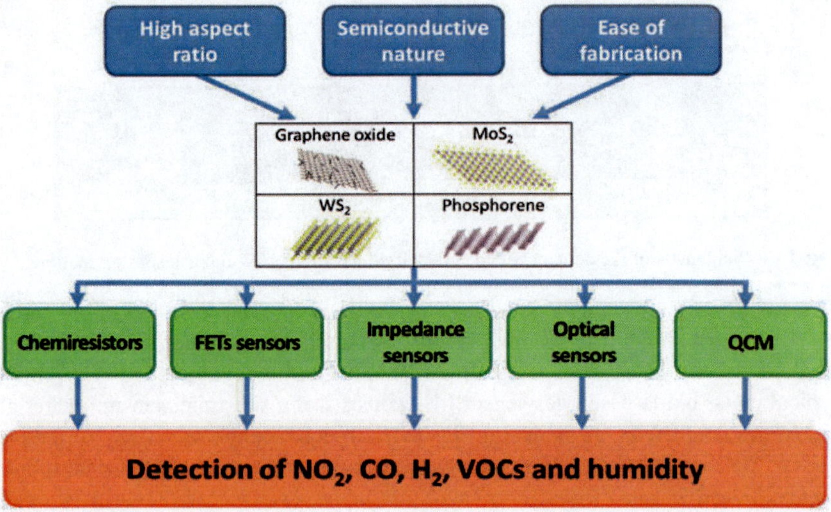

Fig. 15.2 Schematic of few polymer-based gas sensing material showing detection of gases [42]

of the sensor for some specific gases also very difficult, along with the fact that the real applications generally do not involve sensors with all the perfect characteristics, such as for industrial applications, the sensors required do not need a very low detection limit such as at ppb level, but they require very large response time in the seconds. On the other hand, in the case of environmental monitoring systems, there is a slower rate of change of pollutants, so the required detection limit will be higher, but response time could be acceptable.

15.5 Classification of Methods Applicable for Gas Sensing

The classification of gas sensing methods can be done broadly into two groups: methods based on the change of ohmic resistance with different materials and methods based on other kinds of variation parameters. A table for the classification of sensing methods is shown below in Table 15.2.

Several materials can be used for sensing methods. They can be chosen based on their performance, structure, and synthesis properties. A schematic of different materials is shown below in Fig. 15.3.

In addition to that, there are many pros and cons of different materials, and their applications are also limited to some specific areas. Thus, a brief statement of some basic methods of gas sensing [44] are summarized in Table 15.3

Table 15.1 Performance criterion for gas sensing

S. No	Parameter	Description
(a)	Sensitivity	It may be defined as the measurable change in signal response per unit analyte concentration. This analyses a parameter which can detect the lowest concentration of target gas
(b)	Selectivity	It is measured in terms of the selective response of the gas sensor toward a particular target gas among a group of different gases
(c)	Stability	It is defined as the ability of a sensor to sustain its reproducible performance of response through certain repeated cycles or a specific period. It explains that for how much time a sensor can be repeatedly used
(d)	Working temperature	It can be defined as the temperature at which the sensor shows the maximum level of sensitivity for a particular concentration level of the target gas. Power consumption is also directly related to its working temperature. High temperature requires more power consumption
(e)	Response time	It is defined as the time taken by the sensor to obtain 90% of the saturated value of resistance during its change on being exposed to a specific target gas
(f)	Recovery time	It may be defined as the time required to achieve 90% of the baseline value of resistance past the release of the target gas. This criterion also makes a gas sensor responsive and makes it ready to use in a short period ***Both response and recovery times determine the speed of the sensor toward its cyclic operation***
(g)	Detection limit	The lowest concentration of the analyte can be detected by a sensor at a specific value of operating temperature
(h)	Dynamic range	It defines the performance of the sensor through an extended concentration zone of target gas b/w the detection limit and highest value of limiting concentration
(i)	Linearity	The variation of the experimentally determined calibration graph versus target gas concentration
(j)	Resolution	It is defined as the difference of lowest concentration that can be detected by the sensor
(k)	Hysteresis	The maximum difference obtains in output value when the value has reached a level of increase and a decreasing analyte range of concentration
(l)	Life cycle	It can be explained as the time over which the sensor can be continuously operated

15.6 Approach to Improve Sensitivity and Selectivity

For a good and efficient sensor, it must be highly sensitive and selective. It should have good stability with a very fast response and recovery time. These few characteristics make the sensor viable and applicative in different fields of application. To get this

Table 15.2 Classification of methods for gas sensing

Methods of gas sensing	Change of electrical resistance with different materials and their compositions	Metal oxides-based semiconductor
		Polymers
		Carbon nanotubes (CNT's)
		Moisture absorbing material
	Other parameters	Based on optic methods
		Based on acoustic methods
		Based on calorimetric methods
		Based on gas chromatography

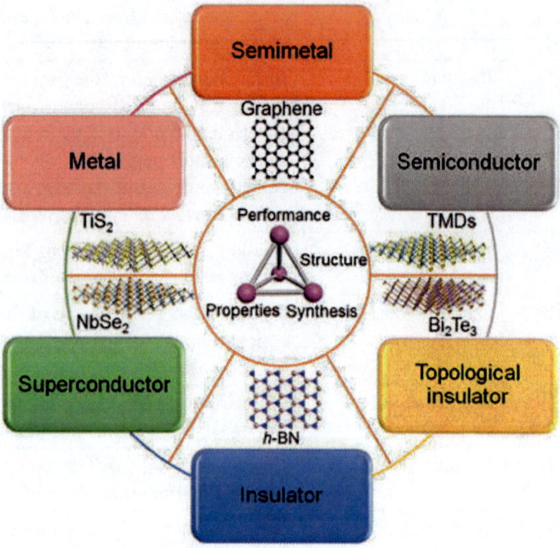

Fig. 15.3 Schematic showing the performance, structure, and properties synthesis of different materials [43]

improvement, some measures need to be taken, which are summarized below in Table 15.4.

Table 15.3 Summary of basic gas sensing methods

Material	Advantages	Disadvantages	Application areas
Metal-oxide semiconductor	Relatively cheaper Short response time Long lifetime	Comparatively less selective and sensitive response High sensitive response to an environmental issue Needs high energy consumption	Industry sector and civil sector
Polymers	Wide range of available target gases Higher sensitivity Short response time Minimal fabrication cost Very simple, compact, and manageable configuration Less power consumption	Large instability Low selective response Irreversibility	Indoor air quality monitoring Best suited to store synthetic products like paints, wax, and fuels Place of work such as chemical industries
CNT's	Highly sensitive Good absorbent Large surface area to volume ratio Fast responsive Lightweight	Expensive Difficult to fabricate Repeatability	Partial discharge (PD) monitoring
Moisture absorbing material	Relatively cheaper Lightweight Highly selective to vapors	Sensitive to friction Irreversibility in absolute humid	Humidity sensor
Optical methods	Highly selective, sensitive, and stable Longer lifespan Not sensitive to environmental change	Difficulty in fabricating and miniaturize Higher cost-value	Air quality monitoring in remote areas Gas leakage detector High-end market product or the most expensive product range
Calorimetric methods	Shows higher stability at ambient temperature Relatively cheaper High sensitive response for industrial gas detection	Prone to catalyst poisoning and explosion Inherent deficiencies in selectivity	Can detect the most combustible gases under industrial environment Petrochemical plants Mine tunnel, Kitchen
Gas chromatograph	Exceptional separation performance Highly sensitive & selective	Higher cost-value Difficult to miniaturization for portable devices	Common laboratory test

(continued)

Table 15.3 (continued)

Material	Advantages	Disadvantages	Application areas
Acoustic methods	Longer lifespan Avoiding secondary pollutants	Low sensitive response High sensitive response to environmental changes	Mechanism of the wireless sensing network

Table 15.4 Summarized approaches to improve sensitivity and selectivity

Approach	Sensitivity	Selectivity
Dielectric resonator	Large surface area Dielectric constant varies with the concentration of gas World Gas Model used in gas generators	Nil
Thermostatic cycle	Highly sensitive to all target gases in each gas cycle Good sensitive response for gases with a quite different sensing temperature	Highly selective of all target gases in each gas cycle For gases with relatively different sensing temperature
Pre-concentrator	Comparative absorption of target gases is enhanced	If the pre-concentrator is selective
Photoacoustic spectroscopy	Cover advantage of both optic and acoustic methods	Nil
Sensor array	Nil	Offer with multidimensional signatures For gases with diverse sensing conditions, the difference is either large or small

15.7 Conclusion

The gas sensors are being constantly focused to lower down the working temperature, improve response time and sensitivity. Efforts have been put by researchers in this field are continuously growing to explore new and novel sensing materials, which would probably bring out the solution to many unsolved technical and industrial issues. The remarkable properties of polymer/CNT nanocomposite-based gas sensing devices have captured the attention of researchers and have been the focus of the sensor research area. The purpose of all the efforts is to bring down the working temperature to reduce the power consumption and thus to improve the other sensing parameters. Also, there is a need for detailed theoretical and experimental examination for the in-depth understanding of the best possible chemical and physical processes in real systems to enhance the specificity of sensor which include metal oxides, doping materials, various analytes, and other molecules which could be possibly present in the atmosphere in specific target applications. This analysis will also contribute to finding solutions for the problems with other issues, such as sensitivity, selectivity, and stability of sensor materials. Our potential approach must be to pay more attention to sensor stability. Higher stability can reduce the frequency needed for verification and re-calibration of sensors. Another significant point is to develop easy and cheaper fabrication methods so that they can be readily available and in reach with the normal population. The development of novel sensitive, selective, and stable gas sensing materials is one of the key factors in the field of nanochemistry and nanotechnology.

References

1. Chen, D., Lei, S., Chen, Y.: A single polyaniline nanofiber field-effect transistor and it's gas sensing mechanisms. Sensors **11**, 6509–6516 (2011)
2. Anderson, T., Ren, F., Pearton, S., Kang, B.S., Wang, H.-T., Chang, C.-Y., Lin, J.: Advances in hydrogen, carbon dioxide, and hydrocarbon gas sensor technology using GaN and ZnO-based devices. Sensors **9**, 4669–4694 (2009)
3. Chaisitsak, S.: Nanocrystalline SnO_2: F thin films for liquid petroleum gas sensors. Sensors **11**, 7127–7140 (2011)
4. Alfeeli, B., Pickrell, G., Wang, A.: Sub-nanoliter spectroscopic gas sensor. Sensors **6**, 1308–1320 (2006)
5. Aldhafeeri, T., Tran, M.-K., Vrolyk, R., Pope, M., Fowler, M.: A review of methane gas detection sensors: recent developments and future perspectives. Inventions **5**, 28 (2020)
6. Chaulya, S.K., Prasad, G.M.: Chapter 3—Gas sensors for underground mines and hazardous areas, sensing and monitoring technologies for mines and hazardous areas. Monitoring and Prediction Technologies, pp. 161–212 (2016)
7. Marr, I., Reiß, S., Hagen, G., Moos, R.: Planar zeolite film-based potentiometric gas sensors manufactured by a combined thick-film and electroplating technique. Sensors **11**, 7736–7748 (2011)
8. Miya, H., Shiina, T., Kato, T., Noguchi, K., Fukuchi, T., Asahi, I., Sugimoto, S., Ninomiya, H., Shimamoto, Y.: Compact Raman Lidar for hydrogen gas leak detection. In: Proceedings of 2009 Conference on Lasers and Electro-Optics/Pacific Rim (CLEOPR), Shanghai, China, 30 July–3 Aug 2009, pp. 1–2
9. Tardy, P., Coulon, J.R., Lucat, C., Menil, F.: Dynamic thermal conductivity sensor for gas detection. Sens. Actuators B **98**, 63–68 (2004)
10. Caucheteur, C., Debliquy, M., Lahem, D., Megret, P.: Catalytic fiber Bragg grating sensor for hydrogen leak detection in air. IEEE Photonics Technol. Lett. **20**, 96–98 (2008)
11. João, B.A., Gomes, J., Rodrigues, J.P.C., Rabêlo, R.A.L., Kumar, N., Kozlov, S.: IoT-enabled gas sensors: technologies, applications, and opportunities. J. Sens. Actuator Netw. **8**, 57 (2019)
12. Fan, Y., et al.: Design and application of toxic and harmful gas monitoring system in fire fighting. Sensors (Basel, Switzerland) **19**(2), 369 (2019)
13. Gonzalez-Jimenez, J., Monroy, J.G., Blanco, J.L.: The multi-chamber electronic nose—an improved olfaction sensor for mobile robotics. Sensors **11**, 6145–6164 (2011)
14. Xiaobo, Z., Jiewen, Z., Shouyi, W., Xingyi, H.: Vinegar classification based on feature extraction and selection from tin oxide gas sensor array data. Sensors **3**, 101–109 (2003)
15. Munoz, B.C., Steinthal, G., Sunshine, S.: Conductive polymercarbon black composites-based sensor arrays for use in an electronic nose. Sens. Rev. **19**, 300–305 (1999)
16. Shrivas, A.G., Bavane, R.G., Mahajan, A.M.: Electronic nose: a toxic gas sensor by polyaniline thin film conducting polymer. In: Proceedings of International Workshop on Physics of Semiconductor Devices 2007 (IWPSD 2007), Mumbai, India, 16–20 Dec 2007, pp. 621–623

17. Kim, S.-J., Hwang, I.-S., Kang, Y.C., Lee, J.-H.: Design of selective gas sensors using additive-loaded in2o3 hollow spheres prepared by combinatorial hydrothermal reactions. Sensors **11**, 10603–10614 (2011)
18. Cubillas, A.M., Lazaro, J.M., Conde, O.M., Petrovich, M.N., Lopez-Higuera, J.M.: Gas sensor based on photonic crystal fibres in the 2ν3 and ν2 + 2ν3 vibrational bands of methane. Sensors **9**, 6261–6272 (2009)
19. Ding, B., Wang, M., Yu, J., Sun, G.: Gas sensors based on electrospun nanofibers. Sensors **9**, 1609–1624 (2009)
20. Bakrania, S.D., Wooldridge, M.S.: The effects of the location of Au additives on combustion-generated SnO_2 nanopowders for co gas sensing. Sensors **10**, 7002–7017 (2010)
21. Hulko, M., Hospach, I., Krasteva, N., Nelles, G.: Cytochrome C biosensor—a model for gas sensing. Sensors **11**, 5968–5980 (2011)
22. Lazik, D., Ebert, S., Leuthold, M., Hagenau, J., Geistlinger, H.: Membrane based measurement technology for *in situ* monitoring of gases in soil. Sensors **9**, 756–767 (2009)
23. Fine, G.F., Cavanagh, L.M., Afonja, A., Binions, R.: Metal oxide semi-conductor gas sensors in environmental monitoring. Sensors **10**, 5469–5502 (2010)
24. Zhang, J., Hu, J.Q., Zhu, F.R., Gong, H., O'Shea, S.J.: Quartz crystal microbalance coated with sol-gel-derived thin films as gas sensor for no detection. Sensors **3**, 404–414 (2003)
25. Xu, X., Wang, J., Long, Y.: Zeolite-based materials for gas sensors. Sensors **6**, 1751–1764 (2006)
26. Tiziana, C.B., Garrett, D.C., Lynford, L.G., Elaine, M.B.: Photonic MEMS for NIR *in situ* gas detection and identification. In: Proceedings of 2007 IEEE Sensors, Atlanta, GA, USA, 28–31 Oct 2007, pp. 1368–1371
27. Madgula, K., Shubha, L.N.: Conducting Polymer Nanocomposite-Based Gas Sensors. In: Thomas, S., Joshi, N., Tomer, V. (eds.) Functional Nanomaterials. Materials Horizons: From Nature to Nanomaterials. Springer, Singapore (2020)
28. Cichosz, S., Masek, A., Zaborski, M.: Polymer-based sensors: a review. Polym. Testing **67**, 342–348 (2018)
29. Liu, X., Cheng, S., Liu, H., Hu, S., Zhang, D., Ning, H.: A survey on gas sensing technology. Sensors **12**, 9635–9665 (2012)
30. Ishiguro, Y., Suzuki, T., Nagawa, Y., Kino, O.: Gas sensors for environment monitoring,. In: Conference Proceedings. 10th Anniversary. IMTC/94. Advanced Technologies in I & M. 1994 IEEE Instrumentation and Measurement Technology Conference (Cat. No.94CH3424-9), Hamamatsu, Japan, vol. 3, pp. 1521–1522 (1994)
31. Lee, D.-D., Lee, D.-S.: Environmental gas sensors. IEEE Sens. J. **1**(3), 214–224 (2001). https://doi.org/10.1109/JSEN.2001.954834. IEEE Xplore
32. Nasiri, N., Clarke, C.: Nanostructured gas sensors for medical and health applications: low to high dimensional materials. Biosensors **9**(1), 43 (2019)
33. Santonico, M., Pennazza, G., Parente, F.R., Grasso, S., Zompanti, A., Stornelli, V., Ferri, G., Bizzarri, M., D'Amico, A.: A gas sensor device for oxygen and carbon dioxide detection. Proceedings, vol. 1, p. 447 (2017)
34. Hunter, G, Chen, L.-Y., Neudeck, P.G., Zhou, H.-J.: Chemical gas sensors for aeronautic and space applications. Source NASA Technical Report Server [NTRS] (1997)
35. Funazakia, N., Hemmia, A., Ito, S., Asanoa, Y., Yano, Y., Miurac, N., Yamazoec, N.: Application of semiconductor gas sensor to quality control of meat freshness in food industry. Sens. Actuators B Chem. **25**(1–3), 797–800 (1995)
36. Yang, J., Chen, B., Zhou, J., Lv, Z.: A low-power and portable biomedical device for respiratory monitoring with a stable power source. Sensors (Basel). **15**(8), 19618–19632 (2015)
37. Fine, G.F., Cavanagh, L.M., Afonja, A., Binions, R.: Metal oxide semi-conductor gas sensors in environmental monitoring. Sensors **10**(6), 5469–5502 (2010)
38. Kuo, C.-G., Huang, C.-W., Chen, J.-H., Liu, Y.-H.: Fabrication of a miniature zinc aluminum oxide nanowire array gas sensor and application for environmental monitoring. Int. J. Photoenergy **2014**, 7. Article ID 515268 (2014)

39. Binions, R., Naik, A.J.T.: Chapter 13—Metal oxide semiconductor gas sensors in environmental monitoring. Semiconductor Gas Sensors. Electronic and Optical Materials. Woodhead Publishing, pp. 433–466 (2013)
40. Bogue, R.: Nanomaterials for gas sensing: a review of recent research. Sens. Rev. **34**(1), 1–8 (2014)
41. Mirzaei, A., Lee, J.-H., Majhi, S.M., Weber, M., Bechelany, M., Kim, H.W., Kim, S.S.: Resistive gas sensors based on metal-oxide nanowires. J. Appl. Phys. **126**, 241102 (2019)
42. Yang, L., Lin, H., Zhang, Z., et al.: Gas sensing of tellurium-modified silicon nanowires to ammonia and propylamine. Sens. Actuators B Chem. **177**, 260–264 (2013)
43. Zhu, Z., Kao, C.T., Wu, R.J.: A highly sensitive ethanol sensor based on Ag@TiO_2 nanoparticles at room temperature. Appl. Surf. Sci. **320**, 348–355 (2014)
44. Li, Z., Li, H., Wu, Z., Wang, M., Luo, J., Torun, H., et al.: Advances in designs and mechanisms of semiconducting metal oxide nanostructures for high-precision gas sensors operated at room temperature. Mater. Horizons **6**(3), 470–506 (2019)

Chapter 16
CFD Study of Two-Dimensional Profile Geometry of an Airfoil

Harshit Bahri, Kaushalendra Kumar Singh, and Harvendra Singh

Abstract In this paper, two-dimensional numerical analysis of airflow over an airfoil has been done. Profile of airfoil has been generated from a fabricated prototype, and the analysis is carried out using computational fluid dynamics software in ANSYS 14.0 workspace. An in-built standard $k - \varepsilon$ model has been used for simulation work and the results show good aerodynamic properties of the airfoil with 0.18 coefficients of lift and a lift to drag ratio as 8.38 at an air speed of 44 m/s at 5-degree angle of attack.

16.1 Introduction

Airfoils are aerodynamic components used in the airplane that produce lift to the airplanes and play a very important role in the efficient design and optimization of airplane wings. The air dynamics over the airplane wings have attracted researchers working in the area since the very early days of aviation. Earlier, it was done through experimentations on different fabricated working models or prototypes in high-precision laboratories. They were highly demanding in terms of cost as well as time. But nowadays, computational techniques have accelerated the study with lower cost and higher accuracy due to a reduction in human intervention in the process. CFD is one of the most efficient tools that enhance designer capability for testing various airfoil shapes with higher accuracy in less time. CFD analysis uses in-built basic governing equations of aerodynamics backed with advanced mathematical tools. Due to these advantages, many works based on computation fluid dynamics have been reported in the last few decades. Many studies have been performed by the CFD tool in order to get the best results and validation, as it is an important tool for analyzing fluid behavior. Modeling of airfoil geometry and analyzing with CFD gives results to find C_D and C_L values. A reduction of 1.7% and 4% in values of C_D and C_L, respectively, were found by Hu and Wong [1]. According to a similar study performed later with different models of the airfoil, it was found that drag was

H. Bahri (✉) · K. K. Singh · H. Singh
G. L. Bajaj Institute of Technology and Management, Greater Noida, UP, India

© The Author(s), under exclusive license to Springer Nature Singapore Pte Ltd. 2022

V. V. Rao et al. (eds.), *Computational and Experimental Methods in Mechanical Engineering*, Smart Innovation, Systems and Technologies 239,
https://doi.org/10.1007/978-981-16-2857-3_17

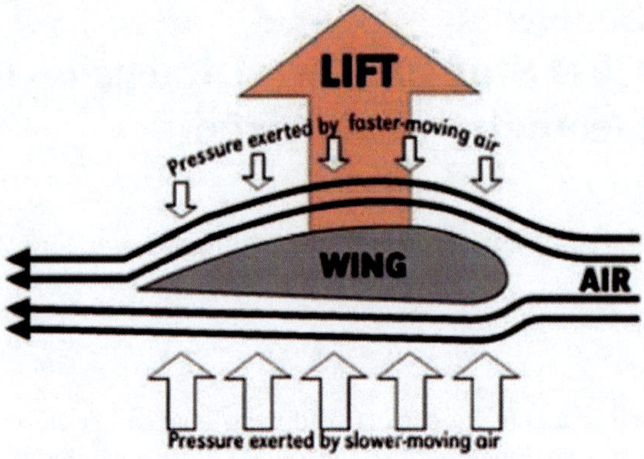

Fig. 16.1 Air dynamics and lift on an airplane wing

reduced by 2.02% while the reduction in the lift was found as 14.06% Ramani and Kumar [2]. Relation between C_D and C_L was established with variable wind speed, and it was found that change in CD was negligible at a certain wind collision angle as speed increases. However, the lift force increases significantly as speed increases Theera-Apisakkul and Kittichaikarn [3]. The study on the comparison of various aspects (likes shape, material) found that magnitude of drag depends on the physical appearance of the profile, such as the shape of the body Alam et al. [4]. It is also observed that drag increases with an increase in lift up to the stalling angle, and the generation of lift affect the flow separation Madhanraj and Shah [5]. At the stalling angle is obtained, the maximum value of lift coefficient gets reduced by flow separation occurring away from the trailing edge Kanimozhi [6].

16.2 Methodology

16.2.1 Airfoil

An airfoil is the cross-sectional shape of airplane wings, blades, or sails; aerodynamic devices are used to give a required force upwards when some fluid flows around it. It also reduces the parallel drag force due to air turbulences and wake formations when the fluid flows around it. The profile of the airfoil greatly influences the lift and drag forces acting on it, which is shown in Fig. 16.1. The basic terminologies used for defining an airfoil geometry are as shown in Fig. 16.2.

16.2.2 Details of the Airfoil Used in the Analysis

Table 16.1 shows the important details of the airfoil used in this study.

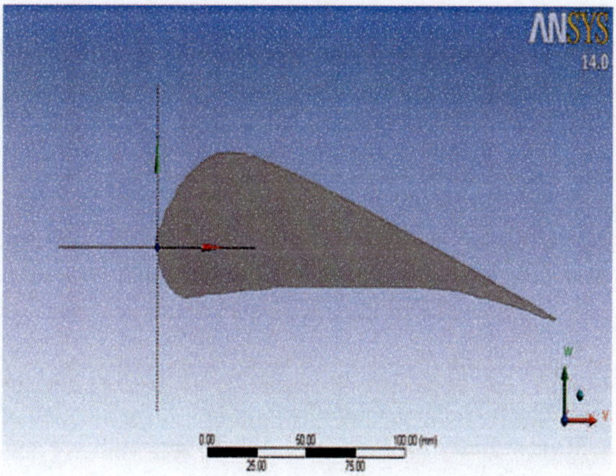

Fig. 16.2 ANSYS 2D model of the airfoil

Table 16.1 Details of the airfoil

Attributes	Details
Way to compute airfoil coordinates with equation and tool utilized	The cross-sectional profile of fabricated airfoil is drawn on a graph paper, and 200 data points are noted down. Thereafter an excel sheet of all the data points is prepared and imported on solid works. The two-dimensional cross-sectional surface model of a fabricated airfoil is generated on solid works imported directly in ANSYS 14.0 Fluent workbench
Name the type of airfoil	NACA2213
Geometrical details of the profile	Chord length (C): 20 cm, Thickness (T): 4.4 cm, Span (S): 20 cm, plan form area (A_p): 0.04 m^2
Calculation lift and drag	$C_D = \frac{2xF_D}{A_p x p_a x V^2}, C_L = \frac{2xF_L}{A_p x p_a x V^2}$

The concepts used to define the shape of the airfoil are:

i. The mean camber line is obtained by the locus of point's midway between the upper and lower surfaces. The thickness distribution along the chord provides it a required shape.
ii. The thickness of airfoil varying along the chord is measured in either of two ways:
   - By measuring thickness perpendicular to the camber line, also described as the "American convention".
   - By measuring thickness perpendicular to the chord line, also described as the "British convention".

The behavior of airfoil's while moving through a fluid is described by:

1. The aerodynamic center.
2. The center of pressure.

16.2.3 Boundary Layers

The boundary layer is the region of the flow of fluid formed due to the effect of viscosity. It is classified into two types based on the flow:

i. laminar, and
ii. turbulent.

In the boundary layer of a laminar, the molecules of the fluid closest to the surface will slow down a great deal and show zero velocity due to the fluid's viscosity. In turn, these surface molecules create a stretch on the upward flowing particles and slow down these particles. The surface effect on the circulation of fluid molecules eventually disintegrates with distance from the surface. The region where these viscous effects are important is called the boundary layer.

In a turbulent boundary layer, eddies are formed, which are larger than molecules. Slow-moving edges close to the surface merging with the rising mass of air. Thus, in the turbulent boundary layer, the motion of air molecules next to the wing surface is faster than that in the case of laminar. The properties of a turbulent boundary layer over a laminar boundary layer are:

i. The higher amount of drag, and
ii. Low susceptibility to flow separation.

16.2.4 Bernoulli's Equation

This equation states that for a steady, incompressible, and frictionless flow, the total head, that is, the sum of pressure head, potential head, and kinetic head remains constant throughout the flow. Mathematically, this equation is discussed as:

$$\left[\frac{P}{\rho}+\frac{V^2}{2}+gz\right]\text{at state 1} = \left[\frac{P}{\rho}+\frac{V^2}{2}+gz\right]\text{at state 2} = \text{Constant} \quad (16.1)$$

where P, pressure of the fluid expressed in Pa; ρ, density of fluid expressed in kg/m^3; v, velocity of fluid relative to airfoil expressed in m/s; z, height at that point expressed in m.

16.2.5 Euler's Equations

The Euler's equation named after Leonhard Euler is derived from the Newton's second law of motion applied to an element in the fluid and is a set of differential terms that express the force analysis on that element.

Mathematically,

$$\frac{dp}{ds} + \rho g \frac{dz}{ds} + \rho V \frac{dV}{ds} = 0 \quad (16.2)$$

16.2.6 Computational Fluid Dynamics (CFD)

Computational Fluid Dynamics (CFD) is the computer-based process to generate flow simulations. CFD involves the numerical solution of the governing laws of fluid in motion. CFD employs the geometrical domain divided into small volumes (mesh) to solve the complex set of partial differential equations. It helps to analyze and simulate the complexity of fluids without any real infra for measurement by instruments.

In this study, ANSYS 14.0 fluent workspace is used to carry out the CFD analysis of a two-dimensional airfoil model. The main features of CFD are:

- The numerical simulation is done in CFD and can also be saved for future references, even after completing the analysis. CFD allows numerical simulation of fluid flows, results of which are available for study even after the analysis is over.
- In CFD, the flow properties can be observed without disturbing the flow, which is advantageous over conventional measuring instruments.
- The flow properties at inaccessible locations (such as inside a combustion chamber) can also be observed using CFD.

16.3 ANSYS Modeling of Airfoil

16.3.1 Two-Dimensional Analysis

For CFD analysis 2D model of the airfoil is first made in solid works, and this model is imported directly in ANSYS Fluent Workbench for 2D analysis. The obtained pressure, velocity contours, and lift drag coefficients are analyzed.

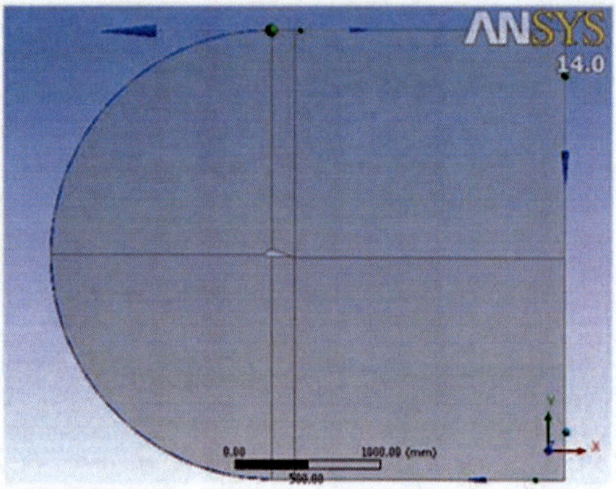

Fig. 16.3 Fluid domain

16.3.2 Geometry

The cross-sectional profile of fabricated airfoil is drawn on graph paper, and 200 data points are noted down. Thereafter an excel sheet of all the data points is prepared and imported on solid works. The two-dimensional cross-sectional surface model of a fabricated airfoil is generated on solid works, which is further imported directly in ANSYS 14.0 Fluent workbench shown in Fig. 16.2. Further steps involved in the modeling of profile in ANSYS 14.0 workbench are as follows.

16.3.3 Fluid Domain

The second step is to make a fluid domain for the airfoil. The domain for spoiler 2d analysis is made C-shaped on the front side and rectangular on the rare side. The airfoil surface is subtracted from the domain resulting in a small hole in the air domain (Fig. 16.3).

16.3.4 Mesh

The third step is to generate the mesh. The mesh is generated so that cells are finer near the spoiler's boundary and coarser in the outer boundary. The front and rare faces of the mesh domain are named as inlet and outlet, and the inner and outer surfaces of the domain are named as airfoil wall and wall (Fig. 16.4).

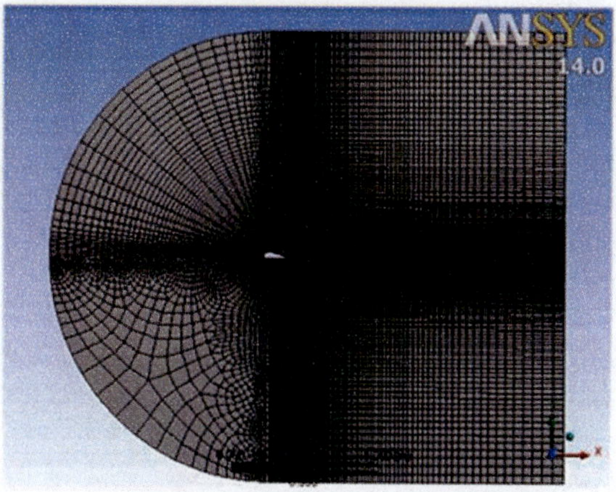

Fig. 16.4 Mesh

Table 16.2 Boundary conditions for airfoil's 2D CFD analysis

S. No.	Description	Values
1	Model	Viscous–Laminar [k-ε Model (2 equations)]
2	Material	Air
3	Inlet velocity	44 m/s
4	Outlet pressure	0 Pa (Gauge pressure)
5	Angle of attack	5°

16.3.5 Boundary Conditions

Table 16.2 shows the boundary conditions used in the analysis.

16.4 CFD Study

A two-dimensional CFD analysis was carried out on the airfoil geometry using ANSYS 14.0 workspace. The governing equations and setting used in the study are as follows:

Transport equations for standard k-ε Models are as follows:

$$\frac{\partial}{\partial t}(\rho k) + \frac{\partial}{\partial x_i}(\rho k u_i) = \frac{\partial}{\partial x_j}\left[\left(\mu + \frac{\mu_t}{\sigma_k}\right)\frac{\partial k}{\partial x_j}\right] + P_k + P_b - \rho\epsilon - Y_M + S_k \tag{16.3}$$

For dissipation ε:

$$\frac{\partial}{\partial t}(\rho\epsilon) + \frac{\partial}{\partial x_i}(\rho\epsilon u_i) = \frac{\partial}{\partial x_j}\left[\left(\mu + \frac{\mu_t}{\sigma_f}\right)\frac{\partial \epsilon}{\partial x_j}\right] + C_{1e}\frac{\epsilon}{k}(P_k + C_{3e}P_b) - C_{2e}\rho\frac{\epsilon^2}{k} + S_e \quad (16.4)$$

Production of K:

$$P_k = -\rho\overline{u'_i u'_j}\frac{\partial u_j}{\partial x_i} \quad (16.5)$$

$$P_k = \mu_t S^2 \quad (16.6)$$

where S is the modulus of the mean rate of the strain tensor, defined as:

$$S = \sqrt{2S_{ij}S_{ij}} \quad (16.7)$$

Type of Mesh: Structured (quadrilateral).

Mesh size: Mesh is generated directly from default settings (scale 1.0), keeping the relevance size as fine. The cells are finer near airfoils's boundary and coarser in the outer boundary.

Convergence criteria: 1e−3.

16.5 Results and Discussion

The solution was converged in 98 iterations, and the following results were obtained. Figures 16.5, 16.6, and 16.7 show the contours of static pressure, dynamic pressure, and total pressure around the airfoil. As can be seen, static and total pressures are highest at the front portion of the airfoil whereas the dynamic pressure is highest at the top portion of the airfoil geometry.

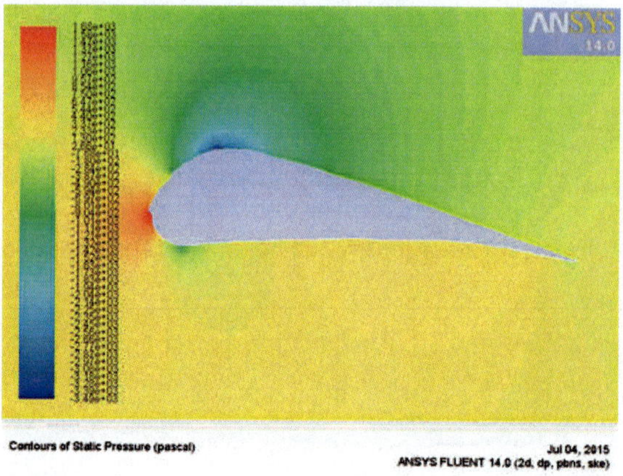

Fig. 16.5 Contours of static pressure

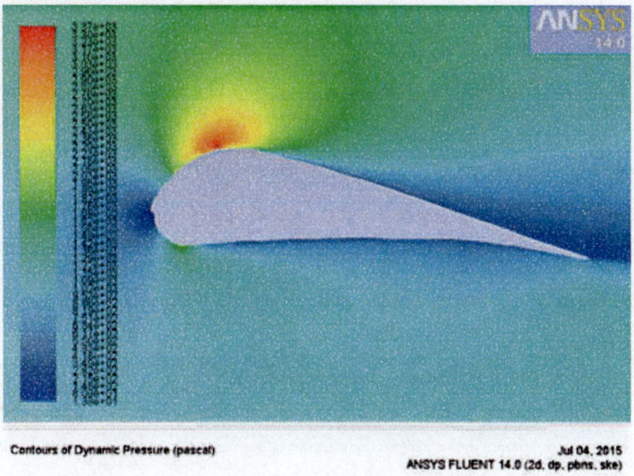

Fig. 16.6 Contours of dynamic pressure

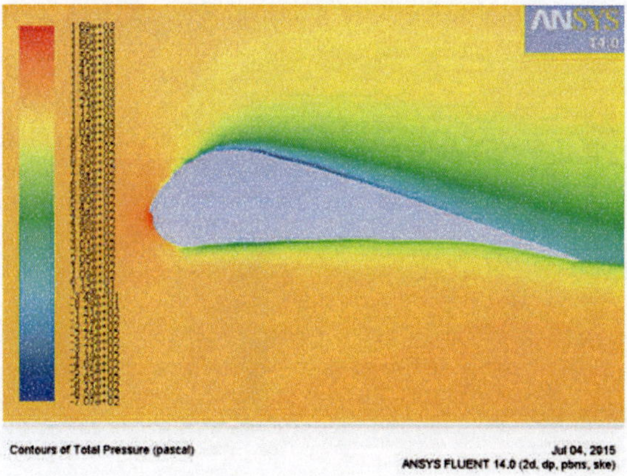

Fig. 16.7 Contours of total pressure

Figures 16.8 and 16.9 show the contours of velocity magnitude and path line around the airfoil geometry. It can be seen that velocity is maximum on the top portion.

Details of Lift, Drag forces, and Coefficients

As represented in Table 16.3, the aerodynamic properties of the fabricated airfoil are good because it creates a nice lift to drag ratio ($C_L/C_D = 1.09/0.13 = 8.38$) up to 8.38. On the other hand, this ratio decreases with an increase in velocity. A good airfoil is supposed to give a greater lift to drag ratios at high speeds. There may be a possibility that the velocity range in which experiments were done does not suit the airfoil, and it can possibly give improved results in some other velocity domains. This is because of the fact that the lift and drag coefficients are both functions of velocity. Hence sometimes they come in a bad proportion for a certain velocity range and sometimes excellent proportion for the other particular velocity domain. All these facts are a

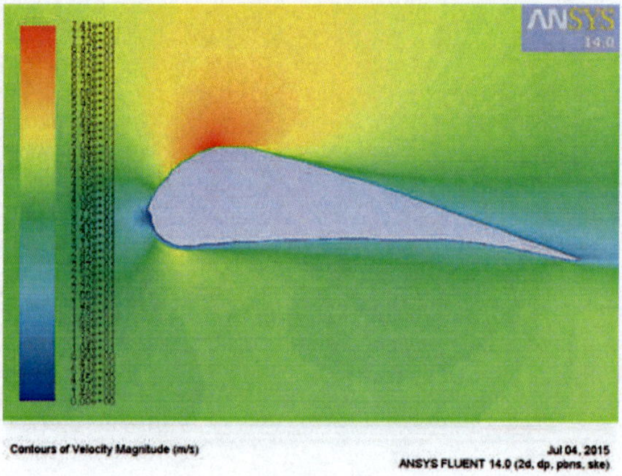

Fig. 16.8 Contours of velocity magnitude

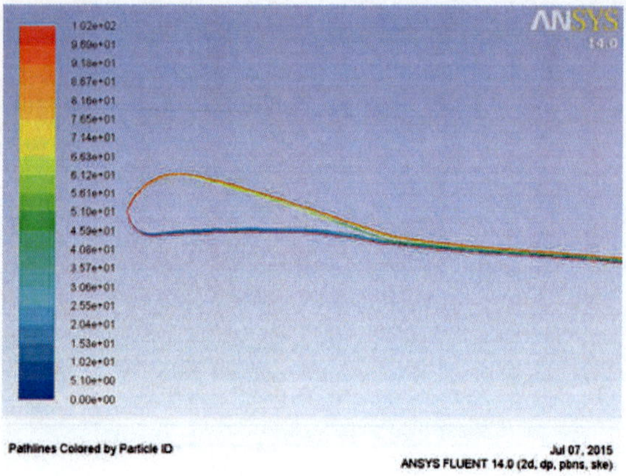

Fig. 16.9 Path line

Table 16.3 Results of spoiler's 2D CFD analysis

Inlet velocity	Angle of attack	Drag force (N)	Lift force (N)	Drag coefficient	Lift coefficient
44 m/s	5°	4.24	12.14	0.08	0.18

matter of optimization of velocity against lift to drag ratio. All above means that the results may be improved by experimenting at different velocity domains and attacks, but that will take excessive time and money.

16.6 Conclusions

Following are the conclusions made on the basis of the results obtained:

i. Static pressure is highest at the frontal portion of the airfoil and lowest at the top of the airfoil.
ii. Dynamic pressure is maximum at the top and minimum at the front.
iii. Velocity is maximum at the top of the airfoil and minimum at the front.
iv. The small changes made in the geometric construction at any stage can help in finding out the required analysis using CFD. This saves much time and fabrication cost in real-time analysis.
v. However, the boundary condition data required for the analysis at boundary conditions can be obtained experimentally.
vi. Sill in the case of turbulent flows, the validation of data through CFD is needed with the data for wind tunnel experiments.
vii. The cross-sectional profile is responsible for the aerodynamic properties of airfoils.

References

1. Hu, X., Wong, E.T.T.: A numerical study on rear-spoiler of passenger vehicle. World Acad. Sci. Eng. Technol. **57**, 636–641 (2011)
2. Ramani, H.B., Kumar, N.: Using shape optimization tool in Ansys software for weight reducation of steel connecting rod. Int. J. Eng. Res. Technol. **2**(2), 1–5 (2013)
3. Theera-Apisakkul, K., Kittichaikarn, C.: Numerical analysis of flow over car spoiler. Kasetsart University, Bangkok (2009)
4. Alam, F., Chowdhurya, H., Guillaumea, E., Yanga, J., Zimmer, G.: On-road and wind tunnel aerodynamic study of human powered vehicles. Procedia Eng. **60**, 473–478 (2013)
5. Madhanraj, V.R., Shah, D.A.: CFD analysis of NACA 2421 aerofoil at several angles of attack. Department of Aeronautical Engineering, Hindustan Institute of Technology and Science, Chennai, Tamil Nadu (2019)
6. Kanimozhi, V.G.: Analysis of airfoil flow pattern using CFD. Department of Aeronautical Engineering, Parisutham Institute of Technology and Science, Thanjavur, Tamil Nadu (2019)

Chapter 17
Tunable Filter at Second Transmission Window Containing 1D Ternary Superconductor/Dielectric Photonic Crystals

Vimal, Sanjeev Sharma, Anil Kumar Sharma, and Rajesh Tiwari

Abstract A tunable optical filter at the second transmission window containing a superconductor/dielectric one-dimensional ternary photonic crystal is designed using the TMM method. The thermal tenability of the structure is controlled by an external temperature of the superconductor Yttrium Barium Copper Oxide (YBCO). The defect layer of the YBCO superconductor plays an important role in designing an optical filter within a bandgap of the structure. The designed structure can be used as a tunable optical filter at the second transmission window in optical fiber communication.

17.1 Introduction

Photonic crystals (PCs) are generally periodic nanostructures of dielectric materials [1, 2]. The dielectric materials possess a photonic bandgap in which we have designed many optical properties like waveguides, ODR mirror, WDM, phase retarders, and filters [3–7]. Both types of photonic crystal conventional like dielectrics and metals and unconventional like magnetic materials, ferroelectric materials, superconductor, semiconductor, and plasma have been studied [8, 9]. The conventional constituents for PC filter having traditional positive index materials but researchers find some drawbacks in conventional PC filter while unconventional constituents have been used to design new types of photonic crystals filters [10, 11]. Some types of conventional PCs are superconductors; they are also called Superconductor-dielectric photonic crystals (SDPCs) [12]. The SDPCs have more advantages compared to

Vimal (✉) · S. Sharma
Department of Physics, GL Bajaj Institute of Technology and Management, Greater Noida, Uttar Pradesh, India

A. K. Sharma
Department of Mathematics, SPC Degree College, Baghpat 250101, Uttar Pradesh, India

R. Tiwari
Department of Applied Science (Physics), ABES, Ghaziabad, Uttar Pradesh, India

© The Author(s), under exclusive license to Springer Nature Singapore Pte Ltd. 2022
V. V. Rao et al. (eds.), *Computational and Experimental Methods in Mechanical Engineering*, Smart Innovation, Systems and Technologies 239,
https://doi.org/10.1007/978-981-16-2857-3_18

metal-dielectric PCs. The metallic loss in these SDPCs is very small [8], and the optical properties of an SDPC can be tunable [13] at the desired wavelength. A defective layer of the superconductor is established into the periodic structure of photonic crystal, called defect mode, and originates narrowband transmission filter [14]. A multi-channel filter has emerged when a single defect layer of dielectric photonic crystal is restored by photonic quantum well structure [15]. The 1D photonic crystal structures are easier to fabricate in comparison to 2D and 3D structures. Xu et al. [16, 17] theoretically design defect modes one-dimensional photonic crystals containing a defect layer with a negative refractive index.

The photonic crystal consists of two different types of single negative materials that can be used to design an omnidirectional (ODR) filter [12]. Some researchers suggested that photonic heterostructures are also used as a terahertz filter; these photonic filters based on the PBGs are used to design various types of optical filters such as wavelength division multiplexers. R. Kumar et al. theoretically design a terahertz tunable filter based on ternary photonic crystal [18]. Recently, Y. Trabeilsi et al. design a narrowband optical tunable filter using superconductor/dielectric Thue-Morse based photonic crystals [19].

We have theoretically designed a narrowband tunable filter using 1DTPCs containing superconductor layers [13]. The superconductor, which is a function of temperature is also depending on the external temperature of the structure, and it is possible to design a thermally tunable optical filter. The complete 1D ternary photonic crystal consists of three layers of dielectric/superconductor materials, and a defect layer of the superconductor is introduced between them in a periodic manner. The photonic crystal consists of superconductors and dielectric materials have attention to a new concept in the field of research [17–19].

17.2 Theoretical and Numerical Method

In this work, a one-dimensional ternary photonic crystal is arranged in a periodic manner. It consists of alternate three layers of dielectric/superconductor materials, and a defect layer of a superconductor is introduced between them, which are shown in Fig. 17.1.

The optical properties of 1D ternary photonic crystal are determined by using the TMM method. Here, the property of the YBCO superconductor is described without using an external magnetic field. The Gorter Casimir two-fluid model [20] well explains the material properties of the superconductor in the absence of external magnetic field. Using two-fluid model, the relative permittivity of a lossless superconductor is defined as [20, 21]

$$\varepsilon_c = 1 - \frac{\omega_{th}^2}{\omega^2} \tag{17.1}$$

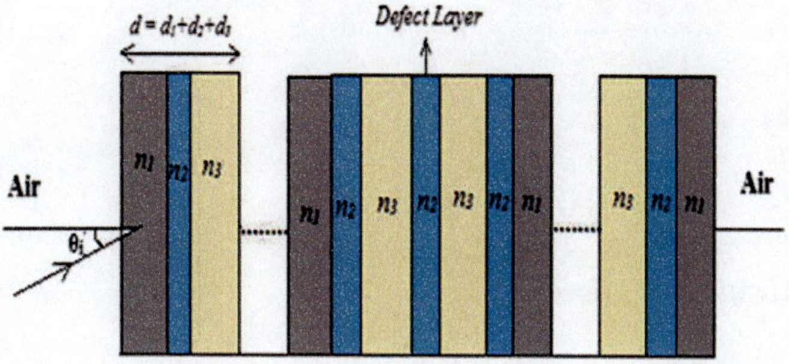

Fig. 17.1 Schematic arrangement of 1D ternary photonic crystal with defect layer

where, ω_{th} is the threshold frequency of the bulk superconductor,

$$\omega_{th}^2 = \frac{c^2}{\lambda_L^2} \tag{17.2}$$

c is the speed of light in free space. λ_L denotes the temperature-dependent penetration depth, which is described by

$$\lambda_L = \frac{\lambda_0}{\sqrt{1 - S(T)}} \tag{17.3}$$

where, λ_0 is the London penetration depth at $T = 0$ K, and $S(T) = \left(\frac{T}{T_c}\right)^p$, where T_c denotes the critical temperature of the superconductor and p depends on the superconductors (for YBCO, it is 4).

The refractive index of this superconductor YBCO related to temperature-dependent penetration depth is defined as [20]:

$$n_S = \sqrt{1 - \frac{1}{\omega^2 \mu_0 \varepsilon_0 \lambda_L^2}} \tag{17.4}$$

At 0 K, the critical transformation temperature and London penetration depth of this superconductor YBCO is 92 K and $\lambda_0 = 140$ nm, respectively [20, 21].

The reflectance and transmittance of this ternary photonic crystal are calculated by using TMM. Applying the transfer matrix method (TMM), the coefficient of reflection and the reflectance of the structure is given by [22],

$$r(\omega) = \frac{(m_{11} + m_{12}p_0)p_0 - (m_{21} + m_{22}p_0)}{(m_{11} + m_{12}p_0)p_0 + (m_{21} + m_{22}p_0)} \tag{17.5}$$

Also, the reflectance and transmittance of the proposed 1D ternary photonic crystal is evaluated by,

$$R = |r(\omega)|^2 \tag{17.6}$$

$$\& \ T = |1 - R| \tag{17.7}$$

17.3 Result and Discussion

In this paper, we have taken the refractive index and thicknesses of Si and SiO_2 materials to be 3.5, 1.45 and 620 nm, 760 nm, respectively. The refractive index of superconductor Yttrium Barium Copper Oxide (YBCO) material depends on temperature. For YBCO materials, the range of refractive index and temperature varies from 1.25–1.4 μm and 20–92 K, respectively. Also, the thicknesses of the YBCO superconductor layer be 93 nm and for the defect layer, it is also 93 nm. Figure 17.2 shows the photonic bandgap of this ternary photonic crystal when the defect layer of superconducting materials is not introduced in the periodic manner of these photonic crystals. When a defect layer of superconducting materials is introduced in the mid of this photonic crystal, it is tuned with temperature. The transmittance of the proposed structure is shown in Fig. 17.3. In this figure, it is clear that the bandgap and defect layer shifted toward a higher wavelength range of the spectrum when the temperature varies from 20 to 90 K. The peak of the defective layer has appeared at wavelength 1297.2 nm with a transmission efficiency of 90% at temperature 20 K. When temperature increases from 20 to 30 K the transmission peak shifted to 1297.8 nm wavelength. Similarly, at temperature 40 K, 50 K, 60 K, 70 K, 80 K, and 90 K, the defect mode layer has been tuned and observed at wavelengths 1298.2 nm, 1298.6 nm, 1301.3 nm, 1304.1 nm, 1307.5 nm, and 1312.3 nm with the transmittance of 0.93, 0.94, 0.98,

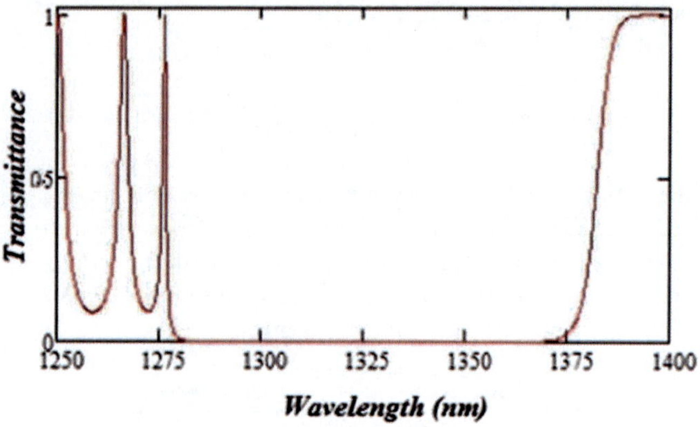

Fig. 17.2 Structure of 1D ternary photonic crystals without defect layer

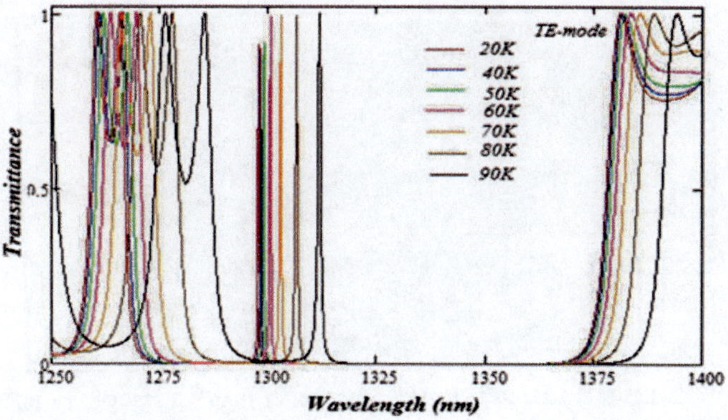

Fig. 17.3 Structure of 1D ternary photonic crystal with defect layer at different temperature

0.99, 0.99, and 1.0, respectively. It is clear that when the temperature of the superconductor increases, the defect mode layer shifted toward the higher wavelength range of the spectrum.

Also, from Table 17.1, it is clear that when temperature increases, the efficiency of the defective layer also increases, and at temperature 90 K, it approaches 100% transmission efficiency. So, it works as a tunable optical filter in the range of photonic bandgap at the second transmission window. In this photonic bandgap, we can design wavelength division multiplexer, optical filter, dense wavelength division multiplexer, etc.

The quality factor of this photonic crystal at various transmission wavelengths is shown in Table 17.1. From Table 17.1, it is clear that when the temperature of the superconductor increases, the efficiency and quality factor of the proposed structure increase. This high-quality factor indicates low attenuation loss in the bandgap of a photonic crystal which is used in optical communication.

Table 17.1 Transmission efficiency and quality factor of ternary PCs at different temperatures

Temperature T (K)	Resonant wavelength (nm)	Transmission efficiency %	Quality factor
20	1297.2	90	3727
30	1297.8	91	3526
40	1298.2	93	3435
50	1298.6	94	3224
60	1301.3	98	3110
70	1304.1	99	2974
80	1307.5	99	2633
90	1312.3	100	2497

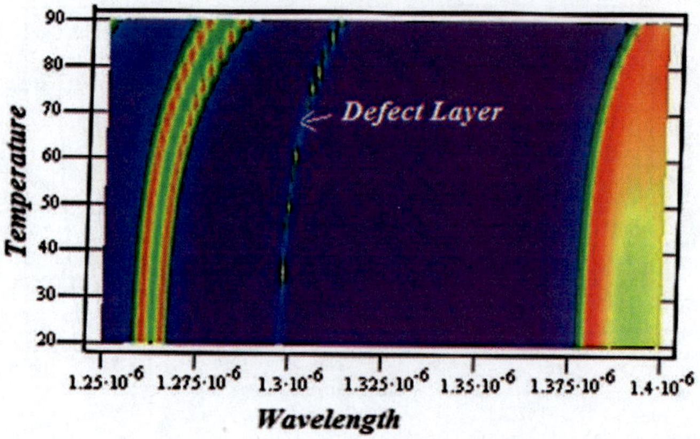

Fig. 17.4 3D View of 1D ternary photonic crystal with defect at different temperatures

The 3-dimensional schematic diagram of this photonic crystal is illustrated in Fig. 17.4. The proposed structure shows the variation of the defect layer with temperature. It is clear that when temperature increases, the defect layer of superconductor material shifted toward the higher wavelength range of the spectrum. Hence the proposed structure works as a tunable filter in optical fiber communication.

17.4 Conclusions

This paper, the optical properties of 1D ternary photonic crystals are analyzed by using the TMM method in this paper. The theoretical result shows that the proposed structure works as a tunable optical filter when the temperature of the superconductor increases with wavelength. In this work, the defect mode layer shifted toward the higher wavelength range of the spectrum, and the efficiency of the layer also increases with the increased temperature of the superconductor. At a certain temperature (92 K), it approaches to 100%. By tuning the wavelength, we design a wavelength division multiplexer or optical filter which is used in optical fiber communication. So, the designed structure can be used as a tunable multi-channel filter at the second transmission window in optical fiber communication.

References

1. Song, B., Noda, S., Asano, T., et al.: Photonic devices based on in-plane hetero photonic crystals. Science **300**(5625), 1537 (2003). https://doi.org/10.1126/science.1083066
2. Li, C., Xue, Q., Ji, Z., Li, Y., Zhang, H., Li, D.: Construction of photonic crystals with thermally adjustable pseudo-gaps. Soft Matter **16**(12), 3063–3068 (2020)

3. Sharma, S., Kumar, A., Singh, K.S.: Design of a tunable DWDM multiplexer using four defect layers of GaAs nonlinear photonic crystals. Optik-Int. J. Light Electron Opt. **212**, 164652 (2020)
4. Kalhan, A., Sharma, S., Kumar, A.: 16-channel DWDM based on 1D defect mode nonlinear photonic crystal. AIP Conf. Proc. **1953**,(2018). https://doi.org/10.1063/1.5032777
5. Zhang, X., Chen, Y.: Broadband phase retarder based on one-dimensional photonic crystal containing mu-negative materials. J. Opt. Soc. Am. B. **29**(10), 2704 (2012)
6. Sharma, S., Kumar, R., Singh, Kh.S., Kumar, A.: Temperature dependence ODR reflection by using 1D binary and ternary photonic crystal. Optoelectron. Adv. Mater.-Rapid Commu. **19**(5–6), 319–324 (2017)
7. Sharma, S., Kumar, A., Singh, Kh.S., Kumar, V., Kumar, A.: Omnidirectional reflector using linearly Graded refractive index profile of 1D binary and ternary photonic crystal. Optik-Int. J. Light Electron Opt. **126**(11–12), 1146–1149 (2015)
8. Bilotti, F., Sevgi, L.: Metamaterials: Definitions, properties, applications, and FDTD-based modeling and simulation. Int. J. RF Microw. Comput. Aided Eng. **22**(4), 422–438 (2012)
9. Jutze, G.A., Foster, K.E.: Recommended standard method for atmospheric sampling of fine particulate matter by filter media—high-volume sampler. J. Air Pollut. Control Assoc. **17**(1), 17–25 (1967)
10. Sharma, S., Gupta, S., Suther, B., Singh, Kh.S.: Design of a tunable transmission mode filter using 1D Ge based nonlinear photonic crystal. AIP Conf. Proc. **2220**,(2020). https://doi.org/10.1063/5.0001128
11. Sharma, S., Kumar, R., Singh, Kh.S., Kumar, V., Jain, D.: Design of a transmission TM mode filter using one-dimensional ternary photonic crystal. AIP Conf. Proc. **1536**, 705 (2013). https://doi.org/10.1063/1.4810422
12. Zhang, H.-F., Liu, S.-B., Yang, H.: Omnidirectional photonic band gap in one-dimensional ternary superconductor-dielectric photonic crystals based on a new Thue-Morse aperiodic structure. J. Supercond. Novel Magn. **27**, 41–52 (2014)
13. Hung, H.-C., Wu, C.-J., Chang, S.-J.: A mid-infrared tunable filter in a semiconductor–dielectric photonic crystal containing doped semiconductor defect. Sci. Direct Solid State Commun. **151**(22), 1677– 1680
14. Dadoenkova, N.N., Zabolotin, A.E., Lyubchanskii, I.L., Lee, Y.P., Rasing, T.: One-dimensional photonic crystal with a complex defect containing an ultrathin superconducting sublayer. J. Appl. Phys. **108**(9), 093117 (2010)
15. Dhingra, M., Shankar, A., Tiwari, B.B.: A review on quantum well structures in photonic devices for enhanced speed and span of the transmission network. Indian J. Phys. **84**(8), 1031–1037 (2010)
16. Carrera, E., Pagani, A., Zangallo, F.: Comparison of various 1D, 2D and 3D FE models for the analysis of thin-walled box with transverse ribs subjected to load factors. Finite Elem. Anal. Des. **95**, 1–11 (2014)
17. Sharma, S., Dwivedi, D., Yadav, A., Sengar, A.S.: Temperature dependence ZnS based 1D photonic crystal. Int. J. Phys. Res. (IJPR) **6**(3) (2016)
18. Kumar, R., Kushwaha, A.S., Srivastava, M., Srivastava, S.K.: Study of tunable ternary layer photonic crystal for terahertz application. J. Nanoelectron. Optoelectron. **10**, 1–7 (2015)·
19. Trabelsi, Y., Ali, N.B., Kanzari, M.: Tunable narrowband optical filters using superconductor/dielectric generalized Thue-Morse photonic crystals. Microelectron. Engg. **213**, 41–46 (2019)
20. Chung-An, Hu., Liu Jia-Wei, Wu., Chien-Jang, Y.-J., Su-Lin, Y.: Effects of superconducting film on the defect mode in dielectric photonic crystal heterostructure. Solid State Commun. **157**, 54–57 (2013)
21. Ji-Jiang, Wu., Gao, J.-X.: Low temperature sensor based on one-dimensional photonic crystals with a dielectric-superconducting pair defect. Optik **126**, 5368–5371 (2015)
22. Yeh, P., Yariv, A.: Optical Waves in Crystals. Wiley, New York (1984)

Chapter 18
A Numerical Method for a Problem Occurring in Conduction of Heat Through a Solid and Other Applications

Shradha Gupta and Sanjeev Sharma

Abstract A singular boundary value problem arises in the conduction of heat through a solid and having other important applications is solved using a novel spline method. The removal of the singularity is done before applying a cubic B-spline and then a B-spline with a free parameter. The numerical examples show that the results have a very close agreement with exact solutions.

18.1 Introduction

Singular boundary value problems have some important applications, including the conduction of heat through a solid, tumor growth problems, and heat sources' distribution in the human head in physiology. Several researchers applied various numerical techniques for this class of differential equations. These methods are difference methods [1, 2], collocation method [3], variational iteration method [4], adomian decomposition method [5, 6], homotopy perturbation method [7], and various spline methods [8–11]. In this paper, the series expansion is first used in a small neighborhood of the singular point to make it a regular problem without singularity. Then, we have first solved the differential equation using cubic B-spline. Further, a method using B-spline with free parameter is also proposed, which improves results further. The detailed elaboration of properties, including the definition of such extended spline, can be found in [12]. Two numerical problems are solved to explain the methods and to demonstrate the advantage of the extended B-spline method when one can choose the free parameter suitably.

S. Gupta (✉) · S. Sharma
Department of Applied Sciences, GL Bajaj Institute of Technology and Management, Greater Noida, Uttar Pradesh, India

© The Author(s), under exclusive license to Springer Nature Singapore Pte Ltd. 2022
V. V. Rao et al. (eds.), *Computational and Experimental Methods in Mechanical Engineering*, Smart Innovation, Systems and Technologies 239,
https://doi.org/10.1007/978-981-16-2857-3_19

18.2 Conversion to Regular Problem and Cubic B-Spline Method

Consider the following differential equation having a regular singular point

$$u''(t) + p(t)u'(t) + q(t)u(t) = r(t) \tag{18.1}$$

with the condition

$$u(a_0) = \alpha, u(b) = \beta, \tag{18.2}$$

where coefficients $p(t),\ q(t)$ are analytic at $t = a_0$. The removal of the singularity is done by expanding in series in a small neighborhood near $t = a_0$ in the interval $[a_0, \delta]$. Now, Eq. (18.1) has the following form of solution

$$u(t) = (t - a_0)^p \sum_{k=0}^{\infty} l_k (t - a_0)^k, a_0 \neq 0. \tag{18.3}$$

Differentiating Eq. (18.3) and substituting in Eq. (18.1) and by comparison of the coefficients of same degrees of $t = a_0$ on LHS and RHS, we obtain the roots of the indicial equation in the form of the value of p.

Now, the solution for (18.1) is

$$u(t) = \sum_{i=1}^{m} \alpha_i R_i(t) + R_{s+1}(t), \quad m \leq 2 \tag{18.4}$$

for $t \in [a_0, \delta]$ where $R_1(t)$, $R_2(t)$ are the independent solution of the homogeneous part and $R_{s+1}(t)$ is PI to (18.1).

Now we derive the condition at the boundary to convert the singular problem to the regular problem on the interval $[\delta, b]$, where δ lies the interval (a_0, b).

$$\alpha_1 R_1(\delta) + \alpha_2 R_2(\delta) = u(\delta) - R_{s+1}(\delta), \tag{18.5}$$

$$\alpha_1 R_1'(\delta) + \alpha_2 R_2'(\delta) = u'(\delta) - R_{s+1}'(\delta). \tag{18.6}$$

On solving Eqs. (18.5) and (18.6) we get

$$\alpha_1 = \frac{\left[u(\delta) - R_{s+1}(\delta)\right] R_2'(\delta) - \left[u'(\delta) - R_{s+1}'(\delta)\right] R_2(\delta)}{R_1(\delta) R_2'(\delta) - R_2(\delta) R_1'(\delta)} \tag{18.7}$$

$$\alpha_2 = \frac{\left[u^{'}(\delta) - R^{'}_{s+1}(\delta)\right]R_1(\delta) - \left[u(\delta) - R_{s+1}(\delta)\right]R^{'}_1(\delta)}{R_1(\delta)R^{'}_2(\delta) - R_2(\delta)R^{'}_1(\delta)} \tag{18.8}$$

Using boundary condition and Eq. (18.4), we get

$$\alpha_1 R_1(a_0) + \alpha_2 R_2(a_0) = u(a_0) - R_{s+1}(a_0). \tag{18.9}$$

Now, Eqs. (18.8), (18.9) and (18.10) give

$$\begin{aligned} \frac{g(\delta)R^{'}_2(\delta) - g^{'}(\delta)R_2(\delta)}{h(\delta)}R_1(a_0) + \frac{g^{'}(\delta)R_1(\delta) - g(\delta)R^{'}_1(\delta)}{h(\delta)}R_2(a_0) \\ = \alpha - R_{s+1}(a_0), \end{aligned} \tag{18.10}$$

here $g(t) = u(t) - R_{s+1}(t)$ and $h(t) = R_1(t)R^{'}_2(t) - R_2(t)R^{'}_1(t)$.

Now, rewriting the equation as

$$\begin{aligned} [R_1(\delta)R_2(a_0) - R_2(\delta)R_1(a_0)]g^{'}(\delta) + \left[R^{'}_2(\delta)R_1(a_0) - R^{'}_1(\delta)R_2(a_0)\right]g(\delta) \\ = \left[\alpha - R_{s+1}(a_0)\right]h(\delta). \end{aligned} \tag{18.11}$$

Equation (18.11) gives

$$Au(\delta) + Bu^{'}(\delta) = D, \tag{18.12}$$

where

$$A = -R^{'}_1(\delta)R_2(a_0) + R^{'}_2(\delta)R_1(a_0), \quad B = -R_2(\delta)R_1(a_0) + R_1(\delta)R_2(a_0)$$

and

$$D = h(\delta)\left[\alpha - R_{s+1}(a_0)\right] + AR_{s+1}(\delta) + BR^{'}_{s+1}(\delta)$$

So BVP without singularity over the interval $[\delta, b]$ is expressed by Eq. (18.1) with the boundary condition given by (18.12) and $u(b) = \beta$.

B-spline formulation: Let us partition $[\delta, b]$ with equidistant points given by $\pi = \{t_0, t_1, t_2, ..., t_N\}$, where $t_0 = \delta$ and $t_N = b$ and $h = \frac{(b-\delta)}{N}$ is the length between two consecutive nodes.

Now, define [13] the basis of spline function for the values of $i = 0, 1, 2, ..., N$

$$B_i(t) = \frac{1}{6h^3}\begin{cases} (t-t_i)^3 & t \in [t_i, t_{i+1}] \\ h^3 + 3h^2(t-t_{i+1}) + 3h(t-t_{i+1})^2 - 3(t-t_{i+1})^3 & t \in [t_{i+1}, t_{i+2}] \\ h^3 + 3h^2(t_{i+3}-t) + 3h(t_{i+3}-t)^2 - 3(t_{i+3}-t)^3 & t \in [t_{i+2}, t_{i+3}] \\ (t_{i+4}-t)^3, & t \in [t_{i+3}, t_{i+4}] \\ 0, & \text{otherwise.} \end{cases} \tag{18.13}$$

Let us interpolate the function $u(x)$ at nodal points by $S(x)$ given as

$$\begin{aligned} S(t) &= \sum_{i=-1}^{n+1} c_i B_i(t), \\ &= c_{-1}B_{-1}(t) + c_0 B_0(t) + ... + c_N B_N(t) + c_{N+1}B_{N+1}(t). \end{aligned} \tag{18.14}$$

Putting Eq. (18.14) in (18.1), we get

$$S''(t) + p(t)S'(t) + q(t)S(t) = r(t). \tag{18.15}$$

Now putting the value of $S(t)$ and its derivatives, we get for ith nodal point

$$\begin{aligned} & c_{i-1}\left(B''_{i-1}(t_j) + p_i(t_j)B'_{i-1}(t_j) + q_i(t_j)B_{i-1}(t_j)\right) \\ & +c_i\left(B''_i(t_j) + p_i(t_j)B'_i(t_j) + q_i(t_j)B_i(t_j)\right) \\ & +c_{i+1}\left(B''_{i+1}(t_j) + p_i(t_j)B'_{i+1}(t_j) + q_i(t_j)B_{i+1}(t_j)\right) = r_i(t_j). \end{aligned} \tag{18.16}$$

Now Eq. (18.12) and 2nd boundary condition give

$$A\left(\sum_{i=-1,0,1} c_i B_i(\delta)\right) + B\left(\sum_{i=-1,0,1} c_i B'_i(\delta)\right) = D \tag{18.17}$$

Together with

$$c_{N-1} + 4c_N + c_{N+1} = \beta. \tag{18.18}$$

Equations (18.16), (18.17), and (18.18) lead to a tridiagonal system. The solution of the system gives the required B-spline solution.

18.3 B-Spline with Free Parameter

In the present section, the method based on B-spline with a free parameter to solve the problem considered above is proposed. $B_i^{\lambda}(t)$ is a spline consisting of a free parameter λ. The definition of its basic function is given by [12]

$$B_i^{\lambda}(t) = \frac{1}{24h^4}\begin{cases} -4\lambda(\lambda-1)(t-t_i)^3, & t \in [t_i, t_{i+1}], \\ (4-\lambda)h^4 + 12h^3(t-t_{i+1}) + 6h^2(2+\lambda)(t-t_{i+1})^2 \\ -12h(t-t_{i+1})^3 - 3\lambda(t-t_{i+1})^4, & t \in [t_{i+1}, t_{i+2}], \\ (16+2\lambda)h^4 - 12h^2(2+\lambda)(t-t_{i+2})^2 \\ +12h(1+\lambda)(t-t_{i+2})^3 - 3\lambda(t-t_{i+2}), & t \in [t_{i+2}, t_{i+3}], \\ -(h+t_{i+3}-t)^3[h(\lambda-4)+3\lambda(t-t_{i+3})], & t \in [t_{i+3}, t_{i+4}], \\ 0, & \text{elsewhere.} \end{cases} \tag{18.19}$$

When $\lambda = 0$, $B_i^{\lambda}(t)$ degenerates into ordinary cubic B-spline basis. B-spline with free parameter function, $S(t)$ is

$$S(t) = \sum_{i=-1}^{N+1} c_i B_i^{\lambda}(t), \quad t \in [t_0, t_N]. \tag{18.20}$$

Using the definition of basis given above and evaluating (18.20) at t_j, we obtain the following

$$\begin{aligned} S(t_j) &= c_{i-1}B_{i-1}^{\lambda}(t_j) + c_i B_i^{\lambda}(t_j) + c_{i+1}B_{i+1}^{\lambda}(t_j), \\ &= c_{i-1}\left(\frac{4-\lambda}{24}\right) + c_i\left(\frac{8+\lambda}{12}\right) + c_{i+1}\left(\frac{4-\lambda}{24}\right). \end{aligned} \tag{18.21}$$

In a similar way, derivatives of order up to two of $S(x_i)$ can be derived into an expression containing c_{i-1}, c_i, c_{i+1} only,

$$\begin{aligned} S'(t_j) &= c_{i-1}B_{i-1}^{\lambda}{}'(t_j) + c_i B_i^{\lambda\prime}(t_j) + c_{i+1}B_{i+1}^{\lambda}{}'(t_j), \\ &= c_{i-1}\left(-\frac{1}{2h}\right) + c_i(0) + c_{i+1}\left(\frac{1}{2h}\right) \end{aligned} \tag{18.22}$$

$$\begin{aligned} S''(t_j) &= c_{i-1}B_{i-1}^{\lambda}{}''(t_j) + c_i B_i^{\lambda\prime\prime}(t_j) + c_{i+1}B_{i+1}^{\lambda}{}''(t_j), \\ &= c_{i-1}\left(\frac{2+\lambda}{2h^2}\right) + c_i\left(-\frac{2+\lambda}{2h^2}\right) + c_{i+1}\left(\frac{2+\lambda}{2h^2}\right). \end{aligned} \tag{18.23}$$

Using Eqs. (18.21)–(18.23), Eq. (18.1) becomes

$$c_{i-1}\left(\frac{2+\lambda}{2h^2}\right)+c_i\left(-\frac{2+\lambda}{2h^2}\right)+c_{i+1}\left(\frac{2+\lambda}{2h^2}\right)$$
$$+p(t_j)\left[c_{i-1}\left(\frac{1}{2h}\right)+c_i(0)+c_{i+1}\left(\frac{1}{2h}\right)\right] \tag{18.24}$$
$$+q(t_j)\left[c_{i-1}\left(\frac{4-\lambda}{24}\right)+c_i\left(\frac{8+\lambda}{12}\right)+c_{i+1}\left(\frac{4-\lambda}{24}\right)\right]=r(t_j),$$

and boundary conditions are

$$A\left(\sum_{i=-1}^{n+1} c_i B_i^{\lambda}(\delta)\right)+B\left(\sum_{i=-1}^{n+1} c_i B_i^{\lambda\prime}(\delta)\right)=D, \tag{18.25}$$

$$c_{N-1}\left(\frac{4-\lambda}{24}\right)+c_N\left(\frac{8+\lambda}{12}\right)+c_{N+1}\left(\frac{4-\lambda}{24}\right)=\beta. \tag{18.26}$$

Now, elimination of c_{-1} from the initial equation of (18.24) and (18.25), we get

$$c_0\left[\left(\frac{2+\lambda}{2h^2}\right)\left(A-\frac{B}{h}\right)+\left(\frac{8+\lambda}{24h}\right)(Bq(t_0)-Ap(t_0))\right]$$
$$+c_1\left[\left(\frac{2+\lambda}{2h^3}\right)B+\left(\frac{4-\lambda}{24h}\right)(Bq(t_0)-Ap(t_0))\right] \tag{18.27}$$
$$=D\left[\left(\frac{2+\lambda}{2h^3}\right)B-\frac{1}{2h}p(t_0)+\left(\frac{4-\lambda}{24}\right)q(t_0)\right]$$

Now eliminating c_{N+1} from first equation of (18.24) and (18.26), we have

$$c_N\left[\frac{2+\lambda}{2h^2}-\frac{8+\lambda}{24h}p(t_N)\right]-c_{N-1}\left[\frac{4-\lambda}{24h}\right]p(t_N)=\beta \tag{18.28}$$

Now (18.24) results in the linear system of $(N+1)$ equations $FX_N = d_N$ in the $N+1$ unknown $X_N=(c_0, c_1, ..., c_N)$ with right side

$$d_N=\left[\left\{\left(\frac{2+\lambda}{2h^3}\right)B-\frac{1}{2h}p(t_0)+\left(\frac{4-\lambda}{24}\right)q(t_0)\right\}D, 0, 0, 0, ..., \beta\right]^T.$$

and the matrix F is given by

$$F = \begin{bmatrix} f_{1,1} & f_{1,2} & 0 & 0 & \cdots & 0 \\ f_{2,1} & f_{2,2} & f_{2,3} & 0 & \cdots & 0 \\ 0 & f_{3,2} & f_{3,3} & f_{3,4} & 0 & \cdots \\ 0 & 0 & f_{4,3} & f_{4,4} & f_{4,5} & \cdots \\ \vdots & \vdots & \vdots & \vdots & \vdots & \vdots \\ 0 & \cdots & 0 & f_{N,N-1} & f_{N,N} & f_{N,N+1} \\ 0 & \cdots & 0 & 0 & f_{N+1,N} & f_{N+!,N+1} \end{bmatrix}, \tag{18.29}$$

where $f_{1,1} = [(\frac{2+\lambda}{2h^2})(A - \frac{B}{h}) + (\frac{8+\lambda}{24h})(q(t_0)B - p(t_0)A)]$,

$$f_{1,1} = \left[\left(\frac{2+\lambda}{2h^2}\right)\left(A - \frac{B}{h}\right) + \left(\frac{8+\lambda}{24h}\right)(q(t_0)B - p(t_0)A)\right],$$

$$f_{1,2} = \left[\left(\frac{2+\lambda}{2h^3}\right)B + \left(\frac{4-\lambda}{24h}\right)(q(t_0)B - p(t_0)A)\right],$$

$$f_{i,i-1} = \frac{2+\lambda}{2h^2} - p(t_i)\frac{1}{2h} + q(t_i)\frac{4-\lambda}{24},$$

$$f_{i,i} = -\frac{2+\lambda}{h^2} + q(t_i)\frac{8+\lambda}{12},$$

$$f_{i,i+1} = \frac{2+\lambda}{2h^2} + p(t_i)\frac{1}{2h} + q(t_i)\frac{4-\lambda}{24},$$

$$f_{N+1,N} = -\left(\frac{4-\lambda}{24h}\right)p(t_N),$$

$$f_{N,N+1} = \left[\left(\frac{2+\lambda}{2h^2}\right) - \left(\frac{8+\lambda}{24h}\right)p(t_N)\right].$$

The matrix F is a tridiagonal matrix. The solution of system $Ft_N = d_N$ together with Eqs. (18.27) and (18.28) give values of c_i, s. Then the required B-spline solution is obtained.

18.4 Numerical Examples

In this section, two numerical examples are given to show the applicability and efficiency of the proposed method.

Example 1 Consider the boundary value problem

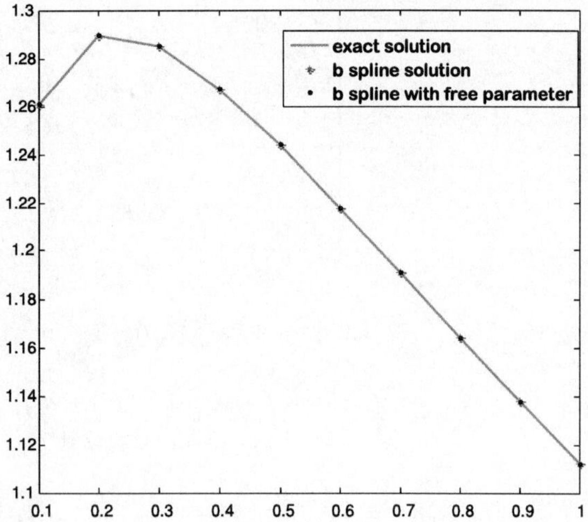

Fig. 18.1 Comparison of B-spline and B-spline with free parameter with the exact solution

$$2t(1+t)\ u''(t) + (1+5t)u'(t) + \ u(t) = 0, \text{ s.t. } u(0) = 1.0, u(1) = 1.112372.$$

The results, for Example 1 for h = 0.1 are shown in Fig. 18.1. The results exhibit that the above B-spline solution gives a good approximation to the exact solution. B-spline with free parameter further improves the results for a suitable choice of parameter.

Example 2 Consider the BVP

$$4t(1+t)\ \ u''(t) + (3+11\ t)\ u'(t) + u(t) = 0, \text{ with } u(0) = 1, u(1) = 1.177701.$$

Numerical results for Example 2 for h = 0.1 are shown in Fig. 18.2. The results show that the above B-spline solution gives a good approximation to the exact solution. B-spline with free parameter further improves the numerical results.

18.5 Conclusion

An effective computational technique based on B-spline to get a solution of singular boundary value problems is proposed. The numerical results attained by cubic B-spline approximate exact solutions very well. The algorithm based on B-spline with free parameter is also formulated. Results show that B-spline with free parameter gives improved for suitable choices of free parameter.

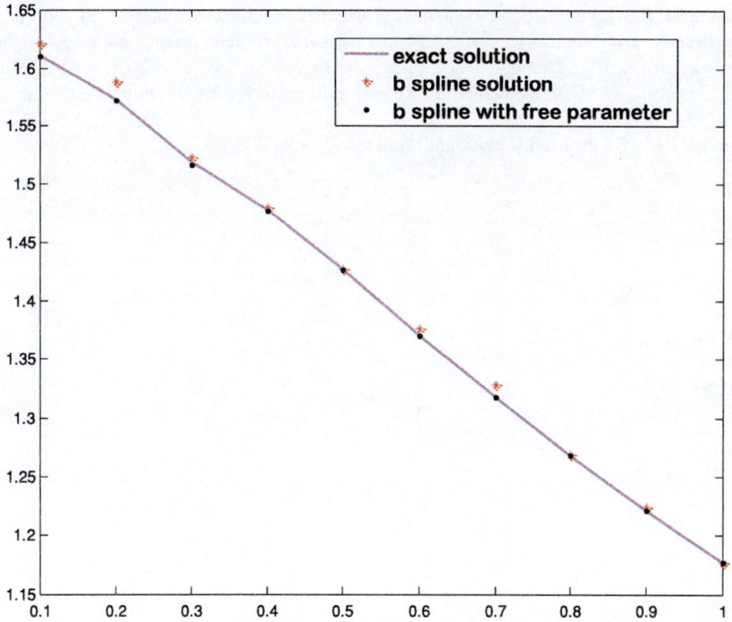

Fig. 18.2 Comparison of B-spline and B-spline with free parameter with the exact solution

References

1. Hoog, F.R., Weiss, R.: Difference methods for boundary value problems with a singularity of the first kind. SIAM J. Numer. Anal. **13**, 775–813 (1976)
2. Roul, P., Goura, V.P., Agarwal, R.: A compact finite difference method for a general class of nonlinear singular boundary value problems with Neumann and Robin boundary conditions. Appl. Math. Comput. **350**, 283–304 (2019)
3. Reddien, G.W.: Projection methods and singular two point boundary value problems. Numer. Math. **21**, 193–205 (1973)
4. Wazwaz, A.M.: The variational iteration method for solving nonlinear singular boundary value problems arising in various physical models. Commun. Nonlinear Sci. Numer. Simul. **16**, 3881–3886 (2011)
5. Kumar, M., Singh, N.: Modified adomain decomposition method and computer implementation for solving singular boundary value problems arising in various physical problems. Comput. Chem. Eng. **34**, 1750–1760 (2010)
6. Roul, P.: A new mixed MADM-collocation approach for solving a class of Lane-Emden singular boundary value problems J. . Math. Chem. **57**, 945–969 (2019)
7. Verma, A.K., Pandit, B., Agarwal, R.P.: On approximate stationary radial solutions for a class of boundary value problems arising in epitaxial growth theory. J. Appl. Comput. Mech. **6**(4), 713–734 (2020)
8. Kumar, M., Gupta, Y.: Methods for solving singular boundary value problems using splines: a review. J. Appl. Math. Comput. **32**, 265–278 (2010)
9. Gupta, Y., Kumar, M.: A computational approach for solution of singular boundary value problem with applications in human physiology. Natl. Acad. Sci. Lett. **35**(3), 189–193 (2012)
10. Gupta, Y., Srivastava, P.K., Kumar, M.: Application of B-spline to numerical solution of a system of singularly perturbed boundary value problems. Math. Aeterna **1**, 405–415 (2011)

11. Chaurasia, A., Gupta, Y., Srivastava, P.C.: Numerical scheme based on non-polynomial spline functions for the system of second order boundary value problems arising in various engineering applications. J. Appl. Comput. Mech. https://doi.org/10.22055/JACM.2020.32435.2012
12. Xu, G., Wang, G.: Extended cubic uniform B-spline and [alpha]-B-spline. Acta Autom. Sin. **34**, 980–984 (2008)
13. Prenter, P.M.: Splines and Variational Methods. Wiley (1989)

Chapter 19
Emotional Intelligence: Assessing Its Impact on Financial Productivity in an Organization

Shivi Mittal, Prabhat Srivastava, and Deepa Gupta

Abstract In any organization the ultimate result for growing its wealth depends upon how wealthy the organization's environment is. Though the main objective of any organization is profit maximization but in further stages of stability and growth, wealth becomes the ultimate objective. No doubt, if interpersonal relations between employer and employee are good, the financial productivity would be maximum. This concept will be better explained with the help emotional intelligence in context with financial outcome. The main objective of this paper is to study the factors of emotional intelligence which are responsible for affecting financial productivity of an organization from the previous studies. The Library research approach is being used to carry this study. It has been observed from the literature review that not only technical skills are important to grow for an organization but the emotional ability of people to connect with their peer group is more crucial. The purpose of the study is to bring previous research on Emotional Intelligence from the past with the current models and explanations in this context. Through this study, we are trying to look for the prospects and trials with which the financial effectiveness through emotional intelligence has been increased through various factors. These factors will be discussed further in detail in order to provide the viewpoint and expected outcome. As the study is conceptual and is being carried out from the previous studies, its critical and crucial aspects would also be examined.

S. Mittal (✉)
Department of Management Studies, G.L. Bajaj Institute of Technology & Management, A.P.J. Abdul Kalam Technical University, Greater Noida, India

P. Srivastava
Shri Venkateshwara University, Gajraula, Uttar Pradesh, India

D. Gupta
Department of Management Studies, G.L. Bajaj Institute of Management & Research, A.P.J. Abdul Kalam Technical University, Gautam Buddha Nagar 201306, India

© The Author(s), under exclusive license to Springer Nature Singapore Pte Ltd. 2022
V. V. Rao et al. (eds.), *Computational and Experimental Methods in Mechanical Engineering*, Smart Innovation, Systems and Technologies 239,
https://doi.org/10.1007/978-981-16-2857-3_20

19.1 Introduction

The thought got a further lift with the release of a book by Daniel Goleman, Emotional Intelligence: Why it can Matter More than IQ (1995) [1]. In another book, Working with Emotional Intelligence, Goleman [2] concentrated on the requirement for passionate knowledge grinding away, a region regularly thought to be more head than heart. The topic "Emotional Intelligence" has a delicate issue in management and psychology disciplines which is still mushrooming and captures the interest of professionals, academia and researchers [3].

The thought isn't constrained to the administrators and pioneers of the association. Any activity that requires managing individuals requires the contribution of enthusiastic knowledge so as to build more proficiency grinding away at work and at home as well. Whereas intelligent quotient (IQ) is relatively fixed, scholars suggest that emotional intelligent quotient (EIQ) needs to be given more attention.

19.2 Objectives

- To comprehend the effect of emotional intelligence on the work performance of an employee.
- To understand the employer–employee relationship.
- To understand the direct relationship and effects between emotional intelligence quotients and financial productivity of the organization.
- To examine the non-intrinsic factors by correlating them with the intrinsic factors.

19.3 Research Methodology

Based on the foregoing discussion, this paper is aimed at exploring the linkage between emotional intelligence and increasing working efficiency of employees in an organization. Using a library research approach, the papers review both empirical and conceptual papers on emotional intelligence, emotional quotient, financial productivity and working efficiency. Secondary data has been used to observe and correlate the views. It examines these two concepts as emotional intelligence of the people really affects the working efficiency and the overall development of any organization as well.

19.4 Emotional Intelligence

Capability in EI is turning into an essential is drawn out in extreme zones of 'emotional work' such as nursing, social work, the service industry and management roles. High EI improves the physical and mental strength of people and supports academic and business execution [4]. Emotional intelligence or emotional quotient is one of the key elements of behavioral finance. The concept was first introduced in the scientific paper of psychology. Emotional intelligence was quickly introduced to the needs of commercial working life. Entrepreneur.com (May 8, 2015) reported that workers with high emotional intelligence are better at working in teams, adjusting to change and being flexible. It is very much important at the workplace to understand the emotions of others and show your emotions to others as well. No doubt, by using this trait one should be more emotionally empowered than others. He has an articulate personality and the ability to understand what others think. When both sender and the receiver think in the same way, of course their aim would be common. Emotional intelligence or emotional literacy is a set of the capability and competitiveness which is being required to understand the intrinsic nature of the person, usually called the soft skills or interpersonal skills that is a counterpart of technical skills, knowledge, work ethics, artificial and general intelligence and professional skills. Goleman (1995) perceived five different classifications of abilities that form the key attributes of EI and recommended that, unlike one's intelligence quotient (IQ), these classified abilities can be learned where absent and improved upon where present [5]. So, after understanding the concept we need to focus upon how EI is related to the financial progress or the working efficiency of any organization. Does it really help? Does it matter when it comes to the overall organizational success?

19.5 Financial Productivity and Work Efficiency

The vast majority of the analysts propose that individuals having high emotional intelligence show signs of better interpersonal relation, work more efficiently and adequately and spend sound time and healthy life than those people having low emotional intelligence.

Emotional intelligence largely shapes the behavior of the leader and determines his or her leadership effectiveness Goleman [6].

Emotional Intelligence gives the definition by which the EI is: "involves a combination of competences which allows a person to be aware of, to understand and to be in control of their own emotions, to recognize and understand the emotion of others, and to use this knowledge to foster their success and the success of others".

The employees are called as the lifeblood of any organization. If the organization better understands the feelings of its employees, the efficiency of the job is being increased and in turn, increases its overall efficiency. Kambiz and Mazid [7] found in their Iranian study that there is a positive correlation between emotional intelligence, job satisfaction and organizational citizenship behavior.

Jung and Hoon [8] in their study prove that counterproductive work behavior (CWB), emotional labor, emotional exhaustion, affectivity are directly related with emotional intelligence (EI). They work on the five factors model and able to find the connectivity between the efficiency of the employees and the productivity of an organization.

Keshav and Mahajan in their paper "Relationship between Emotional Intelligence and Organizational Citizenship Behavior among bank employees" (2017) found that emotional intelligence was positively associated with organizational citizenship behavior [9].

19.6 How Emotional Intelligence and Working Efficiency of an Organization Are Correlated with Each Other

In this fast going world, the benefit of using emotional intelligence along with intelligence quotient is being clearly seen at workplaces. There people find it more useful in order to reduce their stress. They are able to maintain a balance in their family lives a well as at workplace. Psychologically they feel more fit and healthy. As observed from the studied literature, stress is one of the most critical factors in affecting the work efficiency of people. Individual's work efficiency will have direct impact on the financial productivity of any organization. Emotional intelligence became a popular notion after the article by Peter Salovey and John D. Mayer was published. Daniel Goleman educated a broader audience through his book "Emotional Intelligence", where he broke down the concept for everyone to understand. Since then, people have tried inculcating these habits in their daily lives. They then realized its benefits help both at the workplace and at home. It helps them make better decisions and live psychologically healthier lives.

As digital transformation has made work so easier for people, with which they can now concentrate more on their career growth and overall growth of an organization. Emotional intelligence is a vital tool to identify the true leaders as they can now more focus on the decision making roles than paperwork. They are required to be good with number of people. Here, they will have a say in how the company should function. Here emotional intelligence plays a crucial role in understanding the nature, the behavior and their conversation they have. Most importantly, if the leader must have qualitative skill, i.e., emotional intelligence, he will be able to manage and getting things done on time. Additionally, it will also help them turn any situation into a positive solution [10].

In any organization, there is a connect from one department to the other which hugely requires a good working team and it is only possible when that connect is a good one. As a result, it becomes all the more important for the organization to up their game and become emotionally aware.

19.7 Intrinsic Factors Affecting Financial Productivity on Using Emotional Intelligence

It is the need of the hour that every leader must understand that time has changed. Qualitative skills are more important than Quantitative skills. It means emotional intelligence is of much importance along with one's knowledge. Let's discuss the factors which show the relationship between the organizational productive behavior and its soundness depends upon the emotional behavior of its employees. This can be explained through the following model (Fig. 19.1).

- Self-awareness and regulation: it means that the person before understanding the other person should know about himself/herself. If he is short-tempered, he will only see the negative side of the people but on the other hand if he is patient, he will also be patient in listening to others in better way. Basically, a person should know about his strengths and weaknesses first in order to know other's behavior.
- Awareness about others: after knowing about oneself, a person will now be able to understand others behavior in a better way. It is very important to note here that one's mental health is sound only when he is maintaining the balance between his family and work life. Nothing is more important than a family for a human being. So the employer needs to talk to his employee in every aspect of life. To understand their psychological standpoint is essential to create a better work environment and this is the key for motivation in the employees to work for the organizations and hence increase overall productivity.

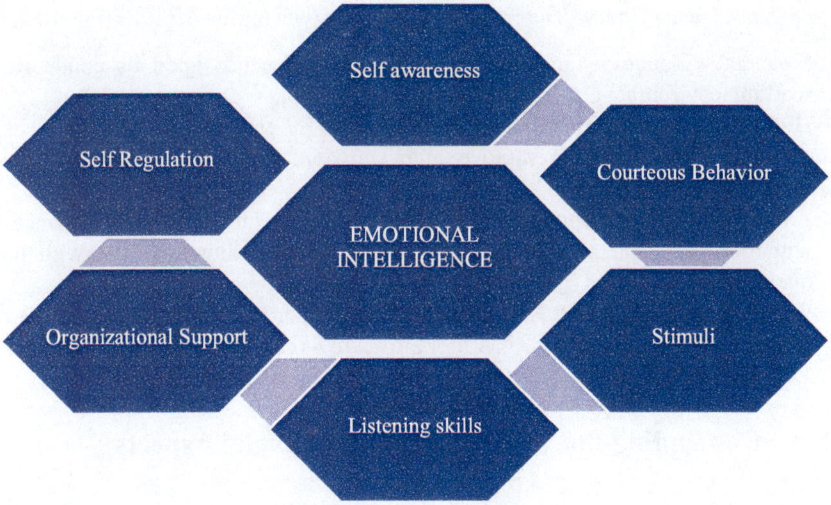

Fig. 19.1 Intrinsic factors and emotional intelligence

- Listening skills: listening to what others say is an important task in any organization though difficult. But to understand their viewpoint will definitely improve the efficiency. Their suggestions and recommendations should be welcome.
- Courteous behavior: giving the tone of respect is taking the same tone. It means the behavior of the person is so humble that there is no question of saying NO to the seniors. This is one thing that prevails harmonious relationship between employer and employee and in turn increases productivity.
- Stimuli: it means the ability to anticipate reactions and respond effectively. It is same as one understands his own thoughts, how to implement that; he should also understand and respond to his employee or subordinates in the same way only. What else? It helps in increasing the happiness, motivation at work and increases the overall productivity of an organization [11].
- Organizational support: support from any organization reduces the stress of the employees. We know that there are various types of stress and causes due to which stress arises. This stress would be reduced if things will be properly communicated. And here all the factors jointly work.

19.8 Non-intrinsic Factors and Overall Financial Productivity: A Correlation

Not only intrinsic factors are important to increase the overall efficiency of an organization but non-intrinsic factors play a motivational role in job satisfaction of an employee which enhances overall productivity of an organization. Let us discuss the non-intrinsic factors that will be crucial to correlate it with intrinsic factors (Fig. 19.2):

- Financial productivity of any organization also depends upon its conducive working environment.
- Higher the transparency is there in the working of the environment, higher will be the work satisfaction of an employee and in turn will surely affect the productivity in a positive manner.
- As far as salary and appraisals are concerned, financial productivity enhances with motivation and motivation comes from finance. It is directly related with the quality work produced by employees.
- Working hours and good HR practices on the other hand prove to secure best place in one's mind to put his/her best efforts to work.

19.9 Concluding the Viewpoint and Its Crucial Aspects

- Emotional intelligence plays a vital role in strengthening, not only the employees relationship with the organization but it also helps organization to understand the employees well and which always leads to better understanding, sound and effective relationship, motivation, confidence, sincerity, enthusiasm, belongingness

Fig. 19.2 Non-intrinsic factors and organizational environment

toward the organization by the employees and which in turn gives more strength and develop healthy, cordial and productive environment within the organization.

- It is need of time for every organization irrespective of the size and number to involve people into various activities which not only satisfy their emotional things but also give stress-free environment for better productivity.
- Both intrinsic and extrinsic factors can lead to job satisfaction; a pleasurable emotional state which a person can only enjoy at his/her workplace if both these factors can correlate with each other.
- For instance, if a person gets promoted, he/she will be able to balance both the personal and professional life because of monetary gain and motivation as well.
- It has also been observed that EI and EIQ both are important to enhance the financial productivity of the organization. There is a direct relationship of financial productivity with the productivity of the employee [12].
- If the employee is enjoying the stress-free environment, emotionally stable, satisfied workplace and their personal and professional interest both are well taken care, certainly they will enjoy the workplace and once the employee starts enjoying the workplace it always leads to increase in wealth of the employees because of long-stay and wealth of the organization because a happy and satisfied person gives more productivity then distress and demotivated person [13].
- Employee's self-satisfaction and high morale always lead to better prospects both for the employees and organization as a whole.
- Therefore, it is found that emotional intelligence and emotional intelligence quotients both play a vital and important role in the growth of an individual as well as organization as a whole because in today's scenario financial satisfaction

has taken a secondary place as compared to happy, healthy and stress-free environment which is only possible when we understand the employee as a mentee and organization has a whole becomes its mentor.

References

1. Goleman, D.: Emotional intelligence: why it can matter more than IQ. Bantam, New York (2005)
2. Goleman, D.: What makes a leader? Harv. Bus. Rev. 93–102 (1998b)
3. Goleman, D.: Working with Emotional Intelligence. Bantam Books, New York (2006)
4. Bar-On, R.: Emotional and social intelligence: insights from the emotional quotient inventory. In: Bar-On, R., Parker, J. (eds.) The Hand book of Emotional Intelligence. Jossey-Bas, San Francisco (2000)
5. Goleman, D.P.: Emotional intelligence: why it can matter more than IQ for character, health and lifelong achievement. Bantam Books, New York (1995)
6. Goleman, D.: Leadership: The power of emotional intelligence: selected writings. More Than Sound LLC, Northampton, MA (2011)
7. Kambiz, H., Majid, M.: A survey on impact of emotional intelligence, organizational citizenship behaviors and job satisfaction on employees performance in Iranian hotel industry (2013)
8. Jung, H.S., Yoon, H.H.: The effects of emotional intelligence on counterproductive work behaviors and organizational citizen behaviors among food and beverage employees in a deluxe hotel. Int. J. Hosp. Manag. **31**(2), 369–378 (2012)
9. Keshav, Mahajan: Relationship between emotional intelligence and organisational citizenship behaviour among bank employees, Jammu (2017)
10. Brown, F.W., Bryant, S.E., Reilly, M.D.: Does emotional intelligence—as measured by the EQ-I influence transformational leadership and/or desirable outcomes? Leadersh. Organ. Dev. J. **27**, 330–351 (2005)
11. https://positivepsychology.com
12. Razi, S., Atif, H.: Relationship between emotional intelligence and employee turnover rate FMCG organization. Pakistan (2013)
13. Sunil, K., Rooprai, K.Y.: Role of emotional intelligence in managing stress and anxiety at workplace. In: ASBBS Annual Conference, Las Vegas, vol. 16, p. 1 (2009)

Chapter 20
Acquiring FG Homogeneous Composite Shell Structure Using Finite Element Analysis

Aman Sharma, Rajat Yadav, and Vikas Kumar Sharma

Abstract Covered composites have different preferences over customary materials, for example, high explicit solidness and lightweight. The significant disadvantage is a shortcoming of boundaries among contiguous films, known as the de-lamination phenomenon. Because of de-lamination, a disappointment for the structure of composite can be seen because of the impacts of bury laminar burdens developed, that is once overlapped composites are exposed to outrageous temperatures. If the unforeseen difference in material properties of the covered composite structure is anticipated these issues can be decreased. along with specific bearings in a controllable manner, the individual material organization differs consistently practically reviewed material (FGM) is that sort of material. Consequently, this article presents the limited component displaying and examination of practically reviewed (FG) shell structures under various stacking, for example, mechanical and warm. Moreover, the introduction of investigation of free vibration of structure of FG circular shell. Further, various kinds of shells are considered to contemplate the impacts of significant boundaries on the reactions of FG shell structures. The reactions were acquired for standardized shells of unadulterated metal (EN 31 steel) shells and FG Functionally graded shells which are contrasted. Besides, both modular and static investigations are done to decide the characteristic frequencies, miss happening, strain, longitudinal pressure, and circumferential pressure, individually.

A. Sharma · R. Yadav (✉) · V. K. Sharma
IET Department of Mechanical Engineering, GLA University, Mathura, India
e-mail: rajat.yadav@gla.ac.in

A. Sharma
e-mail: aman.sharma@gla.ac.in

V. K. Sharma
e-mail: vikash.sharma@gla.ac.in

© The Author(s), under exclusive license to Springer Nature Singapore Pte Ltd. 2022
V. V. Rao et al. (eds.), *Computational and Experimental Methods in Mechanical Engineering*, Smart Innovation, Systems and Technologies 239,
https://doi.org/10.1007/978-981-16-2857-3_21

20.1 Introduction

In different fields of designing, for example, common, mechanical, aviation and atomic building fields the slender walled tube-shaped shells find more extensive applications as essential auxiliary individuals [1, 2]. The hardened and unsolidified shells comprised of metallic and overlaid composite materials (huge width to thickness proportion) are broadly utilized in a submerged, surface, air and space vehicles just as in development of weight vessels, stockpiling vessels, stockpiling canisters and fluid stockpiling tanks [3, 4]. The mathematical defects because of assembling measures takes a prevailing function in diminishing the clasping heap of tube-shaped shells. Clasping is frequently seen as the controlling disappointment method of these structures because of its moderately little thickness of these basic individuals [5, 6]. Reaction of composite structures exposed to blast has been a field of extreme action of specialists in ongoing decades. So composite plates and shells structure one of the essential components of the structures, in this way, considering the shoot reaction of such structures helps to understand and improve their impact opposition [7, 8]. A shell structure by prudence of its calculation can convey applied loads fundamentally by its immediate anxieties lying in their plane with small bowing. The shell activity is a blend of membrane action due to in-plane direct stresses and bending action due to flexural stresses, which attribute to high strength of shell structures [9]. A pre-twisted conical shell is a special case wherein the curvature of mid-surface in one orthogonal direction is absent and the curvature of the mid-surface in other orthogonal direction and the curvature due to twist are of non-zero magnitude. Thus, the resulting surface of such a pre-twisted shell is conical helicoids [3]. The pre-twisted composite shells are structural elements of immense engineering significance. Due to its high strength and stiffness, extensive uses of composites can be found in aerospace, mechanical, marine, civil and other high-performance weight-sensitive applications. Most of the turbomachinery blade configurations used in practice like blades of a fan, compressor, gas turbine, steam turbine, water turbine, marine propeller, windmill and helicopter and flow guide vane are typically very complicated because of their geometry [10]. The disc and its attached blades rotate about an axis perpendicular to the plane of the disc. The failure of a blade in turbomachines The result of blade vibration problems also occurs. Knowledge of these frequencies is important. The blades are also subjected to centrifugal body forces arising out of the rotation. Due to centrifugal force, the initial stress system affects natural frequencies appreciably [11].

The conduct of composite structures against the touchy marvel has been researched utilizing a limited component strategy. Some composite shells, for example, composite plates and sides of the equator with various layer-increasing have been examined utilizing LS-DYNA programming. The impact stacking is reenacted by the blast's weight versus time bends and is unequivocally characterized in LS-DYNA programming [12]. The Tsai-Wu disappointment rule is utilized to foresee the conduct of the composite structure. In this paper, the impact of layer-increasing on the shoot obstruction of the structure is explored. The outcomes show that side of the equator composite has better execution against the shoot stacking than plate and disappointment happens under a more noteworthy burden [13]. Additionally, it is demonstrated that edge utilizes composite structures have great obstruction in examination with cross employs one. Creators examined the impacts of the length, division edge and diverse limit conditions on the clasping burden and post clasping conduct of CK20 tube-shaped boards utilizing exploratory and mathematical strategies. The trial tests have been performed utilizing the INSTRON 8802 servo water-driven machine and for mathematical investigation. Abaqus limited component bundle has been utilized [14]. Default panels and films are refused [15]. Shut shells and bended boards of isotropic, composite coated, piezo-electric, virtually assessed and hyper-flexible materials and nonlinear vibrations of shells exposed to normal and flat excitations are particularly considered. Hypothetical, mathematical and research research is tender to handle dynamic problems, including parametric vibrations, power, dynamic clasping, non-stationary vibrations and disturbing vibrations. Besides, a few unique parts of nonlinear vibrations of shells and boards [16, 17].

20.1.1 Advantages and Disadvantages of Shell Structures

- High-quality regarding its self-weight. (Significant rules to gauge proficiency).
- Aesthetic esteem.
- Lowers the development cost.
- Large space covered.
- Very High Stiffness.

Disadvantages

- Leakage might be seen/prompted because the concrete is the permeable material.
- To include another floor above it, it is unimaginable to expect, because of the mathematics behind this. Essentially, it is the covering structure for rooftop.

20.1.2 Applications

Shells are fundamental auxiliary components of present-day innovation and regular day-to-day existence. Instances of shell structures in innovation incorporate vehicle bodies, water and oil tanks, pipelines, storehouses, wind turbine pinnacles, and nanotubes. Nature is brimming with living shells, for example, leaves of trees, sprouting blossoms, shells, cell layers or wings of creepy crawlies. In the human body supply routes, the eye shell, the stomach, the skin and the pericardium are generally shells too.

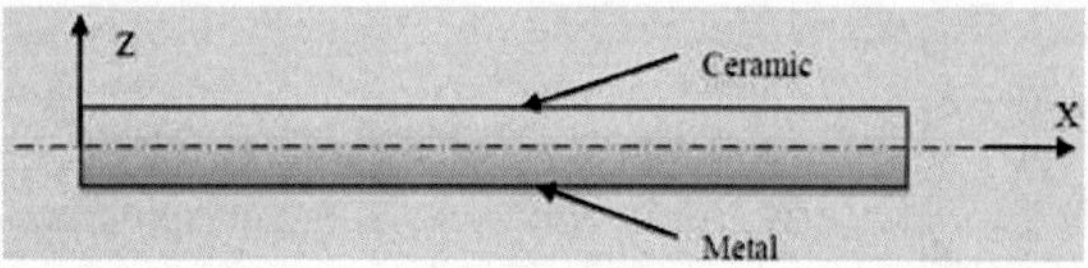

Fig. 20.1 Schematic representation of FG beam

20.2 Functionally Graded Materials

Functionally Graded Materials (FGMs) are a gathering of inhomogeneous materials made out of at least two materials designed to have constantly differing material properties along favored headings. The FG materials are minutely heterogeneous and are produced using a blend of at least two materials that are fitting to accomplish the ideal targets. The general material properties of the FGMs are interesting and not quite the same as the individual material that structures it. Figure 20.1 shows a bar made of such FG material made out of two materials, artistic and metal. The top and base surfaces of this FG shaft are viewed as clay and metal-rich, separately, and the material properties change over its thickness in a smooth and persistent-way. Such a bar can withstand high-temperature inclination over its thickness while keeping up the basic quality and crack sturdiness.

The support in composites utilized as auxiliary materials in numerous aviation and car applications is commonly appropriated consistently. Practically reviewed materials (FGMs) are being utilized as an interfacial zone to improve the holding quality of layered composites. They have accordingly pulled in impressive consideration as of late. One of the benefits of FGMs over overlays is that there is no pressure develops at sharp material limits because of nonstop material property variety to wipe out potential auxiliary.

20.3 Review of Shell Theories

By and large, shell structures are portrayed as the three-dimensional bodies limited by two, generally close, bended surfaces. The greater part of the shell speculations (flimsy and thick, profound and shallow) decrease the three-dimensional flexibility conditions to the two-dimensional portrayal. This is commonly done by dispensing with the arrange ordinary to the shell surface in the advancement of the shell conditions. The precision of flimsy and thick shell speculations can be built up if these hypotheses are contrasted with the three-dimensional hypothesis of flexibility. In this part, the progressive course of shell research is introduced. For clear appreciation of the orderly and sequential improvement of the various parts of shell research, this segment is additionally separated into particular parts referenced beneath.

20.4 Objective

The pressure dissemination of FGM-made round and hollow shell for given burden conditions must be investigated. Henceforth in this work, demonstrating and basic investigation have been done on a reviewed barrel-shaped shell with shifting essential synthesis from internal surface to the external surface. Dispersion of volume part of the components is determined utilizing power law for barrel-shaped shell displaying. Static auxiliary investigation of tube-shaped shell which is presented to inner weight performed and results are confirmed with explanatory arrangements (Fig. 20.2).

20.4.1 Modeling and Analysis

CATIA is a computer-aided 3D interactive application abbreviation. It is one of the major 3D programming used in various airline, car and buyer businesses by associations. CATIA is a multi-stage 3D programming suite created by Dassault Systems, incorporating CAD, CAM just as CAE. Dassault is a French building monster dynamic in the field of flying, 3D plan, 3D computerized models and item life-cycle the executives (PLM) programming. CATIA is a strong displaying device that joins the 3D parametric highlights with 2D instruments and furthermore addresses each plan to-assembling measure. Dassault Systems normally offers new updates, deliveries and bug fixes for every variant. The CATIA programming is written in C++ . It runs on both UNIX and Windows. CATIA provides several workbenches which can be called modules roughly (Figs. 20.3 and 20.4).

- **Part Design**: For strong displaying, the most basic workbench is needed. From drawing in a group environment to iterative nitty–gritty planning, this CATIA

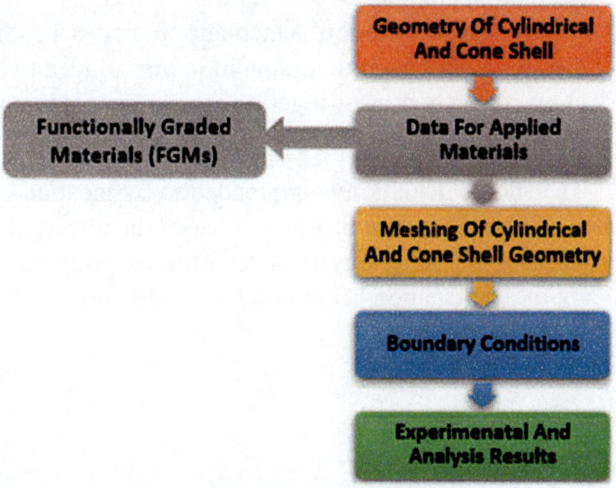

Fig. 20.2 Proposed methodology

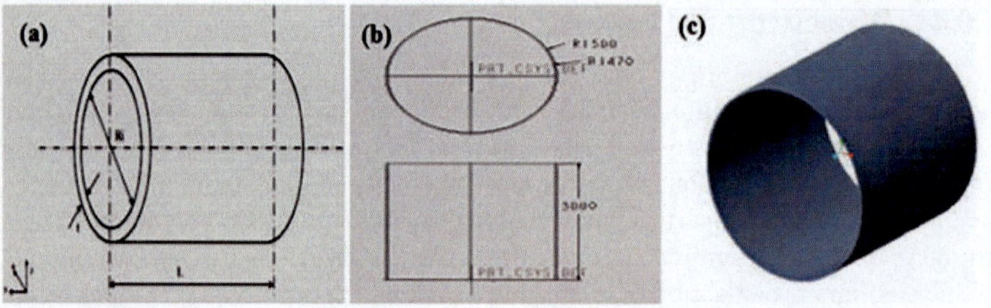

Fig. 20.3 **a** Schematic diagram of cylindrical shell, **b** 2D model of cylindrical shell, **c** Model of cylindrical shell

Fig. 20.4 **a** 2D model of cone shell, **b** Model of cone shell

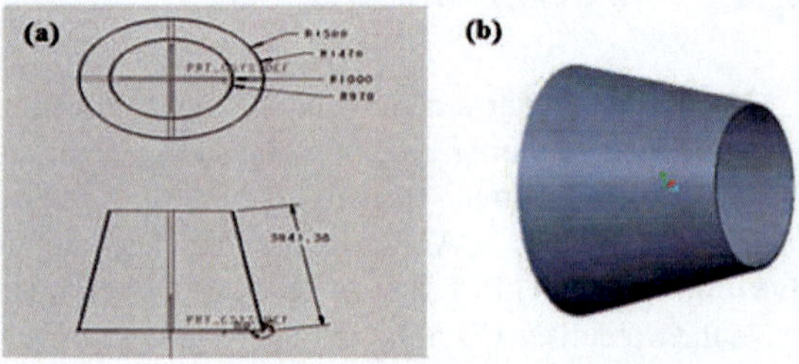

module makes it possible to plan exact 3D mechanical parts with an instinctive and adaptable UI.

- **Generative Shape Design**: Allows both simple and complex shapes with wireframes and highlights on the surface to appear quickly. It provides a huge range of equipment for the production and modification of shapes.
- **Assembly**: Structure, needs, moving congregations and components of nuts and bolts can quickly be adapted. This workshop allows all the components in the form of a computer or section to be linked.
- **Kinematic Simulation**: Kinematics includes a gathering of parts that are associated together by a progression of joints, alluded to as an instrument. These joints characterize how a get-together can perform movement. It tends to the plan survey condition of computerized models.

The shell calculation being tended to by most analysts is the shut barrel-shaped shell wherein the hub of transformation is corresponding to the middle line. The broad use and simplicity of assembling of round and hollow shell are without a doubt the explanation behind such consideration.

20.5 Results and Discussions

Conical shell

In a weight-sensitive application such as aircraft engine turbomachinery that demands extremely high fuel their light-weight, high stiffness and strength make composite materials advantageous. Thus, composite blades can be designed to have a predetermined static and dynamic behavior. The composite blades of pre-twisted shallow conical shells can be manufactured in a mold. Therefore, it is possible to produce complex geometries that result inefficient aircraft turbo propellers. Hence, as a pre cursor to the application of composites in the critical parts of aero-engines or turbo machineries (Table 20.1).

CASE: 1 Cylindrical shell

Material: EN 31

See Figs. 20.5, 20.6 and 20.7.

CASE: 2 Coneshell

Material: EN 31 steel Directional deformation

See Figs. 20.8, 20.9 and Table 20.2.

Structural Functionally Graded Material Ceramic and Metal Aluminum Shell Structures

CASE: 1 Cylindrical Shell Layers

See Figs. 20.10 and 20.11.

Table 20.1 Functionally graded material properties (aluminum alloy and ceramic) for k = 2

Layer number (z)	Young's modulus E (MPa)	Density ρ (Kg/m^3)	Poisson ratio υ
5	220,880	5536	0.31
4	320,720	4830	0.309
3	445,520	4164	0.308
2	595,280	3702	0.307
1	770,000	3307	0.306
−1	96,080	2812	0.304
−2	71,120	2711	0.303
−3	71,120	2711	0.302
−4	96,080	2812	0.301
−5	146,000	2383	0.3

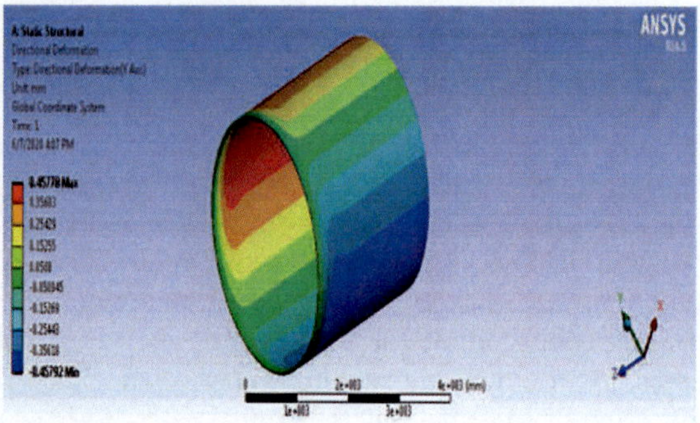

Fig. 20.5 Steel directional deformation

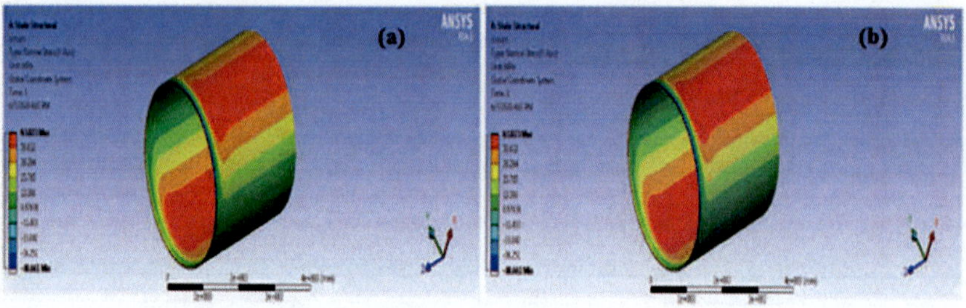

Fig. 20.6 **a** Circumferential stress, **b** Longitudinal stress

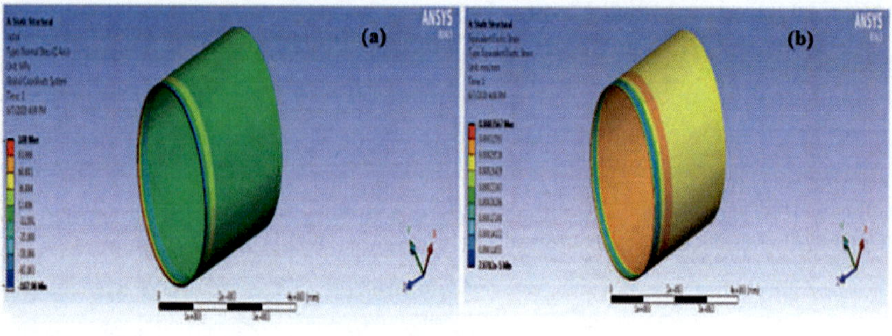

Fig. 20.7 Strain

20.6 Conclusion

For FGM-made structure and single material structure, static auxiliary examination for a barrel-shaped and cone shell was completed. By utilizing power law, volume

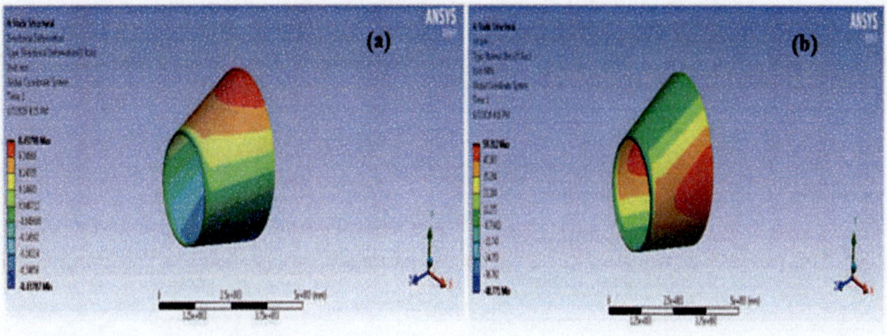

Fig. 20.8 **a** Circumferential stress **b** Longitudinal stress

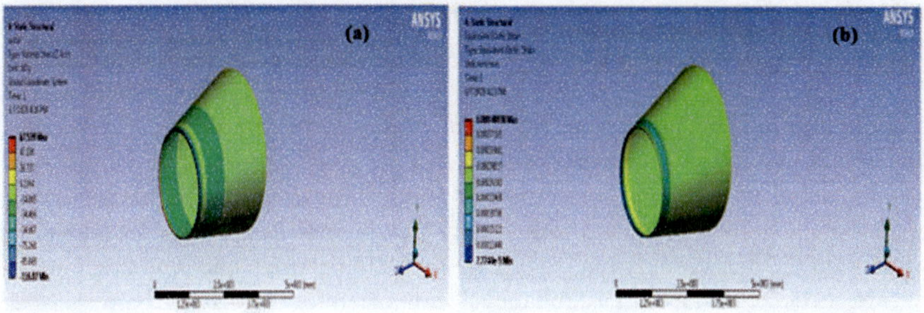

Fig. 20.9 Strain

Table 20.2 Deformation, stress and strain

Shell structures	Directional deformation (mm)	Circumferential stress (N/mm^2)	Longitudinal stress (N/mm^2)	Strain
Cylindrical	0.45778	63.023	108	0.00356
Cone	0.43798	59.312	67.539	0.000408

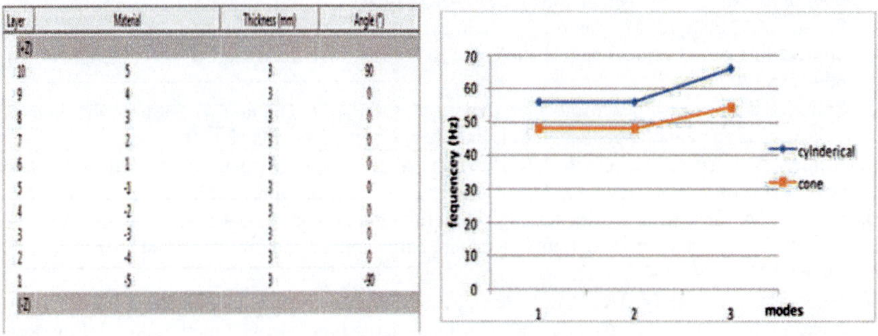

Fig. 20.10 Frequency values of cylindrical and cone shell structures

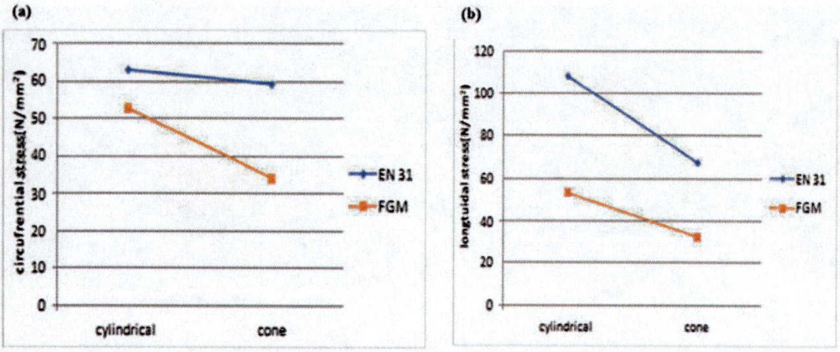

Fig. 20.11 **a** Circumferential stress values of cylindrical and cone shell structures with EN 31 and FGM materials, **b** Longitudinal stress values of cylindrical and cone shell structures with EN 31 and FGM materials

portions and the evaluated structure material properties are determined. To reproduction model, volume division material properties are given for each layer. For the evaluated material structure and single material, static basic examination was performed of cone shell, hollow and round with interior weight load. Rate mistake is found less when with the diagnostic arrangement, reproduction consequences of single material are checked. Henceforth a similar cycle embraced for FGM-made barrel-shaped and cone shell. In FGM-made shell, recreations performed to assess longitudinal and circumferential pressure conveyances. It was seen that over the single material structure at cone shell structure, the reviewed structure shows preferred execution.

References

1. Haghi, R., Behjat, B., Yazdani, M.: Numerical investigation of composite structures under blast loading. J. Mater. Environ. Sci. **8**(6), 2231–2237 (2017)
2. Shariati, M., Sedighi, M., Saemi, J., Allahbakhsh, H.R.: A numerical and experimental study on buckling of cylindrical panels subjected to compressive axial load. J. Mech. Eng. **56**(10), 609–618 (2010)
3. Singh, P.K., Sharma, K.: Molecular dynamics simulation of glass transition behaviour of polymer based nanocomposites. J. Sci. Ind. Res. **77**(10), 592–595 (2018)
4. Kumar, A., Sharma, K., Dixit, A.R.: Carbon nanotube-and graphene-reinforced multiphase polymeric composites: review on their properties and applications. J. Mater. Sci. 1–43
5. Shariati, M., Sedighi, M., Eipakchi, H.: Experimental and numerical studies on buckling and post-buckling behavior of cylindrical panels subjected to compressive axial load. Strength Mater. **43**(2), 190–200 (2011)
6. Narayana, Y.V., Reddy, P.R., Markandeya, R.: Buckling analysis of laminated composite cylindrical shells subjected to axial compressive loads using finite element method. Int. J. Eng. Res. Technol. **2**(1) (2013)
7. Singh, P.K., Sharma, K., Kumar, A., Shukla, M.: Effects of functionalization on the mechanical properties of multiwalled carbon nanotubes: a molecular dynamics approach. J. Compos. Mater. **51**(5), 671–680

8. Singh, P.K., Sharma, K.: Mechanical and viscoelastic properties of in-situ amine functionalized multiple layer grpahene/epoxy nanocomposites. Curr. Nanosci. 14(3), 252–262
9. FarbodAlijani, M.A.: Non-linear vibrations of shells—a literature review from 2003–2013. Int. J. Non-Linear Mech. **58**, 233–257 (2014)
10. Shukla, M.K., Sharma, K.: Effect of functionalized graphene/CNT ratio on the synergetic enhancement of mechanical and thermal properties of epoxy hybrid composite. Mater. Res. Express **6**(8), 085318 (2019)
11. Yadav, R., Chaturvedi, R., Sharma, V.K.: Enhancing and exhausting conventional power module development using magnetic levitation. Eur. J. Mol. Clin. Med. **7**(4), 691–699 (2020)
12. Farhatnia, F., Sharifi, G.A., Rasouli, S. Numerical and analytical approach of thermo-mechanical stresses in FGM beams. In: Proceedings of the World Congress on Engineering, p. 2 (2009)
13. Yadav, A., et al.: Investigating the effects of amine functionalized graphene on the mechanical properties of epoxy nanocomposites. Mater. Today Proc. **11**, 837–842 (2019)
14. Kargarnovin, M.H., Faghidian S.A., Arghavani, J.: Limit analysis of FGM circular plates subjected to arbitrary rotational symmetric loads. In: World Academy of Science, Engineering and Technology, vol. 36, pp. 133–138 (2007)
15. Bambole, A.N., Desai, Y.M.: Hybrid-interface finite element for laminated composite and sandwich beams. Finite Elements Anal. Des. **43**, 1023–1036 (2007)
16. Ray, M.C., Sachade, H.M.: Exact solutions for the functionally graded plates integrated with a layer of piezoelectric fiberreinforced composite. ASME J. Appl. Mech. **73**, 622–631 (2006)
17. Gunes, R., Aydin, M.: Elastic response of functionally graded circular plates under a drop-weight. Compos. Struct. **92**, 2445–2456 (2010)

Chapter 21
An Optimal Control Scheme for Thermal-Hydro System with Distributive Energy Sources

Nagendra Kumar, Brijesh Prasad, Kailash Sharma, Rajat Mehrotra, and Vinamra Kumar Govil

Abstract Wind, solar, fuel cell, small hydro, etc., are emerging technologies for renewable energy and can be built in the future as feasible electricity generation options. This paper introduces the load frequency control (LFC) in the presence of renewable energy resources. The study considers Wind turbine generator (WTG), Solar thermal power system (STPS), Photovoltaic (PV), Diesel source (DEPS), Fuel cell (FC), Battery storage system (BESS), Flywheel (FWES), Ultra capacitor (UC) and Aqua electrolyzer (AE), for sudden load/generation shifts, or both, the power system frequency deviates. The addition of a renewable system also results in a difficult task for an efficient controller design to be realized. With random load variance, the complexity to design an efficient and optimal control scheme is further enhanced. In this study, as an LFC control scheme, a PID controller is used. Two evolutionary approaches named as Big Bang Big Crunch (BBBC) and Harmony Search (HS) have been used to obtain optimal parameters. The performance of the control scheme has been checked two area and three area thermal-hydro system in the presence of other generation and energy storage sources. A comparative assessment of various quantities has been carried out on the basis of different time-domain parameters.

N. Kumar (✉) · B. Prasad · K. Sharma · R. Mehrotra
G. L. Bajaj Institute of Technology & Management, Greater Noida, UP, India
e-mail: nagendra.kumar@glbitm.ac.in

B. Prasad
e-mail: brijesh.prasad@glbitm.ac.in

K. Sharma
e-mail: kailash.sharma@glbitm.ac.in

R. Mehrotra
e-mail: rajat.mehrotra@glbitm.ac.in

V. K. Govil
EE Department, I.E.T Lucknow, Lucknow, India

© The Author(s), under exclusive license to Springer Nature Singapore Pte Ltd. 2022
V. V. Rao et al. (eds.), *Computational and Experimental Methods in Mechanical Engineering*, Smart Innovation, Systems and Technologies 239,
https://doi.org/10.1007/978-981-16-2857-3_22

21.1 Introduction

Currently, a rising trend is the incorporation of distributed generation (DG) into existing power systems to meet increasing power needs [1]. The resulting hybrid power system aims to provide its customers with efficient and high-quality service. In addition, the load demand is highly volatile and it is therefore difficult to preserve the equilibrium of generation and load. If there is some mismatch between generation and load and if it persists for a long time, the device frequency can be affected by this. A control mechanism known as automatic generation control (AGC) can achieve this equilibrium [2]. Therefore, the key contribution of AGC is to sustain system frequency and power within limits by tie-line in the event of any system disruption [3, 4]. In AGC, to maintain equilibrium, the controller plays an important role. Many methods have been used in literature, such as classical, stable, adaptive, optimal, nonlinear, modern [5]. A conventional system comprising thermal/hydro or mixture of both has been studied in the past. The PI controller is used in hybrid system studies to control the output powers of the distributed generation system to achieve power balance conditions due to sudden generation/load variations. The frequency is adversely affected in a hybrid system by load deviations and wind power shifts, so the power balance needs to be maintained. In the literature, small-signal study of a hybrid renewable system with energy storage such as battery, pump storage, to provide rapid response and less oscillation was discussed [6]. The power markets need to be restructured and the power supply policies need to be reformed in the new deregulation climate in the electricity market [7, 8]. In response to the deregulation and re-structuring of the power industry, new business opportunities have led many private sectors, from an economic point of view, to participate in small-scale power generation [7]. The PV integration hybrid power system with FC provides promising advantages over independent service [9]. In hybrid energy systems, the conservation of energy plays an essential role in storing and releasing energy at the right time. In addition to the energy storage device combinations such as FWES-BESS and FWES-UC, the hybrid mixtures of all generation sources such as WTG, PV, FC and DEPS provide effective energy management to meet the power balance requirements [6, 9, 10]. Due to low computing costs and high convergence speed over PSO and differential evolution (DE) optimization methods, PID parameters are obtained using BBBC in this paper [7, 8]. This paper also discusses the use of controller architecture based on HS optimization to achieve minimum deviation in frequency. The performance of both the designed control approaches has been tested on two and three-area power systems with the inclusion of various DG sources. The research examines different profiles of parameters induced by a sudden shift in the demand for generation and load. All potential disruptions, including the erratic nature of wind speed and other uncertainties, have been checked under the proposed control scheme.

21.2 Modelling of the Test System

A basic block diagram of the system comprising various sources like WTG, PV, STPS, etc. is shown in Fig. 21.1. It is seen that these all sources are working in area-1 of the considered test system. The other area does not have any DG source to compensate load deviation. It is also evident that the total power supplied in area-1 is the sum of powers of different DG sources and the conventional thermal generators' power to satisfy the load demand. Figure 21.2 shows the first-order transfer function model of different sources used in DG system. Three WTGs have been used. The output of WTGs and STPS works as an input to AE, which absorbs the fluctuations of wind and generate hydrogen, which further works as an input for FC. The output

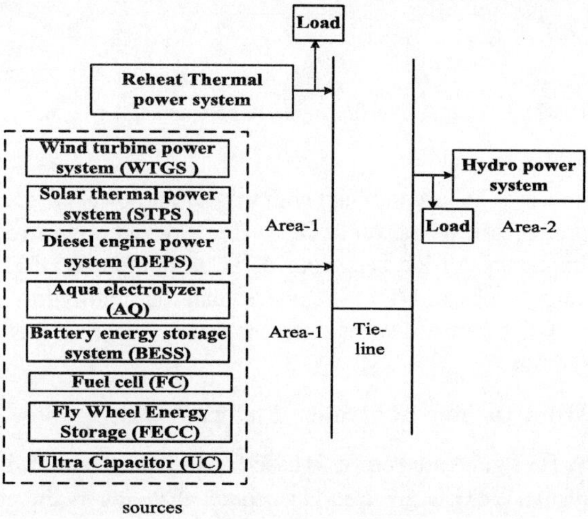

Fig. 21.1 Hybrid system

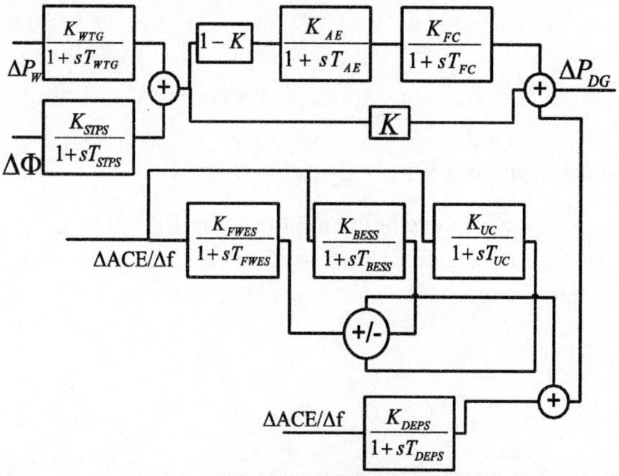

Fig. 21.2 Transfer function model of DG system

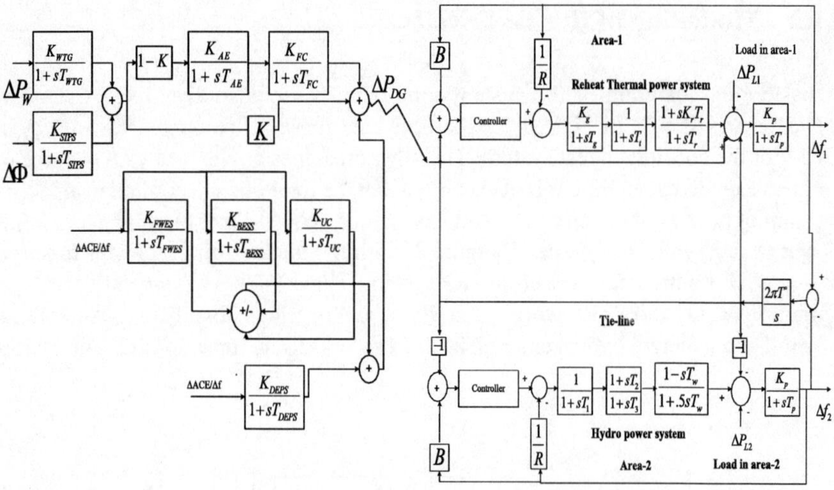

Fig. 21.3 Transfer function model of complete test system

power of FC is combined with the output powers of DEPS, BESS, UC and FWES to produce the net power of DG system. Deviation in Area control Error (ΔACE) or in frequency Δf has been considered as an input to DEPS, BESS, FWES and UC. The complete hybrid (DG + conventional thermal-hydro) system is shown in Fig. 21.3.

Brief description of different sources used in this study has been given in this section.

Wind Turbine Generation System (WTG):

WTG's generated power is based on the speed of the wind. A simple transfer function of the WTG is given in [1], where, ΔP_{WTGk} is the power output and ΔP_W is the input power to the WTG system. Three WTGs systems have been considered in the study.

$$G_{WTGS}(s) = \frac{K_{WTG}}{1 + sT_{WTG}} = \frac{\Delta P_{WTGk}}{\Delta P_W}, k = 1, 2, 3 \ldots$$

Solar Thermal Power Systems (STPS):

Recently, STPS are being explored. In this study ΔP_{STPS} output power of STPS and ΔP_{WTGk}, i.e., output power of WTGs has been added to produce input to the AE system.

$$G_{STPS}(s) = \frac{K_S}{1 + sT_s} \frac{K_T}{1 + sT_T} = \frac{\Delta P_{STPS}}{\Delta \Phi}$$

Fuel-Cell Power Generation (FC):

Fuel cells convert hydrogen directly into electrical energy and considered an important resource due to high efficiency, low pollution, etc., the output power of AE (ΔP_{AE}) is used as an input to FC in the given study, which in turn produced the ΔP_{FCk}.

$$G_{FC}(s) = \frac{K_{FC}}{1 + sT_{FC}} = \frac{\Delta P_{FCk}}{\Delta P_{AE}} k = 1, 2, ...$$

Diesel Engine Power Source (DEPS):

DEPS can be used to supply deficit power to the system to meet load/generation balance.

$$G_{DEPS}(s) = \frac{K_{DEPS}}{1 + sT_{DEPS}} = \frac{\Delta P_{DEPS}}{\Delta ACE}$$

Flywheel Energy Storage System (FWES)/Battery Energy Storage System (BESS):

FWES is a kinetic energy storage device that behaves just like batteries. Another storage device like BESS is also being used to control both active/reactive power output.

$$G_{FWES}(s) = \frac{K_{FWES}}{1 + sT_{FWES}} = \frac{\Delta P_{FWES}}{\Delta ACE}$$
$$G_{BESS}(s) = \frac{K_{BESS}}{1 + sT_{BESS}} = \frac{\Delta P_{BESS}}{\Delta ACE}$$

Ultracapacitors (UC):

UCs are used to store energy during excess generation and produce during peak demand with a capacity of 100 times larger than normal capacitors.

$$G_{UC}(s) = \frac{K_{UC}}{1 + sT_{UC}} = \frac{\Delta P_{UC}}{\Delta ACE}$$

Power Deviation and Frequency Deviation:

The balance of the power frequency is preserved by careful regulation of various resources for power generation, as shown in Figs. 21.2 and 21.3. The balance of power is achieved as given in (24.1),

$$\begin{aligned} \Delta P_{\text{Total}} &= \Delta P_{Conventional} + \Delta P_{DG} \\ \Delta P_{error} &= \Delta P_{Conventional} + \Delta P_{DG} - \Delta P_{Load} \end{aligned} \tag{21.1}$$

Table 21.1 Selected parameters of hybrid system

Gains	Time constant (s)
$K_{WTG} = K_T = K_S = 1$. $K_{AE} = 1/500$, $K_{DEG} = 1/300$, $K_{FC} = 1/100$, $K_{BESS} = -1/300$, $K_{FWES} = -1/100$, $K_{UC} = -7/10$	$T_{WTG} = 1.5$, $T_T = 0.3$, $T_S = 1.5$ $T_{AE} = 0.5$, $T_{FC} = 4$, $T_{DEG} = 2$ $T_{BESS} = 0.1$, $T_{FWES} = 0.1$, $T_{UC} = 0.9$
$K_p = 120$, $K_r = 5$, $K_g = 1$, $K_t = 1$	$T_p = 20$, $T_r = 10$, $T_g = 0.08$, $T_t = 0.3$,
$R = 0.416$, $B = 0.425$	$T_3 = 48.7$, $T_2 = 5$, $T_1 = 0.513$, $T_w = 1$, $T^0 = 0.545$

The net power of different sources utilized in DG system can be written as given in (24.2),

$$\Delta P_{DG} = \Delta P_{WTG} + \Delta P_{STPS} + \Delta P_{FC} - \Delta P_{AE} \pm \Delta P_{FWSS} \pm \Delta P_{BESS} \pm \Delta P_{UC} \tag{21.2}$$

The transfer function of the complete system in respect to frequency deviation can be represented as (24.3),

$$G_{PS}(s) = \frac{\Delta f}{\Delta P_{error}} = \frac{K_p}{1 + sT_p} \tag{21.3}$$

The given equations may be extended, modified as the number of areas or the sources increase. Further to model several sources and conventional system, parameters, i.e., gains and time constants have been chosen from [1] and are given in Table 21.1.

21.3 Controller Design Using Evolutionary Techniques

Since an optimal control scheme is necessary to keep generation and load at balance. Therefore, a PID control scheme using BBBC and HS algorithms has been used in this study. The step-by-step design is given as following.

Due to quick and easy implementation, PID controllers are by far the most common industrial controllers. In this study, A PID control scheme given in (4) has been used as a supplementary controller

$$G_{PID}(s) = K_{Prop} + \frac{K_{Int}}{s} + K_{Dev}s \tag{21.4}$$

To design this control scheme, the mean square of area control error given in (5) has been taken as the optimization problem.

The objective of the control scheme is to eliminate this mean square of error and to bring back the frequency and the tie-line area interchange at the schedule values with less overshoot/undershoot, i.e., oscillations in minimum possible settling time.

$$J_{opt\_fun} = \frac{1}{k}\sum_{i=1}^{k}\left[(ACE_i)^2\right]$$

$$J_{opt\_fun} = \frac{1}{k}\sum_{i=1}^{k}\left[(B_i\Delta f_i + \Delta Ptie_{i\_error})^2\right] \quad (21.5)$$

The constraints have been considered in search of the optimal parameters and are given in [2, 3]. After designing the optimization problem, the optimal parameters have been obtained using two search algorithms namely as BBBC and HS. The brief of both the algorithms is given in as the following:

21.3.1 PID Controller Using BBBC

Step 1: One PID controller is considered for each area. Population for each parameter (K_{Prop}, K_{Int}, K_{Dev}) of the considered PID controllers is generated in this step. This is called Big Bang.

Step 2: It involves in determination of fitness function.

Step 3: Center of mass is evaluated in this step using step 2.

Step 4: New parameters have been generated.

Step 5: Fitness function of newly generated parameters has been determining in this step.

Step 6: This step gives optimum parameters of the controllers.

The complete algorithm with step-by-step modelling can be referred from [3].

A 30-population, 3-variables and 10-parameter limiting size have been used as BBBC parameters to obtain optimal PID control scheme.

21.3.2 PID Controller Using Harmony Search (HS)

PID control scheme coefficients were also obtained using HS algorithm given by Geem et al. [12]. HS is based on the performance process of natural music that occurs when a musician looks for a better state of harmony. HS comprises harmony memory, pitch change and randomization to produce an optimal solution.

In this study, 30-HS memory, 0.9-HS rate, 0.5-pitch rate, (0.0001–1)-BW and 3-variables have been selected to generated optimal parameters of PID control scheme.

21.4 Results and Discussion

The simulation analysis is discussed in this section with the controller mentioned above in the hybrid power system. The area frequency, as well as the tie-line power flow between the control areas, should be regulated by the AGC scheme in an integrated hybrid power system. Therefore, it becomes necessary to investigate the response variation, when existing thermal with different sources (area-1) are interconnected with hydro system (area-2). The PID control scheme is used to eliminate the deviations in frequency and to keep the quantities like frequency, tie-line power, ACE, generation powers at their scheduled value after being affected by a sudden load disturbance. Optimal parameters using both algorithms have been given in Table 21.2.

21.4.1 Case 1: A Sudden Change in Input Power of WTGs and STPS Along with a Sudden Change of Load in the Thermal-Hydro Systems

This case deals with the sudden/step variation in wind input, i.e., $\Delta P_{WTG} = 0.6-0.4$ pu, STPS input, i.e., $\Delta\Phi = 0.38$–0.20 pu, load in thermal area, i.e., $\Delta P_{L1} = 1.0-1.2$ pu and load in hydro area, i.e., $\Delta P_{L2} = 1.0 - 1.1$ pu. These various responses like input powers, frequency deviations, tie-line power and area control error, various generation changes of different generating systems are given in Fig. 21.4. At time t = 30 s, it is seen that once the load is deviating the frequency gets disturbed. It is also seen that the deviation in frequency is settling down at the desired value in steady state. The dip in frequency is more in hydro area than thermal area since it is being fed by DG which is helping to keep balance in generation/load. It is also seen that generation change in thermal areas and others areas are settling to the desired values at steady state. It is also evident that that tie-line power and ACE settle down to the specified values in steady state after being disturbed. It is seen that most of the load demand is taking up by DG sources in thermal system, and by using BESS FWES and UC, the magnitude of oscillations of deviation in frequency and ACE, etc., can be reduced up to a greater amount.

Table 21.2 Optimum values for PID controller

BBBC			HS		
K_{Prop}	K_{Int}	K_{Dev}	K_{Prop}	K_{Int}	K_{Dev}
−1.327	−0.481	−0.627	−1.1	−0.2	−0.4

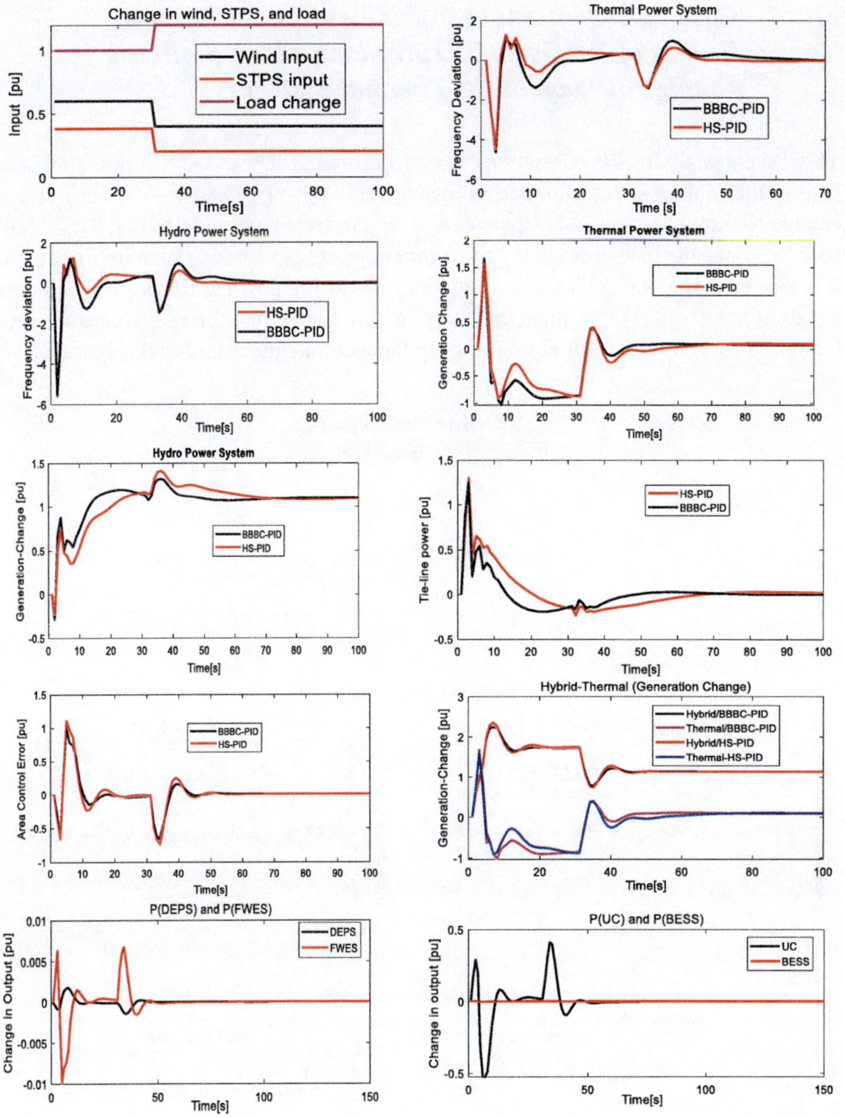

Fig. 21.4 Various responses [pu] of case 1

21.4.2 Case 2: A Continuous (Variable) Change in Input Powers of WTGs and STPS Along with a Staircase Change of Load in the Thermal System

This case deals with the continuous variation in wind input, STPS input, staircase load in thermal area and step load in hydro area, i.e., $\Delta P_{L2} = 1.0 - 1.1$ pu. These changes in input of wind, solar-thermal and load in area-1 are given in Fig. 21.5a. It is seen that once the load is deviating the frequency gets disturbed as seen in Fig. 21.5b. It is also seen that the deviation in frequency is settling down at the desired value in steady state. The dip in frequency is more in hydro area than thermal area since it is being fed by DG which is helping to keep balance in generation/load. Figure 21.5c

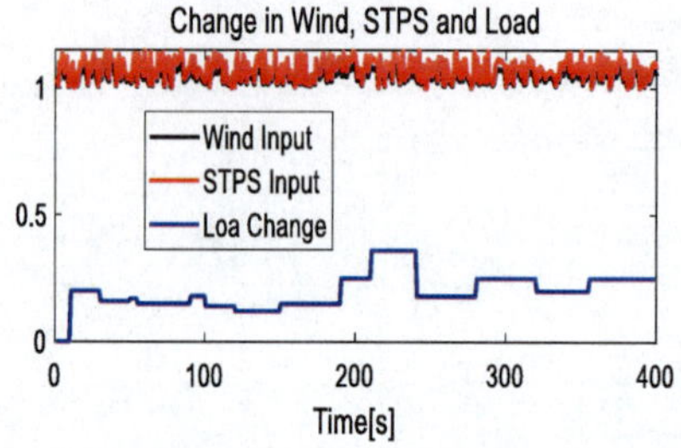

Fig 5(a). Various inputs.

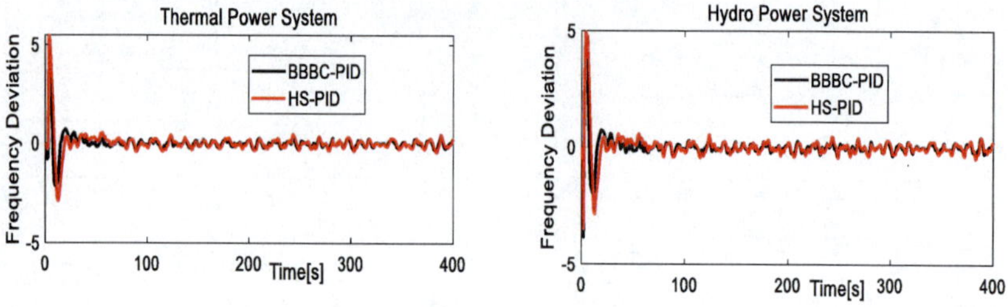

Fig 5(b). Frequency deviations in thermal and hydro areas.

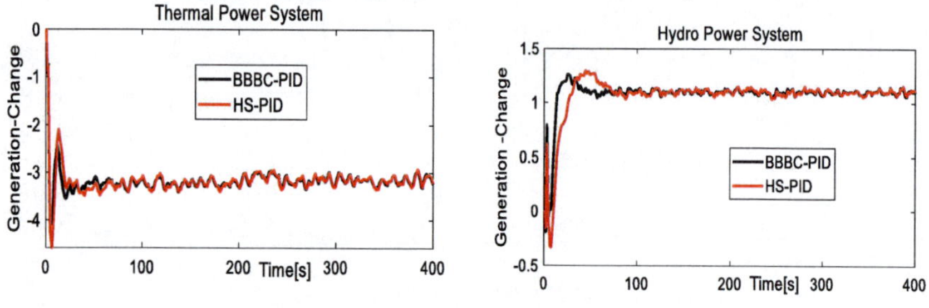

Fig 5 (c). Output of thermal and hydro plants.

Fig. 21.5 Various responses [pu] for case 2

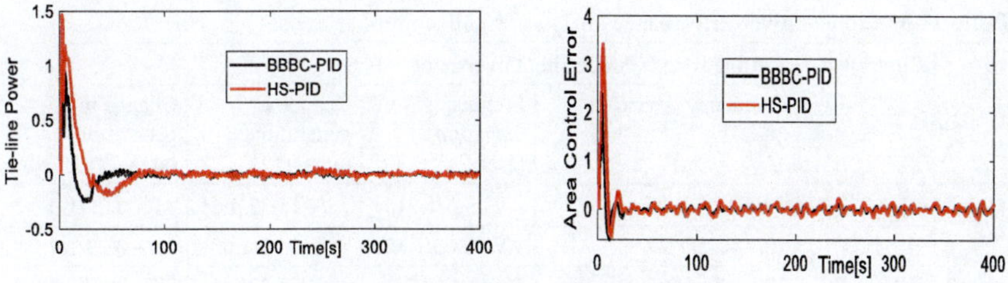

Fig 5 (d). Tie-line power and ACE.

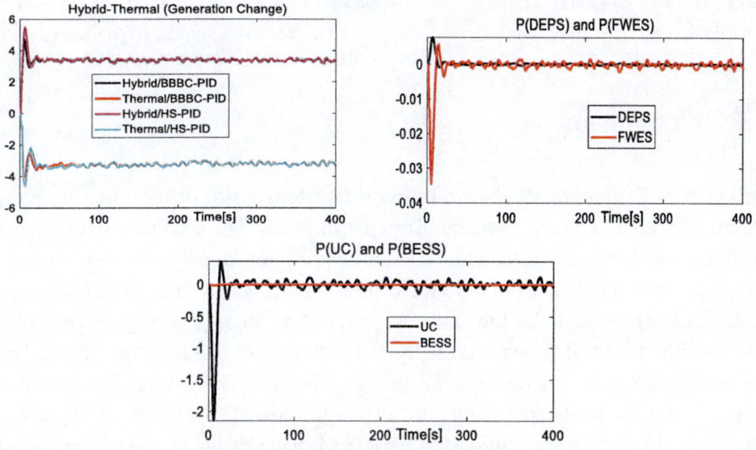

Fig 5 (e). Output of various sources.

Fig. 21.5 (continued)

shows the generation change in thermal-hydro areas using only conventional energy sources. It is evident that the generators in hydro area are settling to the 1.2 pu, as required in steady state. Figure 21.5d shows that tie-line power and ACE settle down to the specified values in steady state after being disturbed. It is seen that most of the load demand is taking up by DG sources in thermal system given in Fig. 21.5e. By using BESS FWES and UC, the magnitude of oscillations of deviation in frequency and ACE, etc., can be reduced up to a greater amount. It is seen that the application of staircase load and continuous input powers of wind and STPS, different quantities having more magnitude in oscillations and are settling at larger settling time, however it is also seen that this condition is also well taken up by both control approaches. However, it is also noted that the consistency of the BBBC-PID controller is almost same and the responses are marked by low overshoot/undershoot, fewer oscillations and faster response. The findings indicate improved output that accommodates the power instability of the WTG and STPS.

Table 21.3 Comparative performance analysis of both control schemes

	Controller	Settling time (s)/undershoot/overshoot			
		Frequency deviation (area-1)	Frequency deviation (area-2)	Change in generation (area-1)	Change in generation (area-2)
Case1	BBBC–PID	65/−4.65/1.05	80/−5.5/1.215	75/−1.032/1.65	75/−0.29/1.3
	HS–PID	70/−4.45/1.27	90/−5.65/1.225	85/−0.92/1.58	90/−0.23/1.4
Case2	BBBC–PID	380/−3.95/2.7	380/−4.55/3	380/−2.07/1.55	380/−0.2/1.38
	HS–PID	405/−4.15/3.3	405/−4.3/2.9	400/−2.09/1.46	405/−1.7/1.48
Case3	BBBC–PID	>400/−2/4.5	>400/−3.8/4.3	>400/−4.2/0.0	>400/−1.8/1.3
	HS–PID	>400/−3/5.5	>400/−3.5/4.9	> 400/−4.6/0.0	>400/−0.35/1.25

21.5 Conclusion

The study of frequency deviation in hybrid systems with different combinations of elements for energy storage is carried out in this paper. The test system is considered to be a two-area thermal-hydro. The DG sources (models of a hybrid generation/energy storage system) have been considered in area-1 of the test system. Wind turbine, diesel set, aqua electrolyzer, solar thermal systems, battery energy storage, flywheels, ultra-capacitor and fuel cell are part of the hybrid system tested. As fuel for the fuel cell, the produced hydrogen from the aqua electrolyzer is used. The wind power, solar thermal and load models are also properly chosen to approximate the substantial performance of the hybrid system being studied. A variety of load change has been considered. The PID controller is constructed to reduce mean square of area control error. BBBC and HS methods have been used to obtain optimum controller parameters. In time-domain simulations, the output of each controller is analysed from the dynamic behaviour of frequency, generation, tie-line power ACE, Output of DG sources. Comparison between the results obtained using both approaches is shown in Table 21.3. It is evident from the comparison that apart from few instances of increased magnitude in oscillations, BBBC-based control.

References

1. Ray, P., Mohanty, S., Kishor, N.: Proportional integral controller based small signal analysis of hybrid distributed generation system. Energy Convers. Manage. **52**, 1943–1954 (2011)
2. Kumar, N., Tyagi, B., Kumar, V.: Multi area AGC scheme using imperialist competition algorithm in restructured power system. Appl. Soft Comput.-Elsevier **48**, 160–168 (2016)
3. Kumar, N., Tyagi, B., Kumar, V.: Optimization of PID parameters using BBBC for multi-area AGC scheme in deregulated power system. Turk. J. Electr. Eng. Comput. Sci. **24**, 4105–4116 (2015)
4. Kumar, N., Tyagi, B., Kumar, V.: Multi-area deregulated automatic generation control scheme of power system using imperialist competitive algorithm based robust controller. IETE J. Res.-Taylor & Francis **64**, 528–537 (2017)

5. Pandey, S.K., Mohanty, S.R., Kishor, N.: A literature survey on load frequency control for conventional and distribution generation power systems. Renew. Sustain. Energy Rev. **25**, 318–334 (2013)
6. Rerkpreedapong, D., Hasanovic, A., Feliachi, A.: Robust load frequency control using genetic algorithms and linear matrix inequalities. IEEE Trans. Power Syst. **18**(2), 855–861 (2003)
7. Tan, W., Xu, Z.: Robust analysis and design of load frequency controller for power systems. Elect. Power Syst. Res. **79**, 846–853 (2009)
8. Debbarma, S., Saikia, L.C., Sinha, N.: Automatic generation control using two degree of freedom fractional order PID controller. Int. J. Elect. Power Energy Syst. **58**, 120–129 (2014)
9. Ghosal, S.P.: Optimization of PID gains by particle swarm optimization in fuzzy based automatic generation control. Elect. Power Syst. Res. **72**, 203–212 (2004)
10. Elmas, C., Yigit, T.: Genetic algorithm based on-line tuning of a PI controller for a switched reluctance motor drive. Electr. Power Comp. Syst. **35**(6), 675–691 (2007)
11. Li, H., Liu, S., Ji, H., Yang, D., Yang, C., Zhao, H.C.B., Hu, Y., Chen, Z.: Damping control strategies of inter-area low-frequency oscillation for DFIG-based wind farms integrated into a power system. Int. J. Elect. Power Energy Syst. **61**, 279–287 (2014)
12. Geem, Z.W., Kim, J.H., Loganathan, G.V.: A new heuristic optimization algorithm: harmony search. SIMULATION **76**, 60–68 (2001)

Chapter 22
Effect of Material Hardness and Operating Conditions on Wear Rate of Sliding Tribopair

Mohammad Hanief

Abstract Wear of a material is one of the most undesirable processes experienced in practical applications. It results in energy loss, material degradation and life-shortening of machine components. This study aims to investigate the behavior of wear rate as a function of different operating parameters. The experiments were conducted on reciprocating tribometer, with three different tribopairs, of ball-on-disc configuration. The 52,100 chromium steel ball was allowed to slide with gun metal, mild-steel and EN-31 steel discs under different conditions. The wear rate for the specimens for each test was calculated by weight loss method. The effect of time, sliding distance and material hardness on the wear rate was examined there off. The Power Law and ANN models were developed for the estimation of wear rate. The model outputs were compared with the observed experimental results. The accuracy of the models was assessed using statistical parameters viz, coefficient of determination (R^2), mean absolute percentage error (MAPE) and mean square error (MSE). It was observed that ANN model was more accurate compared to power law. Further, ANOVA analysis, with 5% p-value, was performed to enunciate the dominant factor affecting the wear rate. It was concluded that the load was the dominant factor influencing the wear rate and was succeeded by the velocity and hardness.

22.1 Introduction

Wear is an undesirable deterioration of a surface of a component by the material removal from surfaces in contact. It is due to the displacement or detachment of material particles from the surfaces. It leads to the sharp reduction in the mechanical properties of the material. The wear of material is a result of the rubbing of surfaces against each other or erosive effect of gaseous or liquid media or scratching of solid surface or any other surface phenomenon.

The wear rate is influenced by a numerous factors as pointed out below:-

M. Hanief (✉)
Mechanical Engineering Department, National Institute of Technology, Srinagar, India
e-mail: hanief@nitsri.net

© The Author(s), under exclusive license to Springer Nature Singapore Pte Ltd. 2022
V. V. Rao et al. (eds.), *Computational and Experimental Methods in Mechanical Engineering*, Smart Innovation, Systems and Technologies 239,
https://doi.org/10.1007/978-981-16-2857-3_23

1. The physical and chemical property of the material
2. Operating condition
3. Characteristics of abrasive media

There is no universal wear mechanism and there exists no simple correlation between the wear rate or surface degradation and values of friction coefficient. This often requires material properties or 'constants' which are obtained by the wear test only. Enormous studies have been performed to investigate both wear during running-in and steady state. Numerous operating parameters have been used for running-in wear characterization. This includes stable surface roughness, coefficient of friction and wear during steady state [1]. Argatov et al. [2] developed an asymptotic wear model with fixed zone of contact and constant acting load for running-in process. Wenlong et al. [3] employed least square support vector machine (LS-SVM) for predicting running-in wear model. Areal surface evaluation parameters were adopted for surface roughness modeling during running-in. Hanief et al. [4] developed a mathematical model for running-in process, assuming the surface roughness as a function of running-in time, explicitly with other process variables (load, velocity, hardness, etc.) constant. Gauss–Newton procedure was employed to find the optimal values of model parameters. A mathematical model was developed by Hanief et al. [5] to determine the relation between wear rate with surface roughness during running-in wear. The correlation of the surface roughness and wear rate was studied by Masouros et al. [6] during the initial bearing operation. The issues in honed surfaces and their effects on performance of ring pack friction were investigated by Mezghani et al. [7]. Al-Samarai et al. [8] investigated the influence of speed and load on performance of alloy of aluminum–silicon on PoD configuration with different loading conditions and three speeds and 70% relative humidity. Uchenna et al. [9] examined the effect of operating parameters on wear characteristics of alloys of AlSi under dry conditions. Haque et al. [10] inquired the influence of load, temperature (pouring) and speed of stirring on microstructure and wear of SiC MMC reinforced with Al6061-Cu. In this study, the dependent variable was wear rate while the independent parameters stirring speed and pouring temperature. In literature, most of the wear models and wear equations were analyzed as to origin, applicability, and content. It is pertinent to mention that not a single model equation or set of equations was noticed for practical and general use. The reason for this is the subjective and erroneous expressions for the wear mechanisms [11]. Singh et al. [12] presented optimum surface roughness model for 316L steel using CNC Lathe pipe with ANN Optimization-ANNO technique. The method was based on ANN and multiple regression analysis (MRA). The main objective was to estimate the most dominant variables and optimal parameters from the set of cutting speed, axial depth, feed rate, and radial depth. Ameen et al. [13] conducted a study on the wear rate for various metals, such as Al, Steel, and Brass under the influence of sliding speed, time and load. The pin-on-disc apparatus was employed to investigate the behavior of the adhesion wear. Singla et al. [14] studied the influence of speed and load on wear characteristics of flu ash-Al 7075 composite. They experimentally investigated the influence of load and sliding speed on the COF and wear characteristics of sliding pin on composite of Al 7075-Fly Ash. Choudhury

[15] worked on design of experiments (DOE) and the ANN for tool wear prediction. The inputs considered were feed, cutting speed and depth of cut while surface finish, flank wear and cutting zone temperature zone were chosen as outputs.

Review of the literature shows that a large amount of research work has been performed by the previous investigators for wear modeling in sliding operations by varying parameters like load, speed, surface coatings, temperatures, compositions, etc. Also, the literature survey shows that the models used for predicting wear in various operation were only regression models. So, in the present study aims to model wear rate in sliding wear operation, by varying parameters like time, sliding distance and hardness while keeping other parameters load, speed, temperatures, etc., constant. Also, the present study uses both power-law and ANN (Artificial neural networks) for the prediction of wear rate.

22.2 Materials and Experiments

Three different materials were used for investigation (i) bronze, which is an alloy of copper, tin, zinc and lead of 85, 5, 5, 5% each, respectively. Bronze was mainly used for making guns, eventually it was overtaken by steel. Gunmetal, machines and casts well. It is corrosion resistant hence used for making hydraulic and steam equipment, valves, gears, castings, statues and numerous objects. It possesses tensile strength of 221–310 MPa, a specific gravity of 8.7, density of 8,719 kg/m^3 and Brinell hardness of 65 to 74. EN8, a medium carbon steel is supplied untreated, usually. EN8 possesses excellent tensile strength. It is generally used in the manufacturing of shafts, stressed pins, gears, studs, keys, bolts, etc., EN8 steel is a very a prominent grade and is easily machinable under different conditions. It may be surface-hardened further to manufacture components with enhanced wear resistance, in the range 50–55 HRC, typically, by induction processes. Mild Steel, also known as plain carbon steel and is now the most common form of steel because it provides material properties that are acceptable for many applications. Mild Steel contains 0.05–0.25% carbon, 0.2% Copper, 98% Iron, 1.03% Manganese, 0.040% Phosphorous, etc. Mild Steel (MS) is malleable and also ductile. Mild Steel having low tensile strength relatively, is cheap and quite easy to form. It is generally used in manufacturing of cams, gears, jigs, automotive and agricultural equipment, frames and machinery parts. The steel possesses density of approximately 7.85 g/cm^3, Young's Modulus of 210 GPa and a Brinell hardness of 120 BHN. In the present experimental study, time, sliding distance and hardness of material were considered as process variables. The Ball material considered for this experiment was AISI 52,100 alloy steel with 60–67 HRC (700–900 HV, 300–320 BHN). The hardness of the materials used are presented in Table 22.1. The test was conducted on Reciprocating Tribometer (Fig. 22.1). The sample was mounted perpendicularly on a stationary holder such that it's one of the faces is forced to press against the ball that is reciprocating on it, then the test parameters both constant and variable are fed in the system. In the present experimental work parameters such as speed, load, temperature, stroke, and frequency were kept constant

Table 22.1 Materials used and their hardness

Specimen	Material	Hardness (BHN)
Disc 1	Gunmetal	65
Disc 2	Mild steel	120
Disc 3	EN8 steel	200
Ball	AISI52100	310

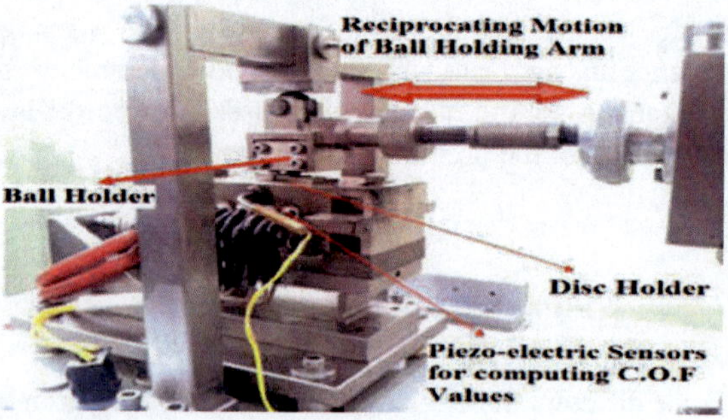

Fig. 22.1 Experimental setup of reciprocating tribometer

while the sliding distance, time and hardness were varied. The wear track so formed on the disc is a straight scar. After each test, the weight loss of the specimen was considered as the wear. The wear rate for each test was estimated from the weight-loss procedure for particular interval of time.

Eight number of sliding wear tests (ball on disc), using different process parameters, were carried on disc samples of same material and each test was repeated at least thrice to have better and accurate results. In total 24 tests were carried out with different process parameters.

22.3 Power Law and ANN Models

In order to predict the wear rate in wear process, the power law equation can be expressed as Eq. 22.1

$$W. = K.H^a.D^b.T^c \tag{22.1}$$

whereas W is estimated wear rate. K, a, b *and* c are constants to be estimated and H, D, T are hardness, sliding distance and time, respectively.

ANN is an advanced computational technique and differs from other computation in vogue. ANN is an information processing structure which is parallel distributed and consisting of interconnected processing elements in unidirectional signal channels known as weights. Although ANN models have been inspired after biological neurons, they are much simplified. Some of the characteristic attributes of ANNs are:

(a) It learns from the data of examples and there after generalizes it on unseen data.
(b) They possess capability to handle the situation where erroneous input exists, incomplete, or are fuzzy.

ANN has attracted the attention of investigators and researchers as it has a capacity to solve problems of non-linear nature. Thus, it has been used for wear rate prediction in the present study. Based on the nature of the problem and its complexity, ANN model can have multiple layers. Generally, it consists of input, hidden and output layers. The input layer receives the input data and are fed to the hidden layers. Thereafter, the next hidden layer evaluates the output. Finally the result is generated in the last layer. The hidden and output layers use transfer functions to compute the final results. In this study, tansig and purelin were was used as transfer functions in hidden and output layer, respectively, as are given in Eqs. 22.2 and 22.3. The typical ANN structure of the model is presented in Figs. 22.2 and 22.3.

$$F = \frac{e^{n} - e^{-n}}{e^{n} - e^{-n}} \tag{22.2}$$

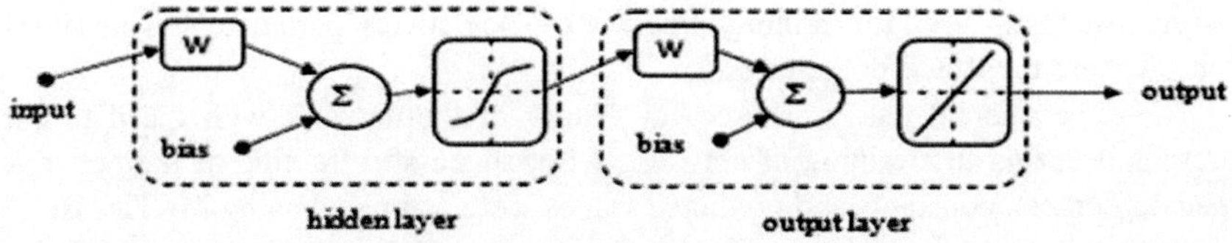

Fig. 22.2 ANN structure and functions used for modelling

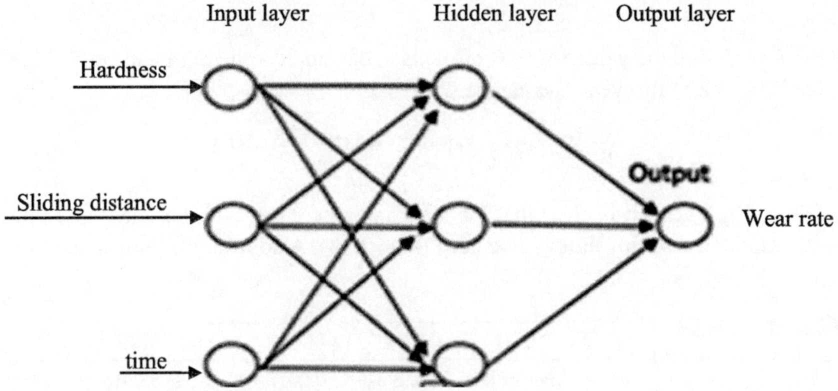

Fig. 22.3 ANN structure used for modelling

Table 22.2 ANN training parameters

No. of neurons	Input-1, hidden-2 and output-1
Initial biases/weight	Between −1 and 1 (randomly)
Activation function (AF)	Tansig
Learning rate (LR)	0.05
Input matrix "X"	H: hardness T: time D: sliding distance
Epochs	1000

$$f = n \tag{22.3}$$

Training constitutes the initial step in an ANN model, where input data is fed to the network along with the desired output data (target). The weights are set initially randomly. The back propagation algorithm (BPA) is used for attaining satisfactory degree of performance by reducing the global error by altering the weights. It is a learning technique in which weights are adjusted by propagating weight changes in backward direction from the output layer to input neurons [9]. The training of the network terminates when the desired level of performance is attained. The decision is taken by the network based on the weights at this stage. In this work ANN-MATLAB toolbox was employed for training of network. The network parameters values used for ANN are presented in Table 22.2.

The experimental data was used for testing of the network with the data not presented during the training of network. After successful training of the network the coefficients, constants and predicted values were estimated using MATLAB.

22.4 Results and Discussion

After calculating the values of the constants a, b, c and coefficient K, the power law model (Eq. 22.1) for wear rate can be written as

$$W = 4.4763H^{-0.4242}.\ D^{1.3328}\ T^{-1.8947} \tag{22.4}$$

From Eq. (22.4) it is clear that the wear rate decreases with hardness and time while as increases with sliding distance (Table 22.3). Also from the Anova analysis,

Table 22.3 Statistical parameters of the models

	R^2	MAPE	MSE
Power law	0.9922	0.0016	0.0018
ANN	0.9860	1.0688×10^{-4}	1.2083×10^{-4}

Table 22.4 ANOVA

	df	*SS*	*MS*
Regression	3	0.50417	0.168057
Residual	21	1.66×10^{-6}	7.92×10^{-8}
Total	24	0.504172	

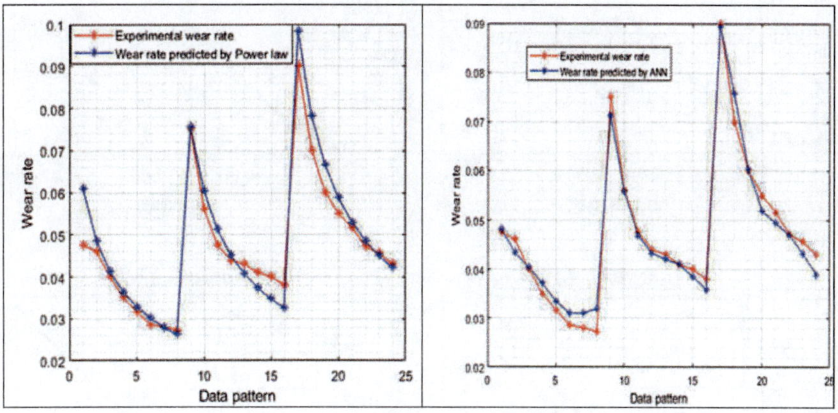

Fig. 22.4 **a** Measured and predicted data of wear rate for power law model and **b** Measured and predicted data of wear rate for ANN model

presented in Table 22.4, it was concluded that wear rate strongly depends on the sliding distance.

The network was trained for 6 input values of three variables, i.e., time, sliding distance and hardness. Out of 24 experimental values, the network was validated for 7 values, while actual wear rate acted as target. The trained networks were finally used for wear rate prediction. Statistical parameters R^2, MSE, and MAPE were determined to find the competence of the models and are presented in Table 22.3. The plot of predictive wear rate and actual wear rate as per the power law and ANN models is shown in Fig. 22.4a, b. It can be seen from Fig. 22.4a, b that the estimated values of wear rate are close to the experimental values for both models. However, ANN model is more accurate. The outputs of the ANN are presented in Fig. 22.5. It can be seen from the Fig. 22.5 that the R^2 is greater the 99% for both training and validation. And for test R^2 is greater than 98%. It can be concluded from Fig. 22.5 that the ANN is competent enough to predict the wear rate with the desired accuracy.

22.5 Conclusions

In this study, Power law and ANN models were used to predict the wear rate. The wear rate with parameters time, sliding distance and hardness was measured by means of

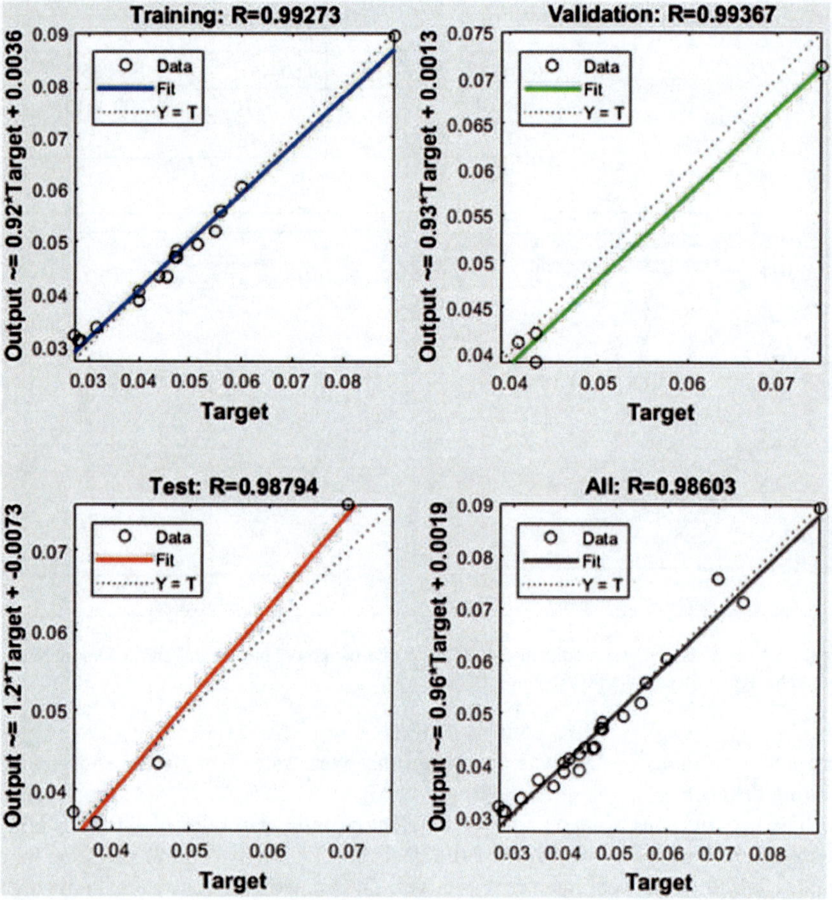

Fig. 22.5 Correlation of parameters for predicting wear rate (training, testing and validation) in ANN model

sliding wear tests conducted on reciprocating tribometer. The data obtained were used to develop the wear rate models. The following conclusions can be drawn from the present study.

The ANN model gives accurate results than Power law model. The developed models were assessed for their prediction capability with known values. Based on the statistical parameters the predicted values were found to be close to measured values.

The proposed models can be used confidently and effectively to predict the wear rate in sliding wear process with high accuracy. The average percentage error in Power law model is 16.8 and 15.6% in wear rate, respectively, while as the average percentage error in ANN model is 4.1 and 4.5% in wear rate, respectively.

References

1. Kumar, R., Prakash, B., Sethuramiah, B.: A systematic methodology to characterise the running-in and steady-state wear processes. Wear **252**, 20445–20453 (2002)
2. Argatov, I.I., Fadin, Y.A.: A macroscale approximation for the running-in period. Tribol. Lett. **42**, 311–317 (2011)
3. Wenlong, Z., Gengpei, X.L., Liping, Z., Liangzhou, C., Xiangqian, J.: Prediction of surface topography at the end of sliding running-in wear based on areal surface parameters. Tribol. Trans. **57**, 553–560 (2014)
4. Hanief, M., Wani, M.F.: Modeling and prediction of surface roughness for running-in wear using Gauss-Newton algorithm and ANN. Appl. Surf. Sci. **357**, 1573–1577 (2015)
5. Hanief, M., Wani, M.F. Effect of surface roughness on wear rate during running-in of En31-steel: modeling and experimental validation. Mater. Lett. (2016)
6. Masouros, G., Dimarogonas, A., Lefas, K.: A model for wear and surfaceroughness transients during the running-in of bearings. Wear **45**, 375–382 (1997)
7. Mezghani, S., Demirci, I., Yousfi, M.: Running-in wear modeling of honed surface for combustion engine cylinder liners. Wear **302**, 1360–1369 (2013)
8. Al-Samarai, R.A., Haftirman, A.K., Al-Douri, Y.: Effect of process parameters on wear rate of Al 6061-Cu reinforced SiCp metal matrix composites. Int. J. Sci. Res. Publ. **2**, 3 (2012)
9. Ozioko, F.U.: Synthesis and study on effect of parameters on dry sliding wear characteristics of AL-SI alloys. Leonardo Electron. J. Pract. Technol. **20**, 39–48 (2012)
10. Meng, H.C., Ludema, K.C. Wear models and predictive equations: their form and content. Wear **181**, 443–457 (1995)
11. Singh, R., Gupta, R., Tripathi, J.: Surface roughness analysis and compare prediction and experimental value for cylindrical stainless steel pipe ss 316l in CNC lathe turning process using ANN method for re-optimization and cutting fluid. Int. J. Eng. Sci. **7**, 58–71 (2014)
12. Hani, A., Khairia, A., Hassan, S., Ethar, M., Mubarak, M.: Effect of loads sliding speeds and times on the wear rate for different materials. Am. J. Sci. Ind. **299**, 106 (2011)
13. Singla, D., Mediratta, S.R.: Effect of load and speed on wear properties of Al7075-Fly ash composite material. Int. J. Innov. Res. Sci. Eng. Technol. **2**, 5 (2013)
14. Feng, C., Jack, X., Wang, X.: Development of empirical models for surface roughness prediction in finish turning. Int. J. Adv. Manuf. Technol. **20**, 348–356 (2002)
15. Choudhury, S.K. Bartarya, G.: Role of temperature and surface finish in predicting tool wear using neural network and design of experiments. Int. J. Mach. Tools Manuf. **43**, 747–753 (2003)

Chapter 23
Fluorescence Characteristics of Coumarin Derivatives with Divalent Metal Ions

Pooja Rani

Abstract The absorption and fluorescence spectra of 3-methyl 7-hydroxyl Coumarin (C4) dye in presence of divalent metal ions (Ni^{+2}, $Co^{+2}$and Sn^{+2}) were investigated. The fluorescence quenching of dye C4 molecule is found to be of second-order and second-order fluorescence quenching rate constant varies in order of $Ni^{+2} > Co^{+2} > Sn^{+2}$ in methanol. The Stern Volmer Plots show that the quenching efficiency increases in $Ni^{+2} < Co^{+2} < Sn^{+2}$. These results are explained based on the paramagnetic and diamagnetic behavior of the said ions.

23.1 Introduction

The main element in the organism is Nickel. It helps lower the blood sugar level, increase the renewal of the red corpuscles and find the anemia. This is a major achievement in the medical field to diagnose the functioning of heart, lungs and asthmatic attack [1]. It is a very advantageous element for individuals. This element also helps to regulate digestion in the human body.

In our neighboring, the amount of Nickel is tremendous and deficient nickel is very rare. In contrast, due to the rich nickel environment, excess nickel creates nose and lung cancer and leukemia [2, 3]. Also, the diseases like nervous breakdown, chronic hepatitis and gall stone can produce by it. Therefore, people need to be more cautious and should take extra care. They are very attentive to health by changing the lifestyle, and the problem of nickel pollution may be paying more devotion. Therefore, the significance of the conclusion of nickel in ecological samples may be overstressed since they have a severe probable risk to the peoples. The chief sources of nickel in marine are after the termination of rocks and stones, natural phases, atmospheric processes, industrial and water disposal [4].

Cobalt has advantages and disadvantages for peoples. For the treatment of anemia, It is used due to it contains vitamin B-12. Due to cobalt exposure, it may cause lung infections that include respiratory problems and asthma. The international cancer

P. Rani (✉)
G L Bajaj Institute of Technology and Management, Greater Noida, UP, India

© The Author(s), under exclusive license to Springer Nature Singapore Pte Ltd. 2022
V. V. Rao et al. (eds.), *Computational and Experimental Methods in Mechanical Engineering*, Smart Innovation, Systems and Technologies 239,
https://doi.org/10.1007/978-981-16-2857-3_24

research categorizes cobalt as 'human carcinogen'. The kidneys, skin problems and lung damage due to cobalt salt exposure.. Without a chest x-ray, these problems cannot observe. This can be mortal [5].

For food preservation, it is recycled for cane coating and tin-plated steel vessels. Tin alloys are utilized in different methods: like the soldering of electric circuits and bell metal. For making superconducting magnets use niobium-tin alloy. Tin oxide is recycled for sensors. The electrical conductivity varies may be observed. Tinfoil is commonly used for packaging foods and drugs. Now, it can be changed by aluminum foil because organic tin is more harmful to people. Regardless of the dangers, they are useful in more industries like paint manufacturing. The more use of tin is still demanding; regard the significances of tin poisoning are known. The effects of organic tin substances may be changes. It depends on the type of substance. Triethylin is a very hazardous organic tin substance for peoples. It has comparable least hydrogen bonds. When hydrogen bonds nurture longer a tin substance will be less harmful to people's health. Peoples can captivate tin bonds by food. The consumption of tin bonds may cause effects like eye and skin problems, headaches, stomachaches, breathlessness, as well as long time effects of depression and brain effects [6].

The development of good, inexpensive, logical methods for the finding of Ni, Co, and Sn in low absorption stages seem. At present, the finding of metal ions is accepted by X-ray fluorescence in electroplating solution carried to find metal [6], flame atomic absorption spectrometry (FAAS) in water solution [7] and gasoline [8], graphite furnace atomic absorption spectrometry (GFAAS) in fingernails and forearm skin [9], gasoline [10] and residual fuel oil [11], electrothermal atomic absorption spectrometry (ETAAS) in aluminum-base alloys [12], marine sediments [13], inductively coupled plasma atomic emission spectrometry (ICP-AES) in plant samples [14], square-wave adsorptive stripping voltammetry (SWASV) in duralumin, iron ore and a river water [15] and flow-injection solid-phase spectrophotometry (FI_SPS) in copper-based alloys [16]. The suggested approaches provide advantages and limitations. However, some limitations like time-consuming mechanisms, advanced instruments, low sensitivity, fast and tactful methods for the find of nickel. The chemiluminescence (CL) investigation is agreeable because fewer finding parameters and extensive linear response ranges may be realized for sure kinds by using modest equipment. Last decades, the techniques for the finding of recommendation metal ions were done by CL Analysis [16].

Coumarins have been widely used as structural units in fluorescence probes which are based on intra-molecular quenching, and as correspondents of radical reactions within the solutions and thin polymeric films. The degree of intra-molecular quenching depends on the structure of coumarin. The fluorescence spectrum of coumarins is majorly used in the area of polymer science [17, 18]. Also they are utilized as photo-initiators [19], for the incorporation into polymer chains by the process of co-polymerization [20], in the assesment of polymer–solvent effects [21], for various structural characterization methods [22]. The Coumarin Period of laser dyes have a solid electron giving substituent (Either hydroxyl,-OH, or an amino, NR2) in the 7-position. Coumarin-1 (7-diethylamino-4-methyl coumarin)

Fig. 23.1 Molecular structure of coumarine-4

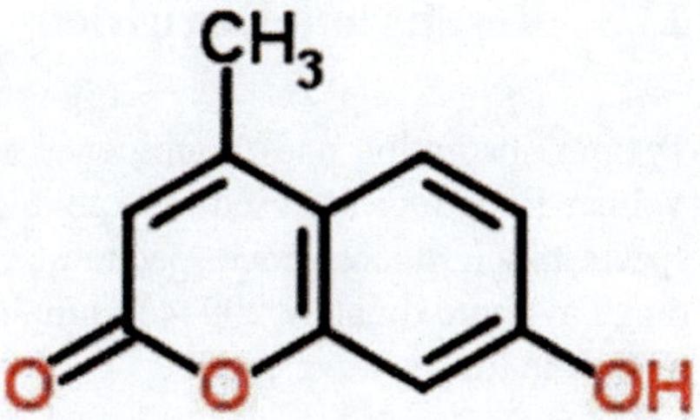

was the first coumarin detected laser action [23]. About 100 coumarin laser dyes are known [24]. The 7-hydroxy coumarins and 7-aminocoumarins with several electron removing substituents in 3rd, 4th and 6th positions give more laser dyes of excellent in the green–blue area [25–27]. In the present study 3-methyl 7- hydroxyl coumarin (C4) dye is used because of certain advantages reported earlier [28]. The molecular structure of C4 is in Fig. 23.1.

23.2 Experimental

The Coumarin dye was obtained from Sigma Chemicals (USA). Without refinement, metal ions can be used. By liquefying the proper quantity of nickel sulfate in water, Nickel (II) ion standard solution (10.0 g/l) can be ready. The required solution ready by dilution of the stock solution in water. 0.1771 g Luminol dissolving with. 0.1 mol/1 sodium hydroxide and diluting with 0.1 mol/1 sodium hydroxide to 1.0×10^{-2} mol/l prepared Luminol stocking solution (1.0×10^{-2} mol/l). The luminol required solution can be ready in a proper concentration of sodium hydroxide. By using the same method other metal ions are used. For quenching one used a solution of comparatively less viscosity to allow fast diffusion of quencher. The quenching method can be beneficial for biological macromolecules. Due to it allow the relaxation rate can be finding lacking temperature confirmation. Methanol can also be used due to similar purpose. Methanol used was of AR grade and find was opaque and non-fluorescent in the kind of excitation and fluorescence emission detected by their absorption spectra. Absorption spectra noted by Shimadzu (260) UV–VIS Spectrophotometer. Aminco-Bowman Spectro- Photofluorometer, connected with Photo-Multiplier tube RJ-758 and 150 W Xenon lamp, was find fluorescence spectra of the dyes beneath observation. For the photo-multiplier tube, the No Correction parameter is required because of the fine response over this range. Excitation curves were the same as the absorption curves of the dyes. So, the excitation curves are selected as the absorption curves to choose the peak wavelength. Dye concentration was low as 2×10^{-5} M/l to ignore self-quenching. Observed fluorescent emission wavelengths assessed to ± 2 nm and absorption wavelength ± 1 nm. Metal ions are mixed in micro-liters as a quencher.

23.3 Results and Discussion

Typical absorption and fluorescence spectra shown in Figs. 23.2 and 23.3 Stern Volmer Equations are used to explain the quenching phenomenon. The absorption spectrum and fluorescence spectra of coumarin-4 exhibits bands in the wavelength range at approximately 200–450 nm. The absorption maxima of the band is found to shift to longer wavelengths.

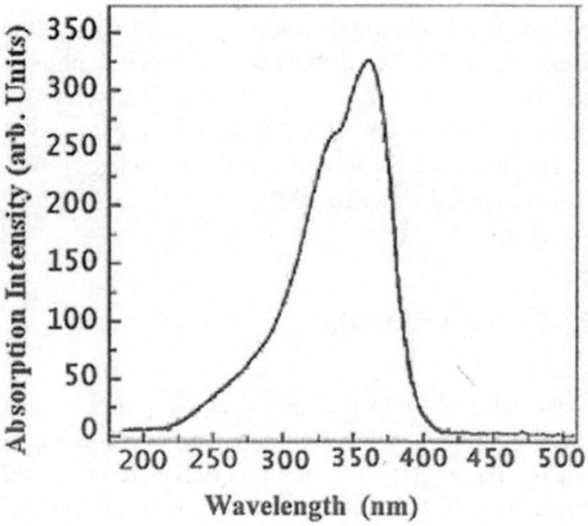

Fig. 23.2 Typical absorption spectra of coumarin-4

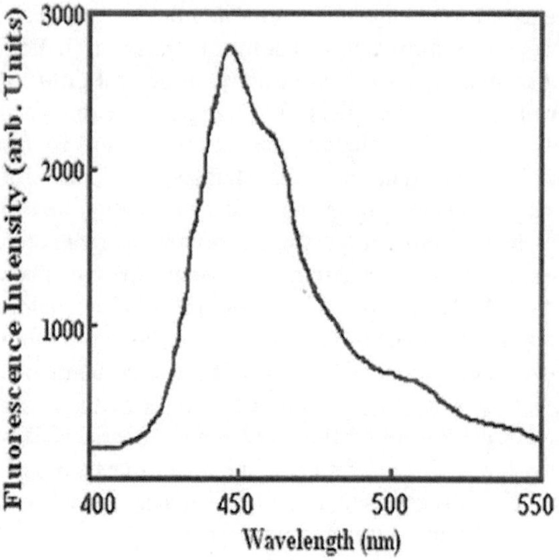

Fig. 23.3 Typical fluorescence spectra of coumarin-4

Fluorescence quenching of the methanol solution of C4 dye in presence of NI++, Co++ and Sn++ metal ions are observed. The second-order quenching rate constant Kq is designed using the Stern Volker relation [28]. It is found that the fluorescence spectral formation does not vary, show the absence of an emitting excited state complex according to experimental conditions. The excited-state lifetime of the C4 in the absence or presence of the Ni++ , Co++ and Sn++ is the same as 3.5 ns calculated theoretically using the method described elsewhere [29].

As shown in Figs. 23.4 and 23.5, the Stern Volmer graph shows that the quantum efficiency increases $Ni^{+2} > Co^{+2} > Sn^{+2}$. The Stern Volmer plot (Fig. 23.4) is linear with a Regression Coefficient of 0.98 and the Stern Volmer constant comes to be 68.5 mol^{-1}. In the case of Sn^{+2} s-order quenching rate constant Kq is 19.57×10^{11} $dm^3\ mol^{-1}\ s^{-1}$.

This is two orders of magnitude more than limiting diffusion rate constants Kdiff in methanol at room temperature of 6.2×10^9 $dm^3\ mol^{-1}\ s^{-1}$ [29]. This study shows that diffusion is not complicated in the quenching mechanism. The stern Volmer graph in Ni++ and Co++ are exponential (Regression Coefficient 0.95 and 0.93, respectively) as depicted in Fig. 23.5. The Stern Volmer Constant is found to be 98.3 mol^{-1} and 106.3 mol^{-1}.

Further, the quenching rate constants rate calculated are 28.09 $dm^3\ mol^{-1}\ s^{-1}$ and 30.38 $dm^3\ mol^{-1}\ s^{-1}$, respectively. In methanol solution, Co++ ions exist in form of the tetrahedral $\{CoCl_4\}^{-2}$ complex [30] which is characteristically deep in the deep and blue and causes efficient fluorescence quenching. Co++ and Ni++ are paramagnetic with electronic structures Ar^3d^7 and Ar^3d^8, respectively. On the other hand, Sn++ is diamagnetic with electronic structure Kr $^4d^{10}$ $5s^2$. Both Co++ and Ni++ are also specified by lesser ionic radii 0.74 and 0.72 Å, respectively. These are

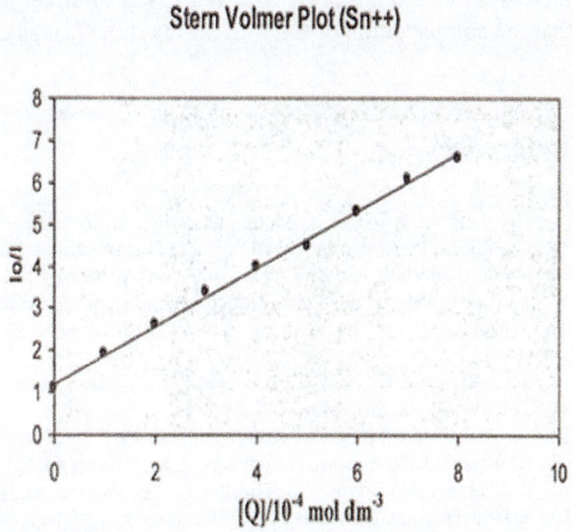

Fig. 23.4 Stern volmer plot of C4 dye in methanol in presence of Sn^{+2} metal ions

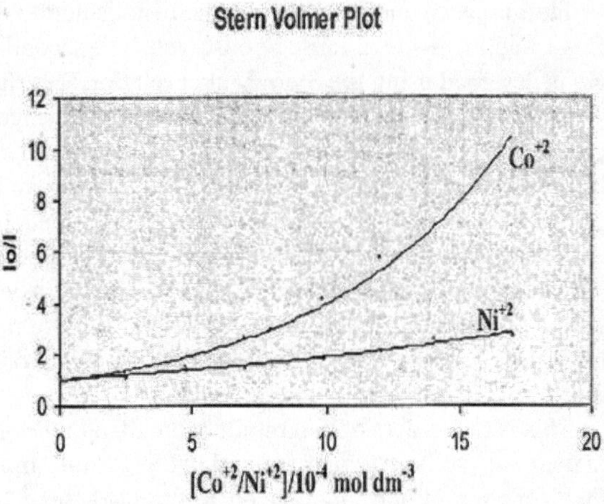

Fig. 23.5 Stern volmer plot of C4 dye in methanol in presence of Co +2 (upper graph) and No +2 (lower graph) metal ions

lesser than the radii of Sn++ (1.40 Å). Complexation of C4 with Sn^{+2} is faster and should be specified by smaller activation energy compared with Ni^{+2} and Co^{+2}.

23.4 Conclusion

From above we concluded that the quenching observed is of dynamic nature. The second-order fluorescence quenching rate constant increases in the order of $Ni^{+2} > Co^{+2} > Sn^{+2}$ in methanol. The Stern Volmer graphs show that quenching efficiencies increases in the order of $Ni^{+2} < Co^{+2} < Sn^{+2}$. Further, the above results also indicate that the complexation of C4 with Sn^{+2} will be fast as compared to Co++ and Ni++.

References

1. Hernburg, S., Kikkane, J., Mellin, G., Lilivs, H.: Arch. Environ. Health **21**(140) (1970)
2. Grandjean, P., Audersen, O., Niellsen, G.D.: Am. J. Ind. Med. **13**(193) (1988)
3. Sanderman, F.W., Morgan, L.G., Andersen, A.: Ann., Clin. Lab. Sci. **19**(44) (1989)
4. Merian, E.: Metals and their compounds in the environment. VCH, New York **22**, 1101 (1991)
5. Sheehan, P.J., Meyer, D.M., Paustenback, D.J.: Toxicology and environmental health **32**, 161 (1991)
6. Schwartz, M.E.L.: Encyclopedia of materials, parts and finishes, 2nd edn. CRC press, Section Tin and Alloys, Properties (2002)
7. Gao, J., Wan, G.F.: Spectrosc. Spect. Anal. **20**, 550 (2000)
8. Khorrami, A.R., Naeimi, H., Fakhari, A.R.: Talanta **64**(13) (2004)
9. Roldan, P.S., Alcantara, I.L., Padilha, C.C.F., Padilha, M.: Fuel **84**, 305 (2005)
10. Kristiansen, J., Christensen, J.M., Henriksen, T., Nielsen, N.H., Menn'e Anal, T.: Chim. Acta **403**, 265 (2000)
11. Campos, R.C., Santos, H.R., Grinberg, P.: Spectrochim. Acta Part B **57**(15) (2002)

12. Vale, M.G.R., Damin, I.C.F., Klassen, A., Silva, M.M., Welz, B., Silvi, A.F., Lepri, F.G., Borges, D.L.G., Heitmann, U.: Microchem. J. **77**, 131 (2004)
13. Carri'on, N., Itriago, A.M., Alvarez, M.A., Eljuri, E.: Talanta **61**, 621 (2003)
14. Barciela-Alonso, M.C., Tubio-Franco, M.C., Prego, R.: Mar. Pollut. Bull. **46**, 1504 (2003)
15. Mestek, O., Tresl, I., Kopl'ik, R., Pavelkov'a, H., Such anek, M., Va'nas, B.: Talanta **55**, 595 (2001)
16. Morfobos, M., Economo, A., Voulgaropoulos, A.: Anal. Chim., Acta **519**, 57 (2004)
17. Soutar, I.: The application of luminescence technique in polymer science. Polym. Int. **26**, 35–49 (1991)
18. Morawetz, H.: On the versatility of fluorescence technique in polymer research. J. Polym. Sci. A Polym. Chem **37**, 1725–1735 (1999)
19. Sarker, A.M., Kaneko, Y., Neckers, D.C.: Photochemistry and photophysics of novel photoinitiators: N,N,N-Tributyl-N-(4-methylene-7-methoxycoumarin)ammonium borates. J. Photochem. Photobiol **117**, 67–74 (1998)
20. Oh, J.K., Wu, J., Winnik, M.A., Croun, G.P., Rademacher, J., Farwaha, R.: Emulsion copolymerization of vinyl acetate and butylacetate in the presence of fluorescent dyes. J. Polym. Sci. Part A Polym. Chem. **40**, 1594–1604 (2002)
21. Kaholek, M., Hrdlovič, P.: Spectral properties of coumarin derivatives substituted at position 3 effect of polymer matrix. J. Photochem. Photobiol. A Chem. **108**, 283–288 (1997)
22. Jones, G., Jimenez, J.A.C.: Azole-linked coumarin dyes as fluorescence probes of domain-forming polymers. J. Photochem. Photbiol. B Biol. **65**, 5–12 (2001)
23. Drexhage, K.H.: Dye Lasers: Topics in Applied Physics, 1, 2nd edn. Schiifer, Springer, New York, F.P (1977)
24. Kagan, M.R., Farmer G.J. and Huth B.G.: Laser Focus, 4(26),(1968).
25. Maeda, M.: Laser Dyes. Academic Press, New York (1984)
26. Song, P.S, Houten, M.L., More, T.A., Herndon, W.C.: J. Photochem, Photobiol. **14**, 521 (1971)
27. Chu, G., Yangbo, F.: J. Chem. Soc. Faraday Trans. I **83**, 2533 (1987)
28. Sharma,V.K., Mohan, D., Sahare, P.D.: Spectrochim. Acta A **67**, 1161 (2007)
29. Sahare, P.D., Sharma, V.K., Mohan, D., Ruspov, A.A.: Spectrochim. Acta Part A **69**, 1257 (2008)
30. Gene, L., Spears, L.G.: J. Chem. **61**, 252 (1984)

Chapter 24
A Novel Method for Performance Enhancement of PV Module Using Graphene

Pragati Tripathi, M. A. Ansari, and Rajat Mehrotra

Abstract This paper aims to improve the output voltage of solar panels by using graphene. The objective of this work is to analyze the absorption coefficient for improvisation in the efficiency of graphene. Various parameters viz. current density, absorption coefficient, wavelength of sun spectrum, Fermi and Dirac point level, and carrier generation rate have been computed and analyzed. The chosen parameters and their characteristics can demonstrate the necessity of solar energy by using graphene. GaAs is used as substrate and graphene layers are used to make junction and gate contact. The obtained graphs indicate that the power conversion efficiency of the solar panel is increased from 10.69 to 11.5%.

24.1 Introduction

These days, Graphene is becoming more popular for solar energy applications as it can also withstand harsh weather conditions. The requirement of an efficient solar cell is becoming important as they are the economical and cleanest form of the power source. The output acquired by solar cells is not considered acceptable if we see both its input and output on the same platform which is obvious from the available statistics in literature. It is observed that solar cell efficiency is merely reported as 10% without employing graphene [1]. Graphene has given tremendous results in terms of efficiency and it is discovered that efficiency is increased to 20% by utilizing graphene [2]. The main difference in their construction is that in place of silicon a layer of graphene can be utilized. The number of layers of graphene

P. Tripathi (✉) · M. A. Ansari
Department of Electrical Engineering, Gautam Buddha University, Greater Noida, India

M. A. Ansari
e-mail: ma.ansari@ieee.org

R. Mehrotra
Department of Electrical & Electronics Engineering, GL Bajaj Institute of Technology & Management, Greater Noida, India
e-mail: rajat.mehrotra@glbitm.ac.in

© The Author(s), under exclusive license to Springer Nature Singapore Pte Ltd. 2022

V. V. Rao et al. (eds.), *Computational and Experimental Methods in Mechanical Engineering*, Smart Innovation, Systems and Technologies 239,
https://doi.org/10.1007/978-981-16-2857-3_25

is inversely proportional to sheet resistance and optical transparency [3]. Talking numerically optical transparency decreases in the following manner according to the increase in the number of the graphene layer. One sheet of graphene gives a visual photograph of 97.7% [4]. Three-layer graphene provides an optical transparency of 90.8% [5]. With the increase of layers of graphene, a decrease of 2.3% is visible in optical transparency [6]. In addition to confining 90% optical transparency to this, a graphene layered sheet can stimulate a sheet with resistance up to 2.1 kΏ sq^{-1} [7]. Most of the time monolayer graphene has 11% less than multilayer graphene layer in terms of damping effect, this happens only because of the presence of a large number of holes, i.e., positive charge carriers are in a high accepting state [8]. The complete property of the graphene sheet can be changed by slightly doping of heteroatoms, wheatear it is photonic or electronic characteristics of the graphene sheet. The first one is p-type and the second most popular doping is n-type doping. When a trivalent atom of metals like aluminum, boron, etc., which has a property to extract electrons and generate a hole in place of it which is a positive charge carrier, this phenomena can be termed as hole doping or in simple language this is known as p-type doping [9]. In the same way, if pentavalent atoms like phosphorus (P), arsenic (As), antimony (Sb) which has the property to donate electron which helps in generating free electrons from pentavalent atoms, this kind of doping is known as electron doping of in simple language it is termed as n-type doping [10]. A variety of carbon compounds (carbon allotropes) transpired to be used in solar cell manufacturing for bringing the basic properties in the domain of cost and efficiency in a pervasive manner [11]. For this, the electron mobility is brought to the conductance band. Many carbon allotropes are meaningless for this context as they are deprived of less no. of free electrons and more in thickness which could be unsuitable for the electronic properties of a solar cell [12]. Graphene which is a co-product of carbon allotrope, graphite tends to withstand all the basic properties for the solar cell preordained thickness and remarkable electron mobility which are considered here [13]. As it is used now a day's for Schottky junction for better outcomes of solar panels efficiency. The upcoming generations looking toward renewable energy sources as the working substitute for cheap and environment-friendly energy sources [4]. This is obvious that the cost and depletion in the availability of non-renewable energy resources are also becoming the crux for people losing their interest in them. That's why more and more research is been carried out on renewable energy resources and embraced everywhere. This paper is covering research on the application of graphene in the solar cell. As we all know that the traditional method of solar energy extraction is not so efficient and still has many aspects that need to be improved especially in the tenet methodology. One small step carried out in many places to enhance the efficiency of the solar cell is to use the graphene layer in it. Graphene is a carbon compound structured honeycomb in a two-dimension lattice [5]. It is a brilliant conductivity, thermally as well as electromechanically [4] Electromobility of graphene at room temperature is 100,000 cm^2/Vs [14]. At room temperature, it is highly flexible and has high mechanical strength [3]. If the temperature is slightly high it affects its current density but still has a current density on an average of nearly 108–109 cm^2/Vs [15]. Graphene can also be used as transparent electrodes as its

Fig. 24.1 Microscope images of graphene nanoribbons formed by a bottom-up process [2]

transmittance rate is 97–98% [16]. There are many more applications of graphene in different fields like carbon nanotubes (CNTs), such as transistors [17] interconnects [18] thermal bumps [16], and gas sensors which are studied in cited papers Most of the times the graphene nanoribbons are used in some cases for efficient outcomes. These are mostly in nanometers. A scanning microscope image of GNRs we formed by the bottom-up approach is shown in Fig. 24.1 [14].

The elementary methodology of the silicon-based solar cell is not very different from graphene-based PV cells. Mostly nowadays all materials are graphene oriented. As by use of graphene operational efficiency could be increased tremendously. The paper here is studying the power absorbed by the solar module concerning the time the Simulink model calibrated by its mathematical model to get the graph reading for our study.

24.2 Material and Methods

This paper is showing MATLAB and Simulink techniques to have a comparison between certain parameters of GaAs Schottky junction field-effect solar cell which is composed of Graphene. This analyzed a graph between carriers generation rate versus time which is compared with previous results analyzed in Reference paper which is a main study in the Simulink and calculated power density, Carrier generation rate and wavelength of sun spectrum by constructing Simulink model by using FDM method (finite difference method) rather than solving drift Diffusion equation by a different method which is a little bit complicated for which some equations are solved using coding and algorithm to remove distortions. The study is made so simple to check and calculate the parameters. It also shows mathematical modeling of graphene with the help of certain equations by constructing models and acquiring graphs between certain parameters which shows the efficacy of the PV module. The paper is so long as nearly methodological opinions for the enterprise and optimization of the graphene PV module with an improved presentation.

24.2.1 Statistical Methods

The crux of the paper is regarding graphene module modelling and simulation so that the operating mechanism also includes the properties of fixing absorption, transporter flexibility, gate oxide thickness, and several junction graphene layers. In Sect. 24.1, a brief Introduction of the properties of graphene is given. In Sect. 24.2 the material used in the whole process and the technique which is applied is given. The geometry of the graphene PV module is also given (Fig. 24.1). In Sect. 24.3, it's the main modeling and simulation approaches are outlined. In Sect. 24.4, results are discussed with their graphs and models are to be taken. In this Section, the MATLAB algorithm result is being compared by the Simulink result. To conclude, the efficiency that has been formed is discussed in Sect. 24.5. Here the complicated Maxwell equations are solved by the GUI technique in MATLAB which gives the optimization of the parameters and gives a brief idea about the design and analysis of the material. Basically, the models are constructed by taking the parameters of the different components as the affinity level of GaAs is set as 4.07 eV, the bandgap of GaAs is 1.42 eV, For obtaining the first graph the values taken are in the Eq. (24.1) the wavelength of the sun spectrum (λ_m) here taken as 0.873 μm, power reflectance of GaAs is defined by the formula: $r = Re^{r\alpha L}$, where $\alpha = 10^{18}$ to 10^{20} cm^{-3}, x is the gate and junction graphene and stated as x = 0 nm, x = 70 nm this is the limited value in which upper and lower limit is defined, L = 500 nm in which recombination rate is obtained by the formula: R = BNp, where $B = 7.2 \times 10^{-16} m^3/s$, $n = 2.24 \times 10^{11}$, $P = 1.85 \times 10^1$ and for obtaining the carrier generation rate at wavelength (λ_m) in the second graph the energy of photon at wavelength (λ_m) is given as 9×10^{17} and for third graph the fit parameter (a,b) is taken as $1 \times 10^{(-5)}$ the absolute temperature (T) here varies from (20–100) K. For obtaining the fourth graph the Dirac point (E_D) of graphene is 4.8 eV Fermi Point (E_F) is 0.5 eV, electrical charge of electron (q) is $1.6 \times 10^{(-19)}$ eV, mobility of holes (μ_h) varies from −5 to 2, critical (T_c) as 300 K and threshold voltage (V_{th}) is taken as −101.3 mV for achieving the mobility of electrons.

24.3 Modelling and Simulation Results

The control density per unit length captivated by GaAs at a sure wavelength.

λm is determined by—[3]

$$P_{abs}(\lambda_m) = \frac{dP(\lambda_m, x)}{dx} = (1 - r)\alpha P_o(\lambda_m)e^{-\alpha x} \tag{24.1}$$

where α is the optical absorption coefficient as shown in the Fig. 24.2. It strongly depends on the wavelength and it can be obtained from this. The parameter r is the power reflectance of GaAs.

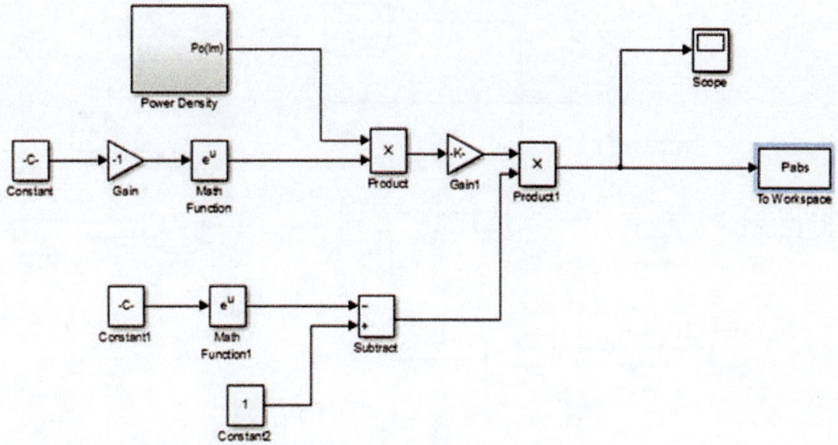

Fig. 24.2 Simulink model of power density absorbed by GaAs

Here in Fig. 24.3, the graph is categorized from the Eq. (24.1) it has been acquired that the power absorbed by graphene solar cell is much higher than the power absorbed by the silicon solar cell because the emissivity and collectively of the graphene is much higher than the silicon which can be seen by the graph in Fig. 24.3. The graph is obtained by constructing the Simulink model of the acquired equation and checked the parameters to acquire the absorption power of the graphene solar cell which is better than the silicon solar cell. Even it has been found that the graphene solar cell can absorb power or conduct electricity in harsh weather conditions in rain or at night due to enhanced collective and high transmittance property.

The carrier generation at wavelength λm can be obtained by—[3]

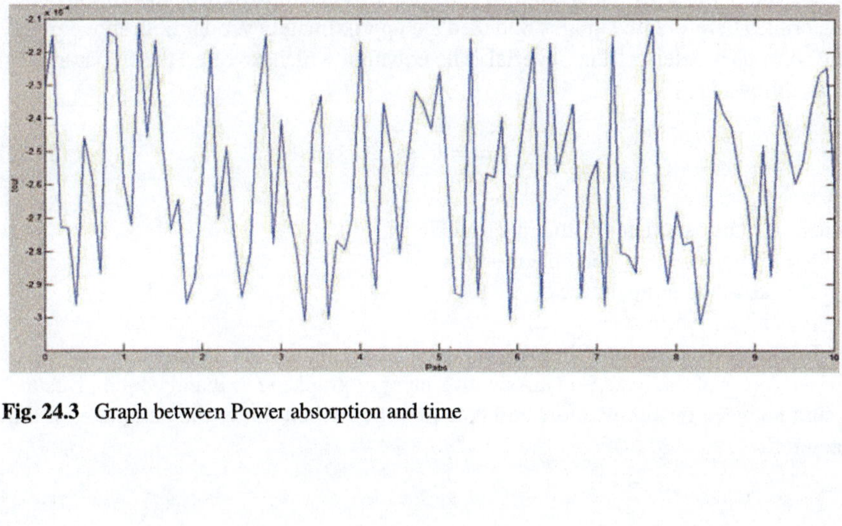

Fig. 24.3 Graph between Power absorption and time

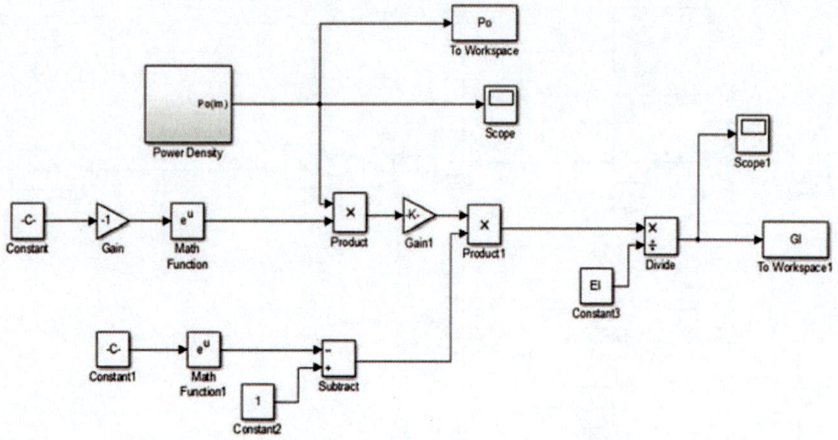

Fig. 24.4 Simulink model of carrier generation

$$G_{\lambda m}(x) = \frac{P_{abs}(\lambda_m)}{E_{\lambda m}} = \frac{(1-r)\alpha P_o(\lambda_m)e^{-\alpha x}}{E\lambda_m} \tag{24.2}$$

where ($E_{\lambda m}$) is the energy of photon at wavelength λ_m,

$G_{\lambda m}(x)$ is the carrier generation rate,

$P_{abs}(\lambda_m)$ is the power absorbed,

α is the optical absorption coefficient, parameter r is the power reflectance of GaAs.

The model of the equation (24.1) is presented in Fig. 24.4 to determine the carrier generation rate of GaAs.

The graph shown in Fig. 24.5 shows that power absorption is randomly varying concerning time. Carrier generation rate is proportional to the power absorption of the material. Therefore, the graphs obtained are approximately similar both showing the different parameters of the material. The equation which revealed the fit parameter is as follows [15]:

$$\mu = a^{*}e^{(T/b)} \tag{24.3}$$

where $a =$ fit parameter of the material GaAs

$b =$ fit parameter of the hole mobility

$T =$ absolute temperature

$\mu =$ hole mobility.

Figure 24.6 shows the Simulink model of the equation (24.3) which revealed the fit parameter of the material GaAs with a layer of graphene in it and which gives the graph between the temperature and hole mobility which shows the parameter of the material.

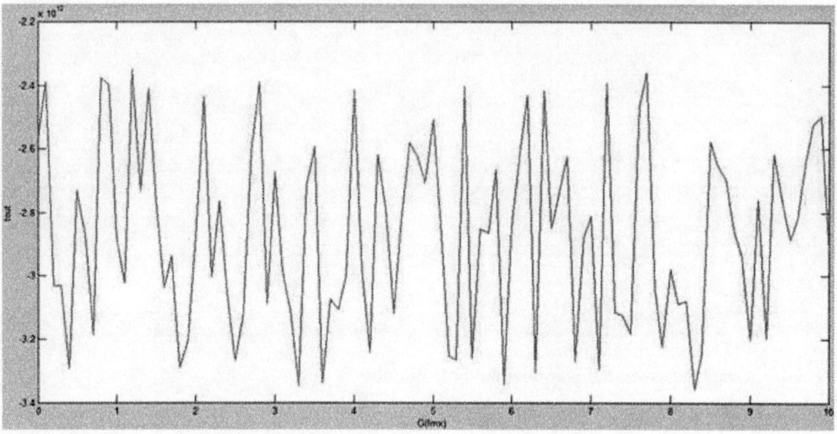

Fig. 24.5 Graph between carrier generation and time

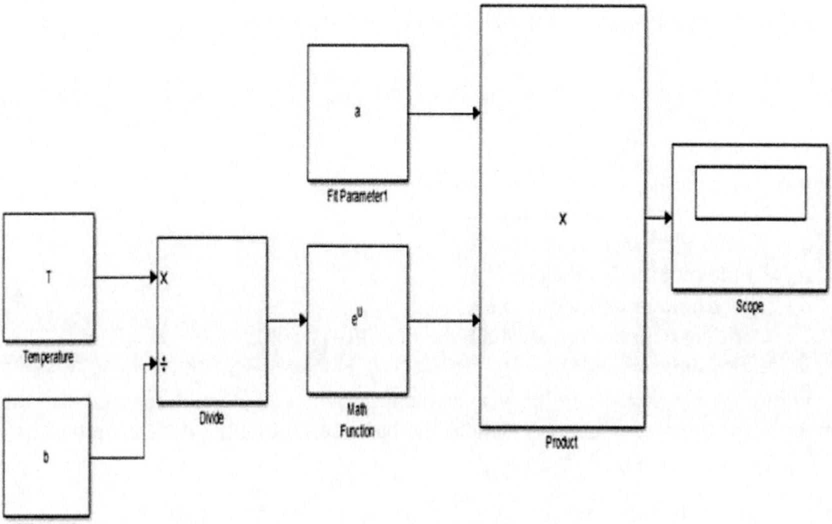

Fig. 24.6 Simulink model of the hole mobility and the temperature

Figure 24.7 shows the hole mobility of PTB7:PC$_{71}$BM as a function of temperature. The hole mobility (μ) depends almost exponentially on the temperature (T) as shown by the red fitting line. The hole mobility gives the relation between the critical temperature and electronic mobility [16]:

$$\mu_h = \frac{q^* T_c}{\mu_n} \tag{24.4}$$

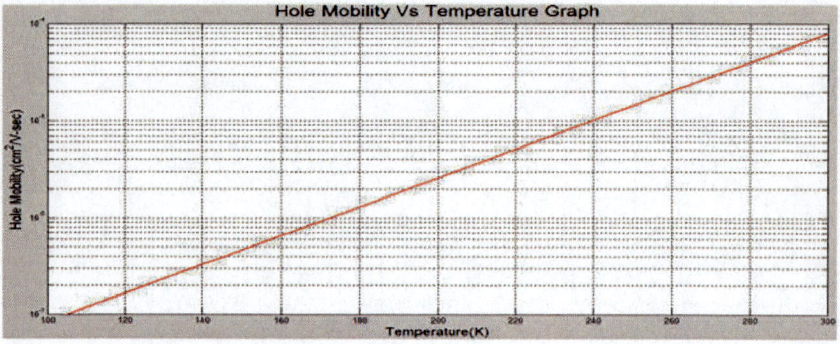

Fig. 24.7 Graph between temperature and hole mobility

$$E_F - E_D = \frac{1}{2}^* \mu_h^* V_{th}^2 \tag{24.5}$$

Substituting the value of Eq. (24.4) in Eq. (24.5), we get

$$E_F - E_D = \frac{1}{2}^* q^* T_c/\mu_n{}^* V_{th}^2 \tag{24.6}$$

where, E_F = Fermi energy of the material
E_D = Dirac energy level of the material
q = electrical charge on an electron
μ_h = hole mobility of the electron
μ_n = electronic mobility of the electron
T_c = critical temperature at which the mobility is stable
V_{th} = threshold voltage.

Figure 24.8 is describing the V_{OC} as the reliability of hole mobility and also the effect in the transfer of charges. Under flat band condition, the difference between

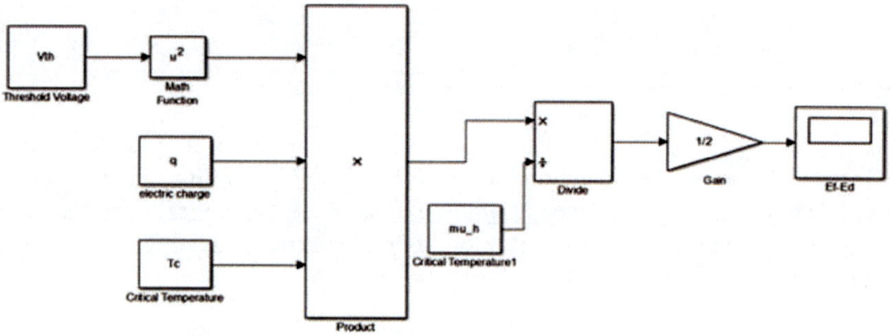

Fig. 24.8 Simulink model of the Fermi level and Dirac point level

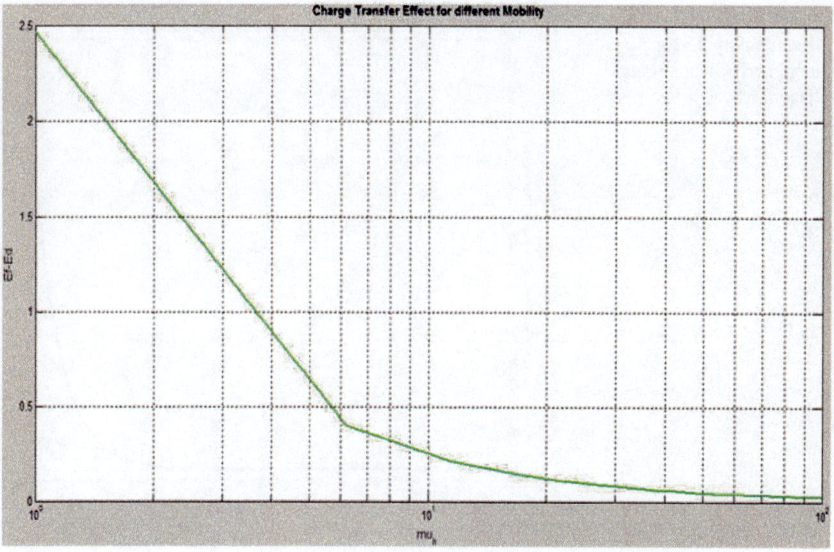

Fig. 24.9 Graph between the Fermi level and Dirac point level versus hole mobility

Dirac point and Fermi level is the value represented by $E_F - E_D$. This value is positive then the graphene will be rich in electrons, or if not, it will be rich in holes. Mainly when the hole mobility is decrementing, this value increases which is because of charge transfer (Fig. 24.9).

24.4 Discussion of Current Density and Power for Different Materials

Here in Fig. 24.10 depicts the graphical representation of current density and voltage. In another graphical representation of power versus voltage, representation has been shown for Silicon material. The comparative analysis has been shown among four materials which are amalgamated with graphene to form an efficient solar module. By the representation, it has been shown that CdTe with graphene is proved to be efficient among all.

Figures 24.10, 24.11, 24.12 and 24.13 are showing the graphs to show the characteristics of Si, CIGS, graphene/CdTe, graphene/a-Si out of which CDTE is showing the best result.

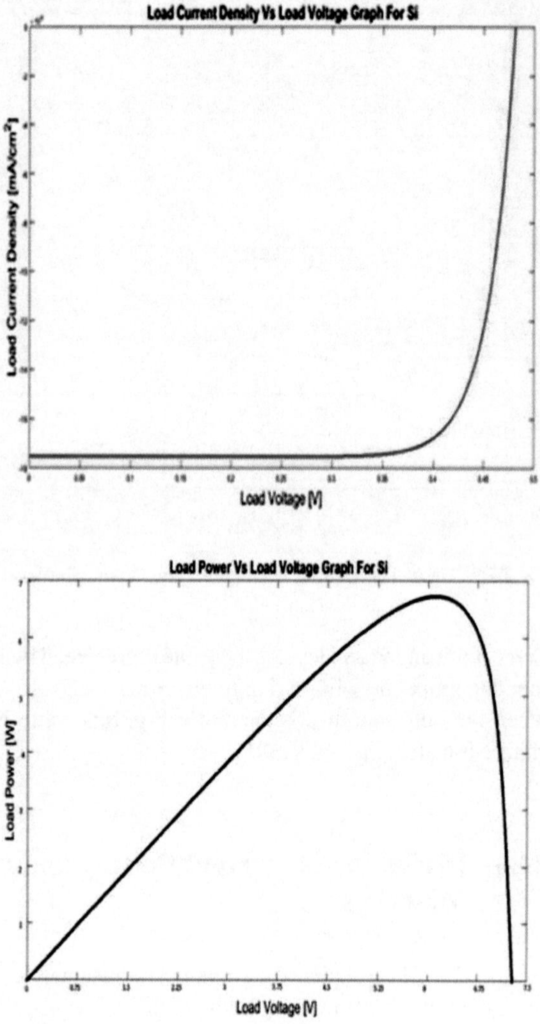

Fig. 24.10 Graph between current density versus voltage and power versus voltage for Si

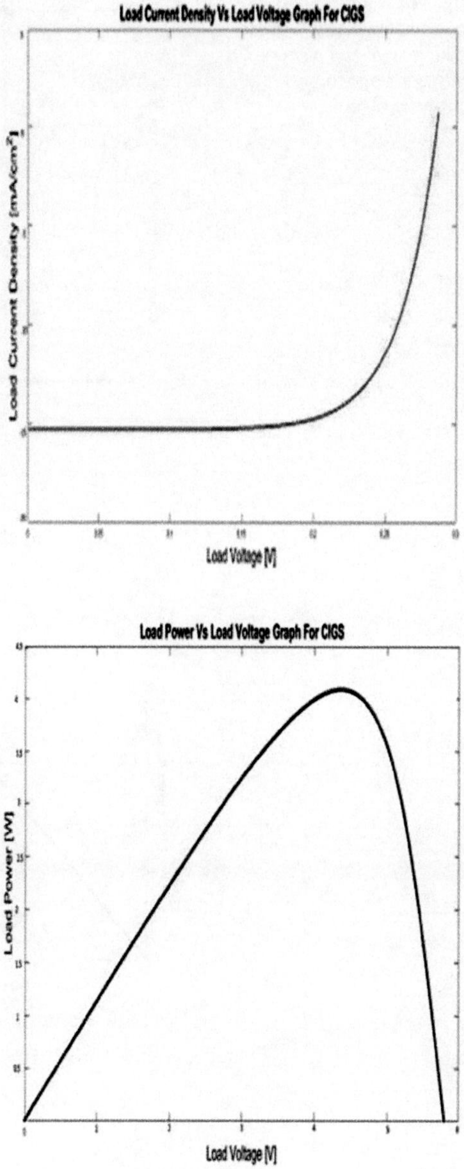

Fig. 24.11 Graph between current density versus voltage and power versus voltage for CIGS

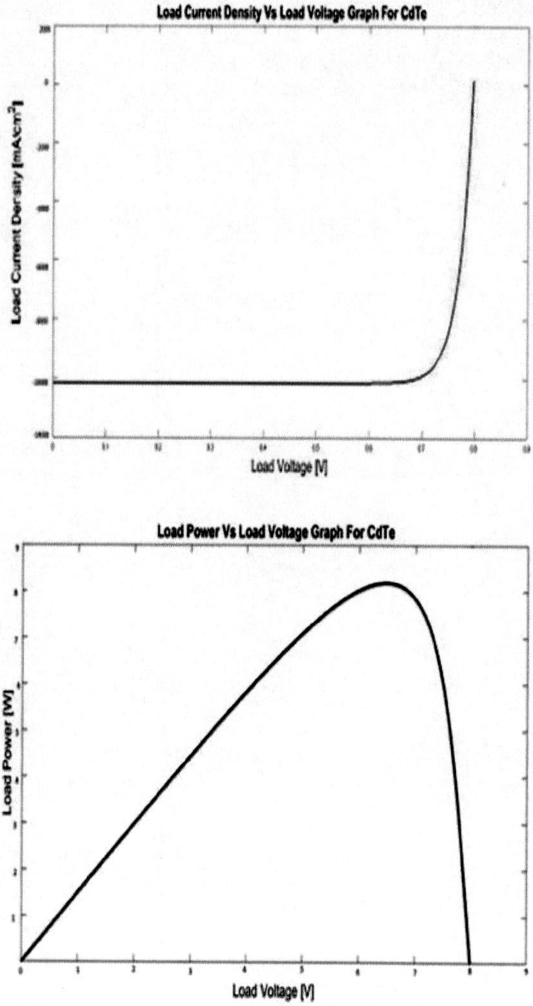

Fig. 24.12 Graph between load current density versus load voltage and power versus voltage for graphene/CdTe

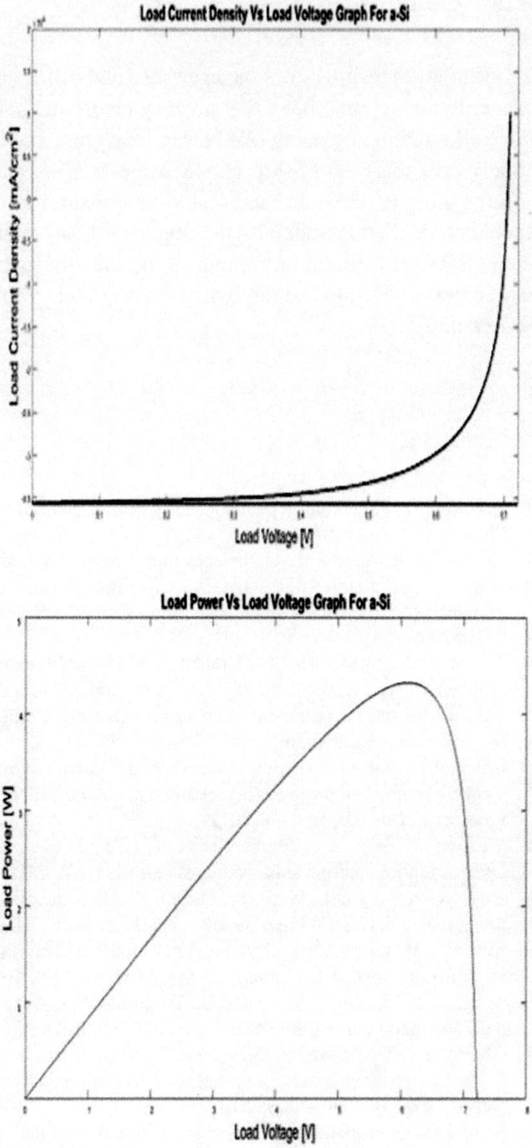

Fig. 24.13 Graph between load current density versus load voltage and power versus voltage for graphene/a-Si

24.5 Conclusion

The simulation result shows its properties and different parameters with their performance. It can be concluded that an easy technique has been introduced to solve the Maxwell equation by using MATLAB. The graphs obtained by the created Simulink models give the relationship between the selected parameters. The reported technique is simple, accurate, and less time-consuming. In future work, the proposed model can be demonstrated by making a hardware model. More parameters can also be taken into consideration for enhancing the efficiency of the solar cells. Other efficient materials can also be sandwiched along with graphene for further improvisation inefficiency.

References

1. Bao, W., et al.: Stacking-dependent band gap and quantum transport in trilayer graphene. Nature Phys. **7**, 948–952 (2011)
2. Rosencher, E., Vinter, B.: Optoelectronics. Cambridge Univ. Press, Cambridge, U.K. (2002)
3. Tongay, S., Lemaitre, M., Miao, X., Gila, B., Appleton, B.R., Hebard, F.: Rectification at graphene-semiconductor interfaces: Zero gap semiconductor-based diodes. Phys. Rev. **X**, 011002 (2012)
4. An, Y., Behnam, A., Pop, E., Ural, A.: Metal-semiconductor-metal photodetectors based on graphene/p-type silicon Schottky junctions. Appl. Phys. Lett. **102**(1), 013110 (2013)
5. Optical absorption coefficient calculator for GaAs [Online]. Available: http://www.cleanroom.byu.edu/OpticalCalc.phtml
6. Chen, W., Seol, G., Rinzler, A.G., Guo, J.: Carrier dynamics and design optimization of electrolyte-induced inversion layer carbon nanotube-Silicon Schottky junction solar cell. Appl. Phys. Lett. **100**(10), 103503 (2012)
7. Wadhwa, P., Seol, G., Petterson, M.K., Guo, J., Rinzler, A.G.: Electrolyte-induced inversion layer Schottky junction solar cells. Nano Lett. **11**(6), 2419–2423 (2011)
8. Hou, J.L., Chang, S.J., Wu, C.H., Hsueh, T.J.: Self-powered ZnO nanowire UV photodetector integrated with GaInP/GaAs/Ge solar cell. IEEE Electron Device Lett. **34**(8), 1023–1025 (2013)
9. Michetti, P., Iannaccone, G.: Analytical model of one-dimensional carbon-based Schottky-barrier transistors. IEEE Trans. Electron Devices **57**(7), 1616–1625 (2010)
10. Tung, L.-T., Mateus, M.V., Kan, E.C.: Tri-gate graphene nanoribbon transistors with transverse-field bandgap modulation. IEEE Trans. Electron Devices **61**(9), 3329–3334 (2014)
11. Gholipour, M., Masoumi, N., Chen, Y.-Y.C., Chen, D., Pourfath, M.: Asymmetric gate Schottky-barrier graphene nanoribbon FETs for low-power design. IEEE Trans. Electron Devices **61**(12), 4000–4006 (2014)
12. Kim, T.G., et al.: Barrier height at the graphene and carbon nanotube junction. IEEE Trans. Electron Devices **61**(6), 2203–2207 (2014)
13. Novoselov, K.S., et al.: Electric field effect in atomically thin carbon films. Science **306**(5696), 666–669 (2004)
14. Hossain, M.B., Muktadhir, M.S., Rana, M.M.: Modelling graphene macroscopic and microscopic conductivity in the subs-shell FDTD method. In: International Conference on Electrical & Electronic Engineering (ICEEE), pp. 53–56 (2015)
15. Anderson, T.J., et al.: Investigation of the epitaxial graphene/p-SiC heterojunction. IEEE Electron Device Lett. **33**(11), 1610–1612 (2012)

16. Chen, W., et al.: Modelling and simulation of graphene-gated graphene-GaAs Schottky junction field-effect solar cell for its performance enhancement. IEEE Trans. (2015). www.ieeexplore.ieee.org
17. Jie, W., Zheng, F., Hao, J.: Graphene/gallium arsenide-based Schottky junction solar cells. Appl. Phys. Lett. **103**(23), 233111 (2013)
18. Fan, G., Zhu, H., Wang, K., Wei, J., Li, X., Shu, Q., Guo N., Wu, D.: Graphene/silicon nanowire Schottky junction for enhanced light harvesting. ACS Appl. Mater. Interfaces **3**, 721–725 (2011)

Chapter 25
Load Frequency Control in Deregulated-Hybrid Power System Integrated with Energy Generation/Storage System

Nagendra Kumar, Mohit Bansal, Shivendra Kaura, and Priyanka Datta

Abstract Wind, solar, fuel cell, small hydro, etc., are emerging technologies for renewable energy and can be built in the future as feasible electricity generation options. This paper introduces the load frequency control (LFC) model in the presence of renewable energy resources for a hybrid system. The study considers hybrid generation system comprising Wind turbine generator (WTG), Solar thermal power system (STPS), Photovoltaic (PV), Diesel source (DEPS), Fuel cell (FC), Battery storage system (BESS), Flywheel (FWES), Ultra-capacitor (UC) and Aqua electrolyzer (AE). For sudden load/generation shifts, or both, the power system frequency deviates. The addition of a renewable system also results in a difficult task for an efficient controller design to be realized. In this study, as an LFC control scheme, a PID controller is used. Two evolutionary approaches named as Big bang big crunch (BBBC) and Harmony Search (HS) have been used to obtain optimal parameters. The performance of the control scheme has been checked in two area and three area thermal-hydro system in the presence of other generation and energy storage sources. A comparative assessment of various quantities like frequency deviation, generation change, etc., has been carried out on the basis of different time-domain parameters.

N. Kumar · M. Bansal (✉) · S. Kaura · P. Datta
G. L. Bajaj Institute of Technology & Management, Greater Noida, U.P., India
e-mail: hod.eee@glbitm.ac.in

N. Kumar
e-mail: nagendra.kumar@glbitm.ac.in

S. Kaura
e-mail: shivendra.kaura@glbitm.ac.in

P. Datta
e-mail: priyanka.datta@glbitm.ac.in

© The Author(s), under exclusive license to Springer Nature Singapore Pte Ltd. 2022

V. V. Rao et al. (eds.), *Computational and Experimental Methods in Mechanical Engineering*, Smart Innovation, Systems and Technologies 239,
https://doi.org/10.1007/978-981-16-2857-3_26

25.1 Introduction

Currently, a rising trend is the incorporation of distributed generation (DG) into existing power systems to meet increasing power needs [1]. The resulting hybrid power system aims to provide its customers with efficient and high-quality service. In addition, the load demand is highly volatile and it is therefore difficult to preserve the equilibrium of generation and load. If there is some mismatch between generation and load and if it persists for a long time, the device frequency can be affected by this. A control mechanism known as automatic generation control (AGC) can achieve this equilibrium [2]. Therefore, the key contribution of AGC is to sustain system frequency and power within limits by tie-line in the event of any system disruption [3, 4]. In AGC, to maintain equilibrium, the controller plays an important role. Many methods have been used in literature, such as classical, stable, adaptive, optimal, nonlinear, modern [5]. A conventional system comprising thermal/hydro or mixture of both has been studied in the past. The PI controller is used in hybrid system studies to control the output powers of the distributed generation system to achieve power balance conditions due to sudden generation/load variations. The frequency is adversely affected in a hybrid system by load deviations and wind power shifts, so the power balance needs to be maintained. In the literature, small-signal study of a hybrid renewable system with energy storage such as battery, pump storage, to provide rapid response and less oscillation was discussed [6]. The power markets need to be restructured and the power supply policies need to be reformed in the new deregulation climate in the electricity market [7, 8]. In response to the deregulation and re-structuring of the power industry, new business opportunities have led many private sectors, from an economic point of view, to participate in small-scale power generation [7]. The PV integration hybrid power system with FC provides promising advantages over independent service [9]. In hybrid energy systems, the conservation of energy plays an essential role in storing and releasing energy at the right time. In addition to the energy storage device combinations such as FWES-BESS and FWES-UC, the hybrid mixtures of all generation sources such as WTG, PV, FC and DEPS provide effective energy management to meet the power balance requirements [6, 9, 10]. Due to low computing costs and high convergence speed over PSO and differential evolution (DE) optimization methods, PID parameters are obtained using BBBC in this paper [7, 8]. This paper also discusses the use of controller architecture based on HS optimization to achieve minimum deviation in frequency. The performance of both the designed control approaches has been tested on two and three area power systems with the inclusion of various DG sources. The research examines different profiles of parameters induced by a sudden shift in the demand for generation and load. All potential disruptions, including the erratic nature of wind speed and other uncertainties, have been checked under the proposed control scheme.

25.2 Hybrid Deregulated Power System

A basic block diagram of the system comprising various sources like WTG, PV, STPS, etc., is shown in Fig. 25.1. It is seen that these all sources are working in area-1 of the considered test system. The other area does not have any DG source to compensate load deviation. It is also evident that the total power supplied in area-1 is the sum of powers of different DG sources and the conventional thermal generators power to satisfy the load demand. Figure 25.2 shows the first-order transfer function

Fig. 25.1 Hybrid system

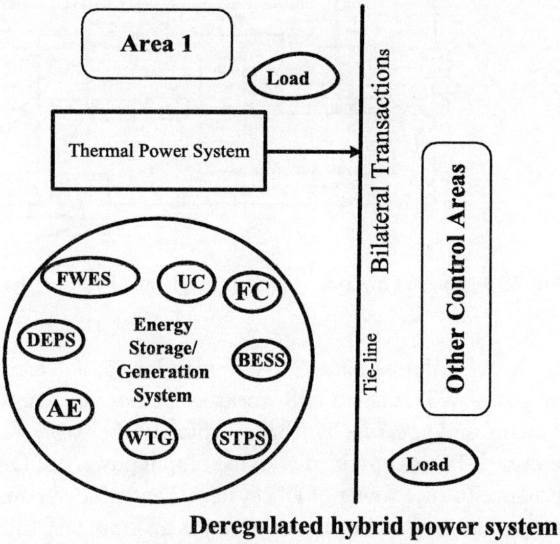

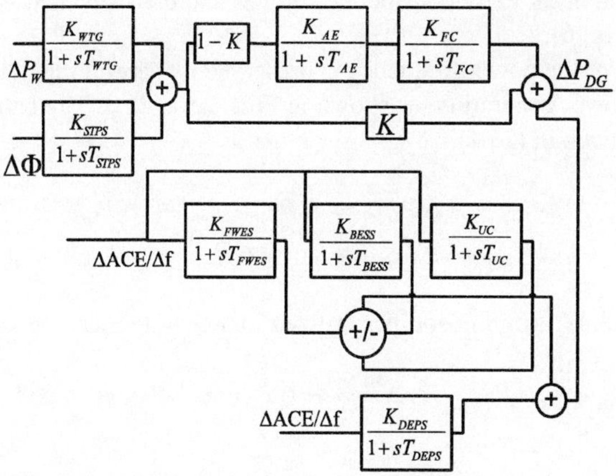

Fig. 25.2 Transfer function model of energy generation system

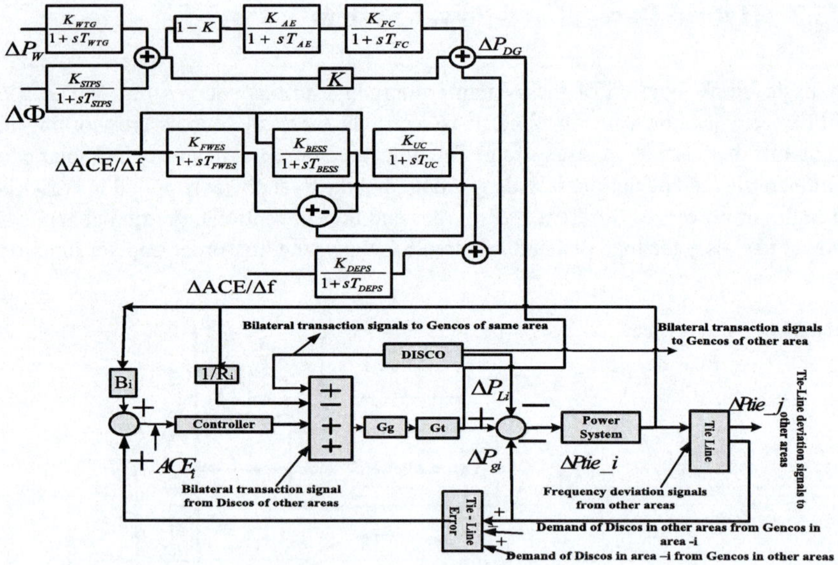

Fig. 25.3 Transfer function model of deregulated-hybrid system

model of different sources used in DG system. Three WTGs have been used. The output of WTGs and STPS works as an input to AE, which absorbs the fluctuations of wind and generate hydrogen, which further works as an input for FC. The output power of FC is combined with the output powers of DEPS, BESS, UC and FWES to produce the net power of DG system. Deviation in Area control Error (ΔACE) or in frequency Δf has been considered as an input to DEPS, BESS, FWES and UC. The complete hybrid (DG + conventional thermal-hydro) system is shown in Fig. 25.3. Brief description of different distributive sources and their values used in this study has been given in [6].

The balance of the power frequency is preserved by careful regulation of various resources for power generation, as shown in Figs. 25.2, 25.3. The balance of power is achieved as given in (25.1),

$$\Delta P_{\text{Total}} = \Delta P_{\text{Conventional}} + \Delta P_{\text{DG}}$$
$$\Delta P_{\text{error}} = \Delta P_{\text{Conventional}} + \Delta P_{\text{DG}} - \Delta P_{\text{Load}} \tag{25.1}$$

The net power of different sources utilized in DG system can be written as (2),

$$\Delta P_{DG} = \Delta P_{WTG} + \Delta P_{STPS} + \Delta P_{FC} - \Delta P_{AE} \pm \Delta P_{FWSS} \pm \Delta P_{BESS} \pm \Delta P_{UC} \tag{25.2}$$

The transfer function of the complete system in respect to frequency deviation can be represented as given in (3),

$$G_{PS}(s) = \frac{\Delta f}{\Delta P_{\text{error}}} = \frac{K_p}{1 + sT_p} \tag{25.3}$$

In a deregulated system, Gencos sells power via poolco or mixed style transactions to various Discos based on economic criteria. In order to achieve the desired generation per DPM, the distribution of ACE among Gencos is proportional to their participation in the AGC. DPM and pu components of Disco load forms demand, i.e., Disco-Genco contract. If P_{ij} is the tie-line power flow from area-i to area-j and k is the total number of areas, the net tie-line power flow from area-i $(Ptie_i)$ in a conventional power system can be written as (25.4)"

$$\text{Ptie}_i = \sum_{\substack{j=1 \\ j \neq i}}^{k} P_{ij} \tag{25.4}$$

In a deregulated situation, because of different transactions, tie-line control shifts. It is possible to describe the net scheduled tie-line power flow from area-i as represented in (25.5)

$$\Delta\text{Ptie}_i\text{\_schd} = \Delta\text{Ptie}_i + \sum_{\substack{j=1 \\ j \neq i}}^{k} Dij - \sum_{\substack{j=1 \\ j \neq i}}^{k} Dji \tag{25.5}$$

where Dji is the Disco demand in area-i from the Genco in area-j, and Dij is the Disco demand in area-j from the Genco in area-i. It is possible to evaluate the tie-line power error as given in (25.6),

$$\Delta\text{Ptie}_i\text{\_error} = \Delta\text{Ptie}_i\text{\_actual} - \Delta\text{Ptie}_i\text{\_schd} \tag{25.6}$$

This error in tie-line is used to generate ACE given in (25.7).

$$\text{ACE}_i = B_i \Delta F_i + \Delta Ptie_i\text{\_error} \tag{25.7}$$

where B_i is frequency bias factor and Δf_i is frequency deviation in area–i. The given equations may be extended, modified as the number of areas or the sources increase.

25.3 Controller Design Using Evolutionary Techniques

Since an optimal control scheme is necessary to keep generation and load at balance. Therefore, a PID control scheme using BBBC and HS algorithms has been used in this study. The step-by-step design is given as follows: Due to quick and easy implementation, PID controllers are by far the most common industrial controllers.

In this study, A PID control scheme given in (8) has been used as a supplementary controller

$$G_{\mathrm{PID}}(s) = K_{\mathrm{Prop}} + \frac{K_{\mathrm{Int}}}{s} + K_{\mathrm{Dev}}s \tag{25.8}$$

To design this control scheme, mean square of area control error is taken as the optimization problem. The objective of the control scheme is to eliminate this mean square of error given in (25.9) and to bring back the frequency and the tie-line area interchange at the schedule values with less overshoot/undershoot, i.e., oscillations in minimum possible settling time.

$$J_{\mathrm{opt\_fun}} = \frac{1}{k}\sum_{i=1}^{k}[(\mathrm{ACE}_i)^2] \tag{25.9}$$

The constraints have been considered in search of the optimal parameters and are given in [2, 3]. After designing the optimization problem, the optimal parameters have been obtained using two search algorithms namely as BBBC and HS. The brief of both the algorithms is given in as the following.

25.3.1 PID Controller Using BBBC

Step1: One PID controller is considered for each area. Population for each parameter (K_{Prop}, K_{Int}, K_{Dev}) of the considered PID controllers is generated in this step. This is called Big Bang. **Step2**: It involves in the determination of fitness function. **Step3**: Center of mass is evaluated in this step using step 2. **Step4**: New parameters have been generated. **Step5**: It produces fitness function of newly generated parameters. **Step6**: This step gives optimum parameters of the controllers. The complete algorithm with step-by-step modelling can be referred from [2–4, 7, 8]. A 30-population, 3-variables and 10-parameter limiting size have been used as BBBC parameters to obtain optimal PID control scheme.

25.3.2 PID Controller Using Harmony Search (HS)

PID control scheme coefficients were also obtained using HS algorithm given by Geem et al. [11]. HS is based on the performance process of natural music that occurs when a musician looks for a better state of harmony [3]. HS comprises harmony memory/pitch change/randomization to produce optimal solution. The complete step-by-step implementation of HS is given in [2]. In this study, 30-HS memory, 0.9-HS rate, 0.5-pitch rate, (0.0001–1)-BW and 3-variables have been selected to generated optimal parameters of PID control scheme.

25.4 Results and Discussion

The simulation analysis is discussed in this section with the controller mentioned above for the deregulated-hybrid power system. The frequency/tie-line power/change in generation, etc., should be managed by the AGC scheme in a deregulated hybrid power system. Therefore, when existing thermal resources with DG resources are linked to other systems, the response variation needs to be examined. Two cases have been considered for two test systems, i.e., two area and three area power system.

25.4.1 *Case1: Two Area Power System*

On a two area power system, the performance of the PID controller was checked. Considered parameters for two area AGC system have been taken from [2–4, 7, 8]. Table 25.1 shows the optimum values of the PID controller parameters computed using BBBC and HS. Both Discos have power contracts with Gencos in compliance with the following DPM.

$$\text{DPM} = \begin{bmatrix} 0.5\ 0.5\ 0\ 0 \\ 0.5\ 0.5\ 0\ 0 \\ 0\ \ 0.0\ 0\ 0 \\ 0\ \ \ 0\ \ 0\ 0 \end{bmatrix}$$

25.4.1.1 A Sudden Change in Input Power of WTGs and STPS Along with a Sudden Change of Load in the Area1

This case deals in the sudden/step variation in wind input, i.e., $\Delta P_{L1} = 0.2 - 0.4$ pu, $\Delta P_{\text{WTG}} = 0.1 - 0.12$ pu and $\Delta P_{\text{STPS}} = 0.2 - 0.1$ pu. Area-2 has a load disturbance of $\Delta P_{L2} = 0$ pu. These changes in input of wind, solar-thermal and load in area-1 are given in Fig. 25.4a. At time $t = 30$ s, it is seen that once the load deviates the frequency gets disturbed as seen in Fig. 25.4b. It is also seen that the deviation in frequency is settling down at the desired value in steady state. The dip in frequency is more in hydro area than thermal area since it is being fed by DG which is helping to keep

Table 25.1 Optimum Values for PID controllers

BBBC			HS		
K_{Prop}	K_{Int}	K_{Dev}	K_{Prop}	K_{Int}	K_{Dev}
−0.999	−0.999	−1	−3.206	−2.411	−1.054

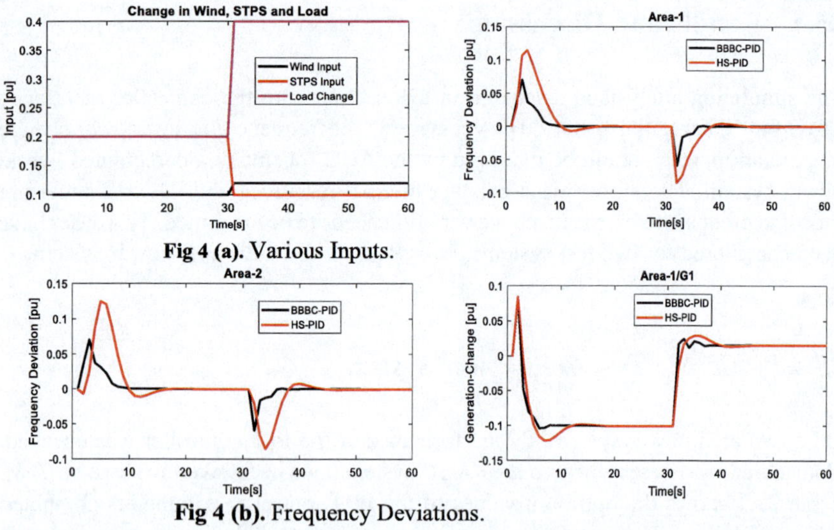

Fig. 25.4 **a** Various inputs. **b** Frequency deviations. **c** Generation change. **d** Tie-line power and ACE. **e** Generation change of different sources. Responses of various quantities for two area system

balance in generation/load. Figure 25.4c shows the generation change in thermal-hydro areas using only conventional energy sources. It is evident that the generators in hydro area are settling to the 1.2 pu, as required in steady state. Figure 25.4d shows that tie-line power and ACE settle down to the specified values in steady state after being disturbed. It is seen that most of the load demand is taking up by DG sources in thermal system given in Fig. 25.4e.

25.4.2 *Case2: Three Area Power System*

For this case, a three area AGC scheme is considered to check the performance of the designed control scheme. Two Discos and two Gencos are in area1 in the three area scheme. One Disco and one Genco are considered in area2 and area3. A GRC of 3 percent/minute (region 2–3) is considered to check the output of the controller. The parameters and their values of three area system have been taken from [2]. PID parameters are given in Table 25.2.

$$\text{DPM} = \begin{pmatrix} 0.4 & 0.3 & 0.2 & 0.1 \\ 0.2 & 0.2 & 0.3 & 0.3 \\ 0.1 & 0.3 & 0.4 & 0.2 \\ 0.3 & 0.2 & 0.2 & 0.3 \end{pmatrix}$$

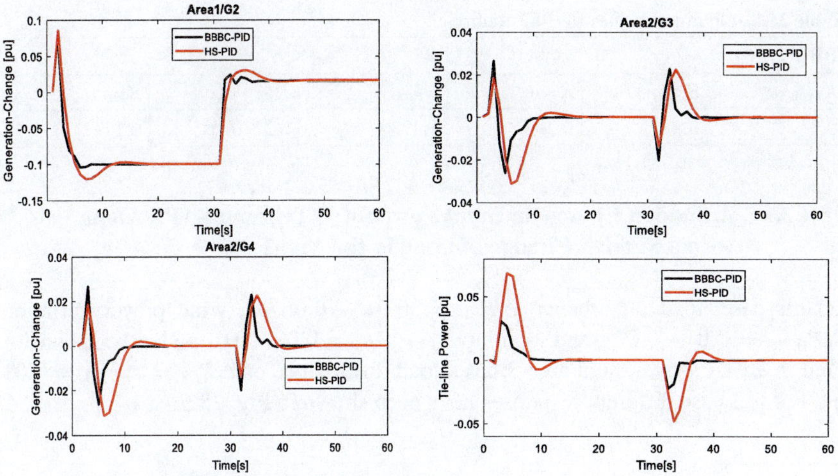

Fig 4 (c). Generation change.

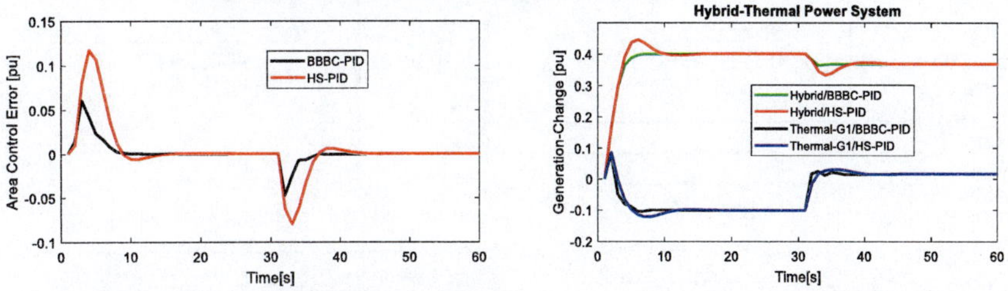

Fig 4 (d). Tie-line power and ACE.

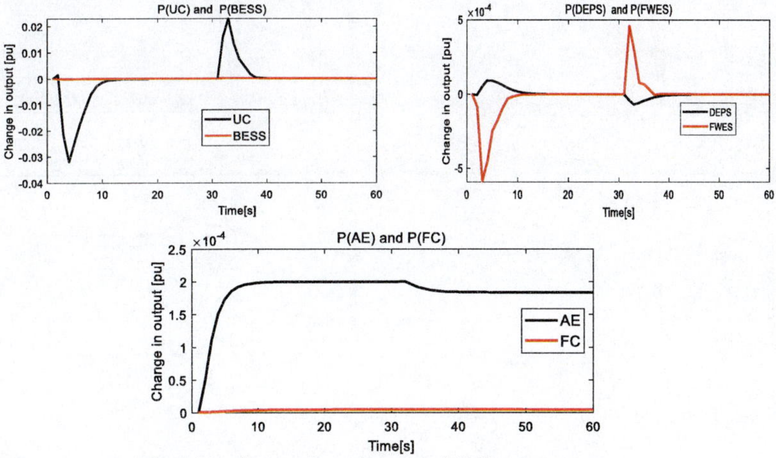

Fig 4(e). Generation change of different sources.

Fig. 25.4 (continued)

Table 25.2 Optimum Values for PID controllers

BBBC			HS		
K_{Prop}	K_{Int}	K_{Dev}	K_{Prop}	K_{Int}	K_{Dev}
0.1	0.2	0.2	1.4788	0.3270	0.1797

25.4.2.1 A Sudden Change in Input Power of WTGs and STPS Along with a Sudden Change of Load in the Area1

In this case, load disturbance $\Delta P_{L1} = 0.01 - 0.06$ pu, wind power variation $\Delta P_{WTG} = 0.01 - 0.02$ p and solar input $\Delta P_{STPS} = 0.02 - 0.01$ pu has been considered in area-1. Area-2 and area-3 has a load disturbance of $\Delta P_{L2} = \Delta P_{L3} = 0.01$ pu. For this case different responses have been shown in Fig. 25.5.

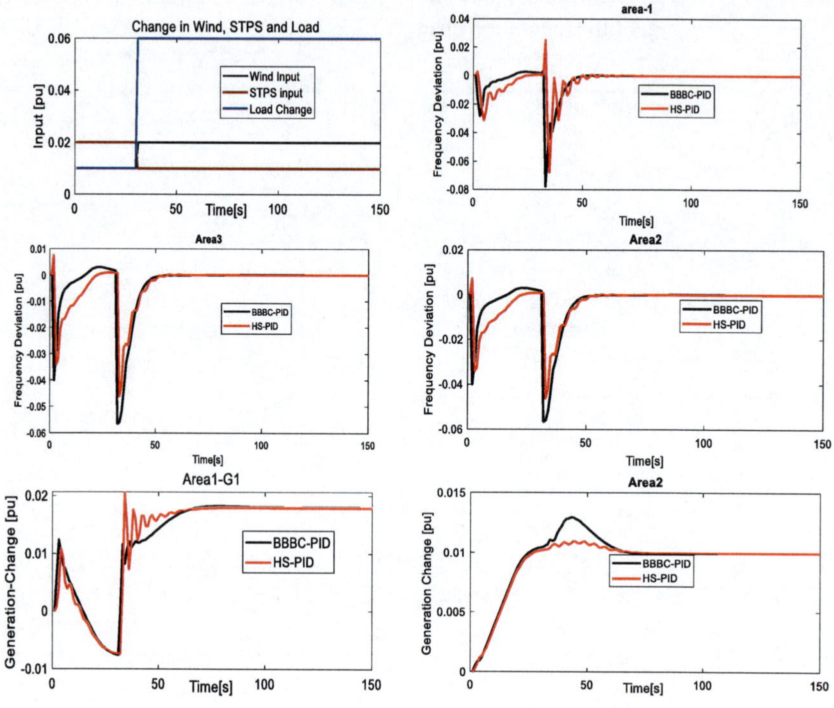

Fig. 25.5 Responses of various quantities for three area system

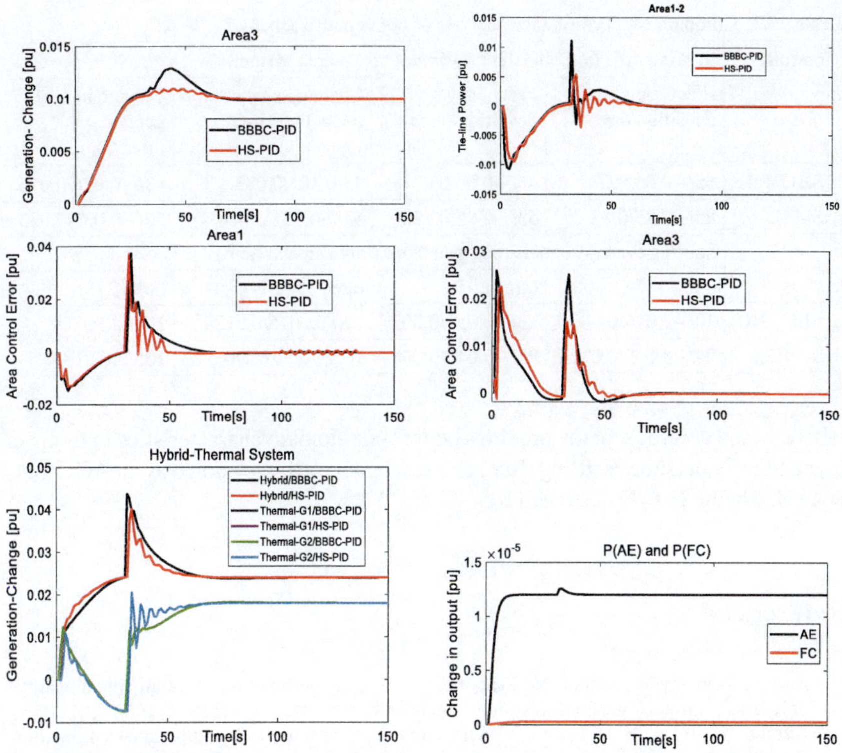

Fig. 25.5 (continued)

25.5 Conclusion

The study of frequency deviation in hybrid systems with different combinations of elements for energy storage is carried out in this paper. The test system is considered to be a two area thermal-hydro. The DG sources (models of a hybrid generation/energy storage system) have been considered in area-1 of the test system. Wind turbine, diesel set, aqua electrolyzer, solar thermal systems, battery storage, flywheels, ultra-capacitor and fuel cell are parts of the hybrid system tested. Hydrogen from the aqua electrolyzer is used as an input to fuel cell. The load model/wind power/solar model is also properly chosen to check the performance of the hybrid system being studied. A variety of load change has been considered. The PID controller is constructed to reduce mean square of area control error. BBBC and HS methods have been used to obtain optimum controller parameters. In time-domain simulations, the output of each controller is analyzed from the dynamic behavior of frequency, generation, tie-line power ACE, Output of DG sources. Comparison between the results obtained using both approaches is shown in Table 25.3. It is evident from the comparison that apart from few instances of increased magnitude in oscillations,

Table 25.3 Comparative performance analysis of both control schemes

Controller	Settling time (s)/undershoot/overshoot (two area system)			
	Frequency deviation (area-1)	Frequency deviation (area-2)	Change in generation (area-1) (G1–G2)	Change in generation (area-2) (G3–G4)
BBBC–PID	45/−0.06/0.07	45/−0.06/0.07	45/0.104/0.083	43/−0.026/0.026
HS–PID	55/−0.85/0.12	53/−0.85/0.128	55/−0.121/0.086	50/−0.031/0.023
	Settling time (s)/undershoot/overshoot (three area system)			
	(area-1)	(area 2–3)	(area-1) (G1–G2)	(area 2–3)
BBBC–PID	70/−0.076/0.003	75/−0.057/0.003	100/0.0075/0.0124	100/0/0.0128
HS–PID	90/−0.069/0.0253	95/−0.046/0.008	105/−0.00748/0.021	105/0/0.0109

BBBC based control scheme provides better time-domain characteristics in terms of (overshoot/undershoot/settling time) and can be an effective choice as an AGC/LFC control scheme in hybrid system too.

References

1. Ray, P., Mohanty, S., Kishor, N.: Proportional integral controller based small signal analysis of hybrid distributed generation system. Energy Convers. Manag. **52**, 1943–1954 (2011)
2. Kumar, N., Tyagi, B., Kumar, V.: Multi Area AGC scheme using imperialist competition algorithm in restructured power system. Appl. Soft Comput. Elsevier **48**, 160–168 (2016)
3. Kumar, N., Tyagi, B., Kumar, V.: Optimization of PID parameters using BBBC for multi-area AGC scheme in deregulated power system. Turk. J. Electr. Eng. Comput. Sci. **24**, 4105–4116 (2015)
4. Kumar, N., Tyagi, B., Kumar, V.: Multi-area deregulated automatic generation control scheme of power system using imperialist competitive algorithm based robust controller. IETE J. Res.-Taylor & Francis **64**, 528–537 (2017)
5. Pandey, S.K., Mohanty, S.R., Kishor, N.: A literature survey on load frequency control for conventional and distribution generation power systems. Renew. Sustain. Energy Rev. **25**, 318–334 (2013)
6. Lee, D.J., Wang, L.: Small-signal stability analysis of an autonomous hybrid renewable energy power generation/energy storage system part I: time-domain simulations. IEEE Trans. Energy Conver. **23**(1), 311–320 (2008)
7. Kumar, N., Tyagi, B., Kumar, V.: Deregulated multiarea AGC scheme using BBBC-FOPID controller. Arab. J. Sci. Eng. Springer **42**, 2641–2649 (2016)
8. Kumar, N., Tyagi, B., Kumar, V.: Application of fractional order PID controller for AGC under deregulated environment. Int. J. Autom. Comput. Springer **15**, 84–93 (2017)
9. Gao, W.: Performance comparison of a fuel cell-battery hybrid power train and a fuel cell-ultra capacitor hybrid power train. IEEE Trans. Veh. Technol. **54**(3), 846–855 (2005)
10. Weissbach, R.S., Karady, G.G., Farmer, R.G.: Dynamic voltage compensation on distribution feeders using fly-wheel energy storage. IEEE Trans. Power Del. **14**(2), 465–471 (1999)
11. Geem, Z.W., Kim, J.H., Loganathan, G.V.: A new heuristic optimization algorithm: harmony search. Simulation **76**, 60–68 (2001)

Chapter 26
Transfer Learning-Based Brain Tumor Detection Using MR Images

Priyanka Datta and Rajesh Rohilla

Abstract One of the most deadly diseases in humans is brain tumor. For clinicians, MRI scan plays a key role in diagnosing and treating tumor. For brain tumor diagnosis, surgical approaches are usually suggested. But the radiologist's analysis of the medical image is time-consuming and also accuracy totally relies upon their skill. Now, Deep learning-based models have gained considerable interest in the diagnosis and treatment of diseases in medical field. As the medical images are limited, so it is a daunting task to train CNN from start and to implement deep learning. In this paper, we develop an automatic brain tumor detection method based on the pre-trained convolutional neural network architectures such as VGG-16, VGG-19, InceptionV3, ResNet50, ResNet101 and EfficientNetB1. The test accuracy achieved with VGG16 and ResNet101 gives highest performance accuracy among all other pretrained network.

26.1 Introduction

Brain tumor is a mass (i.e. benign or malignant) which is produced by tissue besieging the brain or skull within the brain which impacts person's life explicitly. These tumors cultivate irregularly in the brain and put pressure around them [1]. Due to this pressure, different brain disorders are induced in human body. Side effects in patients due to these disorders are dizziness, headache, fainting attacks, paralysis, etc. As stated by WHO, tumor in brain accounts for less than 2% of human cancers in the cancer report; however, extreme bleakness and problems are registered [2]. The Cancer Research Corporation of UK estimated that almost 5200 causalities are recorded per year in UK due to brain disorders and skull tumors [3].

P. Datta (✉)
GL Bajaj Institute of Technology and Management, Greater Noida, U.P., India

P. Datta · R. Rohilla
Delhi Technological University, Delhi, India

© The Author(s), under exclusive license to Springer Nature Singapore Pte Ltd. 2022
V. V. Rao et al. (eds.), *Computational and Experimental Methods in Mechanical Engineering*, Smart Innovation, Systems and Technologies 239,
https://doi.org/10.1007/978-981-16-2857-3_27

Deep learning (DL) has recently been used mainly in medical imaging. Conventional machine learning involves a great deal of domain expertise, human interaction to retrieve the hand-engineered features, i.e. used by classifiers for classification and detection of image patterns. Specialist manual annotation takes a lot of time. DL algorithms, however, maps unprocessed data, i.e. pixel for images directly into outputs, i.e. image classes. With the introduction of AlexNet [4] in 2012, the popularity of DL improved with ImageNet competition, which includes over 1 million images with 1000 different object categories. AlexNet has shown better results in this challenge, as in comparison to other state-of-the-art results obtained from the group of computer vision. DL has advanced fast, thus further significant work has regularly appeared in the area of medical imaging. Different researchers have researched DL in medical imaging [5–9], although some have surveyed individual imaging, i.e. magnetic resonance (MR) imaging, i.e. MRI [10–12], ultrasound (US) [13] and electroencephalogram (EEG) [14]. Convolutional neural networks (CNNs) are the most useful between many DL techniques that were used to actually solve problems in different applications, including detection, segmentation and classification, etc. MRI is a type of medical image modality that is measured by its non-invasiveness as a safe technique and has a reasonable soft-tissue contrast. As attempted by ionizing radiation-based methods, this does not alter the construction, properties and characteristics of particles. The MRI setting does however, offer potential hazards due to 3 magnetic fields that are robust static magnetic fields, gradient-based magnetic fields and pulsed radiofrequency fields that are used to generate 3D images [14]. Eventually, MRI can provide useful information on tissue structures, i.e. shape, size and location. MRI is being categorized as structural and functional imaging. Examples of structural imaging are T1-W MRI, T2-W MRI, Diffusion Tensor Imaging (DTI) and functional imaging is resting-state functional MRI (rs-fMRI) [10]. Nazir et al. [15] categorized brain MRI images into two classes, i.e. benign and malignant. They utilized filter methodology to remove the noise in the MR images as a pre-processing stage for the dataset. Using the normal color moment of every image, they then extracted features. They labeled extracted features by the artificial neural network (ANN). The prediction accuracy obtained was 91.8% in their method. Shree et al. [16] categorized brain MRI images into two classes, i.e. benign and malignant. They extracted features using discrete wavelet transform (DWT) and gray-level co-occurrence matrix (GLCM) preceded by morphological process. The probabilistic neural network was used as a classifier to detect location of tumor in brain MR images. Kanmani et al. [17] categorized brain MR images into two classes, i.e. normal and abnormal. To improve the efficacy of classification accuracy, they utilized the threshold-based region optimization technique along with segmentation. Ahmed et al. [18] also categorized brain MRI images into two classes, i.e. normal and abnormal. For this, they introduced a combination of Artificial neural network technique and a gray wolf optimization technique. Five distinct CNN models have been used by Abiwinanda et al. [19] and their one of the model attained the maximum accuracy. El-Dahshan et al. [20] also categorized brain MR images into two classes, i.e. normal and abnormal using the Discrete Wavelet Transform (DWT) technique for the extraction of features and the Principal component analysis method to reduce the features. Then, they used two classifiers (i.e. feedorward ANN and the k-Nearest neighbor for the identification of images.

In this paper, we diagnose the brain tumor in MR images by extracting features using DL based transfer learning technique. For this, we use six pre-trained models, i.e. VGG16, VGG19, InceptionV3, ResNet50, ResNet101, EfficientNetB1. This framework makes it a lot easier for radiologists to intervene, helps them solve the problem of classification of brain tumors, i.e. whether the tumor is present or not, and helps to develop an appropriate treatment.

Further, we organized the paper into different sections as follows: Sect. 26.2 depicts the information about material and different models used. Section 26.3 illustrates the experimental results and discussion. Finally, in Sect. 26.4, conclusion and future scope are specified, respectively.

26.2 Material and Models

26.2.1 Dataset

The data set used in this paper consists of freely accessible 646 T1 weighted MRI images of brain labeled as non-tumored and tumored attained from The Cancer Imaging Archive (TCIA) Publicly Accessible repository [21]. The images were obtained from 20 patients who identified with glioblastoma. The dataset has 548 samples of tumored images and 98 samples of non-tumored images. The images are in JPG/JPEG format. Figure 26.1. displays few images of Non-Tumored and Tumored classes used in the dataset.

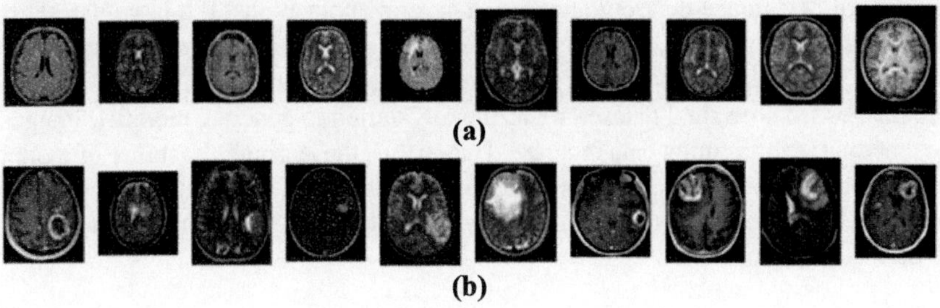

Fig. 26.1 **a** Non-tumored images, **b** Tumored images

Fig. 26.2 Steps involved in data pre-processing

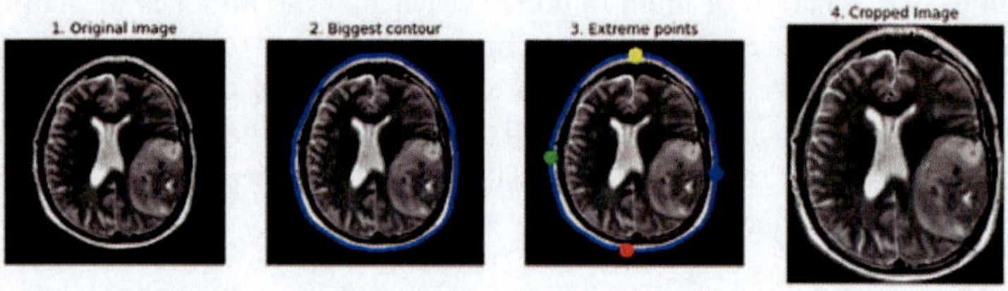

Fig. 26.3 Cropped image after finding extreme points in contour

26.2.2 Data Pre-processing and Augmentation

Data pre-processing can be referred as the transformation of raw data into a form that is more easy to interpret and renders the images more appropriate for any further processing. Figure 26.2 specifies the various steps involved in Data Pre-processing.

The MRI data set contains 253 images, which are divided into 193 training images, 50 validation images and 10 test images. After data splitting the images are cropped to obtain only the portion of brain by using the technique given in Ref. [22]. This method is used for assessing the extreme points within contour lines. Figure 26.3. shows the cropped image after finding extreme points in contour.

However, acquisition processes of MR image are normally costly and complex, the size of MR image dataset is limited in several applications. If a large dataset is present, DL will perform much better. As the dataset used in this work is small, we use data augmentation for artificially increasing the size of training data. For this, the images from original dataset are artificially varied to generate modified images so that the size of training data increase. Due to this, the learning capability of model increase and it became more generalized for unseen data. Then on the other hand, it becomes less susceptible to overfitting. Figure 26.4. shows some samples of images after data augmentation.

26.2.3 Pre-trained CNNs Architectures

DL became prominent with the increasing availability of various datasets and fast gaming graphical processing units (GPUs) nearly a decade ago. DL technique includes numerous layers which learn enormous features from input image and these

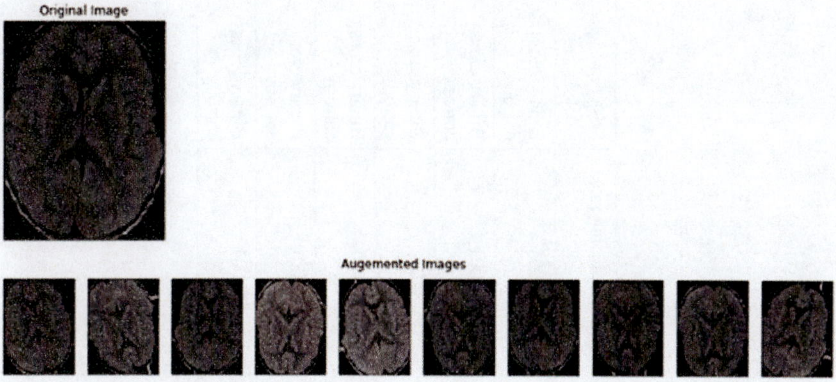

Fig. 26.4 Images samples after augmentation

are used for analyses of various images by providing huge dataset of unlabeled or labeled images [23].

Convolutional neural network (CNN) is frequently utilized DL system architecture in analysis of medical imaging [24]. CNN architectures were built for learning the spatial hierarchies of different features through multiple blocks which includes convolution layers, non-linear layer, pooling layers and fully connected layers. Fully connected layers choose the most effective features and move them to the classification layer. Different pretrained CNN architectures are used in our study. Table 26.1 provides different parameter values used in the different CNN architecture.

In reality, it is unlikely that a person can train a full CNN model from scratch as datasets with adequate sample data are usually not feasible. Persistently, pre-training a CNN on large datasets, e.g. ImageNet seems to have become a common practice. Transfer learning (TL) [25, 26] can be seen as better learning in a novel problem by extracting features obtained from a comparable problem that exists. TL is a method in which features obtained from one data set can be used for other datasets.

VGG16 and VGG19 models [27]—The impact of CNN depth on its performance in computer vision was analysed by Simonyan K. and Zisserman A. Using very small convolution filters, they drive the depth from 11 to 19 weight layers of established VGGNet network. The variations that has 16 and 19 weight layers, referred as VGG16 and VGG19 and they do well. With the increase in depth, the classification error reduces and saturates when the depth exceeds 19 layers. In visual representations, authors affirm the value of depth.

InceptionV3 model [28]—It is a Google Brain Team 48-layer CNN that is trained on the ImageNet database and categorizes objects into 1,000 classes. In comparison to the other inception models and process, it trained much quicker.

ResNet model [29]—Microsoft developed deep residual learning platform, i.e. ResNet, which uses residual learning for simplification of deeper network training and decrease errors through increase in depth. This architecture suggested several

Table 26.1 Parameters of CNN architectures used

Simulation environment	Framework	Mini batch	Data augmentations Parameters	CNN architecture	Input size	Optimization technique	Learning rate	Loss type
			Vertical Flip = True	VGG16	240 × 240	Ftrl	0.001	Binary cross entropy
			Horizontal Flip = True	VGG19	240 × 240	Ftrl	0.001	Binary cross entropy
Python	Keras	16	Random Brightness Range = [0.3 1]	InceptionV3	240 × 240	Ftrl	0.01	Binary cross entropy
			Shear Range = 0.1°	ResNet50	240 × 240	Ftrl	0.001	Binary cross entropy
			Rotate Limit = 10°	ResNet101	240 × 240	Ftrl	0.001	Binary cross entropy
				EfficientNetB1	240 × 240	Ftrl	0.001	Binary cross entropy

structures such as 18-layers, 34-layers, 50-layers and 101-layers framework. This structure is less complexity and more deep in comparison to VGG network.

EfficientNetB1 model [30]—Google Brain Team developed a CNN model named as EfficientNet. These researchers studied the model scaling and identified that carefully balancing the depth, width and resolution of the network can lead to better performance. In order to develop a new model, they scaled neural network to generate more deep learning models, that achieve significantly improved efficacy and accuracy in comparison to prior used CNN.

26.2.4 Proposed Methodology

This proposed diagnosis method aims to improve the accuracy of the detection of brain MRI images by using DL models and TL method. The flow diagram of the suggested work performed is shown in Fig. 26.5.

TL is the process of learning new models generated by new data using the features provided by a pre-trained framework. In order to learn low level features which are utilized to encode medical images, TL is used in which Deep Learning model is pre-trained on a huge data set of images from various medical image modalities

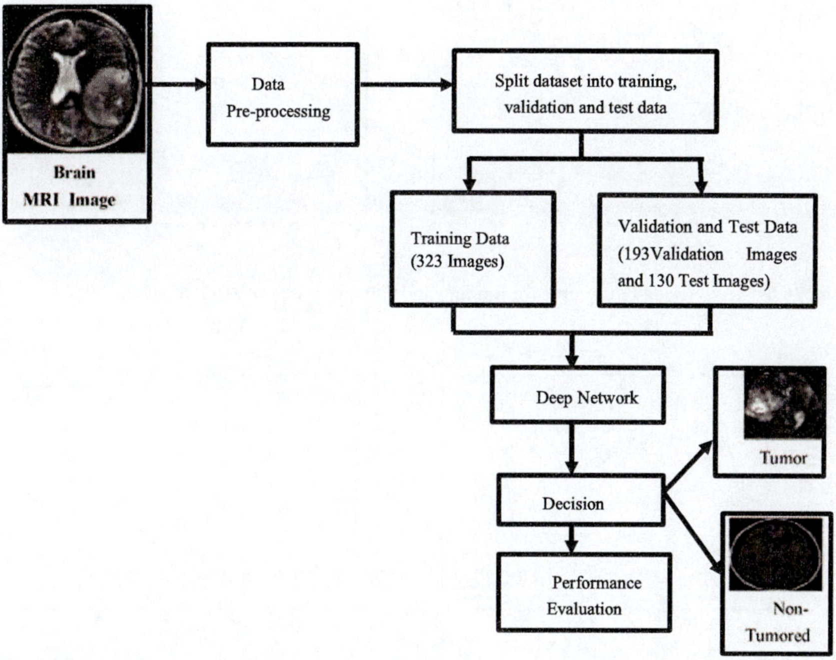

Fig. 26.5 Flow chart of proposed methodology

or different domains. The use of pre-trained DL models allows to learn new tasks quickly. In this work, we first pre-process the MRI data to run the built model and test it. Then, we trained six different pretrained CNN models, i.e. VGG16, VGG19, InceptionV3, ResNet50, ResNet101 and EfficientNetB1 with brain MR Images and then, utilized them to classify tumored and non-tumored images using TL technique.

26.3 Experimental Results and Discussion

Each model has been trained for 50 epochs. Figures 26.6 and 26.7 were attained by training the models using brain MR image dataset for 50 epochs and represents the accuracy and loss curve for training and validation set for VGG16 and ResNet101 models. Table 26.2 displays performance of different model used in our study.

Performance for the different technique was evaluated in terms of different measures such Accuracy, Precision, Recall, F1 score, Kappa and AUC. VGG16 and ResNet101 has highest test accuracy. Training accuracy of ResNet101 is 93%. But, F1 score of VGG16 is more as the precision and recall are more in this case.

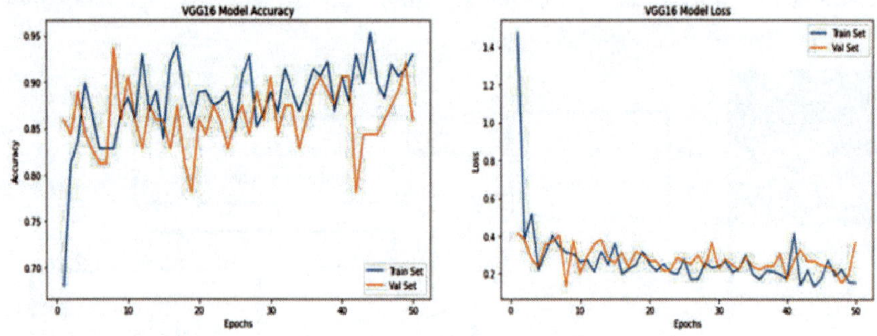

Fig. 26.6 Accuracy and loss curve for training and validation set for VGG16 model

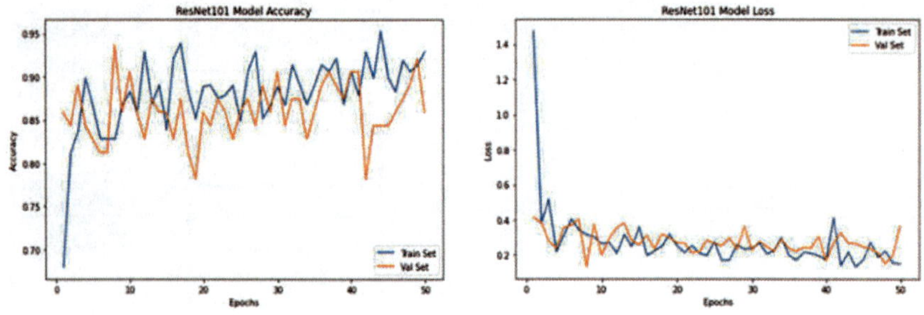

Fig. 26.7 Accuracy and loss curve for training and validation set for ResNet101 model

Table 26.2 Performance of different. Pretrained models

CNN architecture	Accuracy	Precision	Recall	F1 score	Kappa	AUC	Confusion matrix
VGG16	0.92	0.97	0.94	0.95	0.73	0.89	[17 3] [7 103]
VGG19	0.87	0.93	0.91	0.92	0.53	0.78	[13 7] [10 100]
InceptionV3	0.85	0.85	1.0	0.92	0.00	0.50	[0 20] [0 110]
ResNet50	0.91	0.94	0.95	0.95	0.63	0.80	[13 7] [5 105]
ResNet101	0.92	0.96	0.94	0.94	0.69	0.87	[16 4] [7 103]
EfficientNetB1	0.89	0.94	0.94	0.94	0.59	0.79	[13 7] [7 103]

By analyzing the performance measures, we found that both ResNet101 and VGG16 has better performance as compared to other pretrained networks.

26.4 Conclusion

In this paper, a completely automatic system is used for diagnosing tumor in brain MRI images. For this, numerous DL-based pretrained CNN architectures are used. The suggested solution applied the theory of DL employing four pre-trained networks that use the TL approach to enhance diagnosis of brain tumor. The dataset used contains images of 2 classes, i.e. abnormal data which have tumor and normal data which don't have tumor. Although the dataset is not huge, the image data augmentation was relatively good enough to produce excellent performance. As exhibited in Table, it is noticeable that TL through VGG16 and ResNet101 gives highest performance accuracy among all other pretrained network used in this paper. In future work, we will use our model for different medical imaging modalities from different fields and for larger dataset to improve the robustness. Substantial hyperparameter tuning as well as a better pre-processing approach can be conceived that can further improve efficiency of the model. In future work, we are also planning to further classify different types of tumors using larger dataset.

References

1. Grisold, W., Grisold, A.: Cancer around the brain. Neuro-Oncology Practice **1**(1), 13–21 (2014)
2. Stewart B.W., Wild C.P.: World cancer report 2014, IARC. IARC Nonserial Publ, Lyon, France, p. 630 (2014)

3. Brain, other CNS and intracranial tumours statistics Cancer Research UK. (n.d.), 2020 from https://www.cancerresearchuk.org/health-professional/cancer-statistics/statistics-by-cancer-type/brain-other-cns-and-intracranial-tumours?_ga=2.245798211.613033350.1594394057-1698240228.1594394057 (Retrieved July 10, 2020)
4. Krizhevsky, A., Sutskever, I., Hinton, G.E.: ImageNet classification with deep convolutional neural networks, In: Pereira, F., Burges, C.J.C., Bottou, L., Weinberger, K.Q. (eds.) Advances in Neural Information Processing Systems, vol. 25, pp. 1097–1105 (2012)
5. Litjens, G., Kooi, T., Ehteshami, B.B.: A survey on deep learning in medical image analysis. Med. Image Anal. **42**, 60–88 (2017)
6. Shen, D., Wu, G., Zhang, D.: Machine learning in medical imaging. Comput. Med. Imaging Graph 1–2 (2015)
7. Shen, D., Wu, G., Suk, H.-I.: Deep learning in medical image analysis. Annu. Rev. Biomed. Eng. **19**, 221–48 (2017)
8. Suzuki, K.: Survey of deep learning applications to medical image analysis. Med Imaging Technol. **35**, 212–226 (2017)
9. Ker, J., Wang, L., Rao, J., Lim, T.: Deep learning applications in medical image analysis. IEEE Access **6**, 9375–9389 (2018)
10. Liu, J., et al.: Applications of deep learning to mri images: a survey. Big Data Min. Anal. **1**(1), 1–18 (2018)
11. Datta, P., Rohilla, R.: An introduction to deep learning applications. In: MRI Images, 2019 2nd International Conference on Power Energy, Environment and Intelligent Control (PEEIC), Greater Noida, India, pp. 458–465 (2019)
12. Lundervold, A.S., Lundervold, A.: An overview of deep learning in medical imaging focusing on MRI. Z. Med. Phys. **29**, 102–127 (2019)
13. Liu, S., Wang, Y., Yang, X., Lei, B., Liu, L., Li, S., Ni, D., Wang, T.: Deep learning in medical ultrasound analysis: a review. Engineering, 261–275 (2019)
14. Movahedi, F., Coyle, J.L., Sejdic, E.: Deep belief networks for electroencephalography: A review of recent contributions and future outlooks. IEEE J. Biomed. Health Inform. **22**(03), 642–652 (2018)
15. Nazir, M., Wahid, F., Khan, S.A.: A simple and intelligent approach for brain MRI classification. J. Intell. Fuzzy Syst. **28**, 1127–1135 (2015)
16. Shree, N.V., Kumar, T.N.R.: Identification and classification of brain tumor MRI images with feature extraction using DWT and probabilistic neural network. Brain Inf. **5**, 23–30 (2018)
17. Kanmani, P., Marikkannu, P.: MRI brain images classification: a multi-level threshold based region optimization technique. J. Med. Syst. **42**, 1–12 (2018)
18. Ahmed, H.M., Youssef, B.A.B., Elkorany, A.S., Saleeb, A.A., El-Samie, F.A.: Hybrid gray wolf optimizer–artificial neural network classification approach for magnetic resonance brain images. Appl. Opt. **57**, B25–B31 (2018)
19. Abiwinanda, N., Hanif, M., Hesaputra, S.T., Handayani, A., Mengko, T.R.: Brain tumor classification using convolutional neural network. In: Lhotska L., Sukupova L., Lacković I., Ibbott G. (eds) World Congress on Medical Physics and Biomedical Engineering 2018. IFMBE Proceedings, vol 68/1, Springer, Singapore (2019)
20. El-Dahshan, E.S.A., Hosny, T., Salem, A.B.M.: Hybrid intelligent techniques for MRI brain images classification. Digit. Signal Process. **20**(2), 433–441 (2010)
21. Clark, K., Vendt, B., Smith, K., Freymann, J., Kirby, J., Koppel, P., Moore, S., Phillips, S., Maffitt, D., Pringle, M., Tarbox, L., Prior, F.: The cancer imaging archive (TCIA): maintaining and operating a public information repository. J. Digit. Imaging **26**(6), 1045–1057 (2013)
22. Rosebrock, A.: Finding Extreme Points in Contours with OpenCV.PyImageSearch, 11, https://www.pyimagesearch.com/2016/04/11/findingextreme-points-in-contours-with-opencv/ (2016)
23. Schmidhuber, J.: Deep learning in neural networks: an overview. Neural Netw. **6**, 85–117 (2015)
24. LeCun, Y., Kavukcuoglu, K., Farabet, C.: Convolutional networks and applications in vision. In: IEEE International Symposium on Circuits and Systems, pp. 253–256 (2010)

25. Pan, S.J., Yang, Q.: A survey on transfer learning. IEEE Trans Knowl Data Eng. **22**, 1345–1359 (2010)
26. Weiss, K., Khoshgoftaar, T.M., Wang, D.: A survey of transfer learning. J Big Data **3**, 9 (2016)
27. Simonyan, K., Andrew, Z.: Very deep convolutional networks for large-scale image recognition. arXiv preprint arXiv:1409.1556 (2014)
28. Szegedy, C., Vanhoucke, V., Ioffe, S., Shlens, J., Wojna, Z.: Rethinking the Inception Architecture for Computer Vision. arXiv:1512.00567 [cs] (2015)
29. He K., Zhang X., Ren, S., Sun, J.: Deep residual learning for image recognition. In: 2016 IEEE Conference on Computer Vision and Pattern Recognition (CVPR), Las Vegas, NV, USA, pp. 770–778 (2016)
30. Tan, M., Le, Q.V.: EfficientNet: Rethinking Model Scaling for Convolutional Neural Networks. arXiv:1905.11946 (2019)

Chapter 27
Recyclability of Tractor's Engine Component: A Case Analysis of Failure

Bhupendra Prakash Sharma, Rahul Sindhwani, Priyank Srivastava, Pranav Malhotra, Harkirat Singh, Shorya Gupta, and Priyanka Singh

Abstract Structuring items for recyclability is driven by ecological and monetary objectives. A few Design for Assembly (DFA) rules and boundaries can be used to check the recyclability plans. These lists can be utilized for near examination of the recyclability of various items. This helps the designers in settling on structure decisions identified with the item's finish of life. In the present research work, the boundaries will be explored from the existing DFA time gauge tables. The aftereffects of the examination showed the recyclability of the item, as characterized by set up recyclability metrics. A contextual analysis is performed to decide whether DFA boundaries could be used to decide product/part recyclability. A TODIM approach is applied to evaluate the recyclability of a tractor's engine component based on the data obtained from "X" company. Furthermore, TODIM results were examined using MATLAB software. This exploration investigates how designers can utilize the (DFA) boundaries to foresee the recyclability list of an item. The recyclability list will be created utilizing the standards of plan for dismantling and material recyclability. The examination prompted the advancement of a method that an expert could use to decide item recyclability during the reasonable plan period of the fabrication process.

27.1 Introduction

Solid waste is anything that is used and discarded in the form of garbage [1, 2]. This may include anything, from paper, plastic bottles, furniture, food scraps, electronics wastes, etc. Rapid industrialization and urbanization, open markets, increasing population have led to the increase in the generation of solid waste [3, 4]. This situation is most prevalent in developing countries. Though there are many government guidelines and regulation for it, but these are generally ignored while implementation of MSW management [5]. The MSW management is one of the major concerns for governments across the world. The reason being that it is an integrated activity; from

B. P. Sharma (✉) · R. Sindhwani · P. Srivastava · P. Malhotra · H. Singh · S. Gupta · P. Singh
Department of Mechanical Engineering, Amity University, Noida, Uttar Pradesh, India
e-mail: bpsharma@amity.edu

© The Author(s), under exclusive license to Springer Nature Singapore Pte Ltd. 2022

V. V. Rao et al. (eds.), *Computational and Experimental Methods in Mechanical Engineering*, Smart Innovation, Systems and Technologies 239,
https://doi.org/10.1007/978-981-16-2857-3_28

the collection of waste and dumping on defined location for further processing. At the dumping location, the MSW can be disposed-off by using different methods, e.g. landfills, open burning, composting, segregation, fermentation and biological digestion, etc. [6]. These methods are also used in combination, where segregated waste is recycled [7], and then residual waste management is done by landfills or incineration for energy recovery [8]. There is always a concern about landfill method, as the traces of heavy metals [9] or medical waste [10] can lead to groundwater and soil pollution and even raise the health risks for MSW workers or people.

Although there is no textbook definition of recyclability but to frame the process, it can be explained as the ability of a product to be treated using various processes so that it can be reused for the same or different purposes. Along with the same line alteration/modification of a product so that it can be used to perform the same tasks and can be defined as reusability. Usually, various terms such as sustainable development or sustainable manufacturing are associated with recyclability to solidify its vague definition [11–13]. Sustainable manufacturing is defined as meeting the need of the future without sabotaging the requirements of future generations. Need designing products in such a way that their production requires a minimum number of non-renewable sources and at the same time their output/productivity is not hindered. India alone suffers 5 lakh automobile accidents annually, which means around 5 lakh engines must be reused or recycled per annum and the composition of the engine materials should be such that they are easily disintegrated and recycled [14–16]. In this paper, we holistically cover the recyclability of engines of agricultural automobiles focusing mainly on tractors. Engine is one such component, which can always be recycled and if not completely mechanically damaged, can be reused. In the modern world the 3 R's (Repair, Reuse and Recycle) of recyclability plays an important role in the life of any product [17, 18] (Fig. 27.1).

The world is one that has limited resources and serious environmental impacts. It is becoming increasingly evident that a friendly approach needs to be taken toward the environment and a severe lifestyle change is vital. Material products, like staplers, mobile phones or automobiles, are some of the products, which can be made up of one single or up to thousands of different materials [19]. Studies show that industry must process probably over 1000 commercial materials on the market keeping in mind the great number of variations that come with them. Extraction, refinement, transports, product use cycles, recycling or deposition are some of the complex departments that can have various impacts on the environment. During 1900s, these environmental problems were considered local or restricted to a single area due to the impact of a certain commodity. However, today, it is looking increasingly clear that the problems are more complex and related to all the parts in a product's life cycle from extraction of material to deposition of the used product [20]. Therefore, it is essential to investigate the matter for optimizing the MSW by recycling products or components as maximum as possible. In this direction, this research presents a possibility through a case analysis of engine components using TODIM approach and further validated with MATLAB.

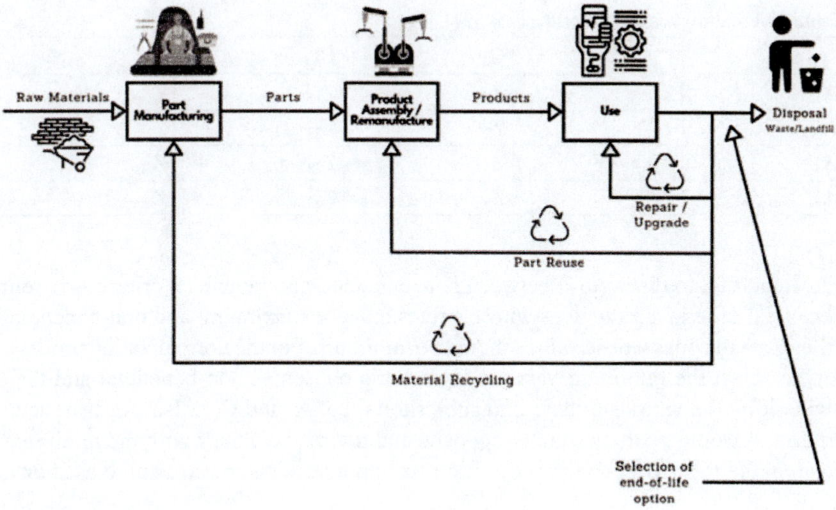

Fig. 27.1 Life of products [17, 18]

27.1.1 TODIM

TODIM is a sub-criterion of multi-criteria decision-making technique (MCDM), which has gained a lot of momentum in today's scientific world. TODIM (TOmada de Decisao Interativa Multicriterio) abbreviates the Portuguese term for interactive multi-criteria decision-making technique [21]. TODIM has taken over the scientific and the analytical world by surprise as it performs a lot of functions and solves a wide array of problems with ease. This method was initially introduced by Gomes and Lima [22, 23]. TODIM method has very well impressed the scientific and data analytics community, which can be clearly seen by the perfect solutions, it has provided to various real-life problems such as solving the problem of finding gas plant over a newfound reserve in Brazil [24] and ranking projects with economic and social consequences in Rio de Janeiro state, Brazil [25].

27.1.2 Steps of TODIM

A = {A1,..., An} is taken as a predefined set of criteria as inputs and C = {C1,..., Cm} as predefined set of sub-criteria as inputs in an MCDM problem. These predefined sets of indices are denoted by N and M, respectively. Various assumptions are taken as A_i as criteria and sub-criteria as C_k. The next very crucial step in this process is the normalization of various values. This is done as values of various criteria and sub-criteria are denoted using various units, and it is necessary to bring them to a comparable value for pairwise comparisons. In the normalization process, it

Table 27.1 Attributes and criteria matrix [21]

	C_1	C_2	C_3	C_m
A_1	Z_{11}	Z_{12}	Z_{13}	Z_{1m}
A_2	Z_{21}	Z_{22}	Z_{23}	Z_{2m}
A_3	Z_{31}	Z_{32}	Z_{33}	Z_{3m}
A_n	Z_{n1}	Z_{n2}	Z_{n3}	Z_{nm}

gets important to distinguish between beneficial and non-beneficial criteria wherein beneficial criteria are the ones whose values must be maximum, and non-beneficial criteria are the ones whose values should be minimum. For the normalization process, it is to select the minimum value in the criteria chosen as non-beneficial and then divide it by the various criteria and sub-criteria, i.e. A_i and C_k. Also, for beneficial criteria, it would be done exactly opposite and divide the values with the minimum value in the row. After this process, the decision matrix can be structured as shown in Table 27.1.

After the formation of decision matrix (DM) of size $\mathbf{L = (L_{ik})_{nxm}}$, a weightage will be assigned to each criterion as $W_k > 0$ in such a way so that sum of all weightage must be equal to 1 i.e. 100%. TODIM method relies on allotting dominance degree to various criteria and then rating them according to their importance in the final matrix formation. The paper focuses on calculating the data provided by a company "X" on the various failures of engine components using TODIM method. In algorithmic form, the implementation of the TODIM method entails the following main steps, Gomes [26], Fan [23].

STEP 1: The first step is to create the decision matrix by deciding the beneficial and non-beneficial criteria (Table 27.2).

For beneficial: $L_{ij} = \frac{Z_{ij}}{\text{MAX}(Z_{ij})}$; for non-beneficial: $L_{ij} = \frac{\text{MIN}(Z_{ij})}{Z_{ij}}$

STEP 2: The weightage thus calculated is now multiplied by the criteria and sub-criteria (Table 27.3).

$\mathbf{F_{ij} = W \times L_{ij}}$ and a new table will be formed.

STEP 3: The data from the table are now added and final values are obtained. The output with the highest value will be the most significant output.

$$\sum A_n = O_n.$$

Table 27.2 Decision matrix table [21]

	C_1	C_2	C_3	C_m
A_1	L_{11}	L_{12}	L_{13}	L_{1m}
A_2	L_{21}	L_{22}	L_{23}	L_{2m}
A_3	L_{31}	L_{32}	L_{33}	L_{3m}
A_n	L_{n1}	L_{n2}	L_{n3}	L_{nm}

Table 27.3 Weightage multiplication [21]

	C_1	C_2	C_3	C_m
A_1	F_{11}	F_{12}	F_{13}	F_{1m}
A_2	F_{21}	F_{22}	F_{23}	F_{2m}
A_3	F_{31}	F_{32}	F_{33}	F_{3m}
A_n	F_{n1}	F_{n2}	F_{n3}	F_{nm}

27.1.3 A Case Application of TODIM on Engine Component from Company 'X'

Finding the solution to problems containing various options with very closely related attributes is something that confuses a lot of analysts and mathematicians in today's world. Even a simple task of choosing a banana in a supermarket or choosing a phone can be confusing if the relevant attributes are very similar to each other. To solve such problems, the formulation of concepts like TODIM was done, which helps in the comparison and calculation of such assumption/problems. A common problem of environmental pollution and the problems faced by the recycling industry struck one's mind and how companies can easily know about the parts and their recyclability. To work on this problem, data were obtained from a tractor production company "X" related to engine disassembly and chosen each of the disassembled components as an assumption and various factors as criteria. Furthermore, a TODIM approach is applied to evaluate the least recyclable and the most recyclable component of the disassembled engine. The same technique can be used for various components of the tractor and further for various models of tractors. Obtaining these data can help R&D department of the company to develop and improve the recyclability of the various components.

27.1.4 Selected Attributes for TODIM Application

Criteria for choosing the attributes.

27.1.5 Part Cost

It is the cost of the Original Equipment Manufacturer (OEM) part that needs to be replaced, refurbished or recycled, by the disassembler/manufacturer.

27.1.6 Oil Cost

- The cost of the oil needs to be changed while changing the part or while taking that certain part out of the assembly for replacement, refurbishment or to be remanufactured.
- Oil is a perishable item that needs to be overhauled from time to time, either way it must be changed.
- Engine oil is the main part of lubrication for the engines, especially in diesel engine where there is a need for a continuous supply of lubricant, else it can lead to an engine seize. Cost analysis of the whole engine assembly is also important as it is not repairable after the engine seizure.

27.1.7 Labor Cost

- This is the cost for the person who is disassembling the part from the whole machine, for that person it is a very complex task as the fasteners may be at non-assessable locations or may corrode with time or damage.
- Another alternative is mechanical robots, which can be programmed as per the application of disassembly. They are fast as compared with the biological counterpart, but they may not be able to access complex locations of fasteners and will be very expensive in operating as well as buying.

27.1.8 Working Hours

Working hours can be described as the total number of times taken by workers to disassemble a certain part/product [26, 27]. Let us take piston rings disassembly, it takes one single worker to disassembly it completely from an unmounted engine. Therefore, it takes 1 work hour to complete this job. If they consider disassembly of a whole engine block, which consists of various components such as pistons, piston rings, cylinder heads, connecting rods, liquid sealant and many more parts, takes four people working successively for 1 h each, therefore it takes four workers for 1 h, contributing to 4 work hours.

27.1.9 Material Group Index (MGI)

It is a type of attribute based on the type of material used to manufacture that certain part or type of different materials used to assemble a component [28]. It can be classified into two categories:

- **Contaminated**: The parts/components that have faced corrosion due to a contaminated working environment or may have been dislodged from the original shape or size, will come under the contaminated category. For example, an Engine Front Oil Seal, it is a rubber part that is used as a seal between the inside and outside of the engine front, over a time it may get wear or may lose its mechanical properties due to high working temperature, therefore it cannot be used again. Similarly, camshaft, it is a component consisting of camshaft, thrust flange, spacer and liquid sealant, which are made up of cast iron, cast iron, steel and silicon adhesive, respectively. Out of which liquid sealant is the only contaminated part as it is made up of silicon adhesive, which must be applied new every time.
- **Non-contaminated**: The parts/components that do not come under the influence of the working environment or get degraded over time are categorized under non-contaminated material/parts. For example, oil pump, crankshaft, bearing, connecting rods are categorized under non-contaminated.

27.1.10 Types of Failures in Engines

- **Broken or burst cases**: This type of failure is caused when an engine is running way above its recommended power limit. This overloading causes the moving mechanical parts in the system to malfunction, which often results in minor cracks to develop in the main cylinder block or even detonating in some cases.
- **Lack of lubrication/oil pump failure**: This is one of the most common types of failure in an engine. This can happen in any engine that is not being maintained properly. When there is a shortage of lubricants such as engine oil in a system, the friction that is generated between the moving parts becomes so high, that over shorter periods, it results in wearing of pistons and block and if this is prolonged, then the entire system fails resulting in a lot of damage.
- **Cylinder head gasket failure**: This is a case where the gaskets are present between the cylinder head and the block to avoid wear out. These gaskets act as sealants for the system, which helps them make airtight. Failure of these components results in the leakage of coolant and engine oil. Due to this, the streak of lubricants may be visible on the engine. This can also result in loss of power in the engine. Changing these gaskets is a part of the preventive maintenance of an engine.
- **Overheat cases**: There can be various factors that lead to overheating of the system. There may be a leak in the cooling system, which causes it to malfunction and not cool the engine as desired. A low amount of engine oil may also lead to heating. This is a very serious issue and needs to be rectified as soon as possible because it may lead to permanent damage to the engine.
- **Camshaft breakage failure**: Sometimes, the camshaft is one of the most important pieces of engine that may break. This can be caused by several reasons such as lack of lubricant, incorrect break-in of the system or due to some play in piston rods. Camshaft breakage brings the entire system to a stop.

- **Rocket cover gasket failure**: Rocket cover may also be referred as valve cover. The gaskets on these tend to wear out over time, and it causes leaks in the system. Engine oil spills result in the engine running low on lubricants. When these gaskets fail, they need to investigate, or they can significantly damage the system.
- **Engine front seal leakage failure**: These seals are also known as crankshaft seals. The seals may dry out over time, which causes them to not function as desired. This results in leakage of oil from the engine. This causes problems such as misfires and oil spills.
- **Engine rear seal leakage failure**: This seal is present at the back of the engine assembly between transmission and engine. This seal leaks sometimes and an indicator for this is engine oil leaking from the front of the transmission. This is something that can easily be rectified simply by the replacement of seal.
- **Intake manifold cracking**: This is the part that directs the right amount of air–fuel mixture into the engine. Cracking in this causes engine misfires and leakage of coolant. This sometimes even causes damage to the pistons. It is imperative to rectify this failure.
- **Engine oil through from exhaust manifold**: In some cases, there may be some engine oil present in the exhaust manifold. This is due to exhaust valve seals wearing out. These seals are made of rubber and may dry out with time. This is a failure that may lead to damaging the exhaust manifold. The seals need to be replaced.
- **Engine cylinder block and cylinder head casting failure**: This is another issue that is associated with overheating of engines. It causes the engine head to crack in some cases. There may be defects in some castings, which may cause this to happen. It can cause damage to the engine block due to foreign particles entering the system. Entire head assembly needs to replace to rectify this [29].

27.1.11 Investigation of Data Obtained by Company X

The data obtained by the company indicating various types of failures of an engine coming to the recycle station underwent and were kept under the assumptions, i.e. A_n. Various criteria that are involved in the recycling process such as the cost part, working hours, oil cost or MGI were selected and kept under C_m. As the MGI is a descriptive attribute, convert it into numerical values assigning numbers 2 to non-contaminated and 1 to contaminated and adding them up as Table 27.4.

Values are taken according to the beneficial or non-beneficial components with MGI as the only beneficial component and calculations are done as shown in Table 27.5. Also, weightage (W_a) is assigned to each criterion. In the present case, the maximum weightage of 30% is given to the part price as it is of the utmost importance to any corporation. The lowest value was given to oil cost and labor hours of 10%. The equal weightage of 25% was given to MGI and working hours keeping in mind the environmental concerns and labor laws. These weights were assigned based on

Table 27.4 Attributes and criteria matrix of engine data

Types of failures	Part cost (Rs.)	Oil cost (Rs.)	Labor cost (Rs.)	Working hours	MGI
Broken or burst cases	18,591	415.09	308	4	12
Lack of lubrication/oil pump failure	11,150	415.09	231	3	10
Engine refurbished	20,323	510.88	308	4	13
Cylinder head gasket failure	205	415.09	77	1	1
Overheat cases	4043	415.09	231	3	9
Cam shaft breakage failure	1273	415.09	154	2	7
Rocker cover gasket failure	40	415.09	77	1	1
Engine front oil seal leakage failure	104	415.09	77	1	1
Engine rear oil seal leakage failure	730	415.09	77	1	4
Intake manifold crack cases	3611	415.09	77	1	6
Engine oil through from exhaust manifold	429	415.09	154	2	4
Engine cylinder block and cylinder head casting failure	15,459	415.09	231	3	7

brainstorming sessions conducted with five experts from the industry and five from academia.

After weightage multiplication in the corresponding column, summation of values of each component is done and final values are sorted from highest to lowest giving the most and the least recyclability as rocker cover gasket failure and the least being the engine refurbishment. In order to conduct the analysis, the ideology adopted was widely used mathematical software MATLAB developed by Mathworks. It is a programming language that is used to do complex computation, visualization and programming. While using the software to get an algorithm based on the factors carefully selected, and the result was verified with the use of conventional methods using the formulas of MCDM approach. In MATLAB, there are various operators or pre-set keywords to obtain desired results and programmed as under:

1. Xval=length(X(:,1));
2. for i= :Xval
3. for j=1:length(W)
4. if Wcriteria(1,j)= = 0

Table 27.5 Decision matrix calculation

Sl. No	Types of failures	Part cost (Rs.)	Oil cost (Rs.)	Labor cost (Rs.)	Working hours	MGI
1	Broken or burst cases	0.002152	0.999783	0.25	0.25	0.923076
2	Lack of lubrication/oil pump failure related	0.003587	0.999783	0.333333	0.333333	0.76923
3	Engine refurbished	0.001968	0.812324	0.25	0.25	1
4	Cylinder head gasket failure	0.195122	0.999783	1	1	0.076923
5	Overheat cases	0.009894	0.999783	0.333333	0.333333	0.692308
6	Camshaft breakage failure	0.031422	0.999783	0.5	0.5	0.538462
7	Rocker cover gasket failure	1	0.999783	1	1	0.076923
8	Engine front oil seal leakage failure	0.384615	0.999783	1	1	0.076923
9	Engine rear oil seal leakage failure	0.054795	0.999783	1	1	0.307692
10	Intake manifold crack cases	0.011077	0.999783	1	1	0.461538
11	Engine oil through from exhaust manifold	0.09324	0.999783	0.5	0.5	0.307692
12	Engine cylinder block and cylinder head casting failure	0.002587	0.999783	0.333333	0.333333	0.538462
	Weightage (Wa)	0.3	0.1	0.1	0.25	0.25

```
Y(i,j) = min(X(:,j))/X(i,j);
else
Y(i,j)=X(i,j)/max(X(:,j));
end
end
end
for i =1:Xval
PWSM(i,1)=sum(Y(i,:).*W);
PWPM(i,1)=prod(Y(i,:).^W);
```

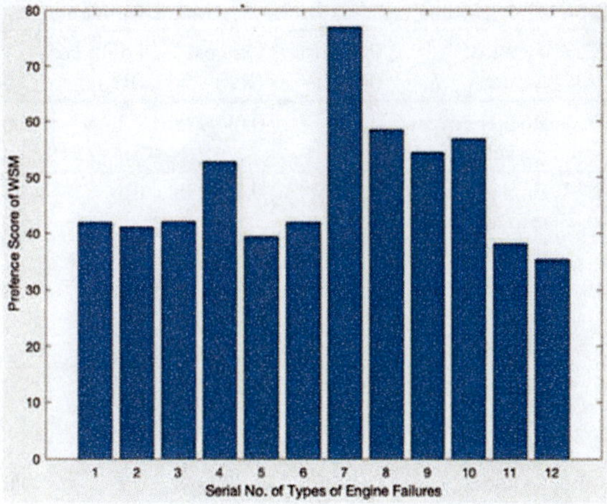

Fig. 27.2 Decision chart of MATLAB

```
end
Preference_Score_of_Weighted_Sum_Model = num2str([PWSM])
figure
bar(PWSM)
```

The operators applied on the given variables considered were namely X, W, W criteria summarized (Fig. 27.2) and hence formed the normalized decision matrix (Table 27.6).

27.2 Conclusion

In the present research work, a detailed study and investigation are conducted for a case of recyclable components of a tractor engine. From the above bar graph shown in Fig. 27.2 and the decision matrix presented in Table 27.6, it can be inferred that the different types of failures that occur in an engine can be represented rank-wise based on their recyclability using a MATLAB-based TODIM approach. Decision-makers will get benefited to adopt this approach to evaluate the recyclability of product, parts or components. This research will propose a guideline for the tractor manufacturers to work on the investigated components. They can work out more with respect to the design for manufacturing and assembly in the rocker cover gasket maximum and in the engine cylinder block and cylinder head casting minimum followed by others. Although, few components are ranked with nearby values such as engine refurbished, camshaft breakage, broken or burst cases and lack of lubrication/oil pump failure-related issues can be taken together into account. This study will help the manufacturers and designers to plan the products in such a way so that they can aim to reduce the costs and energy required. It will give an algorithm to assist in giving quick responses regarding which processes and materials they can use so that

Table 27.6 Normalized rank decision matrix calculation using MATLAB

Sl. No	Types of failures	Part cost (Rs.)	Oil cost (Rs.)	Labor cost (Rs.)	Working hours	MGI	Σ
7	Rocker cover gasket failure	0.3	0.099978	0.1	0.25	0.019231	0.769209
8	Engine front oil seal leakage failure	0.115385	0.099978	0.1	0.25	0.019231	0.584594
10	Intake manifold crack cases failure	0.003323	0.099978	0.1	0.25	0.115385	0.568686
9	Engine rear oil seal leakage failure	0.016438	0.099978	0.1	0.25	0.076923	0.54334
4	Cylinder head gasket failure	0.058537	0.099978	0.1	0.25	0.019231	0.527746
3	Engine refurbished	0.00059	0.081232	0.025	0.0625	0.25	0.419323
6	Camshaft breakage failure	0.009427	0.099978	0.05	0.125	0.134615	0.41902
1	Broken or burst cases	0.000645	0.099978	0.025	0.0625	0.230769	0.418893
2	Lack of lubrication/oil pump failure related	0.001076	0.099978	0.033333	0.083333	0.192308	0.410029
5	Overheat cases	0.002968	0.099978	0.033333	0.083333	0.173077	0.39269
11	Engine oil through from exhaust manifold	0.027972	0.099978	0.05	0.125	0.076923	0.379873
12	Engine cylinder block and cylinder head casting failure	0.000776	0.099978	0.033333	0.083333	0.134615	0.352037

the companies can aim to reduce wastage in the long run. This work addresses to a circular economy of any country directly.

References

1. Moher, D., Liberati, A, Tetzlaff, J., Altmanm, D.G., Altman, D., Antes, G.: Preferred reporting items for systematic reviews and meta-analyses: the PRISMA statement. J. Chinese Integr. Med. **7**(9), 889–896 (2009)
2. Teshome, F.B.: Municipal solid waste management in Ethiopia; the gaps and ways for improvement. J. Mater. Cycles Waste Manag. (2020). https://doi.org/10.1007/s10163-020-01118-y
3. Gundupalli, S.P., Hait, S., Thakur, A.: Multi-material classification of dry recyclables from municipal solid waste based on thermal imaging. WASTE Manag. **70**, 13–21 (2017). https://doi.org/10.1016/j.wasman.2017.09.019
4. Singh, S., Mamatha, K.R., Ragothaman, S., Raj, K.D., Anusha, N., Zacharia, S.: Waste segregation system using artificial neural networks. HELIX **7**(5), 2053–2058 (2017)
5. Priti, Mandal, K.: Review on evolution of municipal solid waste management in India: practices, challenges and policy implications. J. Mater. CYCLES WASTE Manag. **21**(6), 1263–1279 (2019). https://doi.org/10.1007/s10163-019-00880-y
6. Aich, A., Ghosh, S.K.: Application of SWOT analysis for the selection of technology for processing and disposal of MSW. Procedia Environ. Sci. **35**, 209–228 (2016). https://doi.org/10.1016/j.proenv.2016.07.083
7. Abdulla, A.M., Mahrous, F.: Recycling of municipal solid waste in the State of Kuwait. Arab. J. Sci. Eng. **26**(2C), 3–10 (2001)
8. Di Maria, F., Micale, C., Morettini, E.: Impact of the pre-collection phase at different intensities of source segregation of bio-waste: an Italian case study. WASTE Manag. **53**, 12–21 (2016). https://doi.org/10.1016/j.wasman.2016.04.026
9. Kaoser, S., Barrington, S., Elektorowicz, N.: Compartments for the management of municipal solid waste. Soil Sediment Contam. **9**(5), 503–522 (2000). https://doi.org/10.1080/10588330091134374
10. Taghipour, H., Mosaferi, M.: Characterization of medical waste from hospitals in Tabriz, Iran. Sci. Total Environ. **407**(5), 1527–1535 (2009). https://doi.org/10.1016/j.scitotenv.2008.11.032
11. Babaei, R., Abbasi, M., Mirabi, M.: Developing a practical framework for use in the separation of aluminium waste from restaurants and estimation of potential benefits. Detritus **3**, 7–12 (2018). https://doi.org/10.31025/2611-4135/2018.13667
12. Subramanian, P.: Plastics recycling and waste management in the US. Resour. Conserv. Recycl. **28**(3–4), 253–263 (2000). https://doi.org/10.1016/S0921-3449(99)00049-X
13. Awoyera, P.O., Adesina, A.: Plastic wastes to construction products: status, limitations and future perspective. Case Stud. Constr. Mater. **12** (2020). https://doi.org/10.1016/j.cscm.2020.e00330
14. Thomas, L.M., Moosvi, S.A.: Hardened properties of binary cement concrete with recycled PET bottle fiber: an experimental study. Mater. Today Proc. (2020). https://doi.org/10.1016/j.matpr.2020.03.025
15. Parihar, R.S., Ahmed, S., Baredar, P., Sharma, A., Kiran, T.R.: MSWM in Bhopal city: a critical analysis and a roadmap for its sustainable management. Proc. Inst. Civ. Eng. Eng. **172**(2), 83–95 (2019). https://doi.org/10.1680/jmuen.17.00011
16. Marczuk, A., Misztal, W., Jozwiakowski, K., Dach, J., Kowalczyk-Jusko, A.: The research on effectiveness of the electronic and electrical waste selective collection system in Lublin city, Poland. Arch. Environ. Prot. **45**(3), 55–63 (2019). https://doi.org/10.24425/aep.2019.128641
17. Battia , O., Dolgui, A., Tiwari, M., Hegaru, S.S.: Design for manufacturing and assembly/disassembly: joint design of products and production systems. Int. J. Prod. Res. **56**(24) (2018)
18. Sundin, E., Jacobsson, N., Björkman, M.: Analysis of service selling and design for remanufacturing. In: IEEE International Symposium on Electronics and the Environment (2000)

19. de Silva, N., Jawahir, I.S., Dillon, O., Jr.: A new comprehensive methodology for the evaluation of product sustainability at the design and development stage of consumer electronic products. Int. J. Sustain. Manuf. **1**(3) (2009)
20. Berkhout, F., Smith, D.: Products and the environment: an integrated approach to policy. Eur. Environ. **9**, 174–185 (1999)
21. Llamazares, B.: An analysis of the generalized TODIM method. Eur. J. Oper. Res. **269**(3), 1041–1049 (2018)
22. Gomes, L.F.A.M., Lima, M.M.P.P.: TODIM: Basics and application to multicriteria ranking of projects with environmental impacts. Found. Comput. Decis. Sci. **16**, 113–127 (1991)
23. Gomes, L.F.A.M., Lima, M.M.P.P.: From modelling individual preferences to multicriteria ranking of discrete alternatives: a look at prospect theory and the additive difference model. Found. Comput. Decis. Sci. **17**, 171–184 (1992)
24. Gomes, L.F.A.M., Rangel, L.A.D., Maranhão, F.J.C.: Multicriteria analysis of natural gas destination in Brazil: an application of the TODIM method. Math. Comput. Model. **50**, 92–100 (2009)
25. Rangel, L.A.D., Gomes, L.F.A.M., Moreira, R.A.: Decision theory with multiple criteria: an aplication of ELECTRE IV and TODIM to SEBRAE/RJ. Pesqui. Oper. **29**(3), 577–590 (2009)
26. Gomes, L.F.A.M., Machado, M.A.S., Rangel, L.A.D.: Behavioral multi-criteria decision analysis: the TODIM method with criteria interactions. Annals Oper. Res. **211** (2013)
27. Villalba, G., Segarra, M., Ferna´ndez, A.I., Chimenos, J.M., Espiell, F.: A proposal for quantifying the recyclability of materials. Resour. Conserv. Recycl. **37** (2002) 39–53
28. Statistical studies of the costs of six-man versus twelve-man juries. Wm. & Mary L. Rev (1972–1973)
29. Lux, W.: A Reliability Study of Diesel Engines. SAE Technical Paper 640470 (1964)

Chapter 28
Feature Extraction of Face Recognition Techniques Utilizing Neural System as a Classifier

Amrita Rai, C. Shylaja, and Puneet Kumar Mishra

Abstract Face recognition has a large extent of applications from individual recognizable proof and reconnaissance to electronics showcasing and publicizing for chosen clients. There are various advances in facial recognition, for example, preprocessing, feature extraction, and grouping, where feature extraction and grouping are utilized to acquire the greatest precision. In this paper, diverse feature extraction methods, for example, A.A.M, A.S.M, template-based, Gabor-features, and a few are basically surveyed. Aside from these, the various kinds of neural classification systems, for example, backpropagation, convolutional, radial-basis-function, and so on in the space of face recognition, are investigated. The method and calculations created in the present writing are examined, and it is uncovered that every system is one of a kind and has ideal execution. This assessment further makes a relative examination of these frameworks reliant on their focal points and imperatives.

28.1 Introduction

A pattern recognition framework is utilized for distinguishing proof and verification of clients, in light of their special physical properties. Perceiving people on their facial descriptions is a significant marvel, and the assignment of perceiving peers on their appearances easily acts in our day-by-day life. The essential features are spoken to utilize a suitable feature portrayal plot and gain proficiency with the discriminating highlights on which matching algorithms act. Various periods of face recognition frameworks are included. A lot of face recognition methods are structured that perform admirably in controlled situations. It incorporates a portion of predominant techniques, i.e., principal component analysis (PCA), independent component analysis, linear discriminant analysis (LDA), and Fisher face. An assortment of face representation techniques was proposed as of late, and they are arranged into two kinds, such as holistic feature representation and local feature representation. As a typical pattern recognition issue, there are two primary methodologies in a common

A. Rai (✉) · C. Shylaja · P. K. Mishra
ECE Department, G.L. Bajaj Institute of Technology and Management, Greater Noida, UP, India

© The Author(s), under exclusive license to Springer Nature Singapore Pte Ltd. 2022
V. V. Rao et al. (eds.), *Computational and Experimental Methods in Mechanical Engineering*, Smart Innovation, Systems and Technologies 239,
https://doi.org/10.1007/978-981-16-2857-3_29

face recognition framework, i.e., face representation and face matching. Most available local feature descriptors are hand-made, which for the most part, require solid earlier information and are heuristics [1–5].

Among the most common referred techniques for face recognition based on feature extraction are Eigenfaces dependent on PCA, Fisher faces dependent on LDA, and strategies dependent on independent component analysis. A present invariant face recognition framework relies upon a deformable, generic 3D face model that is a mix of an edge model, a shading region model, and a wireframe model for portraying the shape and significant highlights of the face. The initial two models are utilized for image analysis and the third for the most part for face amalgamation. For coordinating the model to face images in a discretionary manner, the 3D model can be anticipated onto different 2D view planes dependent on rotation, interpretation, and scale parameters along these lines producing numerous face picture formats [6–8].

Machine recognition of human face from still pictures and videos has become a functioning examination region in the way of image processing, pattern recognition, neural networks, and computer vision. This has large applications extending from static coordination of controlled organization photos, for example, travel papers, Master cards, driving permits, and the continuous coordinating of surveillance video images introducing various limitations as far as preparing prerequisites. A general and productive plan approach utilizing a "Radial Basis Function (RBF) neural classifier" to adapt to few training sets having high dimensions, which is an issue frequently experienced in face recognition, is introduced in the research directed and hybrid learning algorithm is executed for the study. The PCA is some well-known multivariate statistical procedures that have been generally utilized in the fields of pattern recognition and signal processing. There are various PCA-based techniques utilized for face recognition from one-dimensional PCA to two-directional PCA. The research technique elevates the precision contrasted with 1D PCA, 2D PCA, and two-directional 2D PCA [9, 11].

The significant difficulties are constructing a precise face recognition framework that does not require a large computational expense, and the current researches were led on a real-time face recognition framework by utilizing block processing of local binary patterns of face images caught by N.A.O humanoid. Likewise, a strategy that has included N.A.O humanoid and tried under true conditions was exhibited.

28.2 Review of the Existing Methods

There are a few methods proposed by creators in regards to different feature extraction techniques and a neural system for face recognition. These procedures are investigated and talked about in the below sections.

28.2.1 Geometric-Based Feature Extraction

In the geometric-based method, the local features, i.e., local statistics and locations involving the mouth, eyes, eyebrows, and nose, are at first draw out from the face images. The most significant geometric-based techniques are Active appearance chart models.

Active shape model (ASM): A face recognition model dependent on local coordinating Gabor comprising of three primary modules, one of which is an active state model where image arrangement is performed and utilized for aligning the face. In addition, to choose the shape to draw out features from looking at guides comparative with the stored picture and the module for the spatial situation of the Gabor jet, ASM is used to modify the face present as shown in Fig. 28.1 [1–3]. It is a linear statistical model that uses a linear change in vector space and uses a point apportionment model where a vector portrays the state of the object. A beneficial and sensible calculation for facial landmarks used on most mobile phones is theorized. In this assessment, the first ASM is adjusted to improve its exhibition with three changes, i.e., improving the modified model using the focal point of the eyes by using highlight, which is utilized in the vast majority of the cell phones has been hypothesized. In this examination, the original ASM is altered to improve its exhibition with three changes, i.e., improving the initialization model utilizing focus of the eyes by utilizing feature map of color information, developing a modified model definition and fitting a bigger number of landmarks than the traditional ASM, and last change is developing 2D profile model for identifying faces in the input image, and the technique is assessed on dataset containing more than 1000 face images.

Active appearance model (AAM): A novel technique for deciphering images using this model was shown. The model consists of a factual shape model and the grey-level

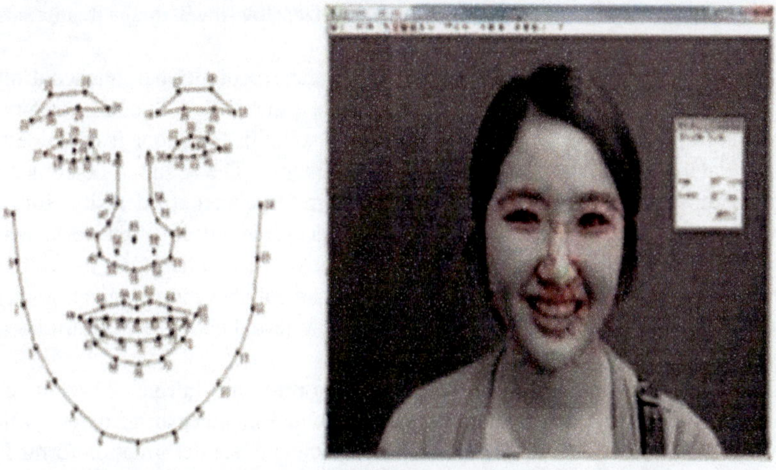

Fig. 28.1 Facial highlights land-marking plan utilized in the exemplary ASM usage

appearance of the image of interest. These models are created by consolidating the shape varieties with a model of appearance variety in a standardized shape structure. In this way, it is inferred that the AAM algorithm will be a significant technique for identifying deformable objects in a huge number of applications. A strategy for arranging facial expression dependent on the extracted features of facial segments was presented. The facial area is distinguished, and afterward, it is prepared to utilize the active appearance model to draw out the important features of the facial parts [1, 10, 11].

Scale-invariant feature transform (SIFT): A model for fuzzy match file for scale-invariant feature transform highlights, which include all of the SIFT vital points in the decision-making process. The SIFT is principally intended for object recognition applications, i.e., face recognition, iris recognition, fingerprint ID, etc. SIFT features are useful for describing the outdoor surroundings. In this, the new fuzzy SIFT classifier is adjusted effectively for robust face recognition from complex-oriented backgrounds with no face editing tool and utilizing a single layout and also the advancement of an exceptionally proficient fuzzy descriptor matching tool. Subsequently, this examination takes into consideration the weak supervision of the face recognition experiment. The SIFT functions admirably just for pose variation and neglects to deliver acceptable outcomes under changing illumination. Furthermore, a hybrid SIFT with a weighing factor in feature matching is suggested, which utilizes a fixed facial landmark localization procedure [2, 4, 5].

28.2.2 Appearance-Based Feature Extraction

This technique is normally used for frontal face detection utilizing color image-based feature extraction and appearance-based organization. Face identification with a certain level of precision and robustness utilizes low-level image features, for example, color and shape.

Local binary pattern (LBP): This technique for face recognition in an uncontrolled condition that works with the LBP of facial images is introduced. The basic technique for separating the LBP surface texture is created, and this technique is examined on various databases that have uncontrolled facial images. The acquired outcomes are demonstrated, i.e., strategy functions admirably in these uncontrolled situations are shown in Figs. 28.2 and 28.3. The real-time face recognition framework, where block processing of LBP examples of the face images taken by NAO, is exhibited. These NAO humanoids are utilized in robot–human collaboration applications. This strategy has been embraced by the NAO humanoid tested under real conditions and defeats the ordinary strategies [6, 7].

Gabor features: Gabor features in face recognition are introduced to increase accuracy. The Gabor-based technique is utilized, which changes the grid, from which the Gabor features are drawn out using mesh, to model face deformation formed by differing pose and statistical-model of the scores, are registered by utilizing Gabor features to increase accuracy.

Fig. 28.2 Training templates from the face database with subject IDs

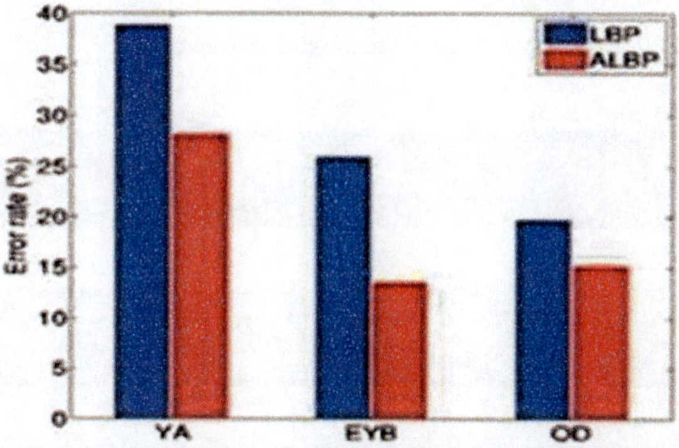

Fig. 28.3 Histogram plot of equal error rate (ERR) showing pitfalls of the local binary pattern (LBP) and augmented local binary pattern (ALBP)

A cloud-based observing framework using face recognition is proposed. This research diminishes Gabor-filter complications by keeping up local features. This research determines techniques for the center symmetric LBP example feature extraction algorithm. To define the presence of detection algorithm author in [9–11] tested, many figures for creating single face dataset and multi-face dataset are shown in Fig. 28.4a, b.

Principal component analysis (PCA): A patch-based PCA technique to manage face recognition is exhibited. The use of relationships between pixels; columns; rows take place; however, the local spatial data are not utilized in this method, and it is seen that patches are more significant fundamental units than pixels for face recognition. This strategy has the most noteworthy precision when contrasted with 1D PCA, 2D PCA, and the two-directional 2D PCA.

For perceiving a human face depending on the features in the image, is shown. The distinguished face in the picture is perceived, utilizing a combination of PCA and FFNS. This technique utilizes the Bio-ID face database as the standard image database as appeared in Fig. 28.5.

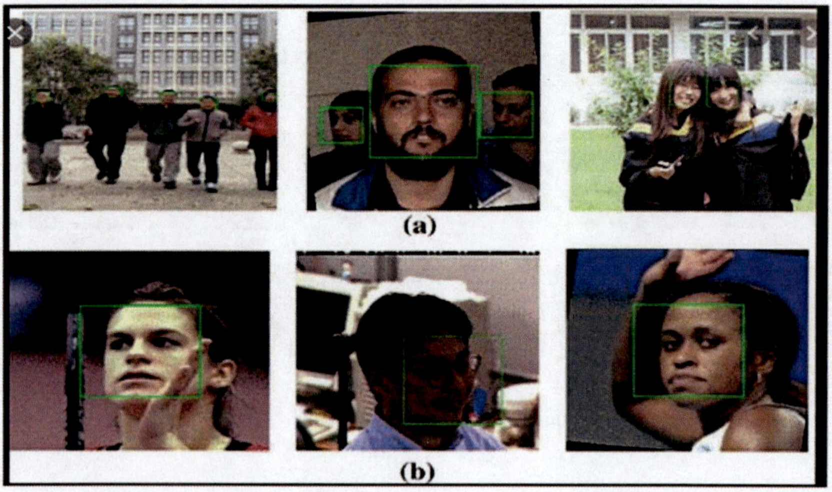

Fig. 28.4 **a** Multi-face dataset, **b** Single-face dataset

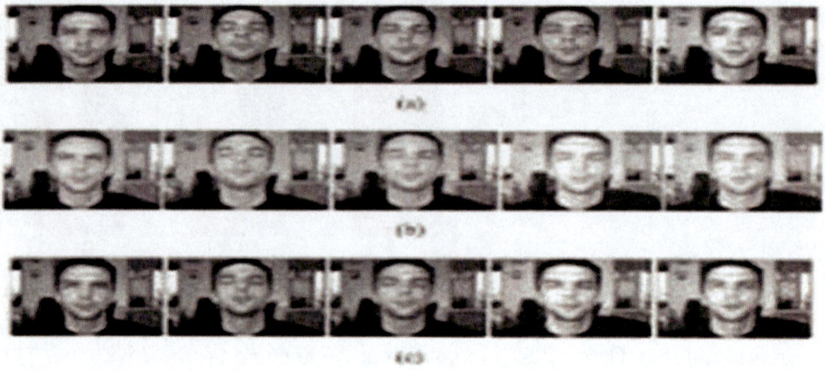

Fig. 28.5 Bio-ID face detection

28.2.3 Template-Based Feature Extraction

The different approaches executed for deciding the "features of faces in recognition" are depicted in the beneath sections.

Deformable templates: A technique for identifying and depicting face features utilizing deformable templates is depicted. This clarifies features-of-interest, for instance, taking eye as an example, and a parameterized template portrays it, and this template connects with the image by changing its parameters. The last parameters can be utilized as descriptors for the feature. In this manner, deformable templates are utilized to highlight original images.

Gravity center template: A gravity center template for the face area by drawing out edges around the body organ of the original picture is prepared, and also, these are checked to distinguish the organs that have been offered. This procedure is snappy, and the condition of the face highlights is depicted. A model to discover and follow the driver's mouth development using a CCD camera for checking and perceiving a driver's yawn was presented. From the start distinguishing driver's appearances using gravity center template and recognizing driver's left and right mouth corners by dim projection and drawing out surface highlights of driver's face corners using Gabor wavelets.

28.2.4 Artificial Neural Networks for Face Recognition

There are a few neural systems proposed by various authors with respect to different feature extraction methods for face recognition.

Backpropagation neural network: This algorithm is the standard and proficient approach to manage machine learning when the data join the complex tangible info. The halfway subsidiary of the cost work concerning any weight or bias is an enunciation for backpropagation. This framework demonstrates how viably the cost changes with change in loads and bias as shown in Fig. 28.6.

Additionally, occlusion identification, where occlusion is a condition where one object is overshadowed by another object, is presented. The research comprises four phases where a backpropagation neural network is utilized to discover the age of the human in the third stage. It demonstrates great execution when contrasted with different artificial neural systems.

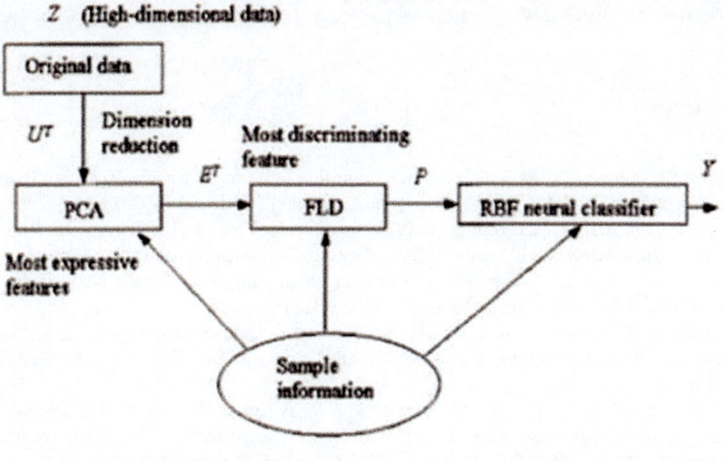

Fig. 28.6 Back propagation neural network

Convolutional neural network (CNN): This strategy is utilized to perceive faces from different points and can manage impediments somewhat. It also demonstrates that the proposed procedure execution can be additionally improved by using better examples, methodologies, and growth strategies. The strategy for dynamic face acknowledgment that acts dependably using human practices in ordinary face acknowledgment situations is outlined.

Probabilistic neural network (PNN): It assists in facial recognition. This system consolidates three stages, i.e., the wavelet parcel deterioration of face picture, planning PNN classifiers with wavelet sub-pictures with low recurrence parts, and finally the blends of the readied PNN classifiers by fluffy basic. Differentiated and four grid subspace calculations, the proposed methodology can get serious precision. It is like manner improves the exactness of the face recognition with less CPU time-span.

28.3 Conclusions

In this paper, different techniques for feature extraction and neural networks, which incorporate dynamic shape, radial basis function, convolutional, gravity center template, were presented. The eight various kinds of feature recognition strategies among which template-based method and active shape-model look progressively productive contrasted with the other methods where these techniques likewise have their constraints. Furthermore, three neural courses of action strategies are discussed, and it is seen that backpropagation neural network and CNN are viewed as more exact in comparison to others. Moreover, they have minor limitations, primarily scaling of the image. The future scope of the work is an improvisation by completing a novel strategy by incorporating the imperatives realized by the two procedures in feature recognition and besides in neural grouping and get more exactness.

References

1. Biggio, B., Fumera, G., Roli, F.: Pattern recognition systems under attack: design issues and research challenges. Int. J. Pattern Recognit. Artif. Intell. 28(07), 1460002 (2014)
2. Klare, B., Jain, A.K.: On a taxonomy of facial features. In: Proc. 4th IEEE Int. Conf. Biometrics Theory, Applications and Systems (BTAS). Crystal City, Washington D.C. (2010)
3. Tao, D., Guo, Y., Li, Y., Gao, X.: Tensor rank preserving discriminant analysis for facial recognition. IEEE Trans. Image Process. 27(1) (2018)
4. Deshpande, N.T., Ravishankar Dr., S.: Face detection and recognition using Viola-Jones algorithm and Fusion of PCA and ANN. Adv. Comput. Sci. Technol. **10** (2017). ISSN 0973-6107
5. Mehra, S., Singh, S.D., Kumari, S., Karatangi, S.V., Agarwal, R., Rai, A.: Design and implementation of biometrically activated self-defence device for women's safety. In: Mathur, G., Sharma, H., Bundele, M., Dey, N., Paprzycki, M. (eds.) International Conference on Artificial Intelligence: Advances and Applications 2019. Algorithms for Intelligent Systems. Springer, Singapore (2020)

6. Bir, P., Karatangi, S.V., Rai, A.: Design and implementation of an elastic processor with hyper threading technology and virtualization for elastic server models. J. Supercomput. Int. J. High Perform. Comput. Des. Anal. Use 1–22 (2020). Publisher Springer Nature. SCI
7. Joo, Er, M., Wu, S., Lu, J., Toh, H.L.: Face recognition with radial basis function (RBF) neural networks. In: Proceedings of the IEEE Conference on Decision and Control, vol. 3, pp. 2162–2167 (1999). https://doi.org/10.1109/CDC.1999.831240
8. Lee, Y.-H., Kim, C.G., Kim, Y., Whangbo, T.K.: Facial landmarks detection using improved active shape model on android platform. Springer Science Business Media New York. (2009)
9. Bolotnikova, A., Demirel, H., Anbarjafari, G.: Real-time ensemble based face recognition system for NAO humanoids using local binary pattern. Analog. Integr. Circuits Signal Process. **92**(2) (2017)
10. Karthigayani, P., Sridhar, S.: Decision tree based occlusion detection in face recognition and estimation of human age using back propagational network. J. Comput. Sci. (2014)
11. Cament, L.A., Galdames, F.J., Bowyer, K.W., Perez, C.A.: Face recognition under pose variation with local Gabor features enhanced by active shape and statistical models. Pattern Recognit. **48**(11), 3371–3384 (2015). ISSN 0031-3203

Chapter 29
Investigating Strategies and Parameters to Predict Maintenance of an Elevator System

Jasmine Awatramani, Gaayan Verma, Nitasha Hasteer, and Rahul Sindhwani

Abstract In this era of automation, our lives are surrounded by machines, be it a mobile phone or an elevator. We humans become careless when it comes to the maintenance of the machine. From the customer's perspective, until an elevator is not working, nobody tends to care. This carelessness, in the long run, can result in loss of human life as well as financial losses. Elevators require maintenance and safety. To overcome both, the machine requires timely maintenance, and it can be executed with the precise product vision with the help of predictive maintenance. It not only predicts future failure but also pinpoints the issues in complex machinery and gives better results in terms of preventive maintenance. The conventional predictive maintenance machine learning techniques are established on feature engineering. It is the manual formation of precise features using domain proficiency and similar methodologies. Due to this, models are hard to reuse because feature engineering is specific to the problem structure and the data available, which can vary from one place to the other. Deep learning methodologies provide better results due to the extraction of new deep features from the dataset compared with the existing features. This work reviews the extant literature as well as showcases the implementation of random forest classifiers on the open-sourced dataset. In our model, an average accuracy of 91.50% was obtained. The dataset consisted of sensor data, which were recorded on the basis of maintenance actions being taken.

29.1 Introduction

In recent years, elevators have become a part of commercial as well as residential complexes. In addition to that, it has emerged as an essential tool in day-to-day life with the growth of the economy and urbanization holding up to 54% of the world's population [1, 2]. Elevator systems tend to lay ease in the lives of people [3, 4].

Recently, the pandemic has impacted maintenance even more severely. It has put a halt to the installations of new equipment as well as has increased pressure to

J. Awatramani (✉) · G. Verma · N. Hasteer · R. Sindhwani
Amity University, Noida, Uttar Pradesh, India

© The Author(s), under exclusive license to Springer Nature Singapore Pte Ltd. 2022

V. V. Rao et al. (eds.), *Computational and Experimental Methods in Mechanical Engineering*, Smart Innovation, Systems and Technologies 239,
https://doi.org/10.1007/978-981-16-2857-3_30

decrease the number of yearly visits. The safety issues of elevators have transformed into a global elevator issue. Therefore, elevator systems require proper maintenance to ensure safety and reliability. Although, there are various obstacles faced during the actual maintenance of elevator systems such as inappropriate maintenance techniques, inappropriate upgrading of elevator instruments, high maintenance costs, and much more. Developing preventive and predictive maintenance arrangements will be the next phase for revamping the security of elevators, which will further escalate the life of elevator systems and decrease the cost of service. Preventive maintenance performs systematic inspections of assets and executes routine maintenance to prevent unexpected equipment downtime or failure. Preventive maintenance can decrease the frequency of equipment failure. Though this maintenance procedure needs a definite cost, there is a possibility that the cost of the whole elevator system might increase if over-maintenance of equipment is being done. If preventive maintenance activities are not useful enough, the rate of equipment failure will be a high rise, and post-maintenance costs of the whole system might also increase.

Predictive maintenance lets you estimate the time-to-failure of a machine. So we can start preventive maintenance and save time and assets from any big issue. Predictive maintenance strategies are nowadays being implemented by elevator production service companies. This implementation comes into use by evaluating the remaining life of the elements that are accountable for faults and remotely tracking faults in elevator systems. Fault detection and diagnosis are needed by elevators to perform a healthy operation.

The traditional predictive maintenance machine learning models are based on feature engineering, which is the manual construction of the right features using domain expertise and similar methods. This usually makes these models hard to reuse since feature engineering is specific to the problem scenario and the available data, which vary from one place to the other.

The most attractive part of applying deep learning in the predictive maintenance domain is the fact that these networks can automatically extract the right features from the data, eliminating the need for manual feature engineering. Deep learning approach provides better results due to the new deep features extracted from the dataset compared with the existing features.

This paper explores the elevator system's optimal maintenance policy to reduce the average maintenance cost. The rest of the work has been structured as follows: Sect. 29.2 provides the in-depth literature review, Sect. 29.3 showcases gap analysis, Sect. 29.4 represents the parameter analysis, Sect. 29.5 represents the implementation of random forest classifier in our model, Sect. 29.6 represents the experimental results on the same and Sect. 29.7 showcases the conclusion along with the future scope.

29.2 Literature Review

Several search engines were used for online library search: IEEE Xplore Digital Library, Springer Link Online Library, Elsevier ScienceDirect and Google Scholar. Most of the authors have implemented deep learning strategies in their models in order to obtain results [1, 2, 5–7]. An effective fault diagnosis and fault prognosis has been observed. Along with which, implementation of decision trees and random forest algorithm has also been observed in order to compare the results. The use of deep learning has helped the authors to extract new deep features from the dataset as well. These new deep features showcased better accuracy in terms of fault diagnosis.

Mishra et al. [1, 2] designed a generic deep encoder model to automatically calculate highly explanatory deep features from the elevator data. Random forest algorithm was applied to determine faults based on statistical attributes to compare results. Strategy with new extracted deep attributes provided great accuracy in determining the faults. In this paper [5], the authors have proposed a conventional framework of intelligent predictive maintenance systems for elevator service consisting of fault prognosis, fault diagnosis, feature extraction, pre-processing signal, IoT and Internet of Service. Gilabert et al. [6] studied predictive maintenance as a result of uncritical machinery. Continuous monitoring along with diagnosis has been done using neural networks, whereas, machine instruments were being observed using vibration systems and applied using Bayesian networks. It also showed the work carried out under the MINICON project that tends to develop cost-effective integrated SPUs. In this paper [7], a consolidated neural-network based on decision support structure for predictive maintenance of rotating equipment. It comprises a vibration-based deterioration database by observing rolling element bearings, developing an ANN model to estimate the life percentile, along with the failure times of roller bearings and cost matrix designing, and probabilistic substitution model that tends to enhance the expected cost per time.

Few papers have showcased the implementation of machine learning algorithms including SVM, genetic algorithms [3, 8, 9]. They focused mainly on reducing the mean cost of the maintenance of an elevator and finding out the most optimal strategy to estimate the time of maintenance. These papers consisted of static datasets only and detection of faults at an early stage.

In this paper [3], the model is combined with a numerical example that is solved by a genetic algorithm with the goal of reducing the mean cost of maintenance. The best equipment maintenance plan of action is found by verifying the practicality and validity of the model. In Langone et al. [8], LS-SVM at an early stage of fault detection was used. Initially, on sensor data coming from the VFFS machine, an unsupervised approach, KSC came into play. After that, the NAR model was illustrated, i.e., a supervised learning methodology, LS-SVM framework. It showed that LS-SVM can successfully assess, predict mechanical conditions based upon sensor data and attain higher performance than basic methods. In this paper [9], a discussion on Multivariate Relevance Vector Machines that examines direct and indirect procedures for forecasting everyday evapotranspiration was being done. The developed models are based on static datasets.

Few miscellaneous studies were mainly based on the research on the mechanical strategies, which included examination of hydraulic oil, MCSA and algorithms and penalty functions to make a degradation and maintenance model [10–13]. These papers were research-based and showcased the impact of predictive maintenance on elevator systems.

Xu et al. [10] drew an outline about the role of predictive maintenance in the epoch of big data and techniques of data-directed fault diagnostics and fault prognostics along with its future extent. In this paper [11], a dynamic predictive maintenance policy for a multi-component approach. It also provides an outline of literature on predictive and condition-based/vanilla maintenance strategies. Flores et al. [12] designed Motor Current Signature Analysis (MCSA) is an identification methodology in which the mechanical structure acts as a transducer, determining short torque variations produced within the mechanical structure. MCSA has been used for determining the condition of various mechanical components such as bearings, motor fan, rotor unbalance and gearboxes, providing information in regards with fault centralization. In this paper [13], various methods have been performed to inspect the percentage of metals and corrosion; sulfur levels in the hydraulic fluid depict the maintenance intervals of an elevator system. The study of zinc, phosphorus, chlorine and calcium verifies that the oil that was used in one elevator is different from the oil used in the other two. Additionally, it releases significant wear to all the elevators as a result of the working environment conditions. This study of the hydraulic fluid can be included in the testing process, which inspects the safety of elevators (Fig. 29.1).

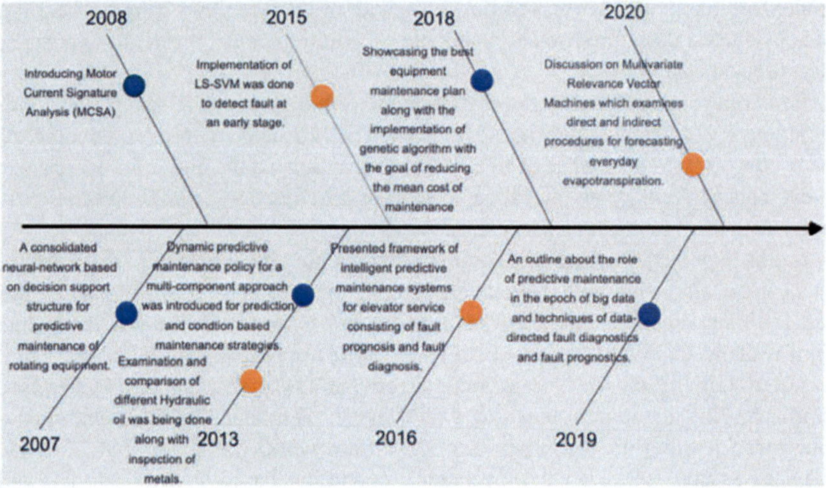

Fig. 29.1 Timeline evolution of techniques to predict the health of elevator systems

29.3 Gap Analysis

In [3], the mathematical explanation was presented, which reduced the mean maintenance cost, but the discussion about the relationship between the two equipment was not discussed. Although the analytical model used in [5] was pretty advanced, still, there is a need for an efficient data-mining approach and analysis model. In this study, the main focus of this work was on diagnostics and prognostics techniques, managing the tasks that are scheduled and with the restricted data, predicting the machine RUL. But, challenges faced by the authors are to improve the efficacy of the model [10]. In [11], to improve the efficacy of maintenance. The work focused more on the component dependencies rather than system dependencies. To look for various techniques in order to model stochastic dependence. The work in [7] showcased the optimization of maintenance of task scheduling. Challenges were faced while recording the readings of bearing when replaced in group. Cost matrix, along with margin life distribution, should also be expanded.

29.4 Parameter Analysis

Elevators work on the principle of the see-saw. The central module of an elevator is the elevator car, in which the commuter or load is carried. The elevator car has a supporting assembly attached to it, that assembly is attached to a traction rope. The vertical motion of the elevator is guided by a pair of rails. The elevator is raised and lowered with the help of traction steel rope and counterweight.

The rope is looped around a pulley, which is known as a driving pulley [14]. The driving pulley is attached to an electric machine. They together form a system known as a traction system. Traction system runs by the elevator's control system present in the elevator, which allows its motion.

There are many mechanical components present in an elevator system, these components are in direct relation with the performance and fidelity of the elevator. The traction system is the driving force of an elevator. The pulley present in the traction system consists of ball bearing and this wears off due to fretting friction caused between the pulley and the steel ropes [15]. Counter-weight balances the elevator car, which puts constant stress on the traction cable, the stress increases at the time of vertical movement of the elevator system. The motor present in the traction system produces short torque vibration, which helps determine the longevity of the elevator system. Distance traveled by the elevator is in direct relation with the elevator as the travels more distance the components wear off. Other components such as motor fans, gearboxes, humidity also affect the health of the elevator systems [12].

Some parameters have more impact on the health of the elevator than others. Ball-bearing present in the driving pulley is responsible for friction-free movement of the elevator system. This component of the traction system wears off as there is constant stress present from the traction steel ropes. Fitting friction among the ropes causes curtailment of the cross-sectional area of the metal, which naturally leads to resistance on the pulley. As a result, the bearing wears off. Distance traveled by an elevator is a clear indicator as to how much an elevator is used. As a consequence, it helps determine the health, wear and condition of the elevator across a timeline [16]. Torque vibration refers to the vibration produced by the mechanical motor present in the traction system, this attribute determines the health of the motor.

29.5 Methodology Used

29.5.1 Dataset and Attributes

The dataset used [4] has been recorded from a variety of IoT sensors used in the elevator industry. It contains the evening usage of an elevator between 16:30 and 23:30. The dataset consists of mainly three attributes: 'Ball-Bearing', 'Vibration' and 'Humidity'. Dataset provides operational data in the form of time series. It showcases the functional life of an elevator system, which is observed to be roughly 27 years. On analysis, it has been observed in Fig. 29.2 that there is a constant decline in revolutions (rpm). A perfectly functional ball bearing has 93 rpm, and it deteriorates as low as 13 rpm in a span of 27 years. The average life of an elevator is 20–25 years. Hence, by observing the graph, it is determined that at a value between 35 and 45 rpm, the elevator is most likely to cause a failure.

In Fig. 29.3, it is observed from the available statistics that in the initial 5–6 years of the elevator system, minor changes occur. The elevator system starts experiencing fluctuation in vibration after 7–10 years of use through the occurrence is less. The major increase in the occurrence of vibration is observed after 15–20 years of use. As traction motors are the source of short torque vibration, the health of the motor can be determined by the intensity of the vibrations.

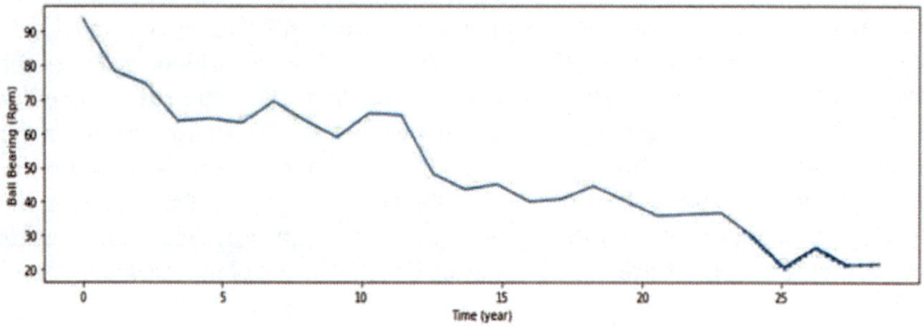

Fig. 29.2 Analysis of revolutions of ball-bearing parameter of an elevator system

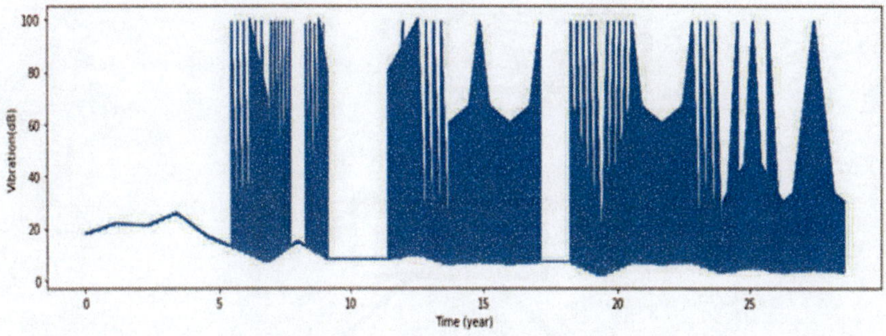

Fig. 29.3 Analysis of vibration parameter of an elevator system

29.5.2 Random Forest

Classification is one of the biggest parts of machine learning and random forest classifier holds the top of the classifier hierarchy [1, 2]. It is basically a bagging technique in which there is a collection of a large number of decision trees. Decision trees have two important properties of low bias and high variance. Random forest classifier creates an array of decision trees that are created from randomly generated subsets of training data. Then, votes from the different decision trees are combined to determine the final category of the test object. Random forests can also apply the weight concept, which takes into account the magnitude of influence of decision tree outcome. Trees with low error values have high weight values and vice versa. This increases the impact of trees that have low error rates. The following parameters have been used in the study:

- Gini Impurity: calculates the purity of the split and it is better than entropy, as it is computationally efficient. The range of gini impurity lies between 0 and 1.
- N_estimators: 20 (total number of trees)
- max_depth: 3 (depth of an individual tree in forest)
- min_samples_split: 2 (minimum samples required to split branch of the tree)
- min_sampels_leaf: 1 (minimum samples required to be at leaf node)

Random forest classifier assigns a random value to the root node sample and classification condition which in our case is "samples = 44408" and "ball-bearing <= 40.045". According to the classification condition, the range of samples is classified into three distinct states: Good (0), Fair (1) and Poor (2). This process continues until all the samples are classified successfully.

In Fig. 29.4, a random forest classifier selects a randomly selected condition value, which is close to the midpoint which in our observation is "ball-bearing <= 40.045". So, all the sample points with ball bearing less or equal to 40.045 are categorized as class 2. In the second level of the tree, the leaf node signifies the number of samples that are categorized as class 2. The non-leaf node on the second level of the tree represents the remaining samples, which are further divided into two subtrees based

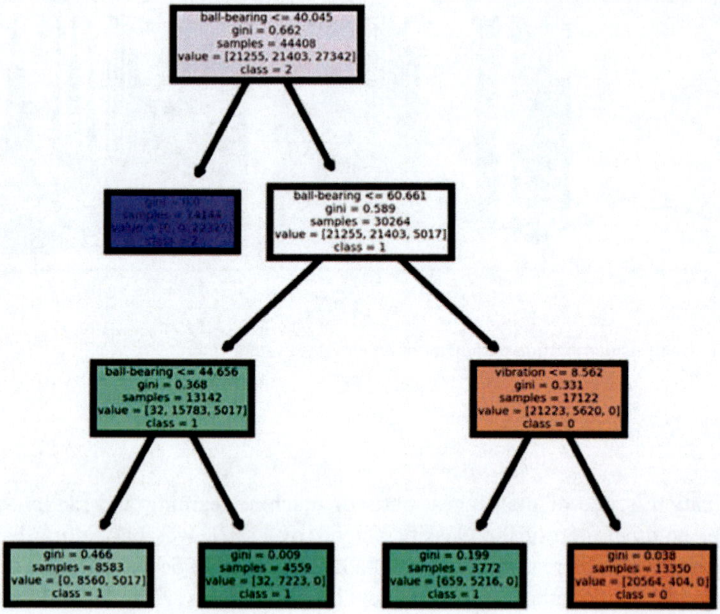

Fig. 29.4 Random forest tree visualization

on the randomly selected condition value selected by the random forest classifier "ball-bearing <= 60.661". This process is repeated until all the samples are classified.

29.6 Experimental Results

An average accuracy of about 91.50% (ranging 91% ± 2%) was achieved using a random forest classifier (Fig. 29.5).

Fig. 29.5 Feature preference score

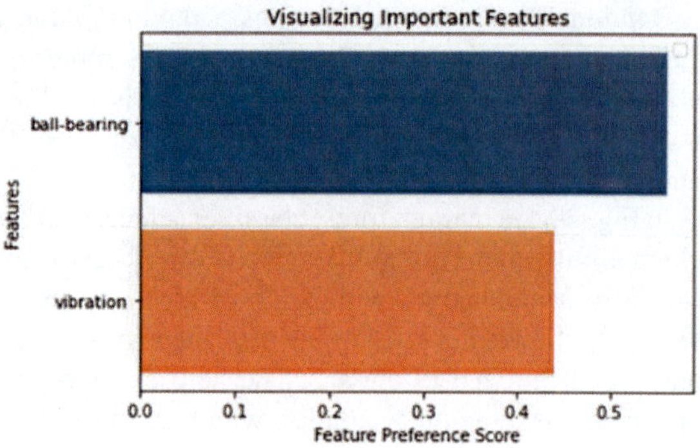

After the implementation of the random forest classifier, it was observed that the ball-bearing feature came out to be of more significance than the vibration feature. This was due to the frequent fluctuations seen in the readings of vibration in our dataset.

29.7 Conclusion

In this study, we investigated strategies and parameters to predict the maintenance of an elevator system and showcased the importance of predictive maintenance, and how it can contribute to better preventive maintenance strategies. With the help of predictive maintenance, it can prevent future accidents, save lives and can be cost-effective. Data analysis being the most crucial part of predictive maintenance is complex too. But, the implementation of deep learning algorithms has contributed in the extraction of new deep features. Thus, this has resulted in better efficacy and analysis of the model used.

In the future, the usage of a predictive maintenance strategy on an elevator system will surely result in optimized maintenance for all elevator components, thus prolonging elevator's overall lifetime. Future issues will be predicted even before the actual customer sees it. The implementation of improvised models along with deep learning in this process will contribute to the generation of new and useful deep features, which will result in more precise results, as well as, proper management of schedules.

References

1. Mishra, K., Huhtala, K.: Condition Monitoring of Elevator Systems Using Deep Neural Network, pp. 381–387 (2020). https://doi.org/10.5220/0009348803810387
2. Mishra, K.M., Huhtala, K.J.: Fault detection of elevator systems using multilayer perceptron neural network. In: 2019 24th IEEE International Conference on Emerging Technologies and Factory Automation (ETFA), Zaragoza, Spain, 2019, pp. 904–909. https://doi.org/10.1109/etfa.2019.8869230
3. Liu, H.J., Wu, J.X.: Research on preventive maintenance strategy of elevator equipment. Open J. Soc. Sci. **6**, 165–174 (2018). https://doi.org/10.4236/jss.2018.61012
4. Axenie, C., Bortoli, S.: Predictive maintenance dataset (2020). https://zenodo.org/record/3653909#.X2cDzz_itPY
5. Kesheng, W., Guohong, D., Lanzhong, G.: Intelligent predictive maintenance (IPdM) for elevator service—through CPS. IOTaS Data Min. (2016). https://doi.org/10.2991/iwama-16.2016.1
6. Gilabert, E., Arnaiz, A.: Intelligent automation systems for predictive maintenance: a case study. Robot. Comput.-Integr. Manuf. **22**(5–6), 543–549 (2006). ISSN 0736-5845. https://doi.org/10.1016/j.rcim.2005.12.010
7. Wu, S., Gebraeel, N., Lawley, M.A., Yih, Y.: A neural network integrated decision support system for condition-based optimal predictive maintenance policy. IEEE Trans. Syst. Man

Cybern. Part A: Syst. Hum. **37**(2), 226–236 (2007). https://doi.org/10.1109/TSMCA.2006.886368

8. Langone, R., Alzate, C., De Ketelaere, B., Vlasselaer, J., Meert, W., Suykens, J.A.K.: LS-SVM based spectral clustering and regression for predicting maintenance of industrial machines. Eng. Appl. Artif. Intell. **37**, 268–278 (2015). ISSN 0952-1976. https://doi.org/10.1016/j.engappai.2014.09.008
9. Leevy, J.L., et al.: Investigating the relationship between time and predictive model maintenance. J. Big Data **7**(1), 1–19 (2020). Received 12 February 2020. Accepted 25 May 2020. Published 05 June 2020. https://doi.org/10.1186/s40537-020-00312-x
10. Xu, G., et al.: Data-driven fault diagnostics and prognostics for predictive maintenance: a brief overview. In: 2019 IEEE 15th International Conference on Automation Science and Engineering (CASE), Vancouver, BC, Canada, pp. 103–108 (2019). https://doi.org/10.1109/coase.2019.8843068
11. Van Horenbeek, A., Pintelon, L.: A dynamic predictive maintenance policy for complex multi-component systems. Reliab. Eng. Syst. Safety **120**, 39–50 (2013). ISSN 0951-8320. https://doi.org/10.1016/j.ress.2013.02.029
12. Flores, A., Carvalho, J., Cardoso, A.J.M.: Mechanical fault detection in an elevator by remote monitoring. (2008). https://doi.org/10.1109/ICELMACH.2008.4800064
13. Kalligeros, S.: Predictive maintenance of hydraulic lifts through lubricating oil analysis. Machines **2**, 1–12 (2013). https://doi.org/10.3390/machines2010001
14. Esteban, E., Iturrospe, A., Isasa, I., Salgado, O.: Vibration-based condition monitoring for residential lifts (2013)
15. Rao, B.P.C., Lei, H., Tian, G., Zhao, H., Mao, Y., Huang, Z.: Health monitoring for coated steel belts in an elevator system (2012). https://doi.org/10.1155/2012/750261
16. Hayati, S.: Prediction and Detection of Abnormal Usage of an Elevator. Tampere University of Technology Master of Science Thesis (2018)

Chapter 30
Groundwater Recharge Using Artificial Filter Mechanism

Deepti Dohare

Abstract Groundwater is one of the most significant assets for giving water for all purposes but due to new trends of living, it is seriously misused for mechanical and homegrown use. In India, groundwater is a basic asset of water, which represents practically 85% of drinking water supplies [1]. The quality and amount of groundwater are being debased at a high rate predominantly because of increment in populace, urbanization, unnecessary abuse and lacking contamination control measures. Groundwater recharge is a hydrologic cycle where water descends from surface toward ground and after getting filtered through various soil media, meets the groundwater. This cycle typically happens in the vadose zone underneath plant roots and is frequently communicated as a motion of the water table surface. Energization happens both normally (through the water cycle) and through anthropogenic cycles (i.e., 'counterfeit groundwater revive'), where water and additionally recycled water are directed to the sub-processes. These exercises can bring about loss of dirt, decreased water penetration, improved surface overflow and decrease in energization. The utilization of groundwater, particularly for water system, may likewise bring down the water table [2]. This study is to provide an efficient arrangement for rejuvenating the waste and polluted water to be used to improve the groundwater table. The study provides an efficient way to effectively remove unwanted particles present in the surface water, which can be further used for various other activities. The experimental study used here is to provide a practical and natural state of water pollution and the efficiency level of the artificial filter.

D. Dohare (✉)
GL Bajaj Institute of Technology and Management, Greater Noida, India
e-mail: deepti.dohare@glbitm.ac.in

© The Author(s), under exclusive license to Springer Nature Singapore Pte Ltd. 2022
V. V. Rao et al. (eds.), *Computational and Experimental Methods in Mechanical Engineering*, Smart Innovation, Systems and Technologies 239,
https://doi.org/10.1007/978-981-16-2857-3_31

30.1 Introduction

Groundwater levels in different segments of India are declining as the nation could not provide enough regenerating springs in shortage-prone territories where it has been utilized for water system, enterprises and drinking water needs of the developing populace throughout the long term. As per the Central Ground Water Board (CGWB) [3], which is an administration organization, around 56% of the wells, which are examined to keep a tab on groundwater level, indicated a decrease in its level in 2013 when collated with the normal conditions of the last 10 years (2003–12) period. It was found that 5,699 wells have indicated a decrease during that period. It additionally inferred that farming part is the greatest client of water followed by homegrown plants and modern area needs. Exhausting groundwater level might be a genuine concern in the event that one glances at the future interest of water in India. It is assessed that the nation would require 1,180 billion cubic meters (BCM) of water every year by 2050. India has, at present, yearly capability of 1,123 BCM of 'utilizable' water with 690 BCM originating from surface water assets and staying 433 BCM from groundwater assets. South has a startling 30% of its groundwater table lower than 60 m beneath the ground (Fig. 30.1).

There has been a 6 rate point plunge in a portion of groundwater wells inside 10 m beneath the ground. This profundity is the edge past in which ranchers need to begin utilizing profound water apparatus, which adds to their difficulty (Fig. 30.2).

The artificial recharge to groundwater focuses on the growth of groundwater reservoir by changing the characteristic development of surface water using reasonable common development procedures [4]. Counterfeit revive methods typically address the following issues

- To improve the economical yield in territories where over-advancement has drained the spring.
- Conservation and capacity of overabundance surface water for future necessities, since these prerequisites regularly change inside a season or a period.
- To improve the nature of the existing groundwater.
- To eliminate bacteriological and different pollutions from squander water so that water is made appropriate for reuse.

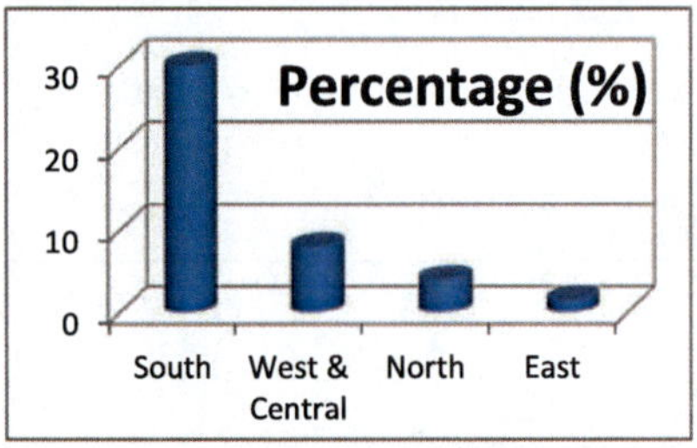

Fig. 30.1 Data show the percentage share of groundwater wells lower than 60 m below the ground

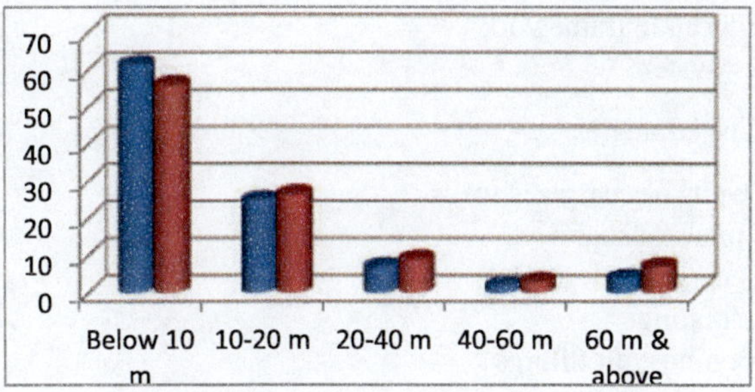

Fig. 30.2 Data show the percentage share of groundwater wells according to depth below the surface

- The essential motivation behind artificial energization of groundwater is to reestablish supplies from springs drained because of exorbitant groundwater improvement.

30.1.1 Sources of Water for Recharge

Before grabbing a revive plot, it is imperative to initially survey the accessibility of satisfactory water for energization. Following are the primary sources, which should be recognized and surveyed for sufficiency:

- Precipitation over the outlined zone
- Large rooftop zones from where water can be gathered and occupied for rejuvenation
- Canals from huge supplies from which water can be made accessible for rejuvenation
- Natural streams from which surplus water can be redirected for revival, without abusing the privileges of different clients.

30.1.2 Artificial Recharge Techniques and Designs

A wide range of methods are known to revive groundwater repository. Like the varieties in hydrogeographical system, the counterfeit revival strategies also change generally. The artificial revival methods can be extensively communicated as follows:-

a. Direct surface strategies

   i. Flooding
   ii. Basins or permeation tanks
   iii. Stream expansion

iv. Ditch and wrinkle framework
v. Over water system

b. Direct sub-surface strategies

i. Injection wells or energizes wells
ii. Recharge pits and shafts
iii. Dug well energize
iv. Bore gap flooding
v. Natural openings, pit fillings

c. Combination surface—sub-surface strategies

i. Basin or permeation tanks with pit shaft or wells

d. Indirect techniques

i. Induced revive from surface water source.
ii. Aquifer adjustment

Other than the above mentioned, the groundwater protection structures like groundwater dams, sub-surface dykes, or particularly named as Bandharas, are very pervasive to capture sub-surface streams. Thus, in hard rock regions, rock breaking strategies including sectional impacting of boreholes with appropriate procedures have been applied to associate the cracks and incremented energization. Concrete fixing of fissures through extraordinarily built, bore well has been used in Maharashtra to preserve sub-surface stream and enlarge bore well yield.

30.1.3 Identification of Area for Recharge

The virtual energy of groundwater is typically taken in the following territories:

- Areas where groundwater levels are declining on standard premise.
- Areas where considerable measure of spring has just been deimmersed.
- Areas where accessibility of groundwater is deficient in lean months.
- Areas where saltiness entrance is occurring.

30.1.4 Advantages of Artificial Groundwater Recharge

Following are the fundamental focal points of misleadingly reviving the groundwater springs [5, 6]:

- No enormous capacity structures are expected to store water. Structures required are compact and practical.
- Enhance the trustworthy yield of wells and siphons.
- Negligible mishaps when contrasted with mishaps in surface stockpiles.

- Improved water quality because of weakening of destructive synthetic compounds/salts.
- No unfavorable impacts like the immersion of huge surface region and loss of harvests. No relocation of nearby populace.
- Reduction in cost of energy for lifting water particularly where upsurge in groundwater level is generous.
- Utilizes the overflow surface spillover which in any case depletes off.

30.2 Methodology

Slow sand filters are the highly efficient filters to improve water quality and remove impurities [7]. Slow sand channels are utilized in water filtration for getting raw water to produce a consumable item. To fulfill the objective of the study, an artificial SSF has been designed so as to replicate higher water purification efficiency in water recharge process. The length and expansiveness of the tanks are controlled by the stream rate wanted by the channels, which regularly have a stacking pace of profound, which can be rectangular or round and hollow in cross-area. These are utilized essentially to 0.1–0.2 m every hour (or cubic meters per square meter every hour). Slow sand channels contrast from all different channels used to treat savoring water as they work by utilizing a complex organic film that develops normally on the outside of the sand. The sand itself does not play out any filtration work yet essentially goes about as a substrate, in contrast to its partners for UV and pressurized medicines. In spite of the fact that they are frequently the favored innovation in many creating nations in view of their low energy necessities and hearty execution, they are additionally used to treat water in probably the most evolved nations, for example, the UK, where they are utilized to treat water [8].

Slow sand channels have the mechanism of having a coagulated layer, which is also called as biofilm or Schmutzdecke in barely millimeters depth above the fine sand layer. The Schmutzdecke is shaped in the initial 10–20 days of activity and comprises of microorganisms, parasites, protozoa, rotifera and a scope of amphibian bug hatchlings. The surface biofilm is the layer that gives the compelling sanitization in consumable water treatment, the fundamental sand is offered as the help vehicle for this natural treatment layer. As water goes through the hypogeal layer, particles of unfamiliar issues are caught in the adhesive network and solvent natural material is also adsorbed. The foreign substances are used by the microorganisms and protozoa for growth. The water delivered from a moderate sand channel is of fantastic quality with 90–99% decreased bacterial cell check [9].

A proper sand channel contains natural movement and is in this way frequently alluded to as a bio-sand channel. As micro-creatures, for example, microbes and parasites, travel through the sand, they slam into it and get adsorbed onto sand particles. The creatures and particles gather in the best thickness in the top layers of the sand, continuously shaping a natural zone. The natural zone is not generally a strong layer, but instead a thick populace that steadily creates inside the top layer

Table 30.1 SSD affectivity

Highly effective for	Somewhat effective for	Not effective for
Bacteria Protozoa Viruses Turbidity Heavy metals (Zn, Cu, Cd, Pb)	Odor, taste Iron, manganese Organic matter Arsenic	Salts Fluoride Trihalomethane (THM) precursors Majority of chemicals

of the sand. The number of inhabitants is essential for a functioning natural pecking order that devours microbes (infection causing life forms) as they are caught in and on the sand surface. 1–3 cm of this organic zone is some of the time alluded to as 'schmutzdecke' or 'channel cake'. This layer is characterized as a layer of particles stored on head of the channel bed or natural development on head of the channel bed. Slow sand channels are generally cleaned by scratching off the biofilm as well as the top sand layer.

30.2.1 Health Aspects

Slow sand filtration is an incredibly proficient technique for eliminating microbial pollution and will generally have no indicator of microorganisms present at the outlet. Slow sand filters (SSFs) are additionally successful in eliminating protozoa and infections. In the event where the turbidity is under 1.0 nephelometric turbidity unit (NTU), a 90–99% decrease in microscopic organisms and infections is accomplished. However, slow sand filtration is commonly not successful for most synthetics. In any such case, it very well may be contended that principle substances for drinking water are of further worry in water gracefully subject to extreme bacterial tainting [10]. For this study, the water sample collected will be made free from the following mentioned health hazards in Table 30.1.

30.2.2 Design of Artificial Slow Sand Filter (SSF)

Filter

It is developed in the utilized electrifies iron oil drum of limit 50 lt. The distance across the cross-segment of the drum is 55 cm and the stature is 90 cm. At the lower part of the drum a tap of size ½″ is plumbed to empty out the separated water. At the stature of 7.5 cm, an iron sheet is set that is punctured to permit water to permeate through it. It is upheld by the assistance of 4 cubical squares of size 7.5 cm. The space accessible between the sheet and the lower part of the drum is utilized to gather the separated water. A layer of appropriately washed rock of sizes 10, 20 and 40 mm

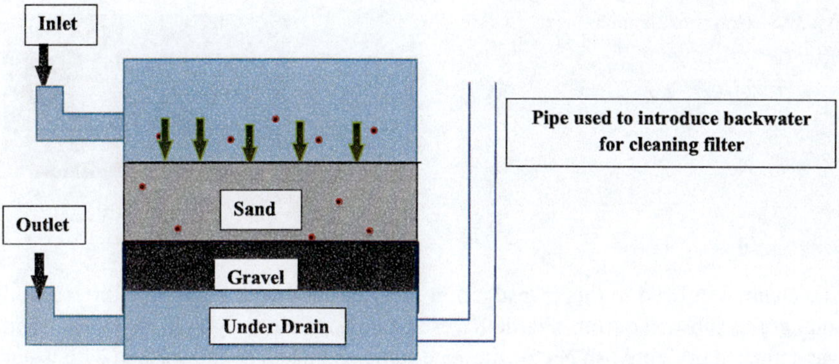

Fig. 30.3 Slow sand filter

up to the stature of 21 cm are set over the sheet, and afterward, the sand layer of 18 cm thickness is set over the rock [11]. The sand utilized in the channel is of Sharda River, which streams close by Pantnagar. The sand layer and rock layer are isolated by a wire work, with the goal that sand layer can be effectively taken out for utilizing sand layer of various thicknesses (Fig. 30.3).

30.2.3 Components of Slow Sand Filter

Gravels

In slow sand filter, we utilized rock of size 10, 20 and 40 mm. The rock is put above iron sheet, which is put at a stature of 7.5 cm from lower part of tank. The thickness of rock layer is kept 21 cm. The layer is isolated into three sections where the base part comprises of rocks of size 40 mm, the center bit comprises of 20 mm rocks and the highest bit of layer comprises of 10 mm rocks. The rock layer holds the sand channel from uprooting during filtration. Before setting in the channel setup, rocks were washed completely with clean water to guarantee that all the foreign materials were liberated to avoid any unfamiliar issue (Figs. 30.4 and 30.5).

Fig. 30.4 Gravel of 40 mm

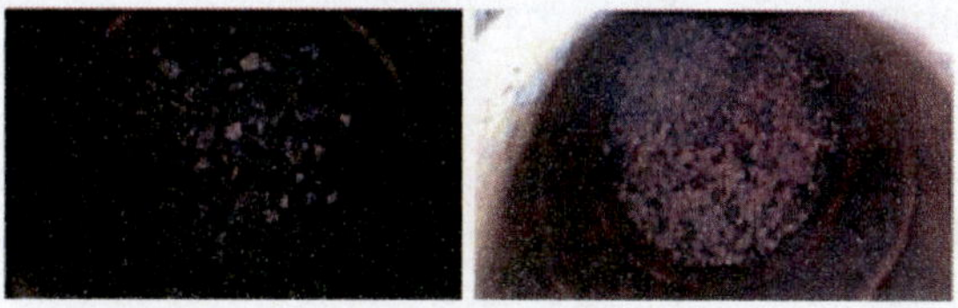

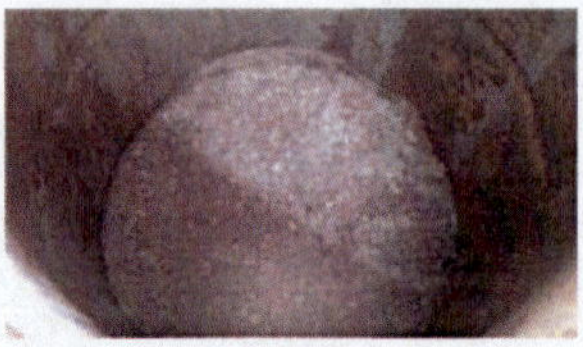

Fig. 30.5 Gravel of 10 mm size

Fine Sand

The channel utilized in this investigation was fine sand with negligible dirt, topsoil and natural substance from Sharda River that streams close by Pantnagar. The sand was screened utilizing fish net to eliminate some greater sizes and washed with water before evaluating to eliminate mud and earth content. The sand was sun-dried and reviewed for the necessary details. Sand particles passing strainer of size 0.3 mm were utilized.

The organic action is upgraded with expanding channel drop (Sand drop). Microorganisms and other suspended particles need to travel more through the sand. In this way, higher expulsion productivity is normal at higher sand depth. The channel drop was changed at 0.18, 0.24 and 0.30 m for coarse rock, fine rock and sand individually. Adequate time was considered for the water particles to go through the bed before the filtrate was gathered at a point for bacteriological examination.

30.2.4 Analysis

The water gathered was preexamined for the underlying properties to get the contaminations level. Additionally, similar water was taken subsequently with different measures for similar boundaries to get the level of filtration by the channel arranged for the examination. Different properties like total solids, suspended solids, settleable solids, pH, total hardness, calcium hardness, magnesium hardness, MPN and turbidity were broken down for the water pollution level [12].

pH

pH of the water was discovered by utilizing the pH meter, which is preadjusted by standard arrangements of pH esteems and can be utilized to straightforwardly peruse pH estimation of the given water sample.

Hardness

Different kinds of hardness of the water were discovered utilizing the reagents like EDTA as titrant, alkali base as support arrangement and EBT as marker. A total of 10 ml test sample was taken to which 1 ml of ammonia buffer and 0.2 g of EBT was added, which will give wine red tone. At this stage, the solution will be titrated against EDTA until blue tone shows up toward the endpoint. When the shading changes to blue, data are noted. Now, utilizing the data, absolute or total hardness is determined

Fig. 30.6 Calcium hardness test and total hardness test

as:

$$\text{Total hardness (as } CaCO_3\text{) mg/l} = C \times 1000/D \tag{30.1}$$

where C = volume of EDTA required by test

D = volume of test taken for titration.

For calcium hardness, again 10 ml sample is taken and 2 ml of NaOH solution and 0.2 g of murexide indicator are added to the arrangement and titrated with EDTA till the shading changes from pink to purple.

From the watched perusing, calcium hardness is determined as:

$$\text{Calcium hardness} = E \times 400.8 \times 2.497/F \tag{30.2}$$

where E = volume of EDTA utilized

F = volume of test taken for titration.

From the above two figurings, magnesium hardness is determined as (Fig. 30.6):

$$\text{Magnesium hardness} = \text{total hardness} - \text{calcium hardness} \tag{30.3}$$

Turbidity

Turbidity is caused in characteristic waters by long-last separated suspended particles of mud, sediment, sand or by some natural material, and by tiny life forms. It is normally communicated in mg/l (or ppm) and might be controlled by optical perceptions, as the obstruction to the section of light through the given water. The standard unit of turbidity is that turbidity that is delivered by blending 1 mg of SiO_2 (called Fullers earth) in 1 L of refined water. Hardness of more than 5 units is observable in the normal water leads to an unacceptable condition for water. Turbidimeter is commonly used to gauge the turbidity of the given water sample.

It is estimated by calibrating the turbidity estimating instrument by utilizing refined water and afterward taking 100 ml of sample and plunging the instrument in the sample to be tested lastly, noticing the readings as shown by the instrument.

Suspended, Dissolved and Total Solids

In this test, the water is tested by gauging the given porcelain dish and recording the mass as M1 gm. At that point, the dish is loaded up with 50 ml of water sample and kept on stream shower so the sample gets dissipated and subsequent to cooling it, it is gauged and taken as M2. Another 50 ml sample and filter paper is taken and weighed as M3. The sample is then permitted to go through the channel paper and oven-dried so that the weight can be taken as M4. From every one of these readings, suspended solids in 50 ml of test can be determined as:

$$\text{Suspended solids in 50 ml test} = (\text{M4} - \text{M3})\ \text{g} \tag{30.4}$$

$$\text{Concentration of suspended solids} = (\text{M4} - \text{M3}) * 1000/50\ \text{mg/l} = \text{x mg/l} \tag{30.5}$$

$$\text{Total solids in 50 mg/l test} = (\text{M2} - \text{M1})\ \text{g} \tag{30.6}$$

$$\text{Concentration of total solids} = (\text{M2} - \text{M1})''1000/50\ \text{mg/l} = \text{y mg/l} \tag{30.7}$$

$$\text{Concentration of settleable solids} = (\text{y} - \text{x})\ \text{mg/l.} \tag{30.8}$$

MPN Test

The total coliform bacteria are generally utilized as the indicator of decision for the quality of drinking water. It is a considerably more explicit indicator of fecal tainting. Escherchia Coli (E-Coli) is the overwhelming individual from the fecal coliform group. Identification of E-Coli in drinking water, accordingly, is taken as proof of late contamination with human or any other creature excrement. The strategy for straightforwardly checking the coliforms is known as the film channel method or layer channel test (Fig. 30.7).

Coliform microorganisms attain age with the development of gas inside a maximum of 48 h at 35 °C. In this test, stock containing lactose and different substances (medium) that hinder non-coliform living beings, is put in a progression of test tubes, three such test tubes are then incubated with 10 ml of water sample

Fig. 30.7 MPN Test setup

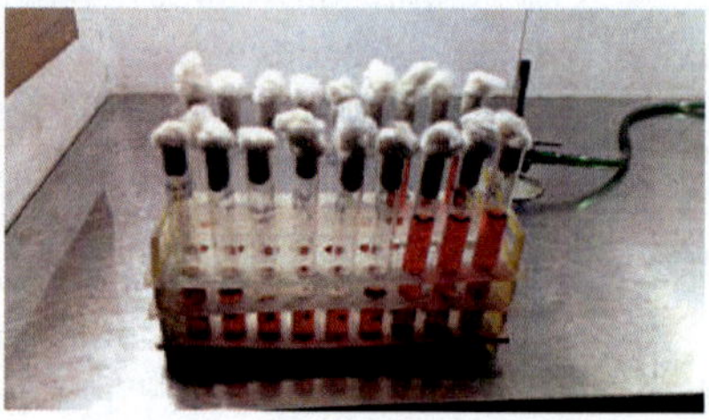

in each cylinder. Another three test tubes are blended in with 1 ml of water test in each cylinder and another three cylinders are blended in with are immunized with 0.1 ml of water sample in each cylinder and these cylinders are at last incubated at 35 °C for 24 h, and development of gas is noted. The presence of gas will give a hypothetical trial of the essence of coliform living beings in given cylinder. In the event that gas is not discovered we may permit further brooding for an additional 24 h, and, may end the test at end of 48 h, which is maximum time for the arrival of gas by coliform microorganisms. In the event that no such gas is developed, it will show a negative test, in which nonappearance of coliform microorganisms is noted.

The test tube indicating positive test results abstract to corroborative test tube, which will wipe out specific microbes of non-clean significance. This is hatched for 48 h, and if gas is framed, it will be a positive corroborative test and if no gas is shaped, it will be a negative test. Presently in the wake of deciding number of positive test tubes, factual techniques are utilized to decide the microscopic organisms thickness that has greatest likelihood with given arrangement of constants, which speak to only the most possible number of the coliform called MPN. The table shows the estimations of MPN per 100 ml of given water sample, when testing is done on nine test tubes (three cylinders each containing 10, 1 and 0.1 ml).

30.3 Results and Discussions

From Table 30.2, it very well may be gathered that the water sample collected for the testing had barely any boundaries like total hardness, calcium hardness, magnesium hardness and pH esteems in charge and underneath as far as possible and as a result of which they are not a matter of concern. Yet, still the channel has indicated decent expulsion productivity for such boundaries. Different properties like total solids, suspended solids, broken up solids, MPN and turbidity demonstrated exceptionally high initial quantities when contrasted with as far as possible. Yet, from the outcomes got in the wake of separating measure, it has been seen that there is a high efficiency appeared from the moderate sand channel. It very well may be seen from the outcomes that above 90% filtration productivity has been accomplished by the channel for suspended solids, MPN and turbidity boundaries. The outcomes for boundaries like total solids, MPN and turbidity have been appeared in Figs. 30.8, 30.9 and 30.10 individually.

From the plots above, it can likewise be seen that for sand depth of 30 cm, the filtration effectiveness is the most extreme for all the boundaries.

Table 30.2 Combined test results

S. No.	Parameters	Permissible standard limits	Water source	Filtered water			Removal efficiency (%)		
				d = 18 cm	d = 24 cm	d = 30 cm	d = 18 cm	d = 24 cm	d = 30 cm
1	Total solids (mg/l)	1000	1200	504	468	456	58	61	62
2	Suspended solids (mg/l)	0	500	27	18	14	94.6	96.4	97.2
3	Dissolved solids (mg/l)	500	700	477	450	442	31.87	35.71	36.86
4	Total hardness (mg/l)	300	70	63	60	58.36	10	14.28	16.63
5	Calcium hardness (mg/l)	75	68.05	58.76	58.76	57.18	10	13.66	15.98
6	Magnesium hardness (mg/l)	30	1.946	1.751	1.24	1.18	10	36.28	39.36
7	MPN per 100 ml	10	1100	49	17	8	95.54	98.45	99.27
8	Turbidity (NTU)	5	117.9	6	5.25	4.95	94.9	95.54	95.8
9	pH	6.5–8.5	7.4	7.1	6.9	6.9	–	–	–

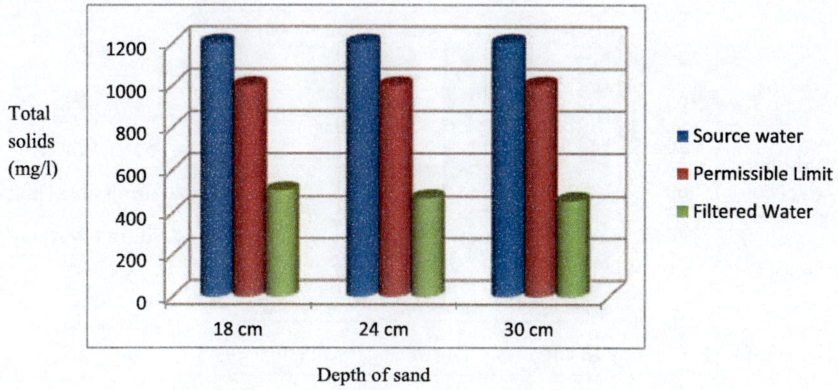

Fig. 30.8 Variation of total solids w.r.t. different depths

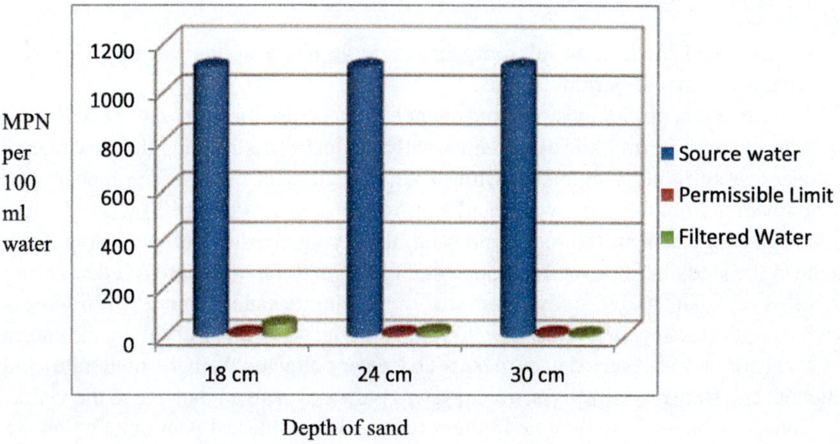

Fig. 30.9 Variation of MPN w.r.t. different depths

30.4 Conclusions

From the above study, it tends to be presumed that purging of well water utilizing sand as a medium in moderate sand channel is doable. The productivity of the channel to eliminate E. coli and lessen turbidity fluctuated was commonly high at various channel depths. This can likewise be credited to the formation of schmutzdecke layer on the sand surface and adsorption properties. Productivity variation of sand channel to eliminate E. coli at 0.18, 0.24 and 0.30 m channel depth is critical. It connotes that moderate sand channel is competent in eliminating E. coli at these depths. Biological action and protozoa group on the top layer of the schmutzdecke alongside adsorption

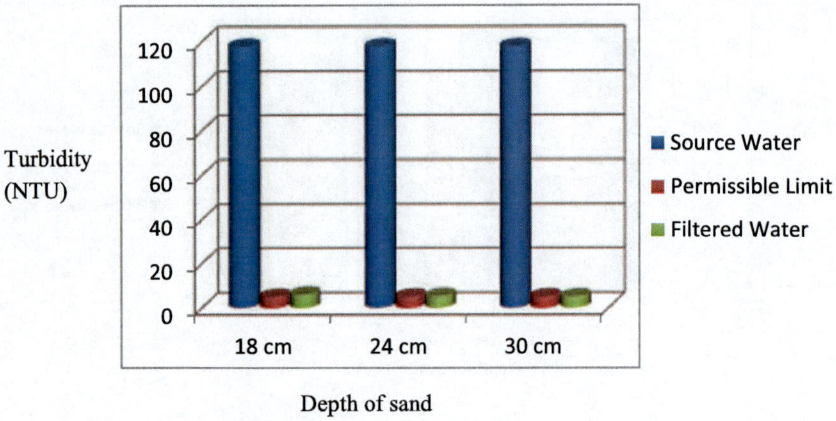

Fig. 30.10 Variation of turbidity w.r.t. different depths

and mechanical catching of microorganisms could likely be the systems of E. coli expulsion in water at various depths.

The variety in productivity of sand channel to decrease turbidity at 0.18, 0.24 and 0.30 m channel profundities is exceptionally critical. This implies the proficiency to decrease turbidity increments with expanding channel depths. The recharge of groundwater from overflow water and stormwater could be planned by providing a similar arrangement of the rocks and sand of varying sizes as per the arrangement done in the study to improve the groundwater level in the areas where it is decreasing at a fast rate. The above study infers that the moderate sand channel is profoundly viable in eliminating the different pollutions from the water and improving the nature of water for various every day purposes. The water collected from the moderate sand channel can be even utilized for water system purposes with no damage to the yields.

The experimental study done in the research shows the real phenomenon taking place in the water and the actual conditions in ground, which are to be analyzed and to be focused upon.

References

1. https://en.wikipedia.org/wiki/Groundwater_recharge
2. Mukharjee, D.: A review on artificial ground water recharge in India. SRG Int. J. Civ. Eng. (SSRG–IJCE) **3**(1) (2016)
3. CGWB, Central Ground Water Board, Ministry of Water Resources, Govt. of India
4. National Drinking Water Mission and Department of Rural Development: Rain Water Harvesting. Government of India, New Delhi (1989)
5. Bhattacharya, A.K.: Artificial ground water recharge with a special reference to India. Artif. Gr. Water Recharge (2010). dlc.dlib.indiana.edu

6. Fadaei, A.: Comparison of the efficiency of physical and biological treatment of slow sand filter in Kahkash (Samaan) treatment plant. Resour. Environ. **5**(4):107–109 (2015). https://doi.org/10.5923/j.re.20150504.01c
7. Gottinger, A.M.: The effectiveness of slow sand filters to treat Canadian rural prairie water. Can. J. Civ. Eng. **38**(4), 455–463. https://doi.org/10.1139/l11-018
8. Eusuff, M.M., Lansey, K.E.: Optimal operation of artificial groundwater recharge systems considering water quality transformations. Water Resour. Manag. (Springer) (2004)
9. Ni'matuzahroh: Behavior of schmutzdecke with varied filtration rates of slow sand filter to remove total coliforms. **6**(4):e03736 (2020)
10. Logsdon, G.S., et al.: Slow sand filtration for small water systems. **1**(5), 339–348 (2002). ISSN 1496-2551 | E-ISSN 1496-256X
11. Pramod, B.N., et al.: Reduction of contamination from water using slow sand filter. Int. Res. J. Eng. Technol. (IRJET) **5**(5) (2018). e-ISSN: 2395-0056
12. Muhammad, N., et al.: Optimization of slow sand filtration. In: 22nd WEDC Conference, New Delhi, India (1996)

Chapter 31
A Computational Technique to Generate Coupler Curve Equation of 6-bar Mechanism

Khalid Nafees, Mohd. Qamar Tanveer, Ajay Mahendru, and Anil P. Singh

Abstract In this paper, a computational technique to generate the equation of coupler point for a prescribed 6-bar, 1-DOF Watt-I mechanism has been described. All the dimensional parameters of the prescribed mechanism, including link length along with orientation, are known. The modeling of the mechanism find out the successive crank rotational positions and, therefore, helps determine the intermediate coordinates of the coupler curve. By applying the mathematical concept of Lagrange's polynomial, the required equation of the coupler curve is generated. The results obtained are supported by graphical plotting, verifying and demonstrating the computational technique for the prescribed 6-bar mechanism. This concept of generating a coupler curve equation is helpful in the analysis of any prescribed mechanism. Moreover, it helps to establish a relationship between prescribed dimensional parameters and desired performance of the mechanism.

31.1 Introduction

Synthesis and analysis are the two major aspects of Kinematics of mechanisms. Synthesis involves developing a mechanism to fulfill required performance, whereas analysis ensures that the prescribed mechanism exhibit properties and motion characteristics to fulfill the requirements. Generally, the output of the analysis of the mechanism is a calculation of forces, accelerations, velocities and displacements of different links of any prescribed mechanism. The analyses of mechanisms have been carried out by researchers with the help of various techniques.

In 1962, Gurney and Tobias [1] carried out a comparative study between variations in chip thickness with the rate of diffusion on stability. The author (s) investigated it for machine tool structure based on the curve of harmonic response. Uicker et al. [2] highlighted displacement analysis for closed kinematic mechanisms using the algebraic method based on matrix equations. Dobrjanskyj and Freudenstein [3] developed

K. Nafees (✉) · Mohd. Q. Tanveer · A. Mahendru · A. P. Singh
Department of Mechanical Engineering, Inderprastha Engineering College, Ghaziabad 201010, India

© The Author(s), under exclusive license to Springer Nature Singapore Pte Ltd. 2022

V. V. Rao et al. (eds.), *Computational and Experimental Methods in Mechanical Engineering*, Smart Innovation, Systems and Technologies 239,
https://doi.org/10.1007/978-981-16-2857-3_32

a computerized method to sketch an automatic graph of a mechanism that is useful in the primary stages of design. The author (s) applied the graph theory concept to determine the structural characteristics of the mechanism. Erdman et al. [4] investigated vigorous error developed due to various strain elements. The author (s) performed synthesis as well as structural analysis for mechanisms that deflect due to external loads. Sheth and Uicker [5] proposed an investigational software technique to mechanize the static and dynamic analyses of arbitrary mechanisms based on network theory and matrix methods. Dubowsky and Gardner [6] worked on planar mechanisms equipped with elastic links having clearances on various connections positions. The author (s) predicted the dynamic behavior of such mechanisms on system elasticity in terms of large impact loads caused due to clearances. Hill and Midha [7] applied the Newton–Raphson technique to analyze and design compliant mechanisms that are subjected to large nonlinear deformations for a prescribed value of the load. Erkaya et al. [8] compared the kinematic and dynamic analyses results of the modified slider-crank mechanism with the usual slider-crank mechanism. The mechanism considered had an additional eccentric link between connecting rod and crank pin. Based on analysis carried out, the author (s) concluded that the output torque of the modified slider-crank mechanism is greater than the conventional slider-crank mechanism. Lv et al. [9] carried out kinematic and dynamic characteristics analysis of the 6-bar mechanism with the help of MATAB software. The author (s) expressed the model of 6-bar mechanism in terms of vector loop equations. Murthy et al. [10] carried out the analysis of spatial 4-link RSCR mechanism to estimate inertia forces in various links of mechanism. The author (s) determines the acceleration and velocities of mechanism links.

So far, analyses have been carried out on different mechanisms in conjunction with various techniques, but still, there exists a research gap to define the characteristics of coupler motion based on dimensional parameters of any prescribed mechanism. For this purpose, in the present research work, a 6-bar mechanism has been considered to generate the equation of coupler tracing point using a computational technique of Lagrange's polynomial. The methodology of the present research work is demonstrated by solving a numerical problem.

31.2 Arrangement of Prescribed 6-bar Mechanism

The prescribed mechanism is a 6-bar Watt-I planar kinematic mechanism. It is a combination of binary links as well as ternary links having revolute pair at each connection. The prescribed mechanism has one binary offset link, two ternary links and three binary links.

The arrangement of connection of various links of the prescribed 6-bar mechanism is shown in Fig. 31.1. Link 1 (binary) is fixed between pivots O_1 and O_2. Link 2 (crank) rotates about pivot point O_1. The input motion is imparted to the mechanism through link 2. Link 3 (ternary link) is connected with link 2, link 4, and link 6 at points A, B, and F, respectively. There is a link 4, which has coupler tracing point C at an

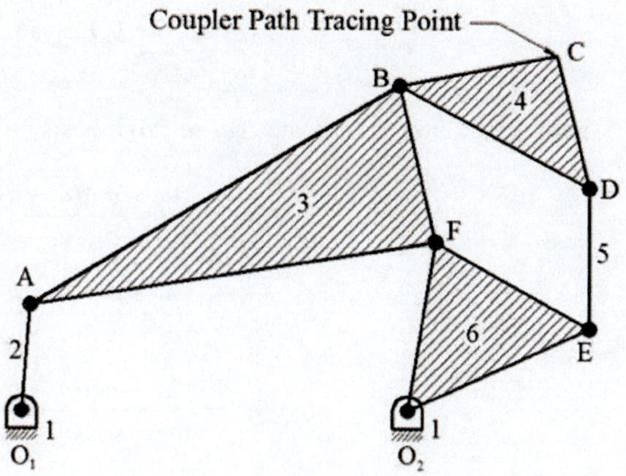

Fig. 31.1 Arrangement of prescribed 6-bar mechanism

offset location. Another link 6 (ternary link) is connected with fixed link 1, link 3, and link 5 at points O_2, F, and E, respectively. Link 4 and link 6 are connected by link 5 at points D and E, respectively. On rotation of link 2, the links of the prescribed mechanism reach at different positions. Link 6 oscillates about point O_2, the tracing point C generates a coupler curve $C_0C_1C_2C_3C_4$ (refer Fig. 31.3).

Link 1 is the binary fixed link O_1O_2; link 2 is binary rotating crank link O_1A; link 3 is ternary link ABF, link 4 is coupler link BDC with offset tracing point that forms the coupler curve through point C; link 6 is ternary link O_2EF; and link 5 is binary link DE.

31.3 Lagrange's Polynomial Interpolation [11, 12]

For a given set of distinct coordinates obtained through AutoCAD modeling of the prescribed mechanism, it is required to form an equation called the coupler curve equation. This process is referred curve fitting. For this purpose of interpolation, a computational technique called Lagrange's polynomial is employed.

To determine the equation of polynomial, the Lagrange's polynomial interpolation computational technique is applied for $(n + 1)$ number of intermediate data points. Suppose the coordinates of intermediate points of the coupler are defined by (a_0, b_0), (a_1, b_1)… (a_n, b_n).

The curve fitting is obtained for basic polynomials based on the constraint condition:

$$L_{n,k}(a_j) = \begin{cases} 1 \text{ when } j = k \\ 1 \text{ when } j \neq k \end{cases} \tag{31.1}$$

The Lagrange interpolating polynomial for nth degree is given by

$$L(a) = \sum_{k=0}^{n} b_k L_{n,k}(x) \tag{31.2}$$

For a set of three coordinates $\{a_0, a_1, a_2\}$, basis polynomials are given by

$$L_{2,0}(a) = \frac{(a - a_1)(a - a_2)}{(a_0 - a_1)(a_0 - a_2)} \tag{31.3a}$$

$$L_{2,1}(a) = \frac{(a - a_0)(a - a_2)}{(a_1 - a_0)(a_1 - a_2)} \tag{31.3b}$$

$$L_{2,2}(a) = \frac{(a - a_0)(a - a_1)}{(a_2 - a_1)(a_2 - a_1)} \tag{31.3c}$$

As per the concept of Lagrange interpolating polynomial, the final 2nd degree polynomial is given by

$$L(x) = b_0 L_{2,0}(a) + b_1 L_{2,1}(a) + b_2 L_{2,2}(a) \tag{31.4}$$

31.4 Numerical Problem to Generate Coupler Curve Equation for Prescribed 6-bar Mechanism

31.4.1 Problem Statement

To generate an equation of coupler curve for a prescribed 6-bar mechanism with following dimensional parameters as shown in Fig. 31.2.

$O_1O_2 = 0.686$; $O_1A = 0.197$; $AB = 0.772$; $AF = 0.741$; $BF = 0.290$;
$O_2E = 0.370$; $O_2F = 0.313$; $EF = 0.313$; $BC = 0.289$; $CD = 0.250$;

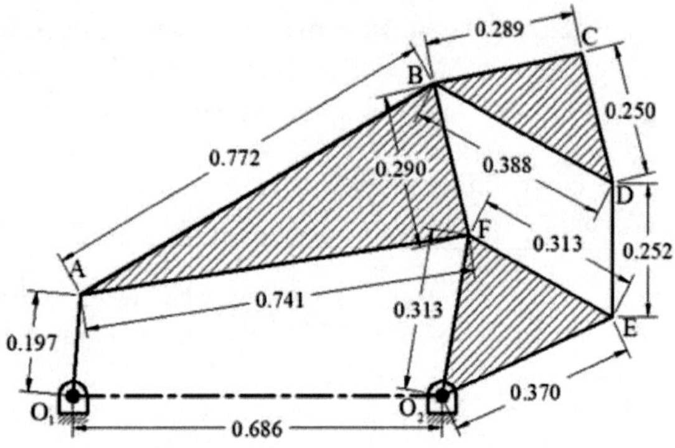

Fig. 31.2 Dimensional parameters of prescribed 6-bar mechanism

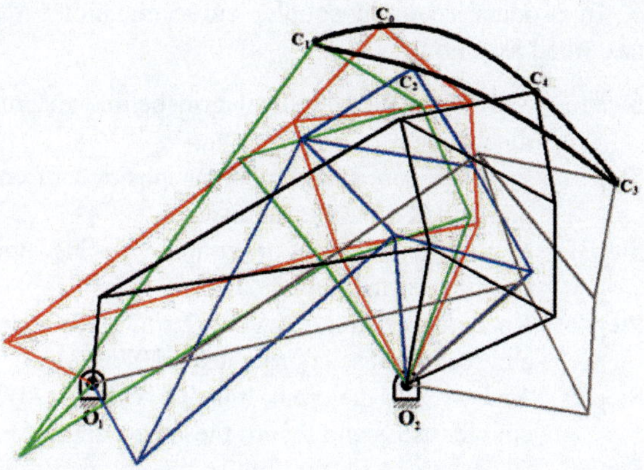

Fig. 31.3 Five crank rotational positions for a coupler of prescribed 6-bar mechanism

Table 31.1 Coordinates of coupler offset tracing point C of the prescribed 6-bar mechanism

Coordinates	C_0	C_1	C_2	C_3	C_4
X	0.950	0.801	1.027	1.423	1.283
$f(X)$	0.941	0.906	0.851	0.621	0.836

BD = 0.388; DE = 0.252.

31.4.2 *Observations*

On each successive rotation of link O_1A by the angle of $2\pi/5$, the coordinates of coupler offset tracing point C are recoded as given in Table 31.1 (Fig. 31.3).

31.5 Methodology for Generation of Coupler Curve Equation

When the crank link of the prescribed 6-bar mechanism rotates, the various joint positions acquire new positions. These positions for coupler point are referred as intermediate points. To generate the equation of coupler curve for a prescribed 6-bar mechanism, the mathematical concept of Lagrange's interpolation polynomial need the coordinates of these intermediate points for the coupler. For this purpose, a model of the prescribed 6-bar mechanism is developed in AutoCAD. By successive rotating the crank link of the model, the coordinates of intermediate points are determined. On substituting these coordinates into Lagrange's interpolation polynomial, the required coupler curve equation is generated.

To produce required coupler curve equations, the step by step procedure is described as follows:

Step1. Switch on the graphical grid behind the model of the prescribed 6-bar mechanism.
Step2. Find the coordinate of home position of coupler tracing point C of the prescribed 6-bar mechanism.
Step3. Mark this position of coupler tracing point as C_0 and record these coordinates in Table 31.1.
Step4. Rotate crank link (O_1A) by $2\pi/5$, mark new position of the coupler as C_1 and record these coordinates in Table 31.1.
Step5. Again rotate the crank link (O_1A) by $2\pi/5$, mark the new position of the coupler as C_2 and record these coordinates in Table 31.1.
Step6. Repeat step4 for two times, thereby marking new coupler positions as C_3 and C_4 and record these coordinates in Table 31.1.
Step7. Substitute all the recorded coordinates value into the equation of Lagrange's polynomial (Eq. 31.4).
Step8. Generate the final equation of the coupler curve for the prescribed mechanism by solving the preliminary equation obtained in step7 for $i = 0, 1, 2 \ldots 4$.

31.6 Results and Discussion

The generated equation of coupler curve for prescribed 6-bar mechanism determined using computational technique Lagrange's polynomial is found to be as:

$$f(X) = 667.64\,X^4 - 2923.63\,X^3 + 3993.21\,X^2 - 1357.03\,X + 883.16$$

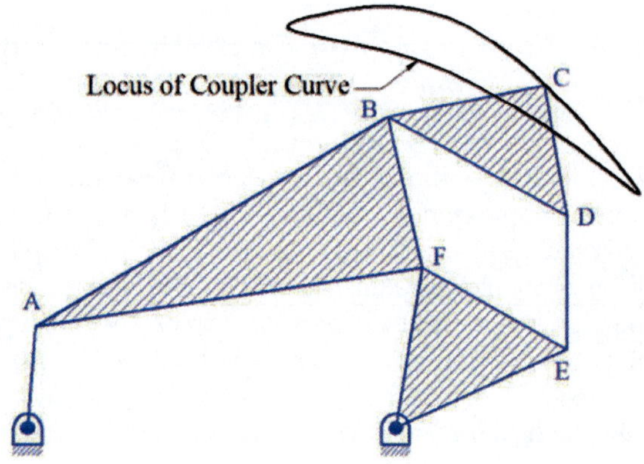

Fig. 31.4 Locus of coupler curve for which equation is being generated for prescribed 6-bar mechanism

The above equation of the coupler curve is graphically shown in Fig. 31.4. This equation for the prescribed 6-bar mechanism helps in comparing the desired performance of the coupler tracing point with actual performance. This analysis is useful to fulfill the accuracy requirements for industrial automation.

31.7 Conclusions

A computational technique to form the coupler curve equation for a prescribed 6-bar, 1-DOF Watt-I mechanism has been discussed in the present research work. The mechanism with prescribed dimensions, including link length with orientations, has been modeled in application software for the purpose of determining successive crank rotational positions. The determination of these positions is helpful in finding out the intermediate coordinates of coupler tracing point, which are subsequently substituted in Lagrange's polynomial. The mathematical concept of Lagrange's polynomial is helpful in the generation of the required coupler curve. This methodology is demonstrated and verified on a numerical problem for the prescribed 6-bar mechanism. The results of the problem are graphically plotted in AutoCAD software. The present work is helpful in the analysis of the prescribed mechanism to generate the equation of the coupler curve for an offset point. Moreover, it helps to establish a relation between the known dimensions and desired performance of a mechanism of this kind which is further useful to enhance its accuracy. The present computational technique of the mechanism analysis is useful in comparing the motion characteristics of various industrial automation processes such as welding robot arms. The generated coupler curve equation is helpful in determining point coordinates for robotic arm motion with continuous motion. This work is further extendable to establish a generalized coupler equation that defines the inter-relationship between coupler curve and dimensions of kinematic link.

References

1. Gurney, J.P., Tobias, S.A.: A graphical analysis of regenerative machine tool instability. J. Eng. Ind. **84**(1), 103–111 (1962)
2. Uicker, J.J., Denavit, J., Hartenberg, R.S.: An iterative method for the displacement analysis of spatial mechanisms. J. Appl. Mech. **31**(2), 309–314 (1964)
3. Dobrjanskyj, L., Freudenstein, F.: Some applications of graph theory to the structural analysis of mechanisms. J. Eng. Ind. **89**(1), 153–158 (1967)
4. Erdman, A.G., Sandor, G.N., Oakberg, R.G.: A general method for kineto-elastodynamic analysis and synthesis of mechanisms. J. Eng. Ind. **94**(4), 1193–1205 (1972)
5. Sheth, P.N., Uicker, J.J.: IMP (integrated mechanisms program), a computer-aided design analysis system for mechanisms and linkage. J. Eng. Ind. **94**(2), 454–464 (1972)
6. Dubowsky, S., Gardner, T.N.: Design and analysis of multilink flexible mechanisms with multiple clearance connections. J. Eng. Ind. **99**(1), 88–96 (1977)

7. Hill, T.C., Midha, A.: A graphical, user-driven newton-raphson technique for use in the analysis and design of compliant mechanisms. J. Mech. Des. **112**(1), 123–130 (1990)
8. Erkaya, S., Su, S., Uzmay, I.: Dynamic analysis of a slider-crank mechanism with eccentric connector and planetary gears. Mech. Mach. Theory **42**(4), 393–408 (2007)
9. Lv, K., Yuan, Y., Zhang, M.J.: Dynamic simulation and velocity adjustment of mechanism based on MATLAB. J. Theor. Appl. Inform. Technol. **51**(1), 85–90 (2013)
10. Murthy, P.S.S., Satyadevi, A., Krishna, A.G., Eswaraiah, K.: Kinematic analysis of a spatial mechanism for estimating shaking effects. J. Phys. Conf. Ser. **662**, 1–6 (2015)
11. Lagrange Interpolation. http://www.math.usm.edu/lambers/mat772/fall10/lecture5.pdf. Last accessed 31 Jan 2021
12. Polynomial Interp. http://www2.lawrence.edu/fast/GREGGJ/Math420/Section_3_1.pdf. Last accessed 31 Jan 2021

Chapter 32
Analysis on Manufacturing Automated Guided Vehicle for MSME Projects and Its Fabrication

Rishabh Chaturvedi, Anas Islam, and Aman Sharma

Abstract Automated Guided Vehicle (AGV) is the material handling equipment that is utilized broadly in a maximum manufacturing company nowadays as it gives more flexibility to the system. The fundamental perception of AGV includes driverless and battery-powered vehicles with programming abilities for path selection and locating. They are equipped to navigate a flexible guide-path network, which might be simply changed and extended. This project may be concentrating on model, unloading and loading mechanism for AGV have been required to define specific criteria, which is automatic operated, lightweight and capable of transporting acrylic. This unloading and loading system worked utilizing ball screw as a development system and suction cup with a vacuum pump as adhering instrument. The outline and fundamental material to manufacture this AGV model would utilize aluminum to decrease the AGV weight. The AGV application in a real-world application is supported human and decreasing price in repetitive movement transportation actions. This manuscript suggests a survey on the control and design of AGV frameworks. We locate many key-related problems incorporating guide-path design, vehicle scheduling, assessing the number of vehicles, battery management, conflict determination, idle-vehicle positioning, and vehicle routing. Furthermore, we suggest a decision system for the design and execution of AGV frameworks, and propose few fruitful study directions. It will be designed for an importantly less price and exorbitant efficiency. The expenditure of the overall build is Rs. 90,000/- which is cost worthy.

R. Chaturvedi · A. Islam (✉) · A. Sharma
IET Department of Mechanical Engineering, GLA University, Mathura, India
e-mail: anas.islam@gla.ac.in

R. Chaturvedi
e-mail: risabh.chaturvedi@gla.ac.in

A. Sharma
e-mail: aman.sharma@gla.ac.in

© The Author(s), under exclusive license to Springer Nature Singapore Pte Ltd. 2022
V. V. Rao et al. (eds.), *Computational and Experimental Methods in Mechanical Engineering*, Smart Innovation, Systems and Technologies 239,
https://doi.org/10.1007/978-981-16-2857-3_33

32.1 Introduction

In general, all the products are produced for specific applications that have to be useful. All the products are created by the fundamental material. That material is known as "Raw material." Without raw materials, there are no products. These raw materials maybe different kinds based on applications. These raw materials are converted into useful products by the industries [1–3]. There are lots of labors or workers for machining these raw materials. Also, there are many workers used for transporting these materials. There are many rooms such as storehouses or warehouses, different workshops, and some inventories for holding or carrying the finished or partly finished goods. A lot of workers cause the high labor cost for the owner of the industry [4–6]. Some workers have to work there to machining these raw materials that are inevitable. But many workers used for transport should be avoided because it reduces the labor cost [7].

The proposed project may be useful for transporting materials from one place to another place without human intervention. The proposed project is made simple and efficient by introducing automation there. The traditional process of transporting the raw materials as well as finished products involves at least 2 or 3 workers in that field [8–11]. Because one person should take the materials from the storehouse and then another person should load this material to the vehicle which carries that materials and another person should drive the vehicle to the appropriate places such as Lathe section, cutting section, drilling section, Milling section, Painting section, Assembly section, Cleaning section, etc. [12–14]. So, the manual method of having many processes involving human and takes a large time and also sometimes, an accident occurs due to carelessness and distractions in the human mind. This is not a cup of cake for everyone. Hence new entrepreneurs will not take up this business. Henceforth, the primary intention of this research work is to automate the material handling process using low-cost automation techniques [15–17]. To design and fabricate a low-cost AGV for handling material handling in industries for handling both finished goods and raw materials with the help of Robotic Arm with Gripper at its ends using Arduino UNO R3 controller. The AGV have been regularly utilized in facilities, such as distribution centers, warehouses, manufacturing plants, and transshipment terminals. AGV might be referred as mobile robots owing to their reprogram capability [18–20]. The AGV reason will be to support decrease prices of manufacturing and expand effectiveness in the manufacturing framework. It also included raw material, tools development, and work in procedure among stations. These developments should be accurately, securely, effectively and without any harm to materials. The project team is isolated into three diverse specs: unloading and loading mechanism, AGV control framework, and AGV mechanical part [21–23].

The vital ability of this AGV is to exchange loads to locality through way under machine control by programming. The AGV will be a material controlling framework utilized for the horizontal movement of materials. The AGVs are suggested in 1955, the utilization of AGVs is enhanced tremendously since their presentation. Many application regions and a variety of kinds have expanded essentially. AGVs might be utilized indoors and outdoors, contributing to tasks, such as distribution, production, and transshipment. The AGVs have been utilized to transport materials of whole kinds related to manufacturing methodology. The cross docks and deposits, engineering centers have been instances of distribution regions. The AGVs have been utilized in these regions for internal transport, for instance, pallets among diverse departments, such as the storage, dispatch, reception, and sorting regions [24–28].

In transshipment schemes, such as AGVs, shipping containers deal with products transport among diverse modes of transport. The very significant variances among traditional and novel application regions have been many AGVs utilized; degree of AGV occupancy, many transport applications, the distances traveled and many pick-ups and delivery points while transport applications become accessible. The panic grades have been utilized to carry fewer requests over short distances among selections, delivery points. Different reinforcement methods, many AGVs are utilized to execute a large number of repetitions, transport tasks to container terminals and exteriors [29]. It was concluded that few methods and novel methods have already proved successful.

32.1.1 AGV (Automatic Guided Vehicles)

AGV vehicles have been utilized in warehouses to transport products. The AGVs appearance was similar to little cars, which might transfer managed in the warehouse. The managed travel will be ensured by special software. There have been numerous kinds of automatic vehicles generated by many industries. Each of these might be utilized for moving pallets [30]. Few of these automatic vehicles might move shelves with goods to be organized. The AGV's central processing framework problems the steering control along with movement speed (Figs. 32.1, 32.2, 32.3 and 32.4).

A common AGV framework comprises basically peripherals vehicle on-site modules, stationary control method. The important modules of the AGV framework have been the oriental way scheme, vehicle, traffic control, and management framework. The AGV will guarantee a secure presentation, load, and surroundings.

Fig. 32.1 Pallet truck AGV

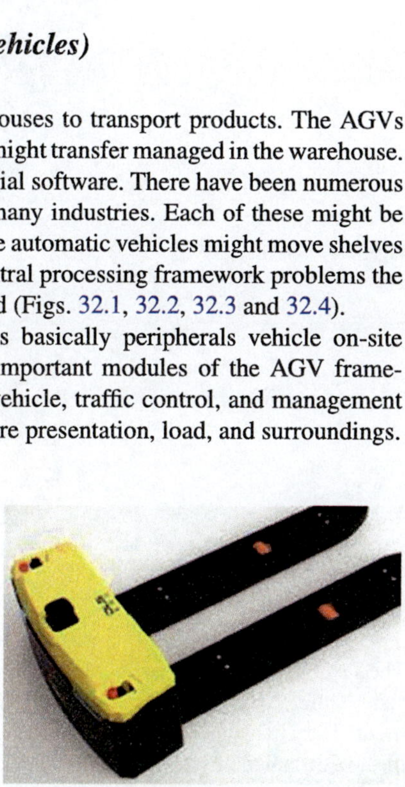

Fig. 32.2 Unit load AGV

Fig. 32.3 Towing AGV

Fig. 32.4 Forklift AGV

32.1.2 Vehicle

The vehicle will be the main component of AGV due to it satisfies the real transport task. The vehicle is distinct reliant on design, determination, and activity environment. They have been utilized in numerous states such as AGV intended to enhance the health care method, AGV established as hotel service robot. Amazon has executed the KWIA robot for automatic storage and products retrieval in the system. The main usual utilizes of these automatic vehicles have been in the production region while providing workstations with raw materials. Similarly, they have utilized while selecting up semi-finished or finished goods and bringing them to the storage region. The automatic forklifts might be deliberated AGVs; nevertheless, the normal concept will be AGVs are the type of cars, which come under the pallet and pick it up. As noted, at that moment, they have been utilized in production environments because of the high prices of buying the systems.

32.1.3 Oriental Way System

The vehicle guidance framework will be a technique by AGVs has been distinct vehicles are managed subsequent predefined ways. The AGV framework utilizes a guide path, and it selects a path based on a programmed path. It utilizes the information given by sensors and will be compared with the value provided by the programmer. While AGV methods a decision point, it only has to choose whether to follow the path. The maximum utilized navigation methods in AGV are:

a. Behavior-based,
b. Landmark-based, and
c. Vision-based navigation.

32.1.4 Traffic Control and Management System

To work effectively and expand AGV output, the vehicle must be well controlled and diminish waiting time at the loading/unloading station. The traffic control might be handled by AGV utilizing onboard vehicle identification and zone control. Numerous destinations are modifying AGV in the adaptable manufacturing environment, utilizing optimization methods, a method for handling traffic in the workspace. The GA will be a method built on the search for natural determination procedure. The ACO method will be utilized to discover the group near-optimal program, which fulfills both loads balancing among AGV, rely on time minimization and travel time. For productive control, two sorts of control framework are utilized inside the workplace:

1. Stationary control framework, and
2. Peripheral control framework.

The AGV efficiency might be calculated by the effective actuation time of AGV from the loading to the unloading cycle. For this, we might assume that AGV transfers at a constant speed through the environment and disregards the deceleration effect of acceleration and other speed modifications. The time for a distinctive delivery cycle will be defined as AGV framework :

1. Upload to pick up station,
2. Travel time to reaching station,
3. Unloading at the output station, and
4. Empty the travel time.

32.2 Proposed Method

32.2.1 *Manual Handling Technique*

In manual handling technique, all the materials are carried and transported solemnly by the workers and not using any vehicle for carrying those materials. This method is the very ancient method for material handling in the world. This method prevails still in small industries such as fireworks, match boxes industries, and this method costs high because lots of humans are required to handle those materials are heavy and a lot, if any.

32.2.2 *Manual Cart (Transporter) Technique*

The manual cart technique possesses an approach in which materials are taken out from the storehouse with the help of humans, and then they place these materials in the vehicle, which carries and transports them to the destination such as workshops etc. One transporter who drives the vehicle which carries the materials that have to be machined. If he loses control, it leads to an accident.

32.2.3 *Conveyor Technique*

These techniques involve some lift or crane type mechanism for lifting the bigger sizes and does not carry the human. This method is mostly used in Automotive and automobile industries such as Audi, Ferrari, Lamborghini industries, particularly in lifting engine components such as engine head, whole crankcase, flywheel, and hood of the vehicle and more are lifted by implementing this way of the procedure. And also, this technique not only lifts the materials but also transports them to its next section. But this forklift method used in the only vast area in industries and also it is an expensive one. By analyzing the existing methods instigated in material handling, the following drawbacks have been identified as follows:

- The existing system takes more time to lift the materials that have to be machined.
- It involves many human labors, which in turn results in high labor cost.

The proposed system entirely focused on, to reduce the humans for materials handling. The primary motto is to automate the entire material handling process. The process is automated by using sensors and an actuator connected to the Arduino UNO R3 controller. This proposed work is not only picking or lifts the raw materials from the storehouse but also transports to the other sections and drops there and also again picks the finished or semi-finished good from the workshop to other painting section or any. This research mainly concentrates on the requirement of small and medium scale enterprises to offer them high reliability, high efficiency, and the most important efficiency and less time consumption.

Advantage of the proposed system:

The advantages of the UGV for material handling in industries project are

- Low price of manufacturing and simple to maintain.
- It is well-suited for picking the materials and dropping them easily with the help of Arm with the Gripper mechanism.
- The machine consumes less power.
- It not only picks up the raw materials and also carries them on the vehicle and transports it to its destination easily by the Line follower concept.
- It is compact and occupies less floorspace.
- It is easily programmable and controllable by using the servomechanism.

32.3 Design and Analysis

Wall Shear Stress for pore size of 0.8 mm Diameter:

As the design is developed in CREO software, it took utmost capability to create this AGV base model. These two diagrammatic representations are clear, which indicates the designation of the model, so with these representations, we can clearly view the AGV structure.

Figure 32.5 represents the maximum WSS in all methods that happened at higher velocity (i.e., 1 mm/s) and WSS is linearly lessening while velocity reduces. Undoubtedly, the shear stress distribution is distinctive and relies on scaffold structure.

32.4 Results and Discussion

And, these AGV model is specialized in ball screw movement. It works on the coding structure and defines how if any object comes across the way in which material handling is done, it stops automatically in which it is programmed, such as sensors. Any environment and inexpensive between autonomous robots might be implemented. There is a main change among the work cycle's practical and theoretical time value being optimized by implementing a diverse technology. Moreover,

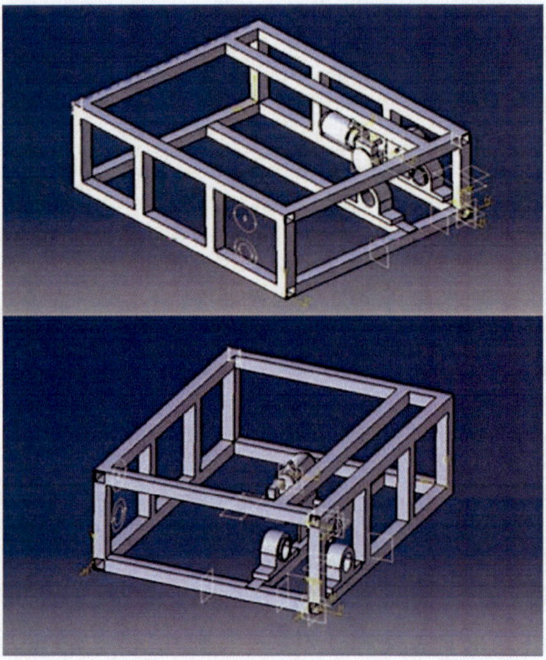

Fig. 32.5 Maximum WSS in all methods happened at higher velocity

one might think of a symbol relaxation holding necessities in traffic control framework; thus, many vehicles might leave diverse crossing areas concurrently; thus, the AGV system presentation might be enhanced. Many researchers neglect AGV battery management issues. In the instance of long travel distances, battery charging becomes a problem and requires being deliberated swell.

32.5 Conclusion

A new system for material handling is designed and established. The whole automation of the framework is attained to be price effective as well. Furthermore, the objective of this system is to pick, to carry and to transport the materials with the help of AGV without any human intervention and reducing processing time is taken by a human is successfully accomplished. The present AGV carries the load with a capacity of up to 3.5 kgs since the capacity of the gripper lifts up to 200 g per lift. Depends on the requirement, the capacity of the gripper and DC motor can be changed. Thus, the model can be used for handling heavyweight materials also. We have deliberated numerous main problems in the AGV method plan and execution. These problems incorporate guide-path design, assessing the essential number of vehicles, battery management, vehicle scheduling (dispatching), idle-vehicle positioning, conflict resolution, and vehicle routing. The main significant problem while designing AGV methods is guide-path design. In normal, we might separate guide-path frameworks into three classes: single-loop, conventional and tandem methods. Everyone has their individual benefits and drawbacks and will be appropriate for particular applications. Attaining the best evaluation of vehicle empty travel under few better dispatching guidelines is a key issue here. A further problem will be how to handle the AGV method effectively. This problem incorporates many subproblems: idle-vehicle positioning, vehicle scheduling, conflict resolution, battery charging, and vehicle routing. The scheduling method also decides the route a vehicle must take to reach its target without collisions.

References

1. Carullo, A., Parvis, M.: An ultrasonic sensor for distance measurement in automotive applications. IEEE Sens. J. **1**(2), 143 (2001)
2. Egbelu, P.J.: Characterization of automated guided vehicle dispatching rules. Int. J. Prod. Res. (1984)
3. Bengu, G.: An assignment algorithm for dynamic picking systems. Inst. Ind. Eng. (1995)
4. Govindaraju, K., Boopathi, S., Parvez Ahmed, F., Thulasi Ram, S., Jagadeeshraja, M.: Embedded based vehicle speed control system using wireless technology. Int. J. Innov. Res. Electr. Electron. Instrum. Control Eng. **2**(8) (2014)
5. Gassmann, H.I., Kristjansson, B.: The SMPS format explained. IMA J. Manag. Math. **19**(4) (2008)
6. Miller, J.K.: Computer assisted circulation control at health sciences library a description of circulation system. State University of New York (1972)
7. Altasa, K.M.: Cycle time models for dual address storage and retrieval systems. In: IEEE International Conference on Systems, Man and Cybernetics (1997)
8. Krishna, R., Ramkumar, R., Lenin, V.R.: Design of three fingered robot gripper mechanism. Int. J. Mech. Eng. Robot. **3**, 18–24 (2015)
9. Singh, P.K., Sharma, K.: Molecular dynamics simulation of glass transition behaviour of polymer based nanocomposites. J. Sci. Ind. Res. **77**(10), 592–595 (2018)
10. Clothier, K.E.: A geometric approach for robotic arm kinematics with hardware design, electrical design, and implementation. J. Robot. (2010)
11. Freedman, M.J.: Circulation system past and present on library automation, School of Library Service, Columbia, New York City, vol. 14/4, p. 279 (1981)
12. Kumar, K., et al.: Experimental investigation of graphene-paraffin wax nanocomposites for thermal energy storage. Mater. Today Proc. **18**, 5158–5163 (2019)
13. Sharma, A., Dwivedi, V.K.: Effect of spindle speed, feed rate and cooling medium on the burr structure of aluminium through milling. IOP Conf. Ser. Mater. Sci. Eng. **998**, (2020)
14. Gohil, M.V., Patel, J.: Design of lead screw mechanism for vertical door wrapping machine. Int. J. Sci. Res. Dev. **2**(2), 185 (2014)
15. Foster, M.R.: Review of programmable logic controllers in control systems education. A review of programmable logic controllers in control systems education (2010) Kumar, A., Sharma, K., Dixit, A.R.: A review on the mechanical properties of polymer composites reinforced by carbon nanotubes and graphene. Carbon Lett. 1–17 (2020)
16. Periyasamy, P., Prince, W.D., Cynthia Christabel, S.: A continuous health monitoring system for photovoltaic array using Arduino microcontroller. Circuits Syst. **7**, 3494–3503 (2010)
17. Gaskins, R.J., Tanchoco, J.M.A.: Flow path design for automated guided vehicle systems. Int. J. Prod. Res. **25**, 667–676 (1987)
18. Gaskins, R.J., Tanchoco, J.M.A., Taghaboni, F.: Virtual flow paths for free ranging automated guided vehicle systems. Int. J. Prod. Res. **27**, 91–100 (1989)

19. Manda, S.: Standard requirements for integrated library system. Asian J. Multidiscip. Stud. **1**(5) (2013)
20. Sharma, A., Sharma, K., Islam, A., Roy, D.: Effect of welding parameters on automated robotic arc welding process. Mater. Today Proc. **26**(2), 2363–2367 (2020)
21. Landers, T., Sadiq, M., Don Taylor, G.: An assignment algorithm for dynamic picking systems. Inst. Ind. Eng. **28**(8) (1996)
22. Yu, T., Sung, A., Srista-an, W., Rothermel, G.: Using property based oracles when testing embedded system applications. In: IEEE International Conference on Software Testing, Verification and Validation, pp. 100–109 (2011)
23. Vaglica, J.J., Gilmour, P.: How to select a microcontroller. IEEE Spectr. IEEE J. Mag. **6**(3) (1990)
24. Araghi, V., Eslaminasab, N., Golnaraghi, M.: Friction induced vibration in lead screw systems. J. Vib. Acoust. **131**(10) (2009)
25. Li, Y., Della Valle, F., Simonnet, M., Yamada, I., Delaunay, J.-J.: High performance UV detector made of ultra-long ZnO bridging nanowires. Nanotechnology **20**(4) (2008)
26. Islam, A., Sharma, S., Sharma, K., Sharma, R., Sharma, A., Roy, D.: Real-time data monitoring through sensors in robotized shielded metal arc welding. Mater. Today Proc. **26**(2), 2368–2373 (2020)
27. Kim, Y.S.: Development of a sensor network based SMPS system, a smart LED monitoring application based on wireless sensor network. Int. Distrib. Sens. Netw. (2014)
28. Chen, Z., Luo, R.C.: Design and implementation of capacitive proximity sensor using microelectromechanical systems technology. IEEE Trans. Ind. Electron. **45**, 886–894 (1999)
29. Vis, I.F.A.: Survey of research in the design and control of automated guided vehicle systems. Eur. J. Oper. Res. **170**, 677–709 (2006)
30. Ganesharajah, T., Hall, N.G., Sriskandarajah, C.: Design and analysis of operational issues in AGV-served manufacturing systems. Ann. Oper. Res. **76**, 109–154 (1998)

Chapter 33
Design and Fabrication of Self-balanced Electric Two-Wheeler

Karanjot Singh, Jaydeep Singh, Amardeep, Shailesh Kumar Singh, and Harshit Kumar

Abstract The current study deals with the self-balancing two-wheeler; those vehicles cannot be staying in adjusting condition with no outer support. This assertion is relevant in each condition when it is very still or gradually moving. It includes an arrangement dependent upon gyroscopic component to settle 2-wheeled vehicles (such as bicycles, bike, mopeds and so forth) in each condition (it is possible that they are very still or moving). This idea likewise can be visualizing fabricating 2-wheeler based vehicles. Adjustment of a two vehicle assumes a huge part inside the muddled transportation. Whenever twist is applied to relate degree hub conventional to the turn hub, causing the gyro to strategy, a moment is made a couple of third pivot, symmetrical to each the twist and turn tomahawks. Since the vehicle slants from vertical, a precession affectation twist is applied to the gyro get together and subsequently the contradicting pivoting component response second can will in general gets back the vehicle in its position. The key arrangement is that movement of the gyro comparative with the body is effectively controlled to concoct dependability in vehicle.

33.1 Introduction

There has been a load of attempts to make a precisely settled vehicle to battle with security and gridlock issues. In 1905, engineers build up a gyroscopic railroad that used a CMG (Control Moment Gyroscope) framework constrained by inactive exertion of numerous instruments and mechanical sensors intended to control the lopsided directions of the railroad. Here, we have made a two-wheeler which might be adjusted by pivoting component. We have also attempted to get a handle on fundamental material science and numerical display. Regardless of which gyro is utilized for 2-wheeler.

K. Singh · J. Singh · S. K. Singh (✉) · H. Kumar
Department of Mechanical Engineering, Dronacharya Group of Institutions, Greater Noida, India

Amardeep
Department of Mechanical Engineering, GL Bajaj Institute of Technology & Management, Greater Noida, India

© The Author(s), under exclusive license to Springer Nature Singapore Pte Ltd. 2022
V. V. Rao et al. (eds.), *Computational and Experimental Methods in Mechanical Engineering*, Smart Innovation, Systems and Technologies 239,
https://doi.org/10.1007/978-981-16-2857-3_34

They are typically utilized for movement or for satisfaction. The first methods of transport in quite a while are two-wheelers. Today most street mishaps are of the two-wheelers. To stay away from such deplorable circumstances, a mechanical gyro will be put inside the vehicles. The two-wheel vehicle might be degreed normally an unsteady framework. Putting such an adjusting framework might be a relentless undertaking. This composition presents a vehicle during which all integrant (mechanical, electrical) is planned from frightfully beginning, develop, incorporate along and tried [1–3]. This vehicle will be seen as natural, battery worked and truly easy to be utilized as a framework. A gyro might be a turning machine or plate during which the pivot of revolution is freed to expect any direction without anyone else. Though turning, the direction of this hub is agitated by shifting or pivot of the mounting, per the oversight of energy. On account of this, gyrators unit accommodating for measure or looking after direction. Spinners likewise are upheld elective operational standards, similar to the electronic, CPU bundled MEMS gyrators found in gadgets, strong state ring lasers, fiber optic whirligigs, and consequently the delicate quantum gyro. Uses of spinners grasp directing frameworks any place attractive compasses would not function (as inside the Hubble telescope) or would not be sufficiently exact (as in overall flight rockets), or for the adjustment of flying vehicles, such as guided helicopters or automated flying vehicles, and sporting boats and business ships. On account of their exactitude, gyrators additionally are used to deal with bearing in passage mining. Spinners additionally utilized in the development of gyrocompasses, which contradiction or supplant attractive compasses (in boats, art and orbiter, vehicles as a rule), to help in dependability or can be utilized as a component of a bearing framework [4–6].

33.1.1 Gyroscopic Impact

The turning system sway is wide used in stream and ships; any place outer turmoil couple is working on the vehicle. In this way, for the dauntlessness of such vehicles, it is fundamental to kill the effect of the outer hefty couple, which might be finished by applying equivalent and inverse couple. Thus, to concoct equivalent and inverse receptive couple, it is fundamental to differ the size and course of the speed of precession. The coordinated turning component couple represents the pace of alteration of energy, and this couple ought to be suing to plate across the pivot of turn to supply it to priority inside the flat plane. When the pivot of turn locomote itself or requirements to locomote the shaft, the circle is mounted sue responsive turning instrument couple [7]. This responsive turning component couple hence made by the gyro is up to the outside interruption anyway it is in alternate manner. Consequently, this couple kills the consequence of disturbance and equilibrium the thing (Fig. 33.1).

Fig. 33.1 3-D model of self-balancing mechanism on 2-wheeler

33.2 Construction

The gyro circle that we have used in this component is a gentle steel flywheel. The wooden casing utilized is locked on the most reduced wood base. Plastic wheels with elastic hold square measure fitted to the wood base at the least, the roundabout openings inside the design square measure prepared to suit the gimbal of the gyro get together, and the DC engine upheld on a U-section is put on the gimbal. The sole pre-imperative for this arrangement to figure is that the mass dissemination, the mass should be predominant on the top side (arrangement should be prime substantial) of the gimbal. The focal point of gravity is so just higher than the gimbal pivot, low-carbon steel plate utilized as gyro square measure fitted to the shaft of the engine. The DC engine is hooked to the U-section, and furthermore, the shaft of the DC engine is going through an opening blessing in the focal point of U-section. The texture used in making the gyro circle, center, and U-section is low-carbon steel. While assortment the entire model, various sizes of deranged and jolts were utilized. One imperative style consideration that we have made during this model is that the gyro plate should be uninhibitedly swinging inside the U-section associated with the construction. We have utilized metal rollers and studs to shape the precise developments and changes free and quick. The circlips square measure set on the inward finishes of the studs to maintain a strategic distance from the studs to move out of the metal rollers, so evading breakdown of model all through the running condition. The model is made in such some manner that the front wheel will move to require turns in order to shift the bearing of movement.

33.3 Experimental Results

The model is controlled by an impact give unit of 12 V yield. When the engine turns over pivoting, the gentle steel plate fitted on the engine shaft begins to turn and, bit by bit, acquires speed. This revolution of the plate winds up in the get together of the pivoting component result; thus when the wheels lose their equilibrium because of the dynamic turning instrument couple, a checking receptive turning system couple is made inside the alternate path on account of pivoting system result, so balancing out the model. This turning component result occurs on each left still as paw perspective. Along these lines, because of the turn of the gyro, a neutralizing responsive pivoting instrument couple winds up in the adjustment of the model. The engine and gimbal shaft gathering are implied so that it is prime huge. This recommends that the focal point of gravity lies higher than the gimbal shaft. That the engine and gyro get together attempt to understand the position such the focal point of gravity of the center can move to descend. Be that as it may, at consistent time the engine and gimbal get together are coordinated among the edge having a bearing response at closes. Thus, the sole expected technique for the engine to understand the unfaltering quality is to one or the other lean forward or in reverse. Thus, when the engine is turned over, the body is close to falling on one or the other viewpoint, and moreover, the engine get together is inclining this causes the precession of turn pivot. Because of this precession, in sync with the paw rule, the receptive pivoting instrument couple follows up on the casing that invalidates the aftereffect of the distressful couple, thus settles the vehicle. When not many turns and motions of the engine, the engine and edge accomplish the fixed position and the gyro is exposed to unadulterated moving movement (Fig. 33.2, Table 33.1).

Figure 33.3a depicts the reasonable idea that after we attempt and increment the sweep of the gyro circle, the moment of Inertia (MI) made by the gyro will increment. In this way, we will say that with a variable sweep of the gyro plate, the moment of Inertia (MI) changes thusly. Figure 33.3b shows the connection between the range of the gyro circle and furthermore the energy made by the gyro. Here, we will see that after we attempt and increment the range of the gyro plate, the energy furthermore will increment because of the variable span and the reverse way around, which recommends that sweep might be a variable issue to get variable force esteems for the spinner. From above Fig. 33.5, we tend to record the readings by keeping range

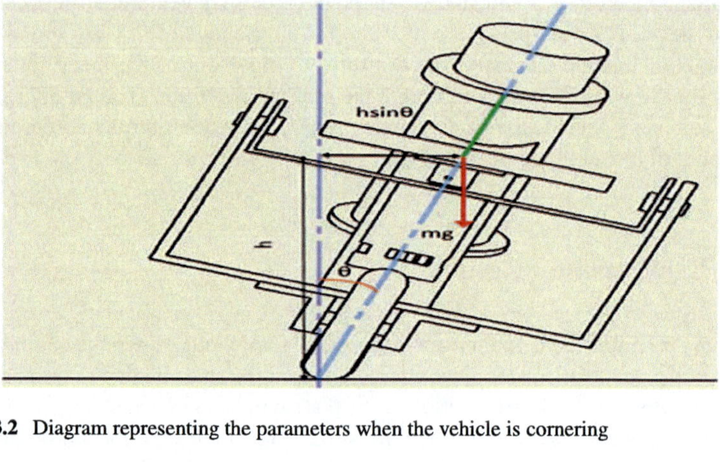

Fig. 33.2 Diagram representing the parameters when the vehicle is cornering

Table 33.1 Change of moment of inertia and angular momentum w.r.t radius

Serial No.	Radius	Moment of inertia	Angular momentum
1	0.01	0.0001	0.197
2	0.02	0.0003	0.559
3	0.03	0.0024	1.297
4	0.04	0.023	2.121
5	0.05	0.031	3.346
6	0.06	0.043	5.979
7	0.07	0.057	7.759
8	0.08	0.075	9.983

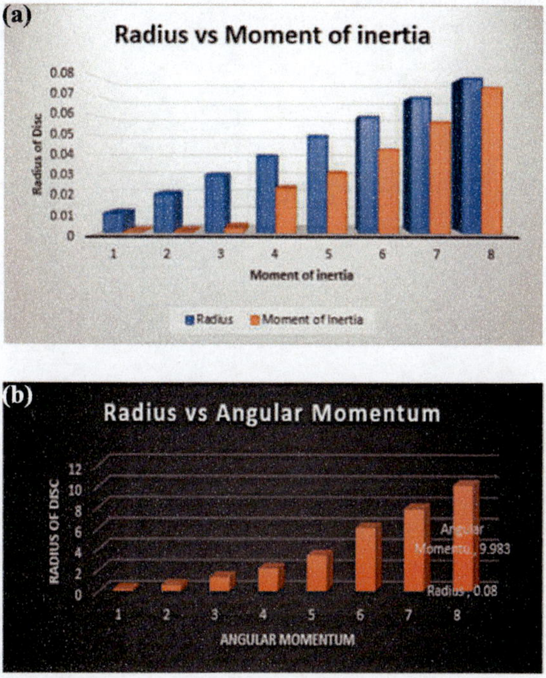

Fig. 33.3 **a** Radius versus moment of inertia, mass-1.2 kg. **b** Radius versus momentum, mass-1.2 kg

on the grounds that the boundary between acknowledging totally various readings of Moment of Inertia and force made by the turning component severally (Table 33.2).

Figure 33.4a shows the connection between the precession point and the Tilt distance of the pivoting component. Here, in the event that we will, in general, fluctuate the point of precession hub, the space of the precession pivots from the conventional shifts subsequently. In this way, we will see that with increment in precession point, the lean distance conjointly will increment, therefore. Figure 33.4b is that the delineation of the outcomes by changed the precession point to get comparing estimations of power to the alteration inside the estimations of the precession point.

Table 33.2 Changes of tilt distance and torque w.r.t precession angle

Serial No.	Precession angle θ	Tilt distance	Torque
1	5	0.0953	1.4578
2	10	0.221	2.247
3	20	0.357	4.512
4	30	0.510	5.752
5	40	0.702	7.482
6	45	0.726	8.659
7	48	0.779	9.219

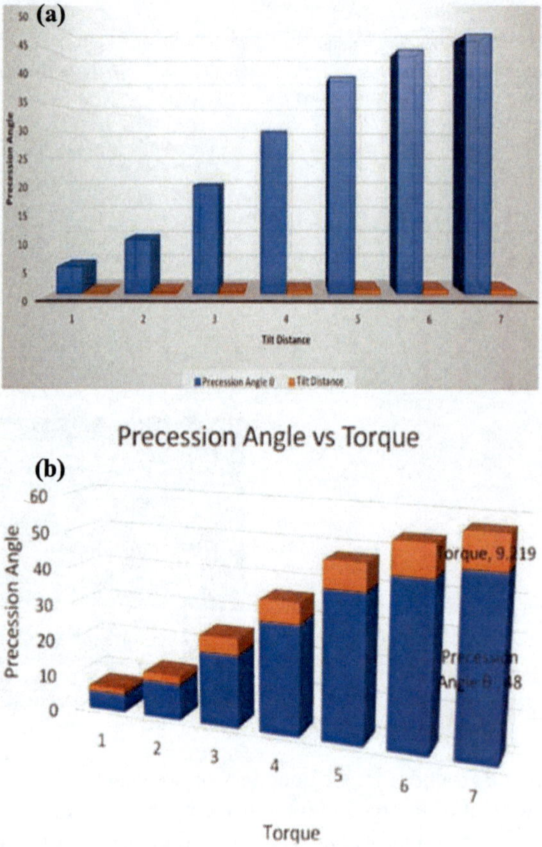

Fig. 33.4 **a** (Precession Angle vs. Tilt Distance). **b** (Precession Angle vs. Torque)

When the precession point builds, we will, in general, see a change inside the value of power created by the turning instrument. From the higher than diagram, we will see that expanding estimations of the precession point winds up in increment inside

Table 33.3 Mass versus speed table

Serial No.	Mass in kg	Speed of disk in RPM
1	1.25	18000
2	1.50	17700
3	2.00	16950
4	2.50	16300
5	3.00	15900

Fig. 33.5 Mass versus speed

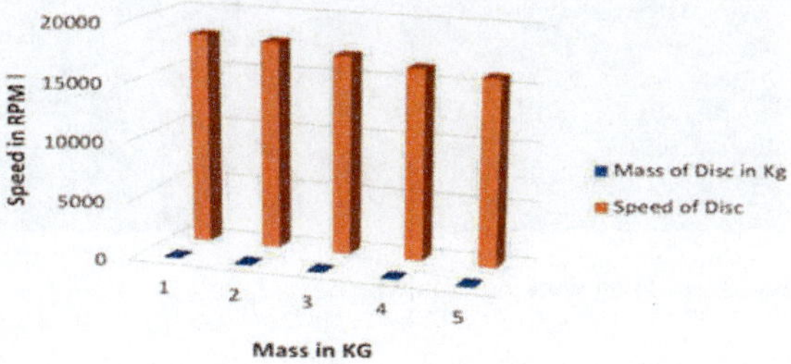

the power produced by the turning instrument. We keep an eye on solid the precession point on the grounds that the boundary that came about inside the change of Tilt distance of the precession pivot and power produced by the turning component severally (Table 33.3).

Figure 33.5 shows the effect of mass on the speed of the turn of the Gyro plate. We tend to should catch from the higher than graphical delineation that once the mass of the Gyro plate is amplified, the speed of the Gyro circle diminishes subsequently. The street on the chart portrays the focuses in connection of each speed of circle and mass once the vehicle is adjusted.

33.4 Results and Discussion

At first, a test was led to see whether the design is regularly adjusted all alone or not. The investigation found that the design is regularly balanced with a high fire up center point engine. At that point, the need was to move it physically once possessions the design lean. With the help of the center engine, we will, in general, may adjust the body. Because of the mounting of the servo engine and its linkage over the body, the undercarriage got huge on one aspect. To get up to speed with the overweight on one feature, the contrary aspect was stacked with the help of batteries. The plate utilized as a turning system was adjusted and focused appropriately to thwart any vibrations and insecure of the circle. The undertaking is up to now has figured out how to discover the standard of self-adjusting. It is figured out how to style AN equation to remain the edge straight unsurpassed. Nonetheless, the edge is not exactly ready to adjust absolutely all alone. It makes a power during a definite bearing of

Fig. 33.6 Final working prototype

the direction of center engines. Nonetheless, the center point engines are a modest quantity overweighed once adding some further weight. This enormous weight (along with along faceting combination with close to along the edge of together with) the power due to direction of the pivoting plate makes such energy to ask the edge to the contrary side as opposed to keeping it straight. Move speedy does not encourage, on the grounds that the underlying effect power is sufficient to make the enormous energy that pushes the edge on the far side our normal cutoff. The circle is slanted to the other way to make up for lost time with this occasion. Nonetheless, till the full casing is inclined toward the purpose of our limit point, it produces bigger energy to reduce the power given by the turning circle and fell on the contrary aspect. The full framework worked totally.

Nonetheless, the electric engine draws in bountiful flow in view of its high power and in this manner the amount of burden it is to deal with. That the engine driver circuit consumes straightforwardly once giving not many yields. The framework is expected to settle the auto through stopping. The elements are, on the whole, totally unique once the wheels of the vehicle are rolling. The higher the speed of the vehicle, the lesser the edge point would be (Fig. 33.6).

33.5 Applications

The on top of pivoting instrument adjustment, thought is used in engine bicycles for cutting edge soundness and wellbeing and conjointly in bikes and option such vehicles for more secure transportation. The pivoting component gathering is set at explicit areas inside the vehicle to encourage the settling sway. Furthermore, to getting utilized in compasses, airplanes, PC educate gadgets, rockets and so on, spinners are brought into customer material science. Since the pivoting system allows the count of direction and turn, architects have fused them into in vogue innovation. The blending of the turning system has taken into consideration extra right acknowledgment of development among a 3D zone than the past solitary estimating instrument among the assortment of cell phones. Spinners in the customer, physical science territory unit, joined with accelerometers (increasing speed sensors) for the extra solid course and movement detecting. Tests of such applications grasp cell phones, game support peripherals, and PC game sets. Voyage ships use whirligigs to level movement touchy gadgets such as self-leveling pool tables. Electrical jumped up controller pivoting system embedded in a very wheel is being sold as an instructing wheel different.

33.6 Conclusion

The last model of oneself adjusting vehicle style has appeared previously. This style has been tried at entirely unexpected RPMs of the circle and conjointly with various loads to watch that the vehicle is adjusting. This paper presents the style and creation of the bike self-adjusting vehicle that is equipped for adjusting itself after the use of outer powers and effects. The vehicle adjusts itself underneath different conditions, such as the constrained slant of the vehicle. Accordingly, the projected framework is a great deal helpful for two-wheel vehicles diminishing mishaps or undesirable falls and expanding wellbeing to the rider. This method diminishes crafted by people additionally as gives eco-accommodating environmental factors. From the undertaking, some definite perceptions are given:

- The higher the fire up, the bigger the counterforce. This means the counterforce is far bigger once the fire up of the center engine is bigger.
- The heading of the turning wheel slant decides the power course of once turning is during an unequivocal bearing.
- The center point engine can be driven even by associating only one battery anyway the fire up of the engine diminished during this case. To remain adjusted the fire up to the evaluated fire up of the engine, it is to be driven by associating two batteries (12 V each) in equal. The edge point up to that the vehicle on shifting, gets steady is 15° from vertical.

References

1. De Filippi, P., Tanelli, M., Corno, M., Savaresi, S.M., Santucci, M.D.: Electronic stability control for powered two-wheelers. IEEE Trans. Control Syst. Technol. **22**(1) (2014)
2. Mohtasib, A.M., Shawar, M.H.: Self-balancing two-wheel electric vehicle (STEVE). In: Proceedings of the 9th International Symposium on Mechatronics and its Applications (ISMA13), Amman, Jordan, 9–11 Apr 2013
3. Corno, M., D'Avico, L., Panzani, G., Savaresi, S.M.: A haptic-based safety-oriented braking assistance system for road bicycles. In: 2017 IEEE Intelligent Vehicles Symposium (IV), pp. 1189–1194 (2017)
4. Selmanaj, D., Corno, M., Savaresi, S.M.: Hazard detection for motorcycles via accelerometers: a self-organizing map approach. IEEE Trans. Cybern. **47**(11), 3609–3620 (2017)
5. Panzani, G., Fiorenti, S., Roselli, F., Corno, M., Savaresi, S.M.: Traction control oriented torque-to-slip identification for powered two-wheelers. In: 2015 European Control Conference (ECC), pp. 836–841 (2015)
6. Corno, M., Panzani, G., Savaresi, S.M.: Single-track vehicle dynamics control: state of the art and perspective. IEEE/ASME Trans. Mechatronics **20**(4), 1521–1532 (2015)
7. Barmpounakis, E.N., Vlahogianni, E.I., Golias, J.C.: Intelligent transportation systems and powered two wheelers traffic. IEEE Trans. Intell. Transp. Syst. **17**(4), 908–916 (2016)

Chapter 34
Use of Response Surface Methodology for Optimization of Received Signal Strength for GSM Signals in Challenging Atmospheric Conditions

Shilpa Choudhary, Abhishek Sharma, Mudita Vats, and Vidit Shukla

Abstract In today's era, mobile communication is the fastest-growing application in the field of wireless communication. In mobile communication received signal strength of radio waves at the receiver end plays an important role in network planning. When the radio waves travel from the transmitter to receivers end through the wireless channel, it has to face many challenges due to varying environment conditions or due to the presence of some obstacles, such as high-rise buildings, some vegetation, water body, etc. All these factors may affect the received signal strength of these radio waves. So for proper mobile network planning, it is required to have complete knowledge of how these parameters effects the received signal strength. So keeping this aspect in mind, this research work was focused on studying the variation in received signal strength for GSM signal with respect to four input parameters, that is, varying atmospheric temperature, Relative Humidity, Air quality index (particulate matter 2.5) and the distance from the base station and optimization of received signal strength was carried out by using response surface methodology.

Nomenclature

Temp	Temperature
RH	Relative Humidity
AQ Index	Air Quality Index
DBS	Distance from Base Station
RSS	Received Signal Strength
RSL	Received Signal Level
MSE	Mean squared error

S. Choudhary (✉) · M. Vats · V. Shukla
Department of Electronics and Communication Engineering, GL Bajaj Institute of Technology and Management, Greater Noida, UP, India

A. Sharma
Department of Mechanical Engineering, GL Bajaj Institute of Technology and Management, Greater Noida, UP, India

© The Author(s), under exclusive license to Springer Nature Singapore Pte Ltd. 2022
V. V. Rao et al. (eds.), *Computational and Experimental Methods in Mechanical Engineering*, Smart Innovation, Systems and Technologies 239,
https://doi.org/10.1007/978-981-16-2857-3_35

MLP	Multilayer Perceptron
RSM	Response Surface Methodology
CCRD	Central composite rotating design
GSM	Global System for Mobile Communications
ANN	Artificial Neural Network
VHF	Very-high Frequency
UHF	Ultra-high Frequency
MCL	Monte Carlo Localization

34.1 Introduction

In today's era mobile plays an important role in every once life. The basic mode of communication for mobile devices is through radio waves. To complete the communication through these mobile devices or between the transmitter and the receiver, the wireless channel plays an important role because it may affect the transmission characteristics of these information-carrying radio waves. These radio waves have to cross the tropospheric layer of the atmosphere while traveling through the wireless channel from the transmitter to the receiver. The refractivity index of air in the tropospheric layer may change due to varying atmospheric conditions, such as temperature, relative humidity, etc., resulting in signal loss or reduced signal strength at the receiving end. As the radio waves are the type of electromagnetic waves, it has similar characteristics as that of a light wave. As the light intensity gets reduced when it passes through the water molecule in a similar way, radio waves get scattered, diffracted and reflected if passes through a highly humid channel and results in reduced received signal strength. Distance between the transmitter and the receiver also affects the received signal strength as the intensity of the radio waves follows the inverse square relation with the distance between the transmitter and receiver. This means the radio wave intensity reduces four times if the distance between the transmitter and receiver get doubled. There are many other factors that may affect radio waves' received signal strength, such as the presence of high-rise buildings, water bodies, vegetation, etc. So for proper mobile network planning, knowledge of behavioral pattern of received signal strength with respect to all these challenging environment conditions is required. Many path loss models have been proposed for the estimation of received signal strength for varying characteristics of the communication channel. Currently, the research work is going on by using the ray tracing method to design the propagation model, and in future, it may become a milestone technology for mobile network planning [1].

Keeping all these issues in mind, this study was carried out for the optimization of received signal strength of GSM signal with four major input parameters, that is, varying atmospheric temperature (Temp), Relative Humidity (RH), user Distance form Base Station (DBS), and Air Quantity Index (AQ Index) for Particulate Matter 2.5 (PM2.5).

34.2 Related Work

Segun I. Popoola et al., in their data article, introduced the path loss data and related information that were required for modeling radio wave propagation in the smart campus environment. They used three different directions to collect the required data set within the same campus to track the record of path loss data. Also focused on some parameters, such as longitude, latitude, elevation, altitude, clutter height and the distance information. This article helped in the prediction of signal coverage, optimize limited network resources, and perform interference feasibility studies [2]. Sreevardhan Cheerla et al. experimentally showed the development essentiality of the path loss prediction model for predicting path loss data to measure the received signal strength. An artificial neural network-based path loss model was used for mobile communication network at 800 and 1800 MHz in South India and found that ANN-based path loss model given more accurate path loss values compared to the other existing path loss models [3]. Eichie Julia Ofure et al. developed ANN-based received signal level determination parametric model with three inputs: atmospheric temperature, relative humidity, and dew point for the calculation of received GSM signal level. In their research work, the MLP network used for the expansion of the model equation, and the Levenberg Marquard algorithm used for the best result in both the hidden layer and output layer with the least MSE value of 0.056 [4]. Segun I. Popoola et al., in their proposed work, measured the received signal strength of transmitted signal at different points in space for radio network planning. They studied three different routes of Lagos-Badagry Highway Nigeria, 900 and 1800 MHz, the operating frequency of the commercial base station. This article helped in the prediction of signal path loss, avoid interference and many other factors [5]. Shih-Hau Fang et al. studied the effect of atmospheric conditions over the GSM network. The whole study was carried out in the central weather bureau in Taiwan, and they studied the effect of distance and wind speed on the received signal strength and according to their findings, it was concluded that with the increase in wind/breeze speed, GSM signal strength decreases [6]. Renato Miyagusuku et al. Suggested a new modeling approach and used the MCL algorithm for creating specific and exact wireless signal strength mappings. Gaussian and path loss models were used to determine the likelihood of locations conditioned to sensor data [7]. Babatunde Adeyemi et al. did their study in Nigeria and monitor the tropospheric radio refractivity by using CM-SAF data [8]. Shilpa Choudhary et al. [9, 10] suggested that with the help of response surface methodology, optimization of output response can be carried out. The main advantage of this methodology is that the optimization through this technique requires a minimum number of the experimental run. Danial Abdorahimi et al. worked over 2.4 GHz to examine the correlation between atmospheric particulate and EM-wave attenuation of the radio signal. In their study, they observed different parameters such as temperature, humidity, and variation of Received Signal level (RSL) with operating condition of low, medium, and high at PM10. They analyzed that Received Signal level has no notable contrast, but polynomial dependency exist in RSL local variance [11]. H. Purohit et al. in their study discussed the Data Interleaving Technique. Their study was more focused on the bandwidth optimization and usage of voice channels in GSM networks. Channel utilization up to 54.22% and bandwidth optimization can be done. All the experiments were performed in Kampala, Uganda [12].

From the related work it is clear that the received signal strength of radio signal get affected with various atmospheric parameters. Much work has already been conducted to monitor the variation in received signal strength for different mobile communication technologies and have many path loss models have already been proposed, but the optimization of received signal strength has still not explored much. So this researched work was carried out to observe the variation in received signal strength with respect to four input parameters, that is, varying air quality index (particulate matter 2.5), Relative Humidity, temperature and the distance from the base station, and the optimization of received signal strength for GSM signals was carried out by using response surface methodology. It is a multi-objective optimization technique and can give the optimized results with the minimum number of experiment runs.

34.3 Materials and Methods

This research work was mainly carried out to observe the variation in received signal strength for GSM signals with varying atmospheric conditions, such as temperature (20–40 °C), relative humidity (25–75%), air quality index (50–500) and distance from base station (0.1–0.3 km). In this experiment Hygro-Thermometer was used to measure the variation in temperature and relative humidity, Variation in received signal strength was observed with spectrum analyzer, Variation in air quality index and distance from base station were measured with AQI meter and GPS installed in smart phone respectively.

In this study, mathematical modeling was done with the help of response surface methodology (RSM). The main advantage of RSM technique is that it requires a minimum number of experiment runs to optimize the response. Various coding level used for input parameters has been shown in Table 34.1. Using these parameters, the experimental design developed through the statistical tool MINITAB 17. After

Table 34.1 RSM range for input parameters

Factors or input parameters	Coded levels				
	−2	−1	0	1	2
Temperature (°C)	20	25	30	35	40
Relative humidity (%)	25	37	50	63	75
Distance from base station (km)	0.1	0.15	0.2	0.25	0.3
Air quality index	50	150	250	370	500

the experiments as per the designed matrix, ANOVA was applied, and good R^2 correlation relation found between input and output responses.

34.3.1 *RSM Methodology*

With the help of response surface methodology, an optimum solution can be obtained. With RSM, optimization of input variables can be done to find the best value of output response and that to with the minimum number of experiment runs. It can be used in any type of application, and it is not application bound.

In this work, RSM architecture can suitable the CCRD matrix, which has five stages with respect to the individual factors. The most fruitful and best among architecture available is the middle composite rotating architecture which is adept by adding two lab experiment level along each coordinate axis at the opposite direction of the origin and at a distance same to the semi-diagonal of the hypercube of the factorial architecture and new extreme values (low and high) for each factor added in this architecture. The same statement has been reported in the previous study.

Under this study, variations in receives signal strength (RSS) for GSM signal was observed for four input variables, that is, Distance from Base Station—DBS, Air Quality index (A.Q Index) for PM 2.5 particle, Relative Humidity (RH) and Temperature (Temp). Variation range of these input parameters has been shown in Table 34.1. Experimental results were obtained after the implementation as per CCRD.

34.4 Result and Discussion

Analysis of Variance was applied and given by Table 34.2, which offers quantitative results about the research model. The maximum level of p-Value is taken 0.05, values

Table 34.2 Analysis of variance

Source	DF	Adj SS	Adj MS	F-Value	P-Value
Regression	4	1170.21	292.551	70.31	0.000
Temp. (C)	1	956.31	956.314	229.83	0.000
RH (%)	1	66.88	66.881	16.07	0.000
DBS (km)	1	28.42	28.419	6.83	0.015
AQ index	1	203.67	203.674	48.95	0.000
Error	26	108.19	4.161		
Lack-of-fit	20	108.19	5.409		
Pure error	6	0.00	0.000		

above 0.05 are referred as an irrelevant. Distance from base station having p-value maximum 0.015 only. ANOVA found satisfactory for this research model.

Model R^2 and R^2 (adj) values are 96.16% and 95.88%, respectively, for RSS (GSM) response. In this work, p-value for mixtures of 0.669 and p-value for clustering of 0.331, both are more than 0.05. Hence, it can be concluded that the experimental data does not have any clusters and mixtures to affect the research results.

MINITAB 17 generated empirical correlations between RSS (GSM) and all input parameters are given by Eq. 34.1.

$$\begin{aligned} RSS\,(\text{dBm}) = {} & -13 + 2.5 * Temp(^\circ\text{C}) + 0.106 * R.H\,(\%) - 83 \\ & * DBS\,(\text{km}) - 0.024 * A.Q\,Index + 0.0151 * Temp^2(^\circ\text{C}) - 0.00061 \\ & * R.H^2 + 14 * DBS^2(\text{km}) - 0.000063 * A.Q\,Index^2 - 0.00312 \\ & * Temp\,(^\circ\text{C}) * RH + 1.64 * Temp\,(^\circ\text{C}) * DBS\,(\text{km}) + 0.00079 \\ & * Temp\,(^\circ\text{C}) * A.Q\,Index - 0.254 * R.H\,(\%) * DBS\,(\text{km}) - 0.000118 \\ & * R.H\,(\%) * A.Q\,Index + 0.064 * DBS\,(\text{km}) * A.Q\,Index \end{aligned} \tag{34.1}$$

In this research work, MINITAB 17 software was used to draw contour and surface plots. All the surface and contour plots were drawn only with two input variables. The rest were kept at some hold value, which is the center value of the input variable, as shown in Table 34.1.

34.4.1 Impact of Input Parameter on Signal Strength for GSM Technology

Figure 34.1a and b shows the surface and contour plots, respectively. These plots show the variations in received signal strength (RSS) of GSM signal with respect to two input variables which are Distance from the base station—DBS (0.10–0.30 km) and Air quality index—AQ Index (50–500) and rest two parameters, that is, temperature (Temp) and Relative Humidity (RH) were kept constant at 30 °C and 50%, respectively. From the surface plot, we can observe that RSS keeps on reducing with increasing DBS. It mainly occurs because radio wave intensity is inversely proportional to the square of distance between transmitter and receiver, which means with an increase in distance between transmitter and receiver, the received signal strength of radio waves reduces. From the same plot, it can also be observed that RSS reduces with increasing AQ Index, but its effect is almost negligible. From the contour plot shown in Fig. 34.1b, the maximum value obtained by RSS is −80 dBm approximately when the DBS and AQ Index are at 0.10 km and 50 respectively.

Variation in RSS with respect to temperature (20–40 °C) and Air quality index—AQ Index (50–500) are shown in Fig. 34.1c and d through surface and contour plots, respectively. At the same time other two input parameters, that is, Relative Humidity (RH) and Distance from the base station (DBS) were kept at a constant value of 50% and 0.2 km, respectively. From the surface plot, it can be observed that RSS

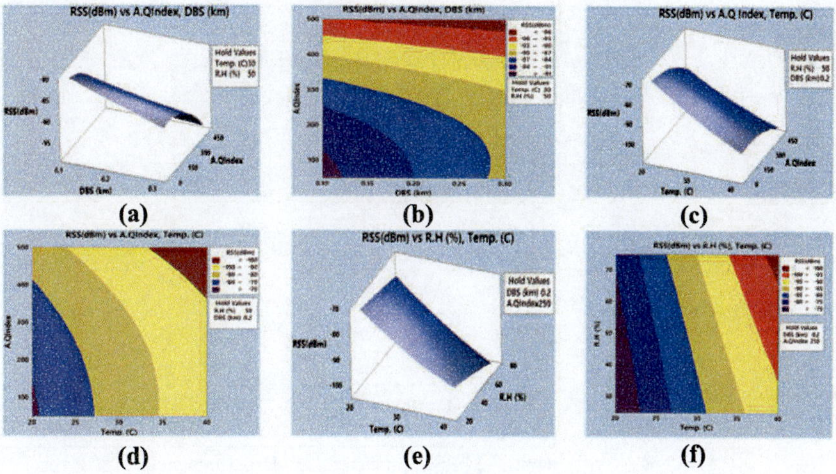

Fig. 34.1 Effect of environment parameters received signal strength for GSM signal

decreases with an increase in temperature, which occurs because the resistivity of medium increases the increase in atmospheric temperature. This results in reduced power and current for the device for a specific voltage. So with increased temperature signal to noise ratio decrease and results in reduces received signal strength. From the same graph, it can be observed that RSS follows the same trend with respect to AQ index as the Temp does, but the effect is almost negligible. Its contour plot shows that the RSS can attain the maximum value of −69.5 dBm approximately and can go as low as −105 dBm approximately. Figure 34.1e shows the Variation RSS with respect to Temp (20–40 °C) and Relative Humidity—RH (25–75%) and at the same time other two input parameters, that is, Distance from the base station—DBS and Air quality index (AQ Index) were kept at a constant value of 0.2 km and 250, respectively. From this figure, it can be observed that RSS keeps on decreasing with an increase in RH, and it occurs because radio waves get scattered, diffracted and reflected in the presence of water molecule, which results in reduced received signal strength. From the same figure, it can be RSS reduces with an increase in temperature Fig. 34.1f shows the contour plot for the same combination of input parameters, and it shows that RSS can attain its maximum value of −71 dBm approximately and can go up to the minimum value of −101.5 dBm approximately.

34.5 RSM Optimization Response and Validation

RSM optimizer is shown in Fig. 34.2. With the help of the RSM optimizer, the best value of response can be found in the optimum values of input variables. The main aim of this study was to maximize the received signal strength for GSM technology.

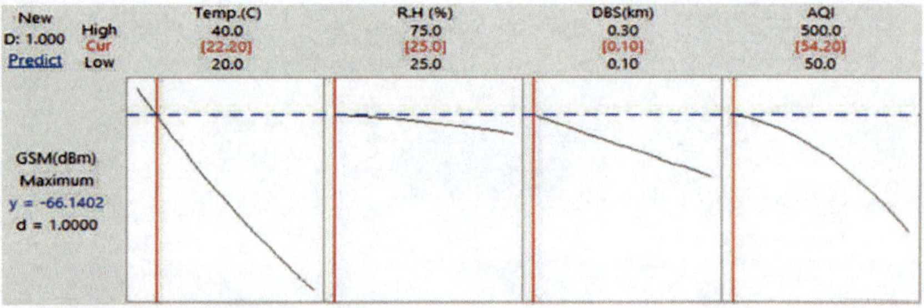

Fig. 34.2 Optimization plot

Table 34.3 Validation RSM result

RSS (GSM) response at 22.2 °C, 0.1 km. DBS, 25% RH, with 55 AQI			
Unit	RSM predicted	Experimental	% Error
dBm	−66.14	−69.42	4.95

With the help of the RSM optimizer, optimum values received for the input parameters were 22.2 °C, 25%, 0.10 km and 54.2 for ambient temperature, Relative Humidity, Distance from the base station and Air Quality Index, respectively. For these optimum values of input parameters, the optimum value received for the response, that is, received signal strength for GSM signal, was −66.1402 dBm. Unity desirability was obtained for the output, which is favorable for the optimization process. For the validation of results obtained by RSM, the experiment was conducted at 22.2 °C Temperature, 25% Relative Humidity, 0.1 km from Base Station at 55 air quality index. Experimental details are as shown in Table 34.3, and it gives the information about the output, that is, received signal strength for GSM signals. Later the comparison was carried out between experiment results and results obtained from the RSM optimizer. Error obtained after comparison was 4.95% which was well within the tolerable limit.

34.6 Conclusions

In this study, the variation in received signal strength for GSM signals was monitored with respect to challenging atmospheric conditions. The main aim was to overcome the effect of these challenging conditions. Later the optimization of response was carried out by using response surface methodology. The main advantage of RSM optimization is that it needs a minimum number of experiment runs to optimize the response. Complete detail of input and output parameters during optimization process is as mentioned below.

- Variation in received signal strength for GSM signal was monitored with respect to four input parameters, that is, Relative Humidity, Air Quality Index for PM 2.5 particle, Distance from Base Station and Temperature.
- The optimum values obtained for input parameters were 22.2 °C, 25%, 54.2, and 0.10 km for Temperature, Relative Humidity, Air Quality Index for PM 2.5, and Distance from Base Station, respectively.
- Received Signal Strength for GSM signal was found to be −66.14 dBm.
- Validation of RSM results was carried out with a confirmation trial, and the error found to be within the range of 5%.

After this study, it is clear that with all these optimized values of input parameters, the best possible value of Received Signal Strength for GSM signals can be obtained even in the presence of unfavorable environmental conditions.

References

1. Yun, Z., Iskander, M.F.: Ray tracing for radio propagation modeling: principles and applications. IEEE Access **3**, 1089–1100 (2015)
2. Popoola, S.I., Atayero, A.A., Arausi, O.D., Matthews, V.O.: Path loss dataset for modeling radio wave propagation in smart campus environment. Data Brief **17**, 1062–1073 (2018)
3. Cheerla, S., Ratnam, D.V., Borra, H.S.: Neural network-based path loss model for cellular mobile networks at 800 and 1800 MHz bands. AEU-Int. J. Electron. Commun. **94**, 179–186 (2018)
4. Ofure, E.J., David, O.O., Oludare, A.M., Musa, A.A.: Artificial neural network model for the determination of GSM Rx level from atmospheric parameters. Eng. Sci. Technol. **20**(2), 795–804 (2017)
5. Popoola, S.I., Atayero, A.A., Faruk, N.: Received signal strength and local terrain profile data for radio network planning and optimization at GSM frequency bands. Data Brief 927–981 (2017)
6. Fang, S., Yang, Y.S.: The impact of weather condition on radio-based distance estimation: a case study in GSM networks with mobile measurements. IEEE Trans. Veh. Technol. **65**(8), 6444–6453 (2016)
7. Miyagusuku, R., Yamashita, A., Asama, H.: Precise and accurate wireless signal strength mappings using Gaussian processes and path loss models. Robot. Auton. Syst. **103**, 134–150 (2018)
8. Adeyemi, B., Emmanuel, I.: Monitoring tropospheric radio refractivity over Nigeria using CM-SAF data derived from NOAA-15,16 and 18 satellites. Indian J. Radio Space Phys. **40**, 301–310 (2011)
9. Choudhary, S., Sharma, A., Srivastava, K., Purohit, H., Vats, M.: Read range optimization of low frequency RFID system in hostile environmental conditions by using RSM approach. EVERGREEN Joint J. Novel Carbon Resour. Sci. Green Asia Strategy **07**(03), 396–403 (2020)

10. Choudhary, S., Sharma, A., Gupta, S., Purohit, H., Sachan, S.: Use of RSM technology for the optimization of received signal strength for LTE signals under the influence of varying atmospheric conditions. EVERGREEN Joint J. Novel Carbon Resour. Sci. Green Asia Strategy **07**(03), 500–509 (2020)
11. Abdorahimi, D., Sadeghioon, A.M.: Review-comparison of radio frequency path loss models in soil for wireless underground sensor networks. J. Sens. Actuator Netw. **8**(2), 35 (2019)
12. Purohit, H., Kaur, P., Mwebesa, D. B., Choudhary, S., Joshi, K.: Channel redundancy analysis and bandwidth utilization in GSM mobile communication network of Uganda. In: 2019 IEEE 10th Annual Ubiquitous Computing, Electronics & Mobile Communication Conference (UEMCON), New York City, NY, USA, pp. 0963–0967 (2019)

Chapter 35
Structural Analysis and Completion of Fatigue Axial-Flow Compressor Using Finite Element ANSYS Technology

Rishabh Chaturvedi, Vikas Kumar Sharma, and Manoj Kumar

Abstract An axial-flow compressor (AFC) will be the flow that enters the compressor in the axial direction and exits from the axial direction and in the gas turbine. The AFC compresses its attempting liquid by initial accelerating the liquid and then dispersing it to acquire a pressure expansion. Presently examination and developmental deliberations in the region of AFC for gas turbine application have been intended to enhancing its operating range without sacrificing effectiveness. An increment in aspect ratio is noticed to have an adverse impact on the execution of single-stage flow compressors. In this manuscript, AFC is planned and simulated in 3D designing software Pro/E. The current patterns get changed on varying aspect ratios. The current work utilized chromium steel, which could be substituted by nickel and titanium alloy. Full compressor methods with steel, nickel and titanium alloy are used for structural testing to validate compressor strength using finite element analysis using software Ansys. CFD examination is also being complete to define fluid performance in ANSYS Fluent.

35.1 Introduction

An axial compressor will be the compressor, which might constantly pressurize gasses. It may be a rotating, airfoil-built compressor in gas or working liquid that primarily flows parallel to the rotation axis or axially. This contrasts from diverse rotating compressors, such as ax centrifugal, mixed-flow, and centrifugal compressors, whereas liquid flow will incorporate a spiral part through the compressor [1–4].

R. Chaturvedi (✉) · V. K. Sharma
IET Department of Mechanical Engineering, GLA University, Mathura, India
e-mail: risabh.chaturvedi@gla.ac.in

V. K. Sharma
e-mail: vikasH.sharma@gla.ac.in

M. Kumar
Greater Noida Institute of Technology, Greater Noida, India

© The Author(s), under exclusive license to Springer Nature Singapore Pte Ltd. 2022

V. V. Rao et al. (eds.), *Computational and Experimental Methods in Mechanical Engineering*, Smart Innovation, Systems and Technologies 239,
https://doi.org/10.1007/978-981-16-2857-3_36

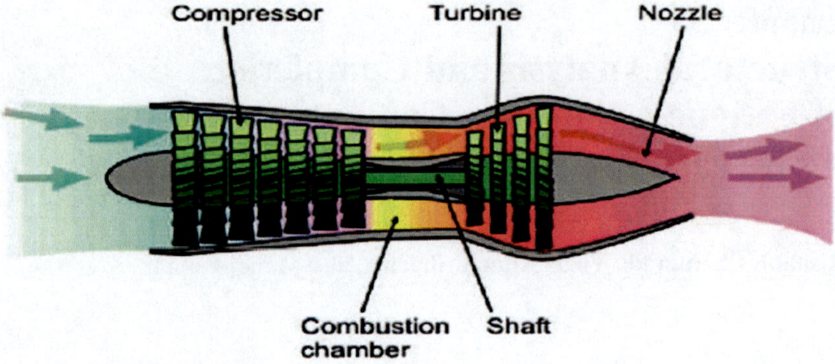

Fig. 35.1 Axial-flow compressor

The energy level of liquid enhances as it streams over the compressor because of rotor blades activity that exerts a torque on liquid [5, 6].

The circumferential part of stream under pressure and moderating the fluid is done by static blades. The electric engine or gas turbine determines the compressors [7–9]. The AFC generate a constant stream of compacted gas and profits of advantages of large mass flow rate, high efficiency, specifically in connection to their cross-section and size [10, 11]. They do but need numerous rows of airfoils to attain an expansive pressure rise, making them unpredictable and unreasonable relative to other plans [12, 13].

The axial compressors have been essential to the plan of large gas turbines, such as high-speed ship engines, plane engines, and little scale power stations. They have also been utilized in mechanical applications, such as blast furnace air, propane dehydrogenation, large volume air separation plants, and fluid catalytic cracking air [14–16]. Because of high reliability, presentation, and adaptable operation throughout the flight envelope, they have also been utilized in aerospace engines [17, 18] (Fig. 35.1).

35.2 Literature Review

The AFC compresses its working stream by initial accelerating liquid and then dispersing it to attain pressure enhances. The pressure increase is insignificant in AFC when air passes from one phase to further, every phase. It will raise the energy level of gas or air streaming by operating the rotor blades that exert torque on the liquid supplied by a gas turbine or electric motor. In this paper, AFC will be planned and simulated in 3D designing software Pro/E. The new plan consists of 30 blades, replaced with 12 blades and 20 blades in this paper. The materials currently used may be chromium steel; nickel and titanium alloy can be substituted with it. The structural investigation may be done on compressor methods to check compressor strength. The CFD investigation is finished to find the air flow.

The AFC is a very significant segment of a gas turbine. In the plan of AFC, the work introduced contains fundamental stream factors and dimensions; this creates the additional design procedure quite easy, and the outcomes are supportive of taking additional variations or development at the time of comprehensive design. The work objective suggested will be to plan AFC by utilizing the mean line technique for provided mass flow rate and essential pressure ratio. The factors described also incorporate thermodynamic assets of stage efficiency, working fluid, stator blades, number of rotor and tip and hub diameters, blade dimensions of both stator and rotor, Mach number, flow and blade angles. Similar factors have also been described for five phases. In any critical number of segments allocated by designers to achieve a smooth twisting profile, the twist of blades can be measured along the length of the blade. Profiles from NACA 65,410 are used to create blade coordinates. Additional, in procedure, the initial phase of the AFC blade is established utilizing Solid works modeling. Similarly, CFD simulation is carried out with the use of Ansys CFX to authenticate the outcomes. Correspondingly Static structural examination is executed to find if the rotor is safe at provided speed.

The structural plan of IND100 AC needs a multistage interrelationship among mechanical design, structural integrity, thermodynamic, and aerodynamic study of factor. These design requirements often oppose, so the equilibrium in engineering is used within the design presentation limits. The text means the conceptual and structural plan of the 16-phase high-pressure shaft IND100 compressor with a full weight flow rate of 310 kg/s/pressure of 12 in ISOSLS situations. Moreover, to estimate the conceptual design examination, fundamental factors such as disk stress study, compressor sizing, rotor dynamics, load and blade mass, conceptual disk design, bearings and material selections have been deliberated utilizing present devices and analytical method. These methods employed the elementary aero dynamic and thermodynamic theory of AFCs to estimate the pressure and temperature for whole phases, velocity triangle, geometrical factors, and stress and weight calculations of compressor disk utilizing Sagerser Empirical Weight Assessment.

This manuscript defines the approaches utilized for computer program generation for examining the off-design presentation of ACs. The determination of survey was to envisage such AC features as efficiency and pressure rise as a rotor speed and mass flow junction, whereas only mechanical geometry of blades and annulus has been provided. The aerodynamic examination is separated into two main parts. The initial part of the examination includes the numerical answer of motion in the compressor to discover temperature, flow field, and pressure rise, and compressor efficiency. The second part concerns the examination of blade fundamentals, in specific their presentation at numerous angles of attack. The numerous effects have been incorporated that have a bearing on this presentation. The axial-velocity ratio and relative Mach number across blade component, together with blade shape and its configuration in the compressor, affect the flow discharging from the blade.

The aspect ratio of the blade is a significant factor and has a strong effect on the presentation of AFC. The presentation of AFCs will be recognized to be influenced by the selection of aspect ratio. A survey is conducted to demonstrate the aspect ratio effect on the presentation by stream analysis of the 1.2 pressure ratio single-stage subsonic compressor using AFC design. The test is performed for a constant diameter of the paddle with an aspect ratio of 1, 2 and 3 and a flow factor and compressor pressure loss. This manuscript presents the finest operating situations happen for aspect ratio among 2 and 3.

35.3 Methodology

The AFC will be the flow that enters the compressor in the axial direction and exits from the axial direction and in the gas turbine. The AFC compresses its attempting liquid by initial accelerating the liquid and then dispersing it to acquire a pressure expansion (Table 35.1).

35.3.1 FEM

The FEM will be the main examination method for determining and replacing difficult issues with simpler ones, attaining estimated answers. FEM being a flexible device, will be utilized in numerous companies to resolve numerous practical engineering difficulties. In finite element technique, it will be possible to produce relative outcomes (Fig 35.2).

35.3.2 Static Analysis Steel Compressor

The maximum deformation of steel having stator angle 24.9°, rotor angle 12.1° is 0.02397 mm (Fig. 35.3).

In the current day, finite element technique will e very broadly and effective utilized devices. By doing a very computational examination, the estimated solution might be enhanced in FEM. In FEM, matrices play a main part in controlling a massive number of formulas. The FEM process will be a disparity method, whereas this perception has contributed considerably in formulating the technique.

Table 35.1 Selection of models, analysis and materials of axial-flow compressor

Models	Analysis	Materials
Case 1: Rotor angle, 12.1°; Stator angle, 24.9°		
Case 2: Rotor angle. 26.4°; Stator angle. 29.0°	Fatigue analysis	Steel Titanium alloy
Case 3: Rotor angle, 39.8°; Stator angle, 33.1°	Static analysis	Nickel alloy
Case 4: Rotor angle, 45.9°; Stator angle, 35.2°		

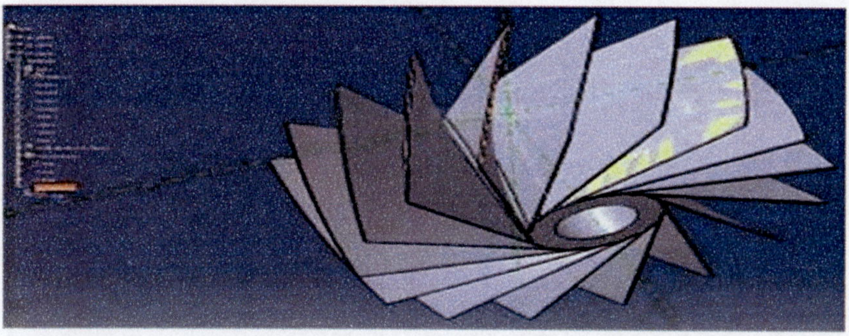

Fig. 35.2 Parametric CAD method

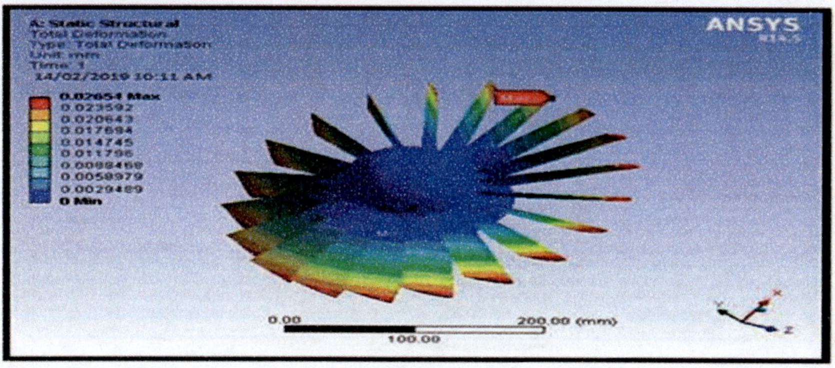

Fig. 35.3 Deformation of steel compressor (stator angle 24.9°, rotor angle 12.1°)

FEM supports in assessing difficult structures in a framework through the planning phase. The design and strength of the method might be enhanced with the support of FEA and computers that justifies the analysis price. FEA has obviously improved the structure's design, which was built many years ago (Figs. 35.4 and 35.5).

The maximum stress of steel having stator angle 24.9°, rotor angle 12.1° is 7.3094 Maps.

The maximum strain of steel compressor having stator angle 24.9°, rotor angle 12.1° is 3.667e–5.

35.3.3 Fatigue Examination of AFC

Case: stator angle 24.9°, rotor angle 12.1°, MATERIAL—TITANIUM ALLOY DAMAGE (Fig. 35.6).

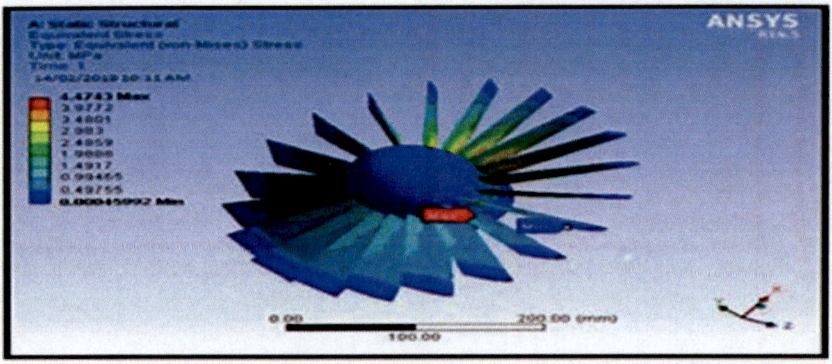

Fig. 35.4 Stress in steel compressor (stator angle 24.9°, rotor angle 12.1°)

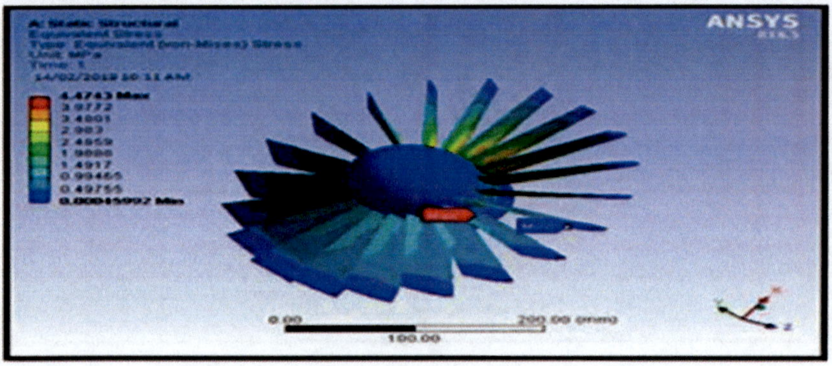

Fig. 35.5 Strain in steel compressor (stator angle 24.9°, rotor angle 12.1°)

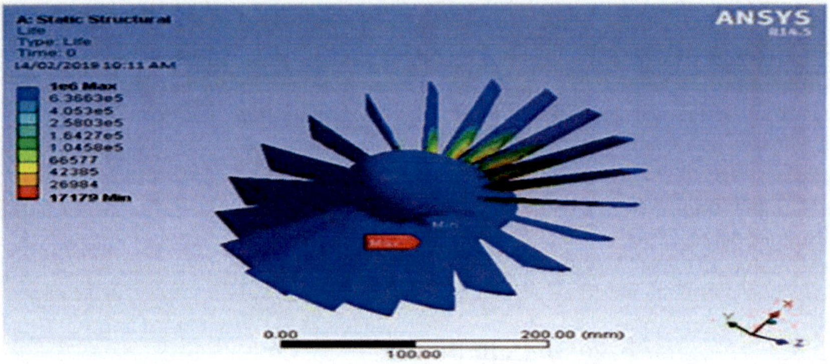

Fig. 35.6 Safety factor

35.4 Results

(See Fig. 35.7).

35.4.1 *Static Analysis*

(See Tables 35.2, 35.3, 35.4 and 35.5).

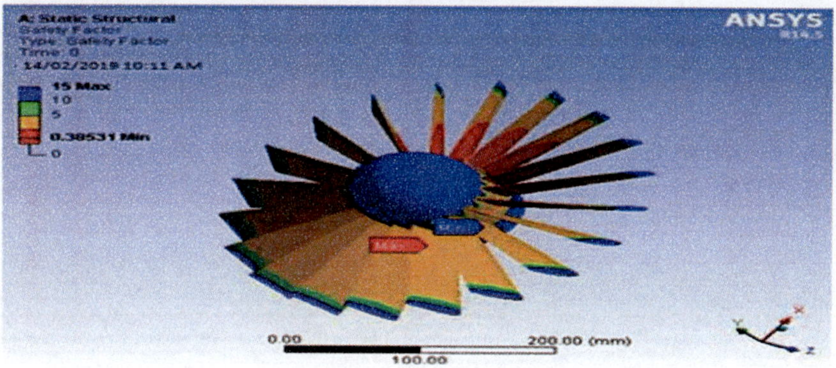

Fig. 35.7 Life

Table 35.2 Static analysis results for stator angle 24.9°, rotor angle 12.1°

Material	Stress (MPa)	Deformation (mm)	Strain
Titanium alloy	4.4743	0.02654	4.0818e–5
Steel	7.3094	0.02397	3.667e–5
Nickel alloy	12.557	0.035879	5.3577e–5

Table 35.3 Static analysis results for stator angle 29.0°, rotor angle 26.4°

Material	Stress (MPa)	Deformation (mm)	Strain
Titanium alloy	5.6591	0.045884	5.1544e–5
Steel	9.2789	0.041448	4.6479e–5
Nickel alloy	16.23	0.062088	6.9165e–5

Table 35.4 Static analysis results for stator angle 33.1°, rotor angle 39.8°

Material	Stress (MPa)	Deformation (mm)	Strain
Titanium alloy	7.326	0.06273	6.6677e–5
Steel	11.996	0.056658	6.0042e–5
Nickel alloy	20.864	0.084821	8.8846e–5

Table 35.5 Static analysis results for stator angle 35.2°, rotor angle 45.9°

Material	Stress (MPa)	Deformation (mm)	Strain
Titanium alloy	7.8331	0.06975	7.1341e–5
Steel	12.837	0.062991	6.4299e–5
Nickel alloy	22.417	0.094202	9.5533e–5

35.4.2 *Fatigue Analysis*

(See Tables 35.6, 35.7, 35.8 and 35.9).

Table 35.6 Fatigue analysis results for stator angle 24.9°, rotor angle 12.1°

Material	Stress (MPa)		Deformation (mm)	Strain	
	Max	Min		Max	Min
Steel	$1 \times e^6$	3574.4	2.7976e5	15	0.23586
Titanium alloy	$1 \times e^6$	17,179	58,212	15	0.38531
Nickel alloy	$1 \times e^6$	798.15	1.25529e6	15	0.1373

Table 35.7 Fatigue analysis results for stator angle 29.0°, rotor angle 26.4°

Material	Stress (MPa)		Deformation (mm)	Strain	
	Max	Min		Max	Min
Steel	$1 \times e^6$	289.01	3.4601e6	15	0.092899
Titanium alloy	$1 \times e^6$	1045.6	9.5638e5	15	0.15232
Nickel alloy	$1 \times e^6$	72.131	1.3864e7	15	0.05311

Table 35.8 Fatigue analysis results for stator angle 33.1°, rotor angle 39.8°

Material	Stress (MPa)		Deformation (mm)	Strain	
	Max	Min		Max	Min
Steel	$1 \times e^6$	150.2	6.6578e6	15	0.071858
Titanium alloy	$1 \times e^6$	534.33	1.8715e6	15	0.11766
Nickel alloy	$1 \times e^6$	40.145	2.491e7	15	0.041315

Table 35.9 Fatigue analysis results for stator angle 35.2°, rotor angle 45.9°

Material	Stress (MPa)		Deformation (mm)	Strain	
	Max	Min		Max	Min
Steel	$1 \times e^6$	126.92	7.87e6	15	0.067148
Titanium alloy	$1 \times e^6$	448.97	2.2273e6	15	0.11005
Nickel alloy	$1 \times e^6$	34.049	2.9369e7	15	0.038452

35.5 Conclusion

The AFC will be the flow that enters the compressor in the axial direction, and exits from the axial direction and in the gas turbine. The AFC compresses its attempting liquid by initial accelerating the liquid and then dispersing it to acquire a pressure expansion. In this manuscript, an AFC is designed and simulated in 3D designing software Pro/E. The existing plans are modified by varying relative proportions. Nickel and titanium alloy swapped chromium steel that is the contemporary utilized material. By detecting static examination, stress, and deformation for nickel alloy and titanium alloy, and incrementing the AFC angles blade and low-stress value of stator angle 24.90, rotor angle 12.10. The examination of the security feature for fatigue detection of titanium alloy is lower than nickel and steel alloy, which is around 24.90 for stator angle and 12.10 for rotor angle. By detecting CFD survey, velocity, a flow rate of mass and pressure drop enhances blade angles of AFC. So, for axial-flow compressor, the titanium alloy is a better option as it ameliorates the performance.

References

1. Srinivas, K., Deepthi, K., Malleswara Rao, K.N.D.: Design and Optimization of Axial Flow Compressor
2. Jaiswal, U.A., Joshi, S.J.: Design and Analysis of Stator, Rotor and Blades of the Axial flow Compressor
3. Aziaka, D.S., Osigwel, E.O., Lebele-Alawa, B.T.: Structural and Conceptual Design Analysis of an Axial Compressor for a 100 MW Industrial Gas Turbine (IND100)
4. Jansen, W., Moffatt, W.C.: The Off-Design Analysis of Axial-Flow Compressors, Author and Article Information
5. Singh, P,K., Sharma, K.: Mechanical and viscoelastic, properties of in-situ amine functionalized multiple layer grpahene/epoxy nanocomposites. Curr. Nanosci. **14**(3), 252–262
6. Kumbhar Anil, H., Agarwal, A.: CFD simulation of single-stage axial flow compressor for varying blade aspect ratio
7 Gaddam Srikanth, A., Srinivas Prasad, S., Mahesh Kumar, A., Mounika, B.: Reddy Design Methodology of a Two Stage Axial Compressor
8. Singh, R.P., Sharma, K., Mausam, K.: Dispersion and stability of metal oxide nanoparticles in aqueous suspension: a review. Mater. Today: Proc. **26**, 2021–2025 (2020)
9. Suthakar, T.: Numerical Investigation of Flow In An Axial Flow Compressor Cascade, Akash, National Institute of Technology, Tiruchirappalli

10. Kumbhar Anil, H., Agarwal, A.: Effect of Variations in Aspect Ratio on Single Stage Axial Flow Compressor Using Numerical Analysis. Associate professor, Technocrats Institute of Technology, Bhopal, M.P, India
11. Gourdain, N., Montagnac, M., Boussuge, J.F.: Numerical Simulation of an Axial Compressor with Non Axisymmetric Casing Treatment CERFACS, Computational Fluid Dynamics Team 31057 Toulouse, France
12. Bharadwaj, G., Sharma, K., Tiwari, A.K.: Performance analysis of hybrid PCM by doping Graphene. Mater. Today: Proc. **26**, 850–853 (2020)
13. Day, J. Stall Inception in Axial Flow Compressors
14. Du, H., Yu, X., Liu, B.: Modeling of the double leakage and leakage spillage flows in axial flow compressors
15. Sieverding, F., Ribi, B., Casey, M., Meyer, M.: Design of Industrial Axial Compressor Blade Sections for Optimal Range and Performance (2004)
16. Chaturvedi, R.: Molding sand nano fluidization role in ceramic substitute. J. Green Eng. **10**(11), 11540–11545 (2020)
17. Yahya, S.M.: Fundamentals of Compressible flow, 81-224-1468-0. Tata Mc-Graw Hill (2003)
18. Hill, P.G., Peterson, C.R.: Mechanics and Thermodynamics of Propulsion, 0201146592, Addison-Wesle (2010)

Chapter 36
Deep Neural Network for Facial Emotion Recognition System

Vimal Singh, Sonal Gandhi, Rajiv Kumar, Ramashankar Yadav, and Shivani Joshi

Abstract Facial emotion recognition is a challenging task to capture and analyze. This paper presents an intelligent approach to detect different facial expressions of a person. In this context, a noval deep learning vgg 19 convolutional neural network architecture is used. Seven different emotions (anger, disgust, fear, happiness, sadness, surprise, and neutral) are considered to recognize emotions. The FER 2013 dataset contains 35,887 different images with these emotions. Results were obtained by using various activation functions. A comparative study of various activation functions has been done.

36.1 Introduction

Intelligent Facial Expression Recognition is a vast and important research topic. Finding emotion from images is difficult and sensitive. The face is the index of our mind. Such a system can understand the situation better to take more fruitful results. The system will also be helpful in human-computer interaction. Since human-computer interaction, interacts with human in uncontrolled environment where the scene lighting, camera view, image resolution, background, user's head pose, gender, and ethnicity can vary significantly.

CNN (Convolutional Neural Network) is a type of neural network which makes an assumption that the input is an image. It contains series of hidden layers that transform the input to output wherein each hidden layer is made up of neurons, and each neuron is connected to all the neurons in previous layers. CNN mainly comprises three types of layers: convolution layer, pooling layer, fully connected layer.

1. Convolution layer
2. Pooling layer
3. Fully connected layer.

The various emotions used in this paper are shown in Fig. 36.1.

V. Singh (✉) · S. Gandhi · R. Kumar · R. Yadav · S. Joshi
G L Bajaj Institute of Technology and Management, Greater Noida, India

© The Author(s), under exclusive license to Springer Nature Singapore Pte Ltd. 2022

V. V. Rao et al. (eds.), *Computational and Experimental Methods in Mechanical Engineering*, Smart Innovation, Systems and Technologies 239,
https://doi.org/10.1007/978-981-16-2857-3_37

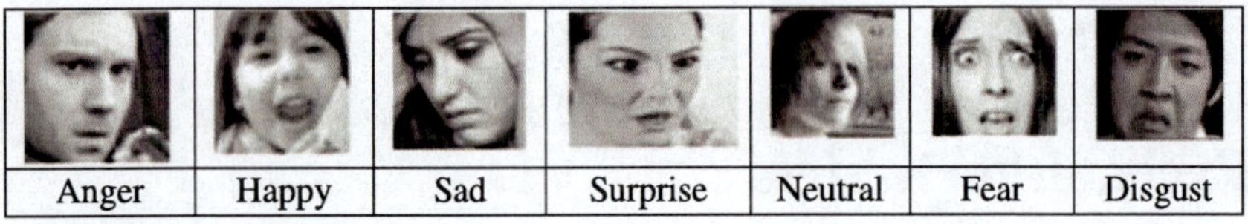

Fig. 36.1 Facial emotion used

36.2 Related Work

Gupta et al. [1] have proposed identification of facial expression using CNN, ResNet, and attention block that gives visual perceptibility. The authors have proposed to use a deep self-attention network for facial emotion recognition. They found that the proposed model outperformed the current CNN-based networks by achieving a higher precision of 85.76. Mellouka and Handouzia [2] performed a survey over multiple research publications that employed deep learning techniques. Authors identified that today researchers are restricted by just knowing the six-basic plus neutral emotion rather than creating a larger database in the future. Mehendale [3] proposed a hybrid of CNN and supervised learning for the detection of facial expression. It has been observed that this model gives better results when it operates for different orientations. The correctness was achieved due to the removal in the background. Zhang et al. [4] performed the facial emotion recognition based on an image of facial expression. Authors use biorthogonal wavelet and fuzzy multiclass support vector machine. This work provides an overall accuracy of 96.7%. Jabid et al. [5] presented an approach for facial emotion by using the local directional pattern(LDP) that takes into consideration the facial geometry. The model also identifies the effectiveness of different reduction techniques such as PCA and AdaBoost in terms of cost and accuracy, which shows the superiority of the LDP descriptor to other features of the descriptors. Mollahosseini et al. [6] proposed a deep neural network architecture for the detection of facial expression which consists of two convolution layers and followed by pooling and then four layers of Inception. It also provides a comparison to several state-of-the-art methods in which engineered features and classifier parameters are usually tuned on a very few databases. The proposed method outperformed over conventional CNN methods in terms of accuracy on both subject independent and cross-database evaluation scenarios. Operto et al. [7] performed a survey to determine ER abilities in children and adults with learning difficulties with different learning disorders incorporated in the study without cognitive disabilities and relates to intelligence. The authors concluded that facial emotional recognition outlays are potentially related to difficulties in cognitive control. Shan et al. [8] presented an approach for facial emotion representation based on the conditions of Local Binary Patterns.

36.3 Methodology

The proposed methodology is implemented as follows:

1. Dataset: The dataset deployed for implementation was the FER2013 dataset from the Kaggle challenge on FER. The dataset contains 35,887 grayscale images out of which 32298 are for training purposes, 3589 for testing. Images in the FER2013 dataset comes under one of the seven categories, namely: neutral, happy, fear, surprise, disgust, angry, and sad.
   Emotion labels in the dataset: 0: −4593 images—angry, 1: −547 images—disgust, 2: −5121 images—fear, 3: −8989 images—happy, 4: −6077 images—sad, 5: −4002 images—surprise, 6: −6198 images—neutral. FER 2013 dataset samples are shown in Fig. 36.2.
2. Preprocessing: Resizing of the image into 48 * 48 grayscale images.
3. Grayscaling: Grayscaling is the process of transforming an RGV image input into a grayscale image whose pixel value from 0 to 255 upon the intensity of light on the image. As the pattern of an image does not depend on color and also the processing of color images requires more processing time and resources. Due to this reason grayscale images are used for processing.
4. Normalization: As neural networks are very sensitive to normalized data, normalization of an image is done to remove illumination variations and obtain an improved face image.
5. VGG19 Architecture:

Fig. 36.2 FER 2013 dataset samples

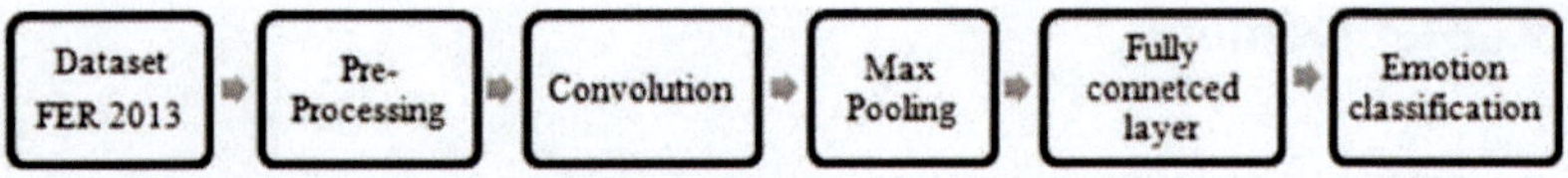

Fig. 36.3 Block diagram of the proposed method

We have conducted extensive experiments to demonstrate the proposed method's effectiveness compared to the most famous classification models, including VGG19 architecture. VGG19 architecture was developed by Simonyan and Zisserman of the University of Oxford with 19 layers, 16 conv, and 3 fully connected. 138 million parameters are eligible for VGG19. VGG19 can train on more than a million images and can classify into 1000 object categories. The detailed methodology is shown in Fig. 36.3.

Block Diagram for Proposed Model:

The VGG19 network consists of sixteen two-dimensional convolutional layers, five max-pooling layers, and three fully connected layers. Max pooling uses the maximum value from each of a cluster of neurons at the prior layer by using a 5 × 5 max-pooling filter. This reduces the dimensionality of the output array. The input to the network is a preprocessed face of 48 × 48 pixels.

36.4 Experimental Results and Discussion

After performing the experiment with the proposed methodology, results are discussed in this section. The experiments are performed on Intel core i5 8400 CPU @ 2.80 GHZ, and python 3. Keeping kernel size 5 * 5 results obtained are shown in Tables 36.1, 36.2 and 36.3. Based on these tables, it is clear that sample size is a very important factor in deep learning. Since the samples of disgust and surprise is low therefore accuracy is poor in these samples. Emotion with happiness performs better compared to others. All others have results that vary in between. Tables 36.1, 36.2, 36.3 and 36.4 show various confusion matrix and results.

Table 36.1 Confusion matrix with Elu activation function

	0	1	2	3	4	5	6
0	333	3	24	13	49	14	59
1	20	20	7	1	3	1	3
2	88	1	207	15	96	48	57
3	15	1	14	785	12	18	54
4	67	0	55	15	300	6	165
5	17	0	35	20	3	317	8
6	28	1	16	35	45	10	485

Table 36.2 Confusion matrix with ReLu activation function

	0	1	2	3	4	5	6
0	332	3	18	14	53	14	61
1	19	21	6	3	2	2	2
2	81	0	187	12	110	52	70
3	8	0	11	791	10	20	59
4	47	0	36	18	342	8	157
5	16	0	34	27	1	315	7
6	16	1	13	26	51	10	503

Table 36.3 Results of Elu and ReLu activation function

Sl. No.	Kernel size (5 * 5)	Activation function	Accuracy
1	5 * 5	Elu	0.682
3	5 * 5	ReLu	0.691

Table 36.4 Comparison with others

References	Methodology	Accuracy(%age)
Mollahosseini et al. [9]	CNN	66.4
Ergen et al. [10]	CNN (ReLu)	57.1
Proposed model	VGG-19 (ReLu, Elu)	69.1

The comparative results on FER 2013 dataset are shown in Table 36.4. Mollahosseini et al. [9] has used a convolution neural network with two convolution layer followed by max pooling and four inception layer and achieved 66.4% accuracy. Tümen et al. [10] achieved 57.1% accuracy, whereas VGG-19 architecture 69.1% accuracy is achieved in the proposed methodology. The proposed architecture is also used in face recognition and pattern recognition [11, 12] applications.

36.5 Conclusions

This paper presents recent development in the facial emotion recognition domain. The paper described VGG 19 architecture with different non-linear activation functions Elu and ReLu. Based on the experimental study, it has been observed that the size of the dataset plays an important role in facial expression recognition. Facial emotions have more samples such as happy, neutral, and angry performed better accuracy than other emotions. Further research based on multimodal deep learning architecture can improve more accuracy in this domain. One challenging issue is recognizing emotions from low-resolution images. It is also observed that the facial expressions are more suitable with the dynamic images in place of static images.

References

1. Guptaa, A., Arunachalama, S., Balakrishnana, R.: Deep self- attention network for facial emotion recognition. Procedia Comput. Sci. Elsevier B.V. **171**, 1527–1534 (2020). https://doi.org/10.1016/j.procs.2020.04.163
2. Mellouka, W., Handouzia, W.: Facial emotion recognition using deep learning: review and insights
3. Mehendale, N.: Facial emotion recognition using convolutional neural networks (FERC). SN Appl. Sci. **2**(3), 1–8 (2020)
4. Zhang, Y.-D., Yang, Z.-J., Lu, H.-M., Zhou, X.-X., Phillips, P., Liu, Q.-M., Wang, S.-H.: Facial emotion recognition based on biorthogonal wavelet entropy, fuzzy support vector machine, and stratified cross validation. IEEE Access **4**, 8375–8385 (2016)
5. Jabid, T., Hasanul Kabir, Md., Chae, O.: Robust facial expression recognition based on local directional pattern. ETRI J. **32**(5), 784–794 (2010)
6. Mollahosseini, A., Chan, D., Mahoor, M.H.: Going deeper in facial expression recognition using deep neural networks. In: 2016 IEEE Winter Conference on Applications of Computer Vision (WACV), pp. 1–10. IEEE (2016)
7. Operto, F.F., Pastorino, G.M.G., Stellato, M., Morcaldi, L., Vetri, L., Carotenuto, M., Viggiano, A., Coppola, G.: Facial emotion recognition in children and adolescents with specific learning disorder. Brain Sci. **10**(8), 473 (2020)
8. Shan, C., Gong, S., McOwan, P.W.: Facial expression recognition based on local binary patterns: a comprehensive study. Image Vis. Comput. **27**(6), 803–816 (2009)
9. Mollahosseini, A., Chan, D., Mahoor, M.H.: Going deeper in facial expression recognition using deep neural networks. In: 2016 IEEE Winter Conference on Applications of Computer Vision (WACV), pp. 1–10 (2016). https://doi.org/10.1109/wacv.2016.7477450
10. Tümen, V., Söylemez, Ö.F., Ergen, B.: Facial emotion recognition on a dataset using convolutional neural network. In: 2017 International Artificial Intelligence and Data Processing Symposium (IDAP), pp. 1–5. IEEE (2017)
11. Kumar, R., Sagar, L.K., Awasthi, S.: Human activity recognition from video clip. Intelligent Computing in Engineering, pp. 269–274. Springer, Singapore (2020)
12. Kumar, R., Joshi, S., Dwivedi, A.: CNN-SSPSO: a hybrid and optimized CNN approach for peripheral blood cell image recognition and classification. Int. J. Pattern Recogn. Artif. Intell. 2157004 (2020)

Chapter 37
An Analytical Study of Partial Replacement of Cement and Quartz Sand by Feldspar

Rajat Yadav, Ravindra Pratap Singh, and Kuwar Mausam

Abstract There are a lot of carbon emissions from cement which demands the usage of other alternatives as its replacement. Research is undoubtedly conveying crazy on the usage from claiming waste items to cement. Likewise, a reinstatement of characteristic sand. Because of that unreasonable utilization for waterway sand Likewise fine aggravator, the availability of the fine aggregate has become scanty. So, it demands the search for other alternative replacements for fine aggregates and cement additives that enhance the properties of fresh and hardened concrete like durability, strength, etc. In our study, we are going to experiment on the effects on mechanical assets of material by moderate replacement of cement by feldspar. Since Quartz is one of the greatest plentiful materials available on earth's crust and being highly siliceous, these materials are being employed as additives for cement and fine aggregate. This study mainly concentrates on the assets of toughened concrete properties, which is the compressive strength of concrete. Here, cement is partially replaced using feldspar up to 25% at regular intervals of 5%. In addition to that, the river sand is completely replaced with quartz sand as fine aggregate. The above materials are being tested for M40 grade of concrete.

R. Yadav (✉) · R. P. Singh · K. Mausam
IET Department of Mechanical Engineering, GLA University, Mathura, India
e-mail: rajat.yadav@gla.ac.in

R. P. Singh
e-mail: pratap.ravindra@gla.ac.in

K. Mausam
e-mail: kuwar.mausam@gla.ac.in

© The Author(s), under exclusive license to Springer Nature Singapore Pte Ltd. 2022
V. V. Rao et al. (eds.), *Computational and Experimental Methods in Mechanical Engineering*, Smart Innovation, Systems and Technologies 239,
https://doi.org/10.1007/978-981-16-2857-3_38

37.1 Introduction

Concrete is a significant versatile and durable creation material that has been in use for centuries now in one form or the other. It is one of the widely accepted due to various properties such as fire resistance, plasticity, chemical resistance, etc. [1, 2].

Cement is the chief compound in the concrete, which is used as a binder. Albeit cement may be one of the most viable and versatile binders that is available and being used, it has many harmful environmental impacts when concrete is directly being exposed and has sabotaging effects on the environment [3, 4]. It produces a lot of pollution, releasing carbon dioxide and greenhouse gases in bulk amounts, harming the environment [5].

The cement industry harvests over 10% of worldwide manmade carbon dioxide emissions, for which 60% may be starting with those substance process. An Chatham house ponders starting with 2018 estimates that the 4 billion tons about bond which may be transformed yearly accounts for 8% of overall carbon dioxide emanations [6, 7]. Due to the excessive use of river sand as fine aggregate, the availability of the fine aggregate has become scanty. Hence other materials can be used instead of river sand for construction purposes [8–10]. So, it demands the search of other alternatives for replacements of fine aggregates and cement additives, enhancing the mechanical assets of concrete such as durability, workability, asset, etc. [11, 12].

Researches proved that mineral powders such as feldspar powder, glass powder, rice husk, etc., have proved that powders, after moderately swapped by cement, increase the compressive asset and reduce carbon emissions [13, 14]. One such material is quartz sand which is used as an Elective for customary waterway sand [15, 16]. These sands need aid utilized within the diversion, looking into golf courses, volleyball courts, baseball fields, children's sandboxes, and beaches. Quartz sand might make discovered Previously, Florida Also indifferent mainland edges [17]. The research demonstrated that use from claiming quartz sand like substitution cost to sand will be a prudent result of settling on the cement safe to weathering.

37.2 Experimental Methodology

37.2.1 Materials

53-grade Regular Portland cement of mark is utilized in research. The assets of cement were tabularized below (Table 37.1).

Table 37.1 Properties of cement

Properties	Magnitude
Specific gravity	2.3
Standard consistency	35%
Preliminary situation time	30 min
Closing situation time	600 min

Table 37.2 Properties of fine aggregate

Properties	Magnitude
Specific gravity	2.53
Bulk density/m^3	1830
Porosity%	29.67
Grading zone	Zone II
Fineness modulus	3.13
Water absorption	1.02%

37.2.2 Fine Aggregate

Quartz sand is used as fine aggregate in this study. Quartz sand is the most abundant silica mineral. The research proved that using quartz sand as a replacement for sand is an economical solution for making the concrete resistant to weathering.

The test procedure is taken from code IS 2720: 1980 part 3 (Table 37.2).

37.2.3 Coarse Aggregate

The well-graded round-shaped crushed aggregates of size 20 mm as per ***IS 2386: 1963-part-3*** is used as coarse aggregate (Table 37.3).

Table 37.3 Properties of coarse aggregate

Properties	Coarse aggregate
Particle shape	Round
Particle size	20 mm
Specific gravity	2.75
Bulk density	1340 kg/m^3
Fineness modulus	3.01%

37.2.4 Feldspar

Feldspar is the most important single group of rock-forming silicate minerals. It is one of the second most abundant minerals available on the earth and makes up about 41% of the earth's crust. In this present study, feldspar is partially replaced with cement up to 25% at regular intervals of 5% (Fig. 37.1 and Tables 37.4 and 37.5).

Fig. 37.1 Feldspar Powder

Table 37.4 Chemical composition of quartz sand

Oxide	Chemical composition (%)
Magnesium oxide	0.006
Ferrous oxide	0.215
Calcium oxide	0.010
Aluminum hydroxide	0.769
Silicon dioxide	99.02
Titanium oxide	0.078
Water	0.020

Table 37.5 Chemical composition of feldspar

Oxide	Chemical composition (%)
Silicon dioxide	64.57
Aluminum hydroxide	18.28
Sodium oxide	0.09
Potassium oxide	16.01
Calcium oxide	–
Barium oxide	0.04

37.3 Experimental Results

37.3.1 Compressive Test

The compressive test was performed when cement was partially replaced with feldspar with different percentages (5, 10, 15, 20, 25%), respectively, for M40 grade of concrete at different ages such as 7, 28, and 60 days, respectively.

It was observed that the compressive strength of concrete decreased with the increase in the percentage of feldspar (Fig. 37.2).

The mean compressive strength of concrete at 28 days is maximum when cement is partially replaced with 5% feldspar. It has then decreased steadily.

The mean compressive strength of feldspar at 60 days is maximum when cement is partially replaced with 5% feldspar. It has then decreased steadily (Figs. 37.3 and 37.4).

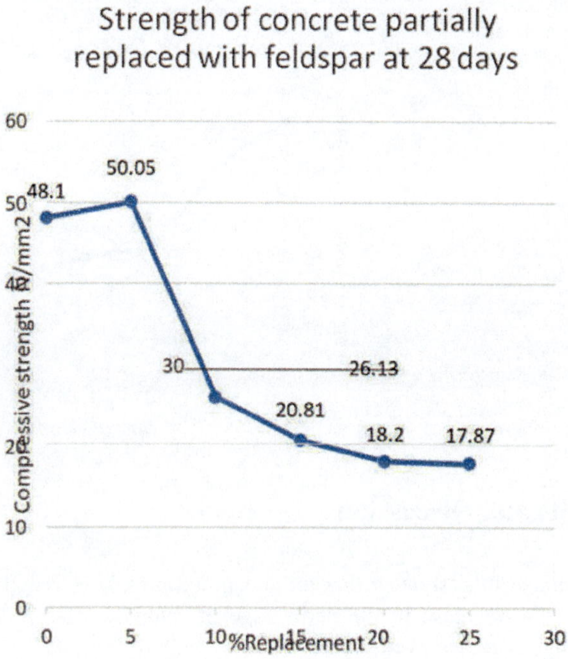

Fig. 37.2 Strength of concrete partially replaced with feldspar at 28 days

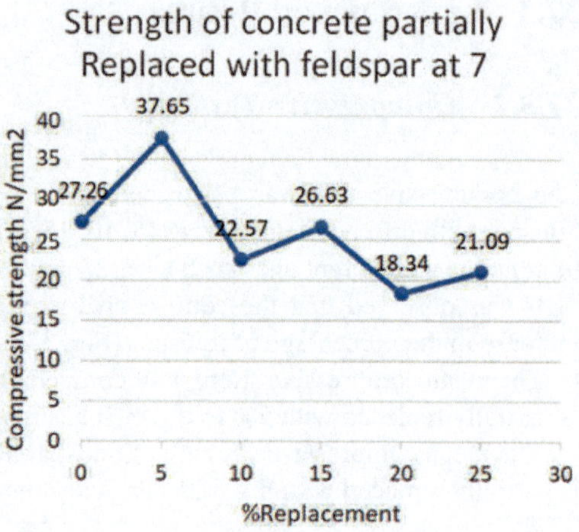

Fig. 37.3 Strength of concrete at 7 days

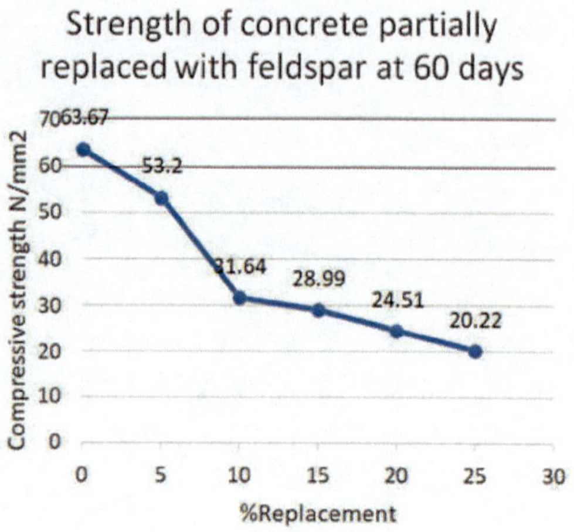

Fig. 37.4 Strength of concrete partially replaced with feldspar after 60 days

37.4 Results and Discussion

The experimental results showed that the mean compressive strength of concrete decreased with the increase in the percentage of feldspar after 7 days, and the maximum strength was achieved at 5%.

Mix design leads to the development of a concrete specification. A mix of M40 grade concrete was used and designed as per relevant Indian Standard specifications in the present investigations (Table 37.6).

Table 37.6 Mix design for M40 concrete

Cement (kg)	Fine aggregate (kg)	Coarse aggregate (kg)	w/c ratio
534.21	508.88	949.67	0.38

References

1. Effect of Quartz Powder.: quartz sand and water curing regimes on mechanical properties of uhpc using response surface methodology—Mohammad Ali Mosaberpanah, OzgirEren
2. Effect of Feldspar.: mica and clay minerals on compressive strength of mortar Atiya Tugrul, SelahattinHasdemir, MuratYilmaz
3. Effect of Mineral powders derived from industrial wastes on selected mechanical properties of concrete—Anna Galinska, Slawomir, CzarBecki
4. Effect of Mineral admixtures and quartz sand on workability of compressive strength of self compacting concrete—Gayatri Komati, Ravi Kumar Garee , Sri Vinay ChowdaryDasari
5. Divya, E., Shantini, R., ArulKumaran, S.: Study on Behaviour of concrete partially replacing quartz sand as fine aggregate
6. Kumar, A., Sharma, K., Dixit, A.R.: A review of the mechanical and thermal properties of graphene and its hybrid polymer nanocomposites for structural applications. J. Mater. Sci. **54**(8), 5992–6026
7. Yadav, A., Kumar, A., Singh, P.K., Sharma, K.: Glass transition temperature of functionalized graphene epoxy composites using molecular dynamics simulation. Integr. Ferroelectr. **186**(1), 106–114
8. Singh, P.K., Sharma, K., Kumar, A., Shukla, M.: Effects of functionalization on the mechanical properties of multiwalled carbon nanotubes: a molecular dynamics approach. J. Compos. Mater. **51**(5), 671–680
9. Singh, P.K., Sharma, K.: Mechanical and viscoelastic properties of in-situ amine functionalized multiple layer Grpahene/epoxy nanocomposites. Curr. Nanosci. **14**(3), 252–262
10. Cementitious composition containing feldspar and pozzalanic particulate material and method of making said composition—AllenPratt
11. Effects of quartz powder on micro structure and key properties of cement paste—Run Sheng Lin, Xiao-Yong Wang , Gui- YuShang
12. Bharadwaj, G., Sharma, K., Tiwari, A.K.: Performance analysis of hybrid PCM by doping Graphene. Mater. Today: Proc. **26**, 850–853 (2020)
13. Singh, R.P., Sharma, K., Mausam, K.: Dispersion and stability of metal oxide nanoparticles in aqueous suspension: a review. Mater. Today: Proc. **26**, 2021–2025 (2020)
14. Broekmans, M.A.T.M.: Failure of greenstone, jasper and CataClasite aggregate in bituminous concrete due to studded tyres: similarities and differences
15. Yang, Q., Zhang, S., Huang, S., Yeb, H.: Effect of ground quartz sand on properties of high- strength concrete in the steam-autoclaved curingExperimental Investigation On Mechanical Properties Of Concrete Containing Quartz Powder And Silica Fume With Steel Fibre - P.Jayanthi, M.Bharath,K.Kanchana
16. Jayanthi, P., Bharath, M., Kanchana, K.: Experimental investigation on mechanical properties of concrete containing quartz powder and silica fume with steel fibre
17. Chaturvedi, R.: Molding sand nano fluidization role in ceramic substitute. J. Green Eng. **10**(11), 11540–11545 (2020)

Chapter 38
A Review of Pick and Place Operation Using Computer Vision and ROS

Aditya Mathur, Chetan Bansal, Sandeep Chauhan, and Omprakash Yadav

Abstract The requirements for more efficient and easily modifiable techniques have provoked the fast advancement in the domain of robotics. The improvement of canny robots prompts the capacity of them to turn into an administrator profoundly proficient and ready to adjust to a wide scope of issues. In any case, notwithstanding the few automated arrangements accessible, most of the current modern robots do not utilize the Robotic Operative System (ROS) and have restrictions as far as self-sufficiently right mistakes during their assignments. Controlling a robotic arm for applications such as object segmentation with the utilization of vision sensors would require vigorous picture processing and calculation to perceive and distinguish the object when using an image processing heavy approach, while a more traditional approach relies on sensors and partial automation in most cases. This paper is coordinated toward compiling the important computer vision techniques for pick and place operation along with the underlying factors that make them better than traditional techniques.

38.1 Introduction

Modern robots, because of their extraordinary speed, accuracy, and cost-effectiveness in redundant errands, presently will, in general, be utilized rather than manual work in computerized creation lines. These amazing machines are not autonomous, as they require manual fundamental operation calibration, for example, adjustment and direction arranging to accomplish characterized assignments [1]. The utmost importance was given in the development of the robots for the future to make the robot fully independent [2].

The issue of trajectory planning is a major one in advanced mechanics and characterizes a fleeting movement law with a given mathematical way, so certain prerequisites are set for the direction purposes [1].

A. Mathur · C. Bansal · S. Chauhan (✉) · O. Yadav
Department of Mechanical Engineering, G L Bajaj Institute of Technology and Management, Greater Noida, India

© The Author(s), under exclusive license to Springer Nature Singapore Pte Ltd. 2022
V. V. Rao et al. (eds.), *Computational and Experimental Methods in Mechanical Engineering*, Smart Innovation, Systems and Technologies 239,
https://doi.org/10.1007/978-981-16-2857-3_39

38.2 Robotic Arm and Their Applications

The mechanical manipulators have pulled in generous interest from various industries. They present key qualities for smoothing out not just daily tasks but even niche applications such as sanitization and disinfection. One of the most important aspects of a robot is the end-effector and is an important aspect in applications like pick and place robot [3]. Instances of utilizations for these pick and place robot include transportation of hardware, pick and place of objects to recently characterize focused on often risky and blocked off undertakings [4]. Industrial robots have been very significant in enhancing the assembling measures. In essence, pick and place robots perform the operation of picking up the object and placing it at a specific place, according to design and requirement parameters [5].

38.2.1 Applications of Robotics Arm

Mechanical robots perform explicit tedious errands such as welding or painting parts in vehicle manufacturing plants. With the improvement of sensors and gadgets for human–robot connection, robotic manipulators are progressively utilized in less controlled climate, such as high-accuracy surgery, putting anything in the same orientation at specified locations [3, 6].

There are numerous fields where robotics is widely used. Some important ones are discussed here:

Robotic arms in manufacturing: The robots are used for spot welding assignments in assembling plants; however, as technology developed, the scope of tasks that robotic arms could perform additionally extended [7]. The advances in technology include the expanding assortment for end-of-arm tooling that has opened up. This implies that Robotic arms can play out a wide scope of errands past welding, relying upon the instruments that are connected to the furthest limit of their arms [8]. Current developments in end-of-arm tools incorporate; 3D Printing apparatus heads and for sheet metal working [9].

Robotic arms in healthcare: Robots are evolving the healthcare sector tremendously and are easily seen in actions in the neighborhood hospitals. Generally, these robots look like R2D2 from Star Wars more than they do a humanoid, yet they have a major impact in the field of medications [7]. Robotic technologies show up in numerous zones that straightforwardly influence patient care. They can be utilized to sanitize patient rooms and working suites, diminishing dangers for patients and clinical staff [10]. They work in labs to take tests and to transport, examine, and store them.

Robotics in the food processing industry: Robots are helpful for food processing by numerous manners. A few robots, for instance, were able to make pizzas with no human mediation [11].

Besides these, some robots were developed that find applications for agricultural purposes [12].

38.3 Robot Operating System

Robot Operating System (or ROS), a Meta Operating system, is a collection of packages built on top of an Operating System [13]. It offers many OS functionalities like low-level device control along with the implementation of common functionalities like hardware abstraction, package management, etc. ROS is not an Operating System in the traditional sense of process management and scheduling; rather, it provides a structured communications layer above the host operating systems of a heterogeneous compute cluster. It particularly supports simulators like the gazebo, Rviz which are very instrumental in making and visualizing robots.

38.4 Pick and Place Operation

Controlling a robotic arm for applications, for example, object arranging with the utilization of vision sensors, would require a vigorous picture preparing calculation to perceive and distinguish the objective object [1, 14]. The technique involves image-processing capabilities and requires coordination toward improving the picture preparing calculation, which is a pre-imperative for the full activity of a pick and spot Robotic arm expected for object arranging task [15]. Toward such kind of assignment, first, the articles are recognized. Furthermore, it is cultivated by including extraction calculations [16]. Next, the separate picture (boundaries in consistence with the classifier) is fed to the classifier to perceive, after which the collected information is used alongside its directions to be prepared for the robotic arm to execute the pick and place task [17].

For the task of pick and place a single object, the pick and place cycle consists of the following tasks:

- Identify and understand the objective item on the rack.
- Plan so that we can perform movement toward the object on the rack.
- Efficiently grasp the object without causing any harm to our surrounding environment.
- Do a clean movement on the desired path to drop-off point.
- Place/Stow the object.

38.5 Pick and Place Operation by Using Computer Vision

Computer vision is a technology in high demand nowadays and brings a revolution in terms of how many industrial operations are utilizing it. It is very scalable, and according to our needs, we can use a camera with better sensitivity [18]. It keeps an eye on the application for distinguishing and examining little parts, directing the regulators. In the event that we take a gander at the manufacturing industry, vision is generally significant. It very well may be utilized for various modern applications, for example, welding, painting, programmed pick and spot objects, military activity, in space, in submerged activity, in agribusiness activity, and so on [10, 17]. The most costly and tedious assignment is to put objects in a particular way known as a palletizer [17]. The fundamental commitment of advanced mechanics is in palletizing.

In a typical vision-based pick and place robot, right off the bat, pictures of objects are captured using image processing tools [1]. Furthermore, an image processing tool is used to determine the operated work pieces' color, size, and shape [18]. Thirdly, after fundamental bodies are discovered, the coordinates of the centroid have to be accurately mapped [19]. Finally, all the arranged items dependent on the shape, size, and color are provided to the manipulator; it is then given the command to fetch the object [8, 20]. To get the edges of the target object in an image, edge detection techniques in OpenCV library are used [21]. This is essential and the first step in finding the object. There are various algorithms for edge detection, and canny is the most used [8, 21]. It detects the edges by looking at the local maxima of the gradient of that particular image pixel value and setting up a threshold value [7]. Once our target image has detected the object, it provides the coordinates, and transform is applied to it to obtain the real-world coordinates [22]. Then a manipulator can go and reach out to get the object and stow/place it at the desired location [16]. A typical computer vision-based technique used in industrial applications is shown through a block diagram in Fig. 38.1 [23].

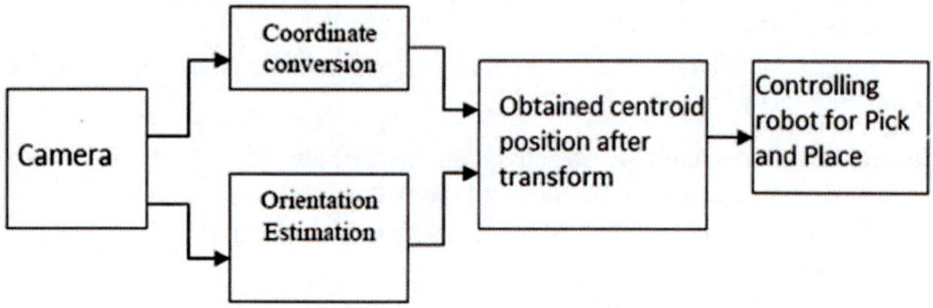

Fig. 38.1 Block diagram of a computer vision-based control system

38.5.1 Artificial Neural Networks (ANNs) and Their Usage in Smart Pick and Place

In an ANN-based system, the images are fed to the neural network to be processed; often convolutional neural networks (CNN) are used for such applications owing to their superior results in dealing with sequential data in the form of images or even audio messages [3]. In the case of pick and place operation, a classifier is made and supervised learning is used to teach the neural network to learn from features and detect and identify various objects instead of using image processing capabilities every single iteration [20].

All the techniques involving only the use of a single camera for pick and place operation require additional components to do successful pick and place operation since a camera can only operate in 2D (i.e., X, Y) and require either sensor or multiple cameras to capture the image in its entirety [1, 3]. On the other hand, systems equipped with 3D (i.e., X, Y, Z) vision can identify objects even if they are randomly posed and correct the orientation [17]. This is revolutionary in its own way as it reduces the need for any additional equipment and helps make the overall system simple and compact. Stereo cameras make use of 3D point cloud for this operation [7]. The system that works on this concept performs activities; range sensing, image segmentation, object recognition, and pose estimation [18]. The work for the range sensor is to estimate the depth of the object.

Classifier for Image segmentation: The use of a classifier, as its name suggests, is to classify the objects into various "classes." It is trained with the help of an Artificial Neural Network (ANN) [24]. The training of a classifier starts with splitting data between test and training blocks, then we need to give it a certain amount of data to learn from, and it comes under the umbrella of Supervised learning because of this reason [7, 14]. After this, we decide the number of layers and the composition of those [3]. In addition, for a classifier, we need to implement the cost function; after that, we define the activation function [25]. In training a classifier, implementing the Feed-Forward Propagation and Cost function is vital, as shown in Eq. 38.1. The cost function has to be regularized. The learning rate, λ, and the number of layers are the factors that determine it [3]. In the final steps of training a classifier, the hyperparameters are tuned according to the application, and it often a task that is very experimental [12, 19]. The general rule of thumb for more accuracy is more number of iterations [3] which would lead to a better model and output.

$$J(\theta) = \frac{1}{m}\sum_{i=k}^{m}\sum_{k=1}^{k}\left[-y_k^{(i)}\log\left(h_\theta\left(x^{(t)}\right)\right)_k - \left(1 - y_k^{(i)}\right)\log(1 - h_\theta\left(x^{(i)}\right))_k\right] \quad (38.1)$$

38.5.2 Implementation in ROS

The pick and place activity can be characterized as a fairly simple assignment that requires a few levels of control. Ordinarily, this approach can be separated into three levels, with an expanded degree of significance [20].

All applications utilizing pick and place procedures have a general environment (or world) just as articles and target areas. Truth be told, a critical advance in pick and place approaches is the distinguishing proof and acknowledgment of that world. Consequently, level one of this methodology is to distinguish the environment by making a 3D model of the world [20]. Making such a model and preparing the acquired 3D model makes it conceivable to recognize all articles in a given scene. In this methodology, we adopt a world in gazebo simulator where we are performing the operations. This permits one to totally distinguish all scenes and all items identified with the scene, saving information identified with every one of them in a simulated environment.

After constructing the 3D model along with the CAD model of the robot, the robot should have the option to react depending on the 3D model yields [19]. This is vital as the robot's ability to move is innate to the pick and place activities, and often times the model fails to pick the desired payload if the conversion has not been done successfully [7]. After the initial processes, the inverse kinematics and the model data are utilized to drive the development of the joint movements of the pick and place robot, which constitutes level two of implementation through ROS [19].

As referenced earlier, the entire pick and place robot application is controlled by a control loop in the majority of techniques, which is level three of ROS implementation. A control loop does the job by providing feedback through forward and inverse kinematic requests in ROS nodes, which help prevent and alter the path and correct the trajectory in pick and place operation [20].

38.6 Future Work

Our efforts are to find the alternative for the traditional bin picking, which is commonly used by many industries nowadays. We have an object whose shape, size, volume, position, and orientation are the same or predefined in traditional bin picking. Depth Camera and computer vision, AI, and ROS can eliminate the constraints mentioned above in bin picking and even deliver better results.

We hope that our work will encourage others to develop their interest in ROS, which we found also has significance in introducing modern robotics concepts with intelligent industrial pick and place robots, leading to the intelligent manufacturing system. Using ROS gives practical exposure to ideas and broadens the thinking because it makes all the things possible in the real world to be implemented inside a simulated environment.

It was also found that Machine Learning could be even more useful with the advent of Reinforcement Learning in the robotics applications such as pick and place robot. Very few research works are available in this area and thus have tremendous scope to explore full autonomy and require no human assistance.

38.7 Conclusions

The traditional pick and place robots do not find applications in the industry these days, but the underlying concepts are vital in improving current pick and place operation techniques. Mostly camera is placed on the upper portion for the pick and place operation, while in some cases, it is mounted near the point of contact. The Neural Networks usage brings about a fundamentally new approach to the whole process and will eventually make the process even smoother with more training data. Depending on the requirement, the techniques can be chosen and be effectively utilized. For example, a neural network approach might be more experimental in nature as compared to the approach of using depth cameras. So depth cameras would be a better choice to work in the industry for quicker and accurate results, while the neural network approach could be more useful where cost is an issue.

References

1. Sam, R., Arrifin, K., Buniyamin, N.: Simulation of pick and place robotics system using Solidworks Softmotion 1–6 (2012). https://doi.org/10.1109/ICSEngT.2012.6339325
2. Schrock, P., Farelo, F., Alqasemi, R., Dubey, R.: Design, simulation and testing of a new modular wheelchair mounted robotic arm to perform activities of daily living. In: 2009 IEEE International Conference on Rehabilitation Robotics, Kyoto, pp. 518–523 (2009). https://doi.org/10.1109/ICORR.2009.5209469
3. Kumar, R., Kumar, S., Lal, S., Chand, P.: Object Detection and Recognition for a Pick and Place Robot (2014). https://doi.org/10.13140/2.1.4379.2165
4. Yang, H., Xu, M., Li, W., Zhang, S.: Design and implementation of a soft robotic arm driven by SMA coils. IEEE Trans. Industr. Electron. **66**(8), 6108–6116 (2019). https://doi.org/10.1109/TIE.2018.2872005
5. Wells, R.L., Schueller, J.K., Tlusty, J.: Feedforward and feedback control of a flexible robotic arm. IEEE Control Syst. Mag. **10**(1), 9–15 (1990). https://doi.org/10.1109/37.50663
6. Hao, W.G., Leck, Y.Y., Hun, L.C.: 6-DOF PC-Based Robotic Arm (PC-ROBOARM) with efficient trajectory planning and speed control. In: 2011 4th International Conference on Mechatronics (ICOM), Kuala Lumpur, pp. 1–7 (2011). https://doi.org/10.1109/ICOM.2011.5937171
7. Lozano-Perez, T., Jones, J.L., Mazer, E., O'Donnell, P.A.: Task-level planning of pick-and-place robot motions. Computer **22**(3), 21–29 (1989). https://doi.org/10.1109/2.16222
8. Ohta, P., Valle, L., King, J., Low, K., Yi, J., Atkeson, C., Park, Y.-L.: Design of a lightweight soft robotic arm using pneumatic artificial muscles and inflatable sleeves. Soft Rob. **5** (2017). https://doi.org/10.1089/soro.2017.0044

9. Bhuyan, A.I., Mallick, T.C., Gyro-accelerometer based control of a robotic ARM using AVR microcontroller. In: 9th International Forum on Strategic Technology (IFOST). Cox's Bazar, pp. 409–413 (2014). https://doi.org/10.1109/IFOST.2014.6991151
10. Bodie, K., Bellicoso, C.D., Hutter, M.: ANYpulator: design and control of a safe robotic arm. In: 2016 IEEE/RSJ International Conference on Intelligent Robots and Systems (IROS), Daejeon, pp. 1119–1125 (2016). https://doi.org/10.1109/IROS.2016.7759189.
11. Örger, N.C., Karyot, T.B.: A symmetrical robotic arm design approach with stereo-vision ability for CubeSats. In: 2013 6th International Conference on Recent Advances in Space Technologies (RAST), Istanbul, pp. 961–965 (2013). https://doi.org/10.1109/RAST.2013.6581353
12. Megalingam, R.K., Vivek, G.V., Bandyopadhyay, S., Rahi, M.J.: Robotic arm design, development and control for agriculture applications. In: 2017 4th International Conference on Advanced Computing and Communication Systems (ICACCS), Coimbatore, pp. 1–7 (2017) https://doi.org/10.1109/ICACCS.2017.8014623
13. Quigley, M., et al.: ROS: an open-source robot operating system. In: ICRA Workshop on Open Source Software, vol. 3. no. 3 (2009)
14. Kumar, R., Lal, S., Kumar, S., Chand, P.: Object detection and recognition for a pick and place robo. In: Asia-Pacific World Congress on Computer Science and Engineering, Nadi, pp. 1–7 (2014). https://doi.org/10.1109/APWCCSE.2014.7053853
15. Jegede, O., Oludele, A., Ayodele, A., Miko, N.: Development of a Microcontroller Based Robotic Arm (2017)
16. Ali, Md., Kashkimbayev, A., K., Yerkhan & Turmakhanov, Zhandos & O., Anuar. : Vision-based robot manipulator for industrial applications. Procedia Comput. Sci. **133**, 205–212 (2018). https://doi.org/10.1016/j.procs.2018.07.025
17. Ghita, O., Whelan, P.: A bin picking system based on depth from defocus. In: Ghita, O., Whelan, P.F. (eds) A Bin Picking System Based on Depth From Defocus. Mach. Vis. Appl. **13**(4):234–244. ISSN 0932-8092. 13. https://doi.org/10.1007/s00138-002-0071-4
18. Reddy, R., Nagaraja, S.R.: Integration of robotic arm with vision system (2015). https://doi.org/10.1109/ICCIC.2014.7238302.
19. Asadi, K., Jain, R., Qin, Z., Sun, M., Noghabaei, M., Cole, J., Han, K., Lobaton, E.: Vision-based Obstacle Removal System for Autonomous Ground Vehicles Using a Robotic Arm (2019)
20. Tavares, P., Sousa, A.: Flexible pick and place architecture using ROS framework. In: 2015 10th Iberian Conference on Information Systems and Technologies, CISTI 2015 (2015). https://doi.org/10.1109/CISTI.2015.7170602
21. arXiv:1901.08180
22. Rahardja, K., Kosaka, A.: Vision-based bin-picking: recognition and localization of multiple complex objects using simple visual cues. In: Proceedings of IEEE/RSJ International Conference on Intelligent Robots and Systems. IROS '96, vol. 3, Osaka, Japan, pp. 1448–1457. https://doi.org/10.1109/IROS.1996.569005(1996)
23. Ganie, A.: Bin picking and machine vision. Ind. Robot: Int. J. **38** (2011). https://doi.org/10.1108/ir.2011.04938baa.002
24. Hao, W., Leck, Y., Hun, L.: 6-DOF PC-Based Robotic Arm (PC-ROBOARM) with efficient trajectory planning and speed control 1–7 (2011). https://doi.org/10.1109/ICOM.2011.5937171.
25. Liang, X., Cheong, H., Sun, Y., Guo, J., Chui, C.K., Yeow, C.: Design, characterization, and implementation of a Two-DOF fabric-based soft robotic arm. IEEE Robot. Autom. Lett. **3**(3), 2702–2709 (2018). https://doi.org/10.1109/LRA.2018.2831723

Chapter 39
Efficient Activated Metal Inert Gas Welding Procedures by Various Fluxes for Welding Process

Aman Sharma, Rishabh Chaturvedi, and Pradeep Kumar Singh

Abstract In the past years, so many developments and innovations are made to increase the performance of the welding machine by reducing the power consumption, cost, labor skills, and many more. In a row, the MIG welding process is one of the welding processes that increase a material's welding properties at the same consumption of power and resources. A-MIG welding is wildly used to optimize the parameter of MIG welding. A steady flow welding powerfully delivers electrical vitality, directed over the bend through a segment of exceptionally ionized gas and metal fumes known as plasma. Metal idle gas (MIG) welding is most regularly used to welder thick segments of tempered steel and non-ferrous metals, such as aluminum, magnesium, and copper compounds. Three sorts of oxides, Fe_2O_3, SiO_2, and $MgCO_3$, were utilized to research the impact of initiating transition helped gas metal bend welding (GMAW) on weld dab math, precise twisting, and mechanical properties. The most noteworthy impact on the inward structure factor was knowledge about an instance of MnO dynamic transition, which diminished the inside structure factor by 20%. If there should arise an occurrence of the outside structure factor, the applied SiO_2 dynamic transition caused the greatest increment, which is 37%. In the hardness dispersion and the microstructure of the joints, including the weld metal and the heath-influenced zone, no critical contrasts were experienced contrasted with the example welded with no transition material.

A. Sharma (✉) · R. Chaturvedi · P. K. Singh
IET Department of Mechanical Engineering, GLA University, Mathura, India
e-mail: aman.sharma@gla.ac.in

R. Chaturvedi
e-mail: risabh.chaturvedi@gla.ac.in

P. K. Singh
e-mail: pradeep.kumar@gla.ac.in

© The Author(s), under exclusive license to Springer Nature Singapore Pte Ltd. 2022
V. V. Rao et al. (eds.), *Computational and Experimental Methods in Mechanical Engineering*, Smart Innovation, Systems and Technologies 239,
https://doi.org/10.1007/978-981-16-2857-3_40

39.1 Introduction

Metal inactive gas welding is known as gas metal arc welding, which uses an expendable conductor to generate arc between workpiece and electrode [1, 2]. The inert gas argon, in rare cases helium (more expensive) gases, is used to protect the weld bead from atmosphere contaminates, these gases are also called shielding gas, and filler materials are normally used [3, 4]. The procedure grants those drivers more amazing control through those welds over contending procedures, for example, protected metal circular segment welding. Furthermore, gas metal circular segment welding, permitting for stronger, higher personal satisfaction welds. MIG welding is utilized to joint ferrous Furthermore non-ferrous materials, for example, steel, copper, aluminum, magnesium, nickel, What's more their alloys [5, 6].

MIG welding procedure has failed to weld a thick section of material in a single pass, and it has low productivity. To recover the performance of MIG welding, triggered flux is used to increase the depth of penetration in a single pass. Activated flux is mixed with the acetone and blinder and applied thin past on the workpiece by brush or spray [7, 8]. A-MIG welding system makes it could reasonably be expected to heighten the accepted MIG welding hones for joining those more than 10 mm thick plate by absolute pasquinade with no edge preparation. The mechanical properties moved forward contrasted with those welds made without any flux material. Over our examination, we investigated the impacts of eight distinctive animated fluxes and mixtures on the point of interest throughout GMAW utilizing animated protecting gas [9, 10]. In the event that of circular segment welding the main impetus of the struck them streams in the weld pool could have a chance to be originated starting with four separate phenomena, those buoyancies, that surface tension (which brought about the with the goal known as Marangoni impact), the high-speed development of the circular segment plasma, and the Lorentz energy [11, 12] (Fig. 39.1).

Investigation on the consequence of flux coated gas tungsten arc welding on 304L. In this investigation, the researcher used SiO_2 flux, and the result shows that the penetration increased up to 200%, the ferrite number increased up to 14%, and the

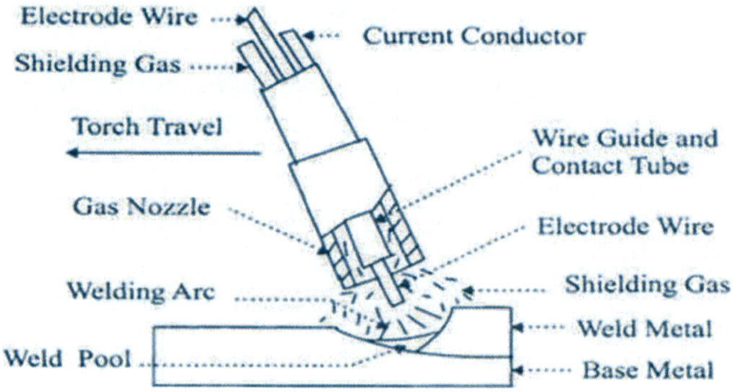

Fig. 39.1 Schematic diagram of MIG welding

hardness value of the material also increased [13–15]. Study on the consequence of ternary fluxes on the depth of penetration in A-TIG welding of AISI 409 ferritic stainless steel. In this study, the flux increased the depth of penetration up to 100% [16, 17]. Xie et al. [18] have researched on impacts about nano-particles fortifying initiating flux on the microstructures. Also, mechanical properties of TIG-welded AZ31 first mass of the magnesium compound joints. Those specialists utilized blended TiO_2 and nano-SiC particles as actuated flux and demonstrated that microstructure, microhardness. On combination zone, extreme elasticity might have been improved [19, 20].

Based on previous research, the present investigation deals with three types of activated fluxes applied to the workpiece prior to welding. The bead geometry is then compared after welding with and without using activated fluxes. The effect of the different fluxes is also studied.

39.2 Experimental Procedure

39.2.1 Base Metal and Its Composition

The material utilized in this learning is austenitic stainless steel 301. The composition of the base metal is given in Table 39.1. The work pieces were cut in the dimension of 100 mm × 50 mm × 10 mm plates. Before welding, the work pieces were grinded with 220 grit emery papers, and it was cleaned with acetone.

Table 39.1 Conformation of material

Component	Percentage weight (%)
C	0.054
Si	0.284
Mn	1.11
P	0.021
S	0.008
Cr	14.25
Mo	0.093
Ni	6.15
Co	0.067
Ti	0.009
V	0.054
Fe	Rest

39.2.2 *Preparation of Flux*

The weld bead geometry of welding should be studied after applying the flux paste on the surface of work piece. The pasted were made with the mixing of acetone and a very small amount of sodium silicate as a binder and apply with the help of brush or spray shown in figure acetone has a tendency to vaporize and leave the flux on the surface of work piece and sodium silicate has a tendency to sticking and bind the flux particle together (Fig. 39.2).

39.3 Experimental Procedure

In the first step, the welding parameters are decided on the basis of several trial experiments and are listed in Table 39.2.

Prior to welding the metal power is mixed with acetone and sodium silicate (as a binder) to make paint-like solution and applied a very thin layer of approximately 0.2 mm on the workpiece surface with the help of brush or spray. Uniformity in

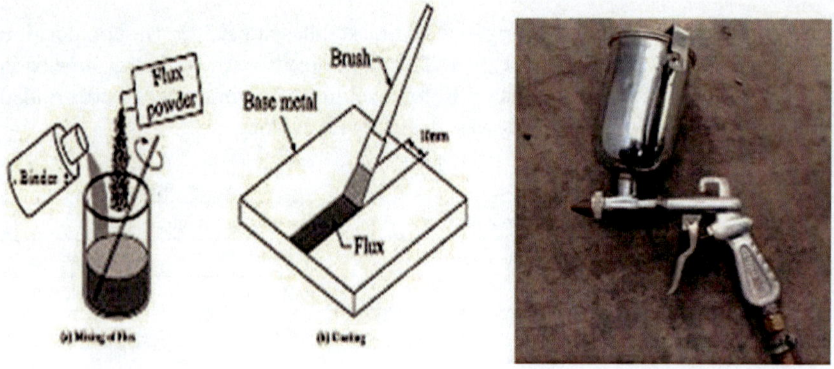

Fig. 39.2 Flux preparation

Table 39.2 Process parameter for welding

Parameters	Value	Units
Welding current	180	A
Total arc voltage evolution	12–18	V
Travel speed	120	mm/min
Arc gap	2.5	mm
Diameter of electrode	2.6	mm
Gas flow rate	14	L/min

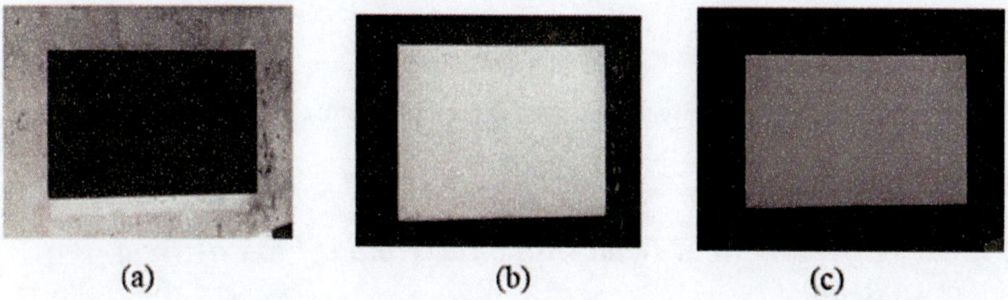

Fig. 39.3 Photograph of applied fluxes **a** MnO_2 flux, **b** TiO_2 flux, and **c** SiO_2 flux

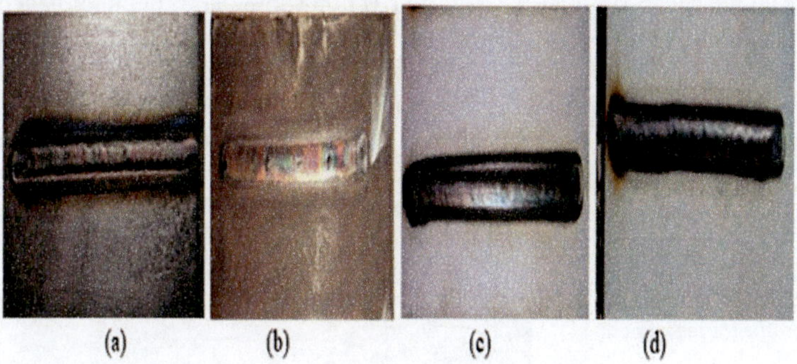

Fig. 39.4 Photograph of weld bead on workpiece **a** without flux **b** MnO_2 flux **c** TiO_2 flux **d** SiO_2 flux

applied flux is most important to get good quality of the weld and more depth of penetration (Fig. 39.3).

TIG welding is done on the work pieces, which are coated with the fluxes and without flux. The pattern of the TIG welding on the workpiece shown below (Fig. 39.4).

39.4 Stereo Zoom Microscope Analysis

The weld bead measurement was done as per UNS SS30100 Standard on stainless steel specimen on the stereo zoom microscope analyzer available at SLIET Longowal (Punjab). For stereozoom microscope analysis, the specimens have been prepared using a standard procedure such as grinding, polishing using successively fine grades of emery up to 3000 grit size and Al_2O_3 power. It helped to remove scratches on the surface that are to be metallographically analyzed. Before analysis, the etchant is used to reveal bead geometry. A stereo zoom analyzer is used to measure the bead geometry like bead width, depth of penetration, and heat-affected zone (HAZ) of the specimens.

39.5 Results

Stereo zoom analysis of the weld bead with and without flux is represented as follows.

39.5.1 *Effect of Weld Bead with and Without Flux of Weld Bead*

In conventional TIG welding, the direction of convection in the weld is toward the edge from the center of the weld. This convection shows a low depth of penetration and wide weld width. But in activated TIG welding, the convection is reverse from edge to center to the weld. The images show the differences between conventional TIG welding and activated TIG welding (Figs. 39.5, 39.6, 39.7 and 39.8).

The results show that the application of fluxes increases the depth of penetration. MnO_2 activated flux has the lowest penetration among the other fluxes up to 4.21 mm and SiO_2 activated flux has the highest penetration depth up to 5.18 mm.

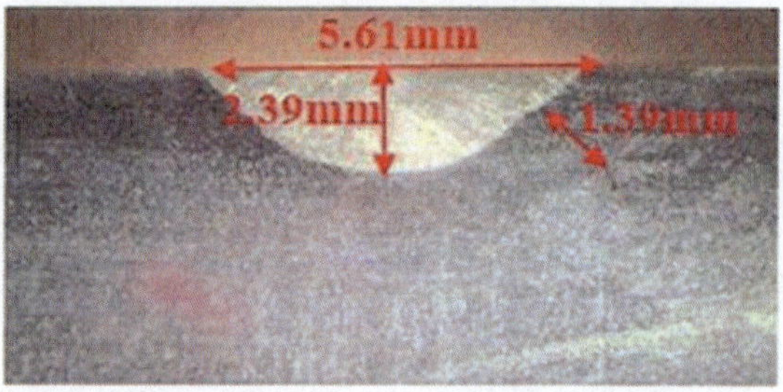

Fig. 39.5 Stereo zoom analyzer images for without flux

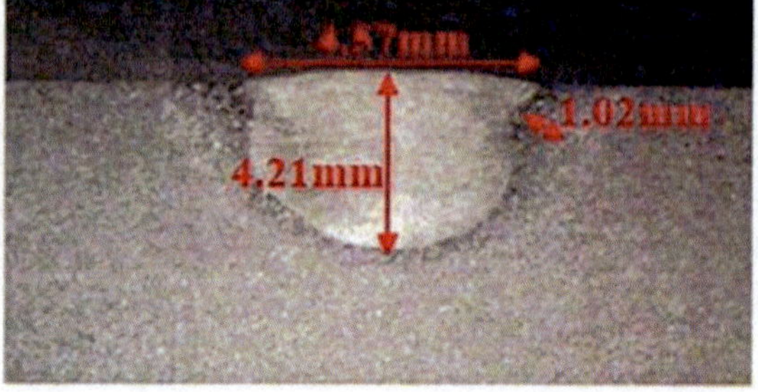

Fig. 39.6 Stereo zoom analyzer images for MnO_2 flux

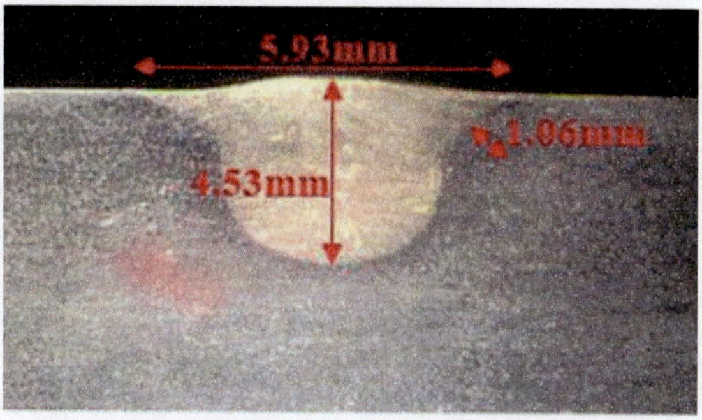

Fig. 39.7 Stereo zoom analyzer images for TiO_2 flux

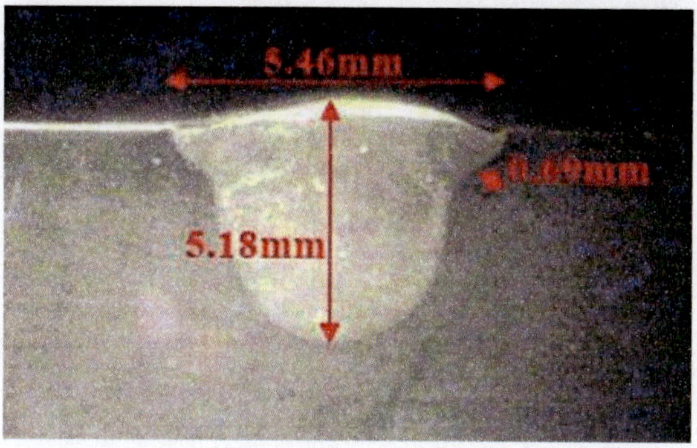

Fig. 39.8 Stereozoomanalyser images for SiO_2 flux

39.5.2 Effect of Aspect Ratio with and Without Fluxes

In conventional welding, the surface tension at the midpoint of the weld is less than the edge of the weld that shows molten metal takes place from center to edge; this phenomenon is called the Marangoni effect. Due to the phenomena of surface tension in conventional TIG, the welding depth of penetration is less than the width of the weld bead [4]. Therefore the aspect ratio in conventional welding is 0.426. The application of flux constricts the arc to the center of the weld bead; this leads the surface tension toward the center of the weld bead, and the penetration is achieved more than the conventional TIG welding process. It is observed that the use of SiO_2 activated flux provides a better aspect ratio as compared to other used flux in study (Table 39.3 and Fig. 39.9).

39.6 Conclusions

- From the study following conclusion are drawn.
- The use of flux is more beneficial as compared to conventional TIG welding.

Table 39.3 Aspect ratio for various fluxes

TIG welding process	Depth of penetration (mm)	Weld width (mm)	Aspect ratio
Without flux	2.39	5.61	0.426
Using MnO_2 flux	4.21	4.57	0.921
Using TiO_2 flux	4.53	5.93	0.764
Using SiO_2 flux	5.18	5.46	0.949

TIG Welding process	Depth of penetration(mm)	Weld width (mm)	Aspect ratio
Without flux	2.39	5.61	0.426
Using MnO2 flux	4.21	4.57	0.921
Using TiO2 flux	4.53	5.93	0.764
Using SiO2 flux	5.18	5.46	0.949

Fig. 39.9 Aspect ratio

- It is observed that the aspect ratio with flux is increased 80–122% for different coated flux.
- The extreme feature ratio experiential is 0.949 for SiO_2 flux. So it is extra advantageous to usage SiO_2 flux.
- It is observed that using SiO_2 flux results in minimum heat exaggerated zone (HAZ), results in unaffected strength of weld and base metal.

References

1. Zuber, M., et al.: Effect of flux coated gas tungsten arc welding on 304L. IACSIT Int. J. Eng. Technol. **6**(3), 177–181 (2014)
2. Venkatesan, G., et al.: Effect of ternary fluxes on depth of penetration in A-TIG welding of AISI 409 ferritic stainless steel. Pocedia Mater. Sci. (Elsevier Ltd.) **5**, 2402–2410 (2014)
3. Bodkhe, S.C., Dolas, D.R.: Optimization of activated tungsten inert gas welding of 304L austenitic stainless steel. Procedia Manuf. **20**, 277–282 (2018)
4. Xie, X., et al.: Effects of nano-particles strengthening activating flux on the microstructures and mechanical properties of TIG welded AZ31 magnesium alloy joints. Mater. Des. (Elsevier Ltd.) **81**, 31–38 (2015)
5. Chaturvedi, R., Islam, A., Sharma, A., Sharma, K., Sharma, R.: Design and analysis of mechanical gripper of aristo-robot for welding. **83**, 23202–23209 (2020). ISSN: 0193-4120

6. Sambherao, A.B.: Use of activated flux for increasing penetration in austenitic stainless steel while performing GTAW. Int. J. Emerg. Technol. Adv. Eng. 220–224 (2013)
7. Nayee, S.G., et al.: Effect of oxide-based fluxes on mechanical and metallurgical properties of dissimilar activating flux assisted tungsten inert gas welds. J. Manuf. Process. (Elsevier Ltd.) **16**, 137–143 (2014)
8. Patel, A.B., et al.: The effect of activating fluxes in TIG welding by using ANOVA for SS 321. ISSN Int. J. Eng. Res. Appl. 41–48 (2014)
9. Jeyaprakash, N., Haile, A., Arunprasath, M.: The Parameters and Equipment Used in TIG Welding: A Review, pp. 11–20 (2015)
10. Vora, J.J., et al.: Experimental investigation on mechanism and weld morphology of activated TIG welded bead-on-plate weldments of reduced activation ferritic/martensitic steel using oxidefluxes. J. Manuf. Process. (Elsevier Ltd.) **20**, 224–233 (2015)
11. Kumar, K., et al.: Experimental investigation of graphene-paraffin wax nanocomposites for thermal energy storage. Mater. Today: Proc. **18**, 5158–5163 (2019)
12. Kumar, A., Sharma, K., Dixit, A.R.: A review on the mechanical properties of polymer composites reinforced by carbon nanotubes and graphene. Carbon Lett. 1–17 (2020)
13. Singh, P.K., Sharma, K., Kumar, A., Shukla, M.: Effects of functionalization on the mechanical properties of multiwalled carbon nanotubes: a molecular dynamics approach. J. Compos. Mater. **51**(5), 671–680
14. Singh, P.K., Sharma, K.: Mechanical and viscoelastic properties of in-situ amine functionalized multiple layer graphene/epoxy nanocomposites. Curr. Nanosci. **14**(3), 252–262
15. Bajpai, A., et al.: Investigations on structure–property relationships of activated flux TIG weldments of super-duplex/austenitic stainless steels. Mater. Sci. Eng. A (Elsevier B.V.) **638**, 60–68 (2015)
16. Sándor, T., Mekler, C., Dobránszky, J., Kaptay, G.: An improved theoretical model for A-TIG welding based on surface phase transition and reversed Marangoni flow. Metall. Mater. Trans. Phys. Metall. Mater. Sci. **44A**, 351–361 (2013)
17. Lu, S., Fujii, H., Sugiyama, H., Tanaka, M., Nogi, K.: Weld penetration and Marangoni convection with oxide fluxes in GTA welding. Mater. Trans. **43**(11), 2926–2931 (2002)
18. Singh, A.K., Dey, V., Rai, R.N.: Techniques to improve weld penetration in TIG welding (A review). Mater. Today: Proc. **4**(2), 1252–1259 (2017)
19. Pamnani, R., Vasudevan, M., Jayakumar, T., Vasantharaja, P.: Development of activated flux, optimization of welding parameters and characterization of weld joint for DMR249A shipbuilding steel. Trans. Indian Inst. Met. **70**(1), 49–57 (2017)
20. Adetunji, O.R., Adegbola, A.O., Afolalu, S.A.: Comparative study of case-hardening and water-quenching of mild steel rod on its mechanical properties. Int. J. Adv. Res. **3**(6), 1–9 (2015)

Chapter 40
Computational Analysis of Heat Transfer Characteristics of TiO and CuO_2 Mixed with Water for Heat Exchanger Applications

Anuj Gupta, Vinod Kumar Yadav, Rishabh Kumar, and Nirbhay Singh Parmar

Abstract In this paper, computational determination of heat transfer features of TiO and CuO_2, mixed with aqua, as a base fluid for heat exchanger application, is conducted using ANSYS® Fluent software. The parameters like Nusselt number, heat transfer and skin friction coefficients are dealt with in detail. The computed results were then compared with the results of Al_2O_3/water mixture. The results revealed that with an increased flow rate, the value of Nusselt number, and consequently the rate of heat transfer, improves significantly by using TiO and CuO_2 particles in aqua. However, the fluid's inlet temperature has insignificant effect on the radiator performance. The TiO concentration in water showed highest rate of heat transfer compared to Al_2O_3 and CuO_2. Furthermore, CuO_2 exhibited highest skin friction coefficient when compared with Al_2O_3 and TiO. In addition, Al_2O_3 executed least Nusselt number and skin friction coefficient compared to CuO_2. The rate of heat transfer gets enhanced by about 4–5 times by using TiO and CuO_2 in place of Al_2O_3 under all operating conditions. On the other hand, the heat transfer characteristics of TiO and CuO_2 do not vary much and are in close match with each other.

A. Gupta
Department of Mechanical Engineering, Vishveshwarya Group of Institutions, Dadri, UP, India

V. K. Yadav (✉) · R. Kumar
Department of Mechanical Engineering, G. L. Bajaj Institute of Technology and Management, Greater Noida, UP, India

N. S. Parmar
Department of Mechanical Engineering, Indus Institute of Technology and Management, Kanpur, UP, India

© The Author(s), under exclusive license to Springer Nature Singapore Pte Ltd. 2022

V. V. Rao et al. (eds.), *Computational and Experimental Methods in Mechanical Engineering*, Smart Innovation, Systems and Technologies 239,
https://doi.org/10.1007/978-981-16-2857-3_41

40.1 Introduction

With the decrease in size of engineering equipment, the need for compact heat exchanger increases. On the other hand, due to the restricted flow area of tubes, the heat transfer coefficient of liquid needs enhancement. Nanofluids containing metal as foreign particles exhibit superior performance. Nanofluids containing metals or oxides of metal particles of the order of less than 100 nm mixed with primary coolant are finding applications in automotive radiators as a substitute for baseline coolant [1]. The metal or metal oxides improve the heat transfer performance of the working fluid and significantly improves the heat transfer rates. The presence of nanoparticles even in a small extent improves the surface area, and consequently the heat transfer rates are improved.

Shah and Kiran Kumar [2] experimentally studied the effect of varying the mass flow rate and air velocity over 15 nm SiO_2 nanofluid using water as base fluid. They reported higher values of effectiveness and heat transfer rates with an increased concentration of nanoparticles compared to water. They also reported that the effectiveness of the heat exchanger reduces with increasing Reynolds number due to the presence of efficient spiral heat exchangers. Tijani and Sudirman [3] analyzed the performance of water and anti-freeze-based nanofluids using ANSYS fluent solver for the automotive radiator. Water and Ethylene glycols with 50–50% concentration was prepared as base fluid. They added nanoparticles of Al_2O_3 and CuO having 0.05, 0.15 and 0.3% concentrations in the base fluid and evaluated the performance keeping constant mass flow rate. They observed that CuO exhibited superior heat transfer characteristics with 0.3% nanoparticle and flow rate of 6 litres per minute (lpm). Contreras et al. [4] evaluated the thermos-hydraulic characteristics of silver and graphene nanofluids (0.01%, 0.05% and 0.1%) on automobile radiators through experimentation in presence of a base fluid containing 50% water and 50% ethylene glycol. The mass flow rates were varied in the range of 0.08–0.11 kg per second (kg/s). The inlet temperature of the coolant was varied from 55 to 85 °C and the air velocity was fixed as 2.1 m/s. They recorded an enhancement of 4.1% heat transfer rate at a specific pumping power and mass flow rate. They further observed that the rate of heat transfer gets enhanced by 4.4% by using silver nanofluids. On the other hand, the grapheme mixture exhibited a drop in thermo-hydraulic performance compared to its original fluid. Selvam et al. [5] experimentally evaluated the heat capacity, density, thermal conductivity, pressure drop, viscosity and convection coefficient of automotive radiators running on graphene nanoplatelets (0.1–0.5% volume concentration) with water-ethylene glycol mixture. The flow rate was varied from 10 g per second to 100 g per second, temperature from 35 to 45 °C and surrounding air velocity was set at 3 m/s and determined the convection coefficient. They reported that by increasing the graphene nanoplatelets concentration, the convective heat transfer coefficient increases significantly. In addition, a rise in the pressure drop of the nanofluids with increased concentration and mass flow rate of graphene was noticed. They reported that the heat transfer coefficient improves due to increased thermal conductivity, reduced boundary layer thickness and particle clustering. Furthermore, they showed that nanofluids having 0.5% (volume) is significant in laminar flow region when maintained at 45 °C in automobile cooling systems, exhibiting less pressure drop and enhanced heat transfer coefficient. Sahoo et al. [6] analyzed the louvered fin-type automotive radiator using water-based CuO, Silver, TiO_2, Cu and Fe_2O_3 in Al_2O_3 nanofluids with 50–50 volume fraction. They found that Al_2O_3-silver/water hybrid nanofluid exhibit superior heat transfer rate, effectiveness, pressure drop and pumping power compared to base fluid (water) followed by Copper, Copper-oxide, Fe_2O_3 and TiO_2. They also reported that rate of heat flow, coolant flow and pumping power gets reduced for Al_2O_3-silver/water-hybrid nanofluids for similar radiator size. On the other hand, the radiator size gets reduced and pumping power increases with Al_2O_3-silver/water hybrid nanofluid compared to the base fluid (water) for the same coolant and heat flow rates. Naraki et al. [7], through experimentation, investigated the coefficient of heat transfer of water/Copper-oxide nanofluids under laminar flow condition (Reynolds number 100-1000) using car radiator. They observed that the overall heat transfer coefficient improves significantly using nanofluids compared to the base fluid. On the other hand, a decrement in overall heat transfer coefficient, with an increase in inlet temperature of the nanofluids (from 50 to 80 °C), was observed. Peyghambarzadeh et al. [8] compared the heat transfer behavior of H_2O and EG, in their pure form, with binary mixtures prepared by them in the laboratory. In addition, appropriate quantities of Al_2O_3 nanoparticle were also seeded to the base fluid to study the change in car radiator's heat transfer performance through experimentation. The mass flow rates of the liquid varied from 2 to 6 lpm and the inlet temperature of the fluid also varied during experiments. Their results revealed that nanofluids significantly improve the rates of heat transfer compared to their base fluids. They reported that the maximum boost in Nusselt number, by about 40%, was noticed for both nanofluids under best-operating conditions. Furthermore, the rate of heat transfer of nanofluids is highly dependent upon the particle concentration. In addition, the flow rates were independent of inlet temperature. Leong et al. [9] studied the behavior of copper nanofluids, based on ethylene glycol, in automobile radiators. Their studies inferred that the heat transfer performance improves with the usage of ethylene glycol nanofluids (using copper) as the base fluid in comparison with pure ethylene glycol. A reduction in fair frontal area was also evaluated.

The literature review reveals that good number of research data is available to judge the performance of automotive radiators using nanofluids like SiO_2, Al_2O_3, CuO, Ag, TiO_2, Cu, Fe_2O_3 and graphene. But the performance analysis using TiO and CuO_2 is seldom reported in literatures. The present work aims to compute the heat transfer behavior and thermal performance of TiO and CuO_2 mixed with aqua base for industrial and domestic heat exchanger applications.

40.2 Materials and Method

The present work aims to find the most appropriate nanofluid, that may serve as a better coolant, for radiators compared to pure water. The performance of the

Table 40.1 Calculated properties of nanofluids

Properties	Titanium oxide	Copper dioxide
Density (kg/m^3)	893.205	1523.8
Specific heat (kJ/kgK)	4172.423	2666.75
Conductivity (W/mK)	0.810	1.23

water-based coolant homogeneously mixed with Titanium oxide, Copper dioxide and Aluminum oxide is investigated numerically.

40.2.1 *Thermo-Physical Properties of Fluids Used in the Present Work*

The properties of the materials (nanofluids) used in the present work are calculated using Eqs. 40.1–40.3. The numerical values of the material properties are presented in Table 40.1.

$$\text{Density,}\quad \rho_{nf} = \varphi\rho_p + (1-\varphi)\cdot\rho_{bf}C_{pbf} \tag{40.1}$$

$$\text{Specific Heat,}\quad C_{pnf} = \frac{\varphi\rho_p C_{pp} + (1-\varphi)\cdot\rho_{bf}C_{pbf}}{\rho_{nf}} \tag{40.2}$$

$$\text{Conductivity, } k_{nf} = \frac{k_p + (\emptyset-1)\cdot k_{bf} - \varphi\cdot(\emptyset-1)\cdot\left(k_{bf}+k_p\right)}{k_p + (\emptyset-1)\cdot k_{bf} + \varphi\cdot\left(k_{bf}-k_p\right)} k_{bf} \tag{40.3}$$

where ρ denotes density, C denoted specific heat, k denotes thermal conductivity and φ for the proportion of nanoparticles in fluid. In subscripts of notations, *nf* represents nanofluid, while *p* is for the nanoparticle and *bf* is for the base fluid.

40.2.2 *Governing Equation and Boundary Conditions*

For incompressible fluids, the assumption of steady-state is applied, with assumption of Newtonian flow inside the tubes. The Newtonian behavior of aqua-based nanofluids for nanoparticle with about 4% concentration was observed. Inlet velocity and temperature of the tubes are uniform. The condition of thermal equilibrium was assured between the base particles and nanoparticles. These assumptions were used for specific heat capacity computation of the working fluid. Particle deposition effect was neglected due to the fact that the concentration and size of nanoparticles were small and presumably may be treated as homogeneous fluid. The momentum and continuity equations are presented using Eqs. 40.4 and 40.5.

$$\nabla . V = 0 \tag{40.4}$$

$$\rho_{nf} \cdot (\nabla \cdot V)V = -\nabla P + (\mu_{nf} + \mu^t)\nabla^2 \cdot V \tag{40.5}$$

The turbulence viscosity was computed using the standard form of k-ε model (Eq. 40.6).

$$\mu^t = C_\mu \frac{k^2}{\epsilon} \tag{40.6}$$

The energy conservation equation is shown in Eq. 40.7.

$$\rho_{nf} + C_{pnf}(V.\nabla)T = k_{nf}\nabla^2 T \tag{40.7}$$

The investigation was performed in laminar flow regime with the condition of no-slip on walls of the tube. Following boundary conditions were subjected to nanofluid:

- Inlet mass flow rate of the fluid: 2, 3, 4, 5 m^3/s.
- Turbulence intensity magnitude: 10%
- Convective heat transfer coefficient: 150 W/m^2K with air at 303 K (Similar to reference [1]).
- Inlet fluid temperature: 35, 45 and 55 °C.

40.3 Results and Discussion

This paper presents the results of computational investigations obtained for three different nanofluids (Aluminum, Titanium and Copper-based) at different mass flow rates and at different temperature conditions. ANSYS Fluent 17.2 Academic Version [10] has been used for the investigation of the present study.

40.3.1 Validation

Delavari and Hashemabadi [1] experimentally evaluated the effect of convective heat transfer, in forced condition, in an aqua-based nanofluids and compared the results with water in an automotive radiator. Nanofluid (0.1–1 volume %), with five concentrations, was prepared by adding Al_2O_3 nanoparticles in water. The prepared fluid was passed via radiator having 34 vertical tubes of elliptical cross-section. The air was passed through cross-flow inside the tubes with fixed speed. The flow rate of the liquid was varied in the range of 2–5 lpm to ensure fully turbulent behavior (9000 < Re < 23000). In addition, the effect of fluid's entrance temperature on the coefficient

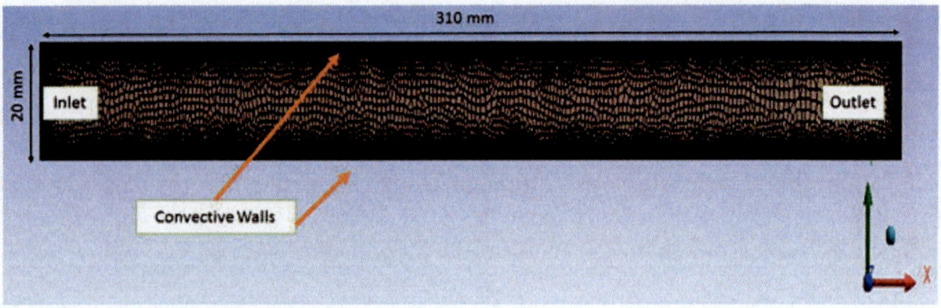

Fig. 40.1 Computational domain and mesh

of heat transfer, through the radiator, was analyzed by varying the temperature from 37 to 49 °C.

To validate the accuracy of the model developed for the present study, the results of Delavari and Hashemabadi [1] was compared. Figure 40.1 shows the computational domain developed for the current problem along with its mesh. At the entry-level inlet and at the exit outlet boundary conditions were prescribed. The following parameters were chosen for analyzing the current problem:

- Mass flow rate of the fluid: 2, 3, 4, 5 m^3/s
- Convection coefficient: 150 W/m^2K
- Pressure outlet: 0 Pa
- Inlet fluid temperature: 35, 45 and 50 °C.

Figure 40.2 compares the experimental results of Delavari and Hashemabadi [1] with numerical results of the present work. It can be observed that similar patterns for the variation in the Nusselt number at various fluid flow rates are encountered. This also indicates that with increased flow rate, the Nusselt number improves, thereby increasing the rate of heat transfer.

The surface heat flux, pressure variation along the channel and variation of fluid temperature along the channel was also studied. Figure 40.3 presents the variation of surface heat flux with distance at different mass flow rates. It can be seen that with increased flow rate, the surface heat transfer increases and gets stabilized while passing through the bundle of tubes.

Figure 40.4 presents the variations in the working fluid's pressure due to skin friction. This also infers that with increased forking fluid's flow rate, pressure loss along the length of the tube increases as a result of which the maintenance cost of the tube may increase due to the presence of metal particles in the working fluid chosen in this study.

Figure 40.5 shows the variation in the working fluid's temperature while flowing through the tube. It can be extracted with the trends that with increased flow rate, the difference in temperature between the inlet and exit decreases. The results support the physics of fluid flow that may be expected in heat exchangers.

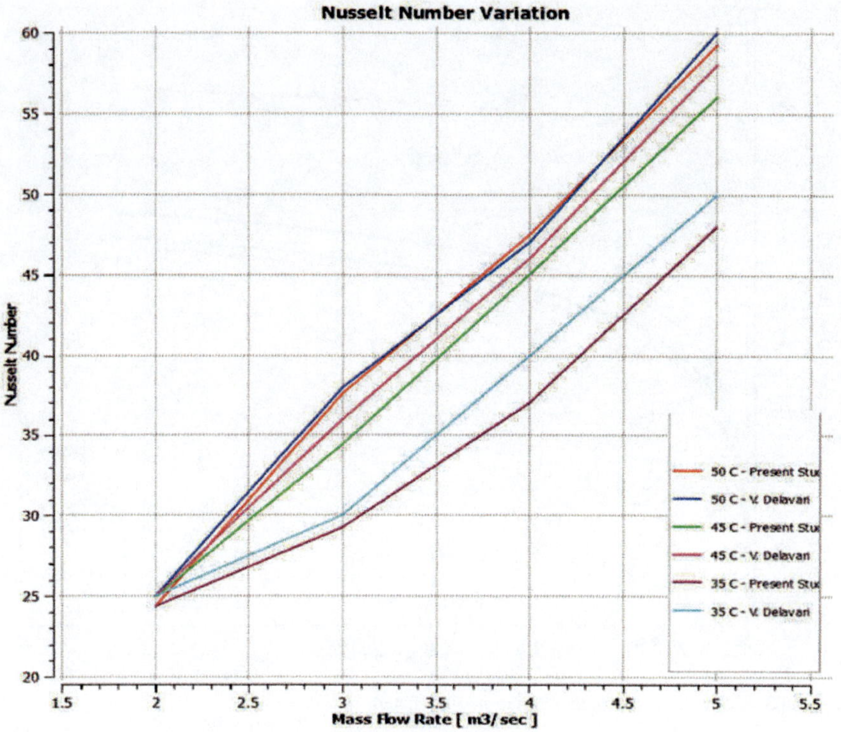

Fig. 40.2 Nusselt number variation with mass flow rate

40.3.2 Heat Flow Analysis

The heat transfer rates from the hot nanofluids through convective heat transfer mode was investigated. Figure 40.6 represents the temperature contours of the nanofluid (TiO) with different concentration (by varying the % volume in water). The results infer that with increased concentration of the nanofluid, the rate of heat transfer improves significantly. To further validate, the dimensionless Nusselt number may prove to be a suitable parameter. Figure 40.7 presents the variation of Nusselt number under different operating conditions with TiO nanofluid. The numerical simulations were conducted for 1% concentration of the working fluid entering inside the domain at various temperatures and mass flow rates. It has been observed that by increasing the working fluid's temperature, the Nusselt number increases thereby enhancing the rate of heat transfer. In addition, with increased flow rate, the Nusselt number improves following a linear trend.

Figure 40.8 presents the variation in the Nusselt number with CuO_2 as working substance. Similar pattern for the temperature change and fluid flow rates were observed.

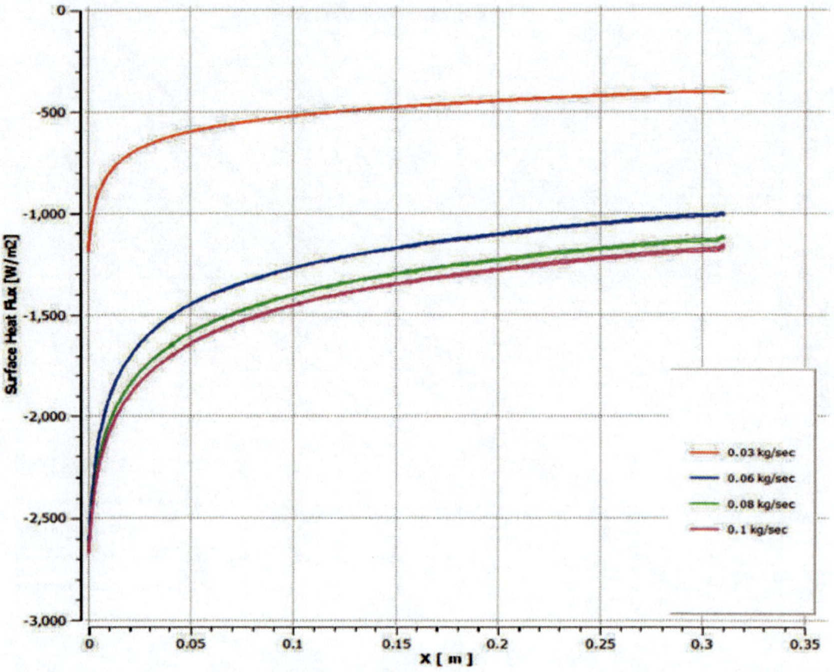

Fig. 40.3 Convection coefficient variation with distance

To find the most appropriate nanofluids, to be used as a cooling medium, the radiator of an automobile, the Nusselt numbers obtained with Al_2O_3 [1], TiO (present work) and CuO (present work) has been compared and plotted in Fig. 40.9. The comparison was made for fluid inlet temperature of 318 K and 1% concentration of respective nanoparticles. The mass flow rates were set as 2, 3, 4 and 5 m^3/h. It is observed that TiO exhibits better heat transfer capability compared to CuO_2 and Al_2O_3 under all mass flow rate conditions. Furthermore, the heat transfer characteristics of Al_2O_3 are inferior to all the fluids studied in this work. The rate of heat transfer gets enhanced by about 4–5 times by using TiO and CuO_2 in place of Al_2O_3 under all operating conditions. On the other hand, the heat transfer characteristics of TiO and CuO_2 do not vary much and are in close match with each other. The preceding section presents the comparison of skin friction coefficient of the studied fluid for a better understanding of the results obtained so far.

40.3.3 Skin Friction Analysis

Figure 40.10 presents the comparison of the skin friction coefficient over the inner side of the tube for Al_2O_3 (Delavari and Hashemabadi [1]), TiO (present work) and

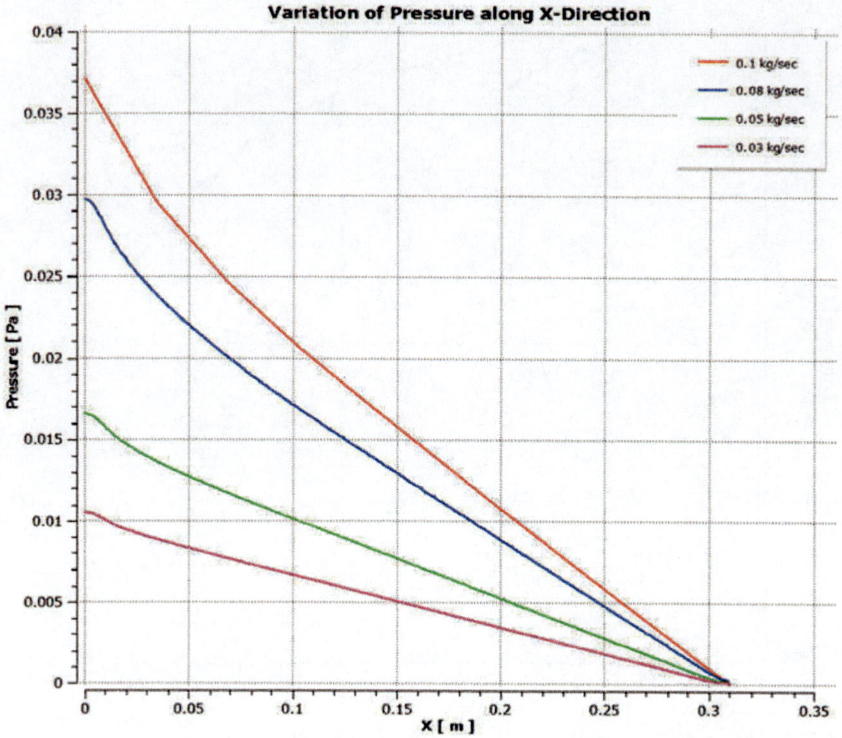

Fig. 40.4 Pressure loss along tube length

CuO (present work). More will be the magnitude of skin friction, more wear and tear rate in the interaction of fluid and the tube. It can be inferred that CuO_2 has maximum skin friction as compared to other two fluids while Al_2O_3 exhibited the least value of skin friction coefficient.

40.4 Conclusions

In the present work, computational analysis using ANSYS Fluent [10] was performed to compute the heat transfer parameters (like skin friction coefficient and Nusselt number) of some nanofluids that may be appropriate as coolant for automobile radiators. The following conclusions were drawn:

(i) TiO exhibited highest heat transfer rate compared to Al_2O_3 and CuO_2.
(ii) CuO_2 exhibited highest skin friction coefficient compared to Al_2O_3 and TiO. This in turn increases tube wear due to enhanced interaction between tube and working fluid.

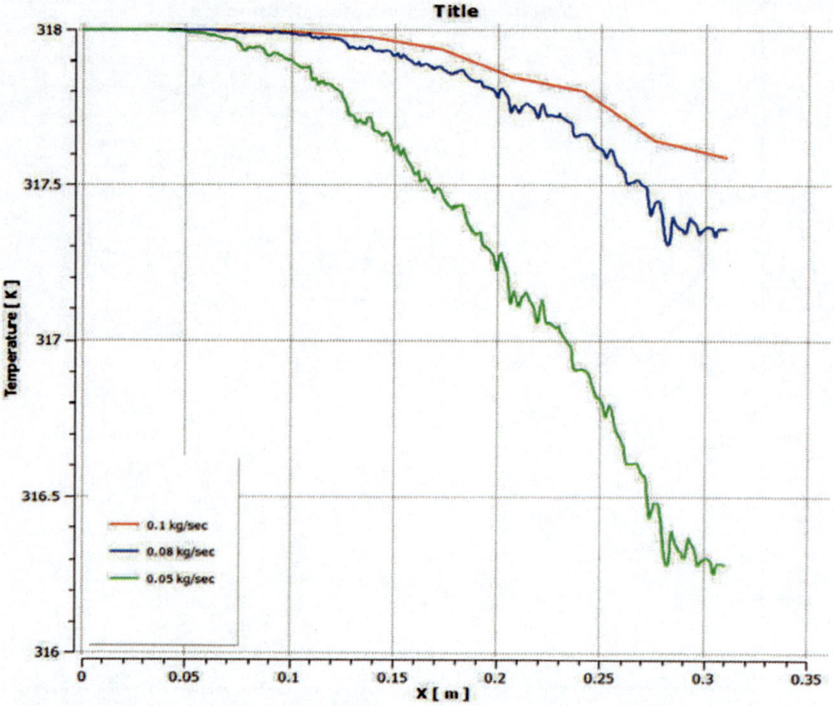

Fig. 40.5 Fluid temperature variation along the length of the tube

(a) (b)

Fig. 40.6 Temperature contour for TiO **a** with concentration of 0.1% **b** with concentration of 0.3%

(iii) Al_2O_3 exhibited least skin friction coefficient and Nusselt number compared to CuO_2 and TiO thereby executing least heat transfer coefficient when used as working fluid in automobile radiators.

From the numerical results of the present work, it can be inferred that water alone in radiators as coolant may be replaced with nanofluids like TiO, Al_2O_3 and CuO_2 that are capable of enhancing the heat transfer rates between the tube and the fluid. Out of these fluids, TiO improves the heat transfer rates to highest extent. These

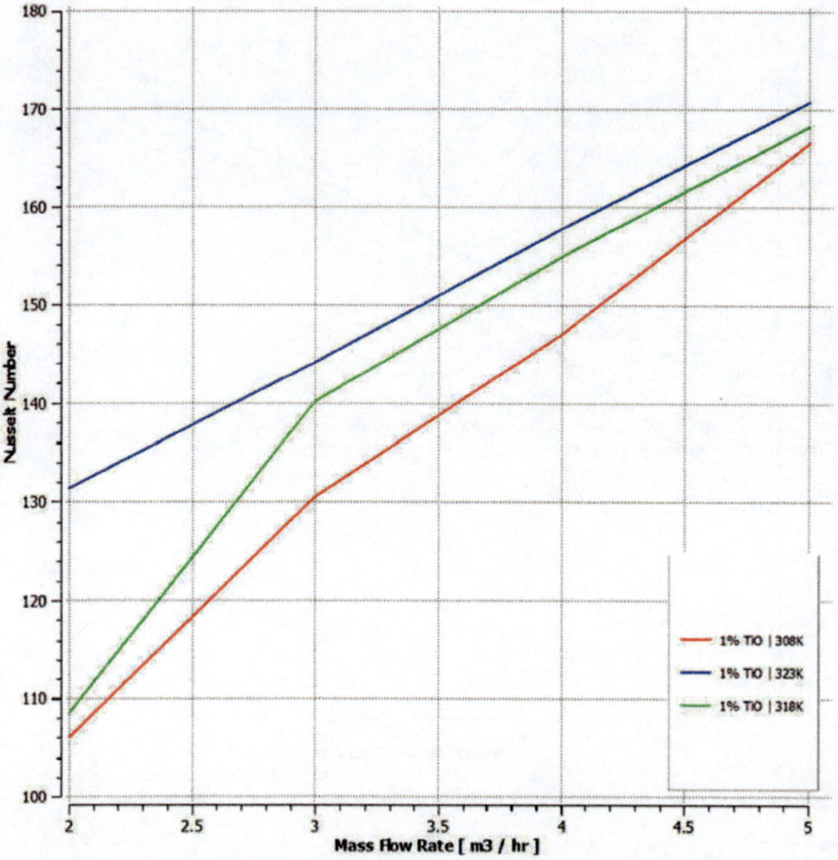

Fig. 40.7 Nusselt number behavior for TiO at different flow rates and temperatures

nanofluids may be suitable as coolants in automobiles used for industrial and residential applications. Inclusion of these nanoparticles to very less extent may enhance the cooling efficiency of the heat exchanging devices used in various applications.

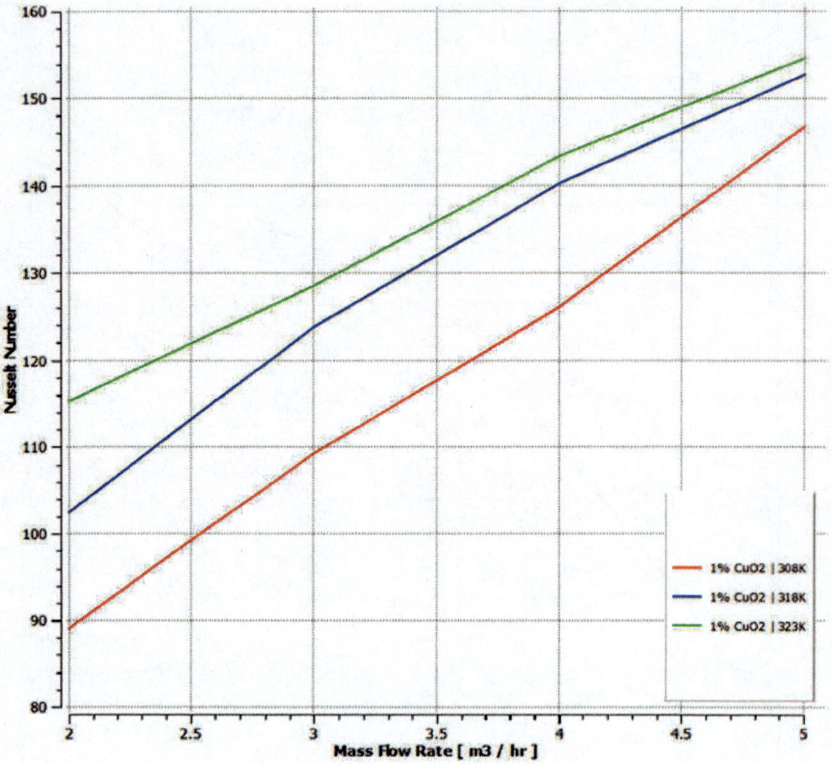

Fig. 40.8 Nusselt number pattern with CuO_2 at different flow rates and temperatures

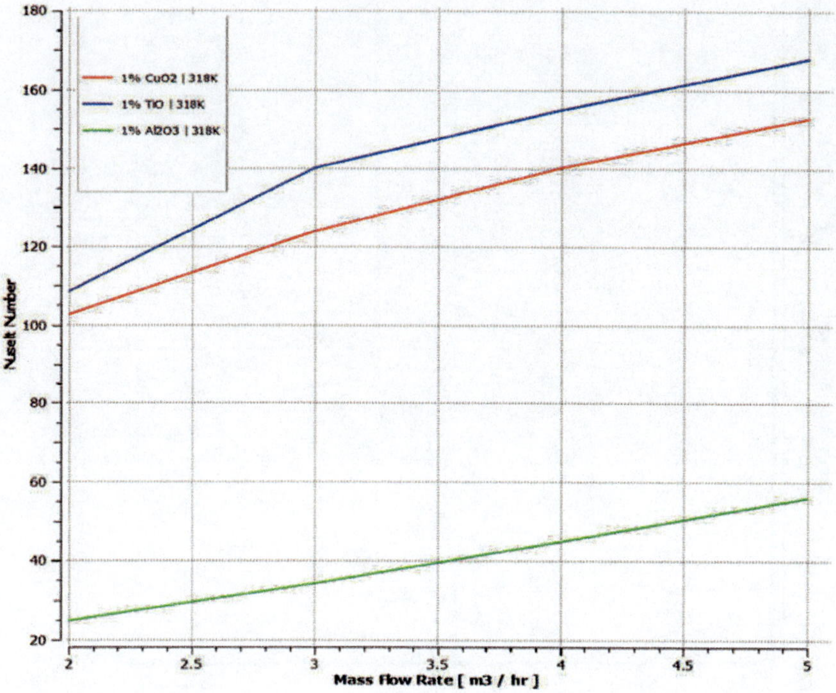

Fig. 40.9 Nusselt number comparison between different fluid at different temperature and mass flow rate

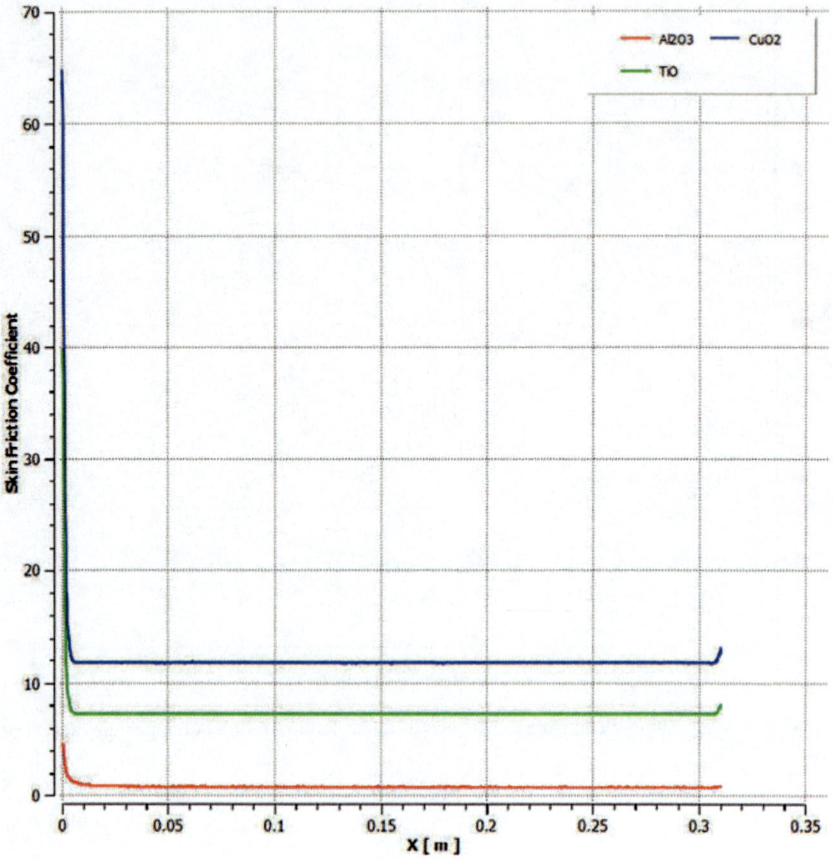

Fig. 40.10 Skin friction coefficient of nanofluids

References

1. Vahid, D., Hashemabadi, S.H.: CFD simulation of heat transfer enhancement of Al_2O_3/water and Al_2O_3/ethylene glycol nanofluids in a car radiator. Appl. Thermal Eng. **73**(1), 380–390 (2014)
2. Shah, S., Kiran Kumar, K.: Experimental study & heat transfer analysis on copper spiral heat exchanger using water based SiO2 nanofluid as coolant. World J. Nano Sci. Eng. **8**, 57–68 (2018)
3. Tijani, A.S., Sudirman, A.S.B.: Thermos-physical properties and heat transfer characteristics of water/anti-freezing and Al2O3/CuO based nanofluid as a coolant for car radiator. Int. J. Heat Mass Transf. **118**, 48–57 (2018)
4. Contreras, E.M.C., Oliveira, G.A., Filho, E.P.B.: Experimental analysis of the thermohydraulic performance of grapheme and silver nanofluids in automotive cooling systems. Int. J. Heat Mass Transf. **132**, 375–387 (2019)
5. Selvam, C., Mohan Lal, D., Harish, S.: Enhanced heat transfer performance of an automobile radiator with graphene based suspensions. Appl. Thermal Eng. **123**, 50–60 (2017)

6. Sahoo, R.R., Ghosh, P., Sarkar, J.: Performance analysis of a louvered fin automotive radiator using hybrid nanofluid as coolant. Heat Transf. Asian Res. (2016)
7. Naraki, M., Peyghambarzadeh, S.M., Hashemabadi, S.H., Vermahmoudi, Y.: Parametric study of overall heat transfer coefficient of CuO/water nanofluids in a car radiator. Int. J. Thermal Sci. **66**, 82–90 (2013)
8. Peyghambarzadeh, S.M., Hashemabadi, S.H., Hoseini, S.M., Seifi Jamnani, M.: Experimental study of heat transfer enhancement using water/ethylene glycol based nanofluids as a new coolant for car radiators. Int. Commun. Heat Mass Transf. **38**, 1283–1290 (2011)
9. Leong, K.Y., Saidur, R., Kazi, S.N., Mamun, A.H.: Performance investigation of an automotive car radiator operated with nanofluid-based coolants (nanofluid as a coolant in a radiator). Appl. Thermal Eng. **30**, 2685–2692 (2010)
10. ANSYS® Academic Research Mechanical, Release 17.2

Chapter 41
Investigation on Automobile Fire and Its Root Causes

Shailendra Singh Chauhan, Aditya Kumar Bhati, Mihir Tomar, Pankaj Kumar Mavi, Siddharth Singh Gurjar, Yash Chauhan, and S. S. Saxena

Abstract This investigation is based on the practical experience reported in the case history, research papers and reported cases of automobile burning at the police and fire stations. Since the fire outbreak in automobiles creates harmful effects to the automobile owner and as well as to other people, it is necessary to detect the fire in the automobile and develop some device for the instant control of the fire.

41.1 Introduction

In the last few years, we all know that thousands of people have died and some got major injuries due to fire accidents in automobiles. According to WHO REPORT ON "ROAD TRAFFIC INJURIES, 2020" [1] and report issued by the Ministry of Road and Transportation, Govt. of India [2] stated that approximately 1.35 million people died in road accident out of which thousands of people became victims of fire accident.

Generally fire takes place in presence of oxygen, heat and fuel, fire is classified into three different classes described as Class A, Class B, Class C. Class A fires are the fires that can be extinguished by water. Class B fires are the fires that are based on oil and Class C fires are based on electricity. In automobiles, the fire outbreak takes place due to Class B and Class C fires [3] and as per fire classes, different kinds of fire extinguishers are invented.

There are various reasons for fire outbreaks in automobiles. Some are due to collision, spillage of fuel, intentionally caused accidents, adopting silly practices in vehicle like smoking, mechanical/electrical causes due to modification in vehicle work by unprofessional technicians which result in short circuit causing heavy avoidable loss in terms of life and material [4–9]. Road Transportation Ministry strictly

S. S. Chauhan (✉) · A. K. Bhati · M. Tomar · P. K. Mavi · S. S. Gurjar · Y. Chauhan
Department of Mechanical Engineering, G.L. Bajaj Institute of Technology & Management, Greater Noida, India

S. S. Saxena
Air Commodore VSM, Auto Wings Training and Consultancy Centre, Greater Noida, India

© The Author(s), under exclusive license to Springer Nature Singapore Pte Ltd. 2022

V. V. Rao et al. (eds.), *Computational and Experimental Methods in Mechanical Engineering*, Smart Innovation, Systems and Technologies 239,
https://doi.org/10.1007/978-981-16-2857-3_42

instructed that modification without permission is illegal but still, this activity is continuously going on [10, 11]. These accidents lead to major losses to the reputation of vehicle manufacturing company and parts suppliers. The customer hesitates to buy the vehicle of that company whose vehicles are catching fire frequently. Fire accident also leads to financial loss to insurance companies.

To reduce fire accident casualties, investigation of this paper is proposed which is beneficial for all citizens, automobile industry, insurance companies as well as the nation. This system is based on microcontroller and can be as additional equipment or compulsory essential kit for all vehicles. The system is intended to keep in mind International Standard and makes it affordable so that every vehicle owner can buy it. The proposed system is equipped with various sensors like smoke, flame, heat, temperature and will detect the fire point in the vehicle and give a warning to the driver to escape from the vehicle as soon as possible and to take precautions.

41.2 General Causes of Fire in Automobiles

The investigation found that Fire in automobiles can take place due to various reasons some of which are listed as the following: [12] (Fig. 41.1).

41.2.1 Overheating Engine

Various reasons cause overheating of the engine; some of them are lack of coolant, blasts due to poor timing or leaking exhaust valves, which can cause missing and

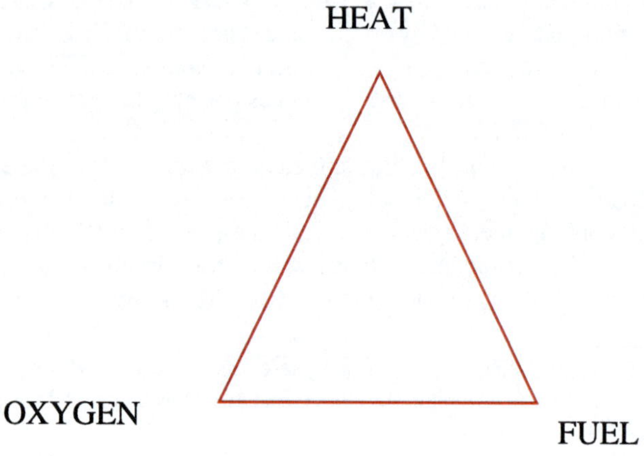

Fig. 41.1 Fire triangle

loud explosions that even burst the air filters. An engine that overheats and causes a car to catch fire is a good example of how one problem can lead to another. A car's engine probably won't overheat enough to simply burst into flames on its ownbut an engine can overheat, making internal fluids, like coolant and oil, and rise to high temperatures and begin to spill out of their designated areas of circulation. Followed by this, they drip, drizzle and spurt throughout the engine bay and onto the exhaust system, landing on other hot parts, where they can easily ignite and spread.

41.2.2 Battery Failure

These are the most common causes of car fires. Car batteries are more prone including normal or hybrid battery packs [13–16]. A typical car's standard battery is capable of causing trouble. The battery's charging cycles can cause the hydrogen gas to explode with the electrical current the battery provides (along with faulty or loose wiring), which can produce sparks that can quickly ignite a fluid drip or leaked vapors. The electrical system's hazards are also present from the engine bonnet to the rear of the vehicle. Electrical wiring runs throughout the entire car; through channels, into doors, under the carpet and through powered and heated seats, just to name a few places where a stray, unnoticed frayed wire could cause the damage.

41.2.3 Leakage in Fuel Line

Leaks in the fuel system are the most common cause of vehicle fires. Number of complicating factors can cause a fuel leak, but they're tricky because fuel leaks can also arise on their own and with very little warning. A fuel system leak is really dangerous. Petrol at a temperature of just 45 °F (7.2 °C) or above can quickly catch fire from a simple spark. Petrol that reaches 495 °F (257.2 °C) will ignite by itself. It's easy to see how fuel dripping onto hot metal and plastic parts can cause a fast-spreading fire. The best way to reduce the chances of a fuel system fire is to make sure the car is properly maintained.

41.2.4 Short Circuit and Wire Damaged by Rats

The damage to the insulation leads to the occurrence of electric failures, which are accompanied by heating of the wire. If the problem cannot be solved fairly quickly, a hole is formed at the point of depression, which causes closure. Such cases of ignition are usually preceded by failures in the vehicle's electronics. Wires bitten by rats also lead to short circuit.

41.2.5 Catalytic Convertor

Overheating catalytic converters are a fire risk that's often overlooked, but think about it: One of the consistently hottest parts of your car runs the entire length of the vehicle is the exhaust system [17–19]. They usually overheat because they are working too hard to burn off more exhaust pollutants than they're designed to process. In other words, if the car's engine isn't operating efficiently (due to worn spark plugs or leaking injectors etc.), it doesn't burn the fuel properly, a lot of extra stuff ends up in the exhaust system. The converter then has to work extra hard to do its job, which makes it even hotter than usual. An overworked (or clogged) catalytic converter can easily go from its normal operating temperature range of about 1,200–1,600 °F (648.9–871.1 °C) to up over 2,000 °F (1,093.3 °C). This causes long-term damage not only to the cat but also to the car's surrounding parts. The car is designed to withstand the cat's normal temps, but it can't consistently cope with temperatures several hundred degrees higher. If the catalytic converter gets hot enough, it could ignite the cabin insulation and carpeting right through the heat shields and metal floor.

41.2.6 Ignition System Fault

The extensive wiring in modern vehicles may be damaged in an accident. A severe collision can sever cables and cause sparks that can ignite fuel. Faulty wiring in ignition systems may short and spark fires. Sometimes mechanical work may result in the ignition of flammable combustibles or an explosion, leading to vehicle fires. The transmission and clutch can overheat and if you have an automatic car and you're holding the brake for too long, it will wear out causing excessive heat.

41.2.7 Smoking (Cigarette)

Smoking is not only harmful to your health but is also one of the reasons for car fire. Ashes from a lit cigarette could burn the upholstery of your vehicle resulting in a car fire. The push-type car cigarette lighter also can lead to a short circuit.

41.2.8 Jammed AC Compressor and Choking Radiator

When an air compressor overheats, the problem is typically related to factors involving irregular suction or discharge pressure [20]. Alternately, the problem could stem from insufficient ventilation, eroded oil or any given number of issues related to the wear and tear of machine parts.

Air compressor overheating issues are often the result of excess discharge pressure, which typically stems from one or more of the following issues:

- Dirty condensing coils
- Ill fitted discharge line
- Blockage of condenser air
- Inconsistent condenser fan
- Undersized condenser
- A failing thermal valve
- Low suction pressure
- Improperly fitted components
- Faulty metering devices
- Loss of refrigerant
- Obstructed strainers
- Pressure drop
- Excess discharge pressure.

41.2.9 Leakage in Fluid Line

The average car or truck has a number of flammable and highly dangerous fluids under the bonnet like petrol or diesel, engine oil, transmission fluid, power steering fluid, brake fluid and even engine coolant [21–24]. All of those fluids are circulating when the car is on, and all of them can catch fire pretty easily if their lines, hoses or reservoirs take the pressure from a hit. So even though one of the car's vital liquids is unlikely to start spewing or dripping out of nowhere generally, something else has to go wrong since all of these fluids are flammable. Combined with another aggravating factor, like a car crash or a failed part, the result could be a fire. Though such a blaze is most likely to start in the engine bay, where all of these dangerous liquids are concentrated, keep in mind that some of them, like the brake oil or petrol, etc., are present in the car from one end to the other while the car is moving.

41.2.10 CNG/LPG Fitting by Un-professional Mechanics

A factory-fitted or a standard CNG/LPG kit comes with an automatic cut-off on the detection of leakage. However, there's no such feature on the cheaper market units and hence, they end up being one of the major car fire causes. In many cases, the leakage of CNG gas is from the damage of pipe connections which is the main cause of fire in the automobile [25, 26]. Also due to traffic collision, CNG components are damaged and such types of fires can be initiated. It is a common scenario in the world that drivers often install aftermarket substandard CNG/LPG kits on their vehicles.

Fitting is mostly done by un-professional mechanics, poor quality cylinders start rusting and corroded after few years. As such, cylinder can't bear the pressure and bursts leading to fire [27, 28].

41.2.11 Collision of Car with Other Solid Rigid Body

Depending on the impact site, a car crash can spark a car fire. Most vehicles' weak points are designed to absorb the impact to prevent it spreading to other explosive parts like battery; and are designed pretty well, so the sheet metal absorbs the force of a blow and protects internal, dangerous spots like the engine, the battery and even the petrol tank [29]. Even then a hard hit is likely to cause fluid leaks and spillage, as well as heat and smoke [30–32]. And, as we know high heat and spilled fluids create perfect conditions for a fire.

41.2.12 Sensors Not Working Properly

When sensors like coolant, fuel, mass air flow, throttle position sensor, cam shaft sensor, etc.) are not working properly, they become major cause of vehicle fire. In 2013 due to fault detected in the battery sensors, Honda recalled 1.2 million Accord midsize cars. The company reported that a 12 V battery sensor was located on the negative battery terminal and it was not properly sealed from the moisture and can get electrical short circuit [33]. It was found that the shorted sensor can heat up and can be the main cause of the fire. Recently, a Chevy volt passenger HEV has reported a fire incident due to the use of a high voltage lithium-ion battery in the car [34]. There is a possibility of fire initiation if the engine compartment temperature and moisture increases. Therefore a battery is one of main sources of fire in the car [34]. Sometimes excessive charging of the battery will also lead to the possibility of fire and must be controlled with the use of a temperature sensor [35, 36]. The role of the sensors used in the battery and engine compartment is to read data. Over a period of time the sensor presents the data in different patterns and depends upon the present condition of the battery and engine compartments. The synthesis of the data must be based on the time and pattern of the data. The temperature rise in the compartment may increase, decrease or may be constant for a period of time so a designer needs to develop an intelligent system which can monitor the change in temperature with the use of sensor [37–40].

41.2.13 Repair/Installation of Add on Equipment by Local Mechanics

It is always advised to never tamper with the electronics on your vehicle which could become a major reason for a short-circuit under the hood. Many incidents have been reported in recent times where people installed extra LED lights, customized horns, etc., on their vehicle resulting in a car fire due to short-circuit caused due to lose connections done to the central electronic system of the car.

41.2.14 Temperature Rise in Cabin Area/Bonnet

In vehicles that are parked, no ventilation and/or air conditioning takes place. If a vehicle is exposed to direct solar radiation, an immediate temperature rise occurs. The high cabin air temperature can threaten children and animals that are left unattended in vehicles [17]. In the USA, lethal heat strokes cause a mean death rate of 37 children per year. In addition, temperature-sensitive goods (e.g. drugs in ambulances and veterinary vehicles, mobile, batteries, UPS) can be adversely affected by high temperature.

41.3 Automobile Accident Due to Fire

Losses during highway vehicle fire accident from 2014 to 2016 are presented in Fig. 41.2, of reported highway vehicle fires and all other fires [41]. The average number of casualty per 1,000 highway vehicle fires was higher than for casualty for all other fires. The average number of injuries per 1,000 highway vehicle fires, as well as the dollar loss per fire, were lower than the same loss measures for all other fires.

In the majority of highway vehicle fires, 83%, occurred in passenger road vehicles, 66% occurred in standard passenger vehicles, 17% occurred in some other type of passenger vehicle, and less than 1% occurred in buses are shown in Fig. 41.3. Only 10% of highway vehicle fires occurred in non-passenger vehicles, 9% occurred in transport vehicles, 1% occurred in other types of vehicles, such as construction vehicles, farm tractors and special purpose vehicles. In the remaining 7% of highway vehicle fires, the type of vehicle was not reported.

In Fig. 41.4 it is easily observed that unintentional actions accounted for 38% of highway vehicle fires. These fires may be the result of either careless behavior or accidental actions. Fires caused by intentional actions constituted only 5% of highway vehicle fires.

Due to the volume of fires, many automobile fires are not investigated. Determining the cause of vehicle fires is often challenging. As a result, in 23% of highway

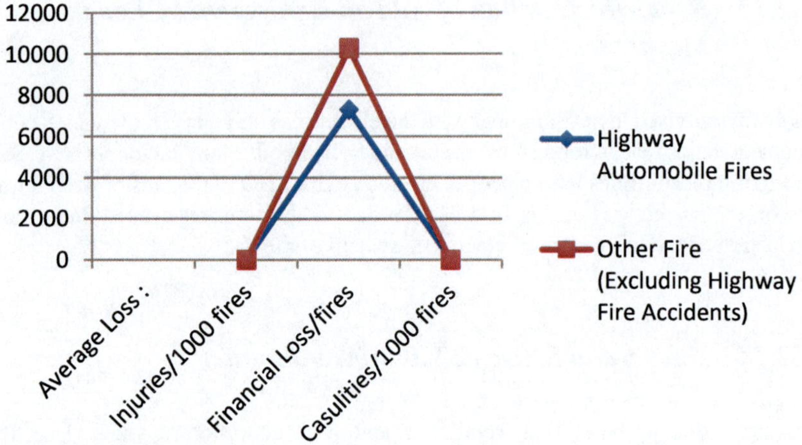

Fig. 41.2 Loss during highway vehicle fire accident (2014–2016)

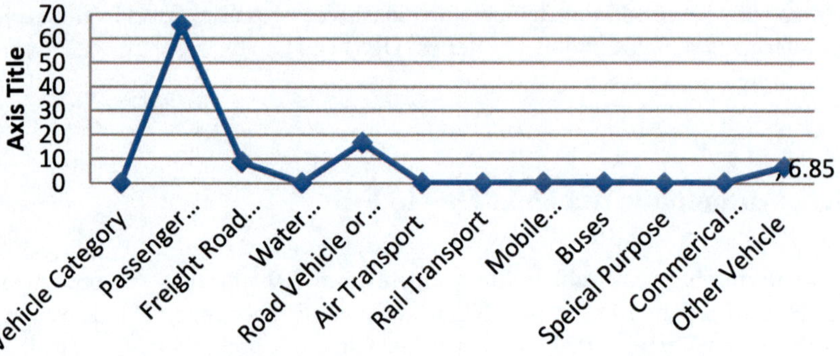

Fig. 41.3 Fire accident in different vehicle class

vehicle fires, no cause was determined after an investigation. These fires are generally a result of mechanical problems, ranging from a faulty design in the vehicle to an improperly installed device.

Figure 41.5 displays the vehicle component area-wise fire accidents, most of the fire originates from the vehicle itself specifically around 61.9% from the engine and wheel. The second most common place is the cabin where fire comes into action it's around 11.9%.

There are various heat sources in the vehicle fire as presented in Fig. 41.6. It was noticed that 65% of highway vehicles get fire due to high ignition temperature and powered equipment used in the vehicle. Whereas other materials like smoking, multiple heat sources and hot objects are fewer regions for the fires.

Figure 41.7 show the fire took place due to the electrical circuits. In most cases, 87% fire originates due to the cable wire causing a great loss. The most second

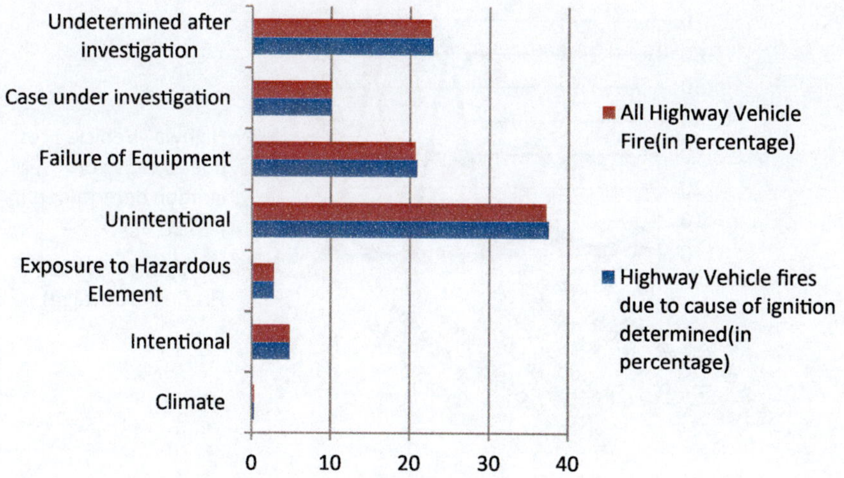

Fig. 41.4 Root causes of fire accident in automobiles

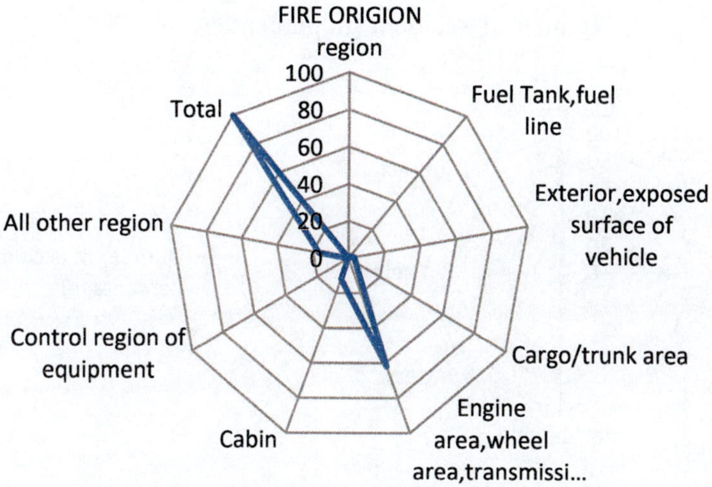

Fig. 41.5 Vehicle components area wise fire accidents origin

common part is the failure or exploding battery in the vehicle. The remaining 8% relates to other causes like the contact of two uncoated wires, etc.

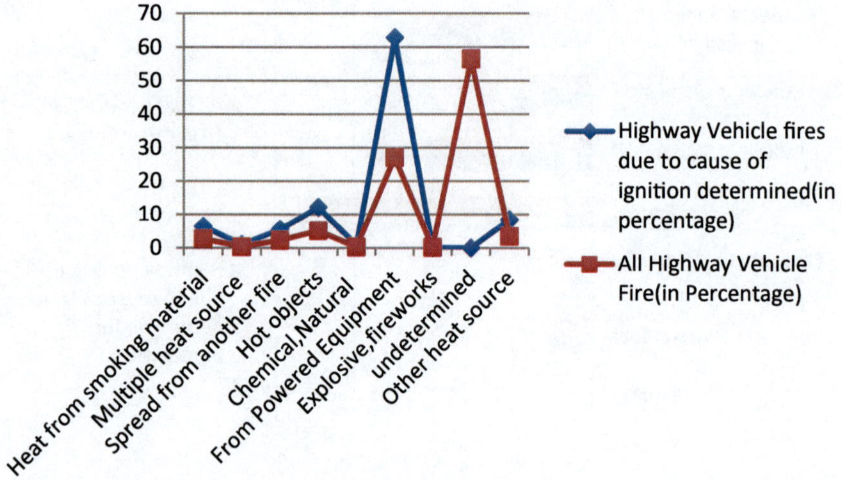

Fig. 41.6 Heat source in vehicle fire (2014–2016)

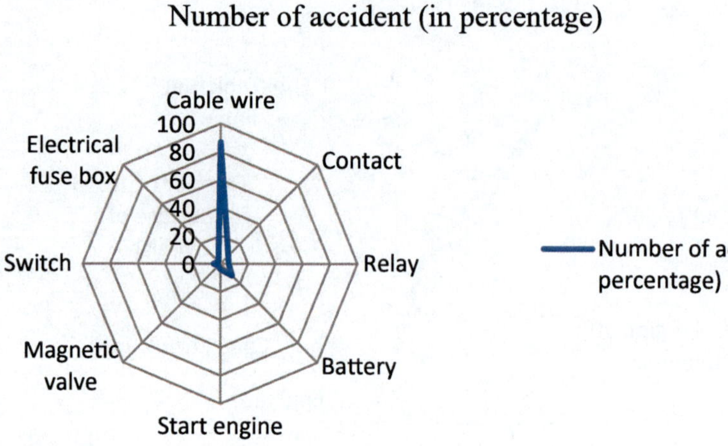

Fig. 41.7 Fire accident reported due to electrical part

Figure 41.8 shows the common surfaces involved at the beginning of vehicle fire. Mechanical failure or malfunctioning of the vehicle leads to the major factor for the fire in vehicles, such as the malfunctioning of the exhaust pipe causes around 24% fires (most common) and the other due to the engine (using low-quality fuel, no proper service) consist of around 16% fires.

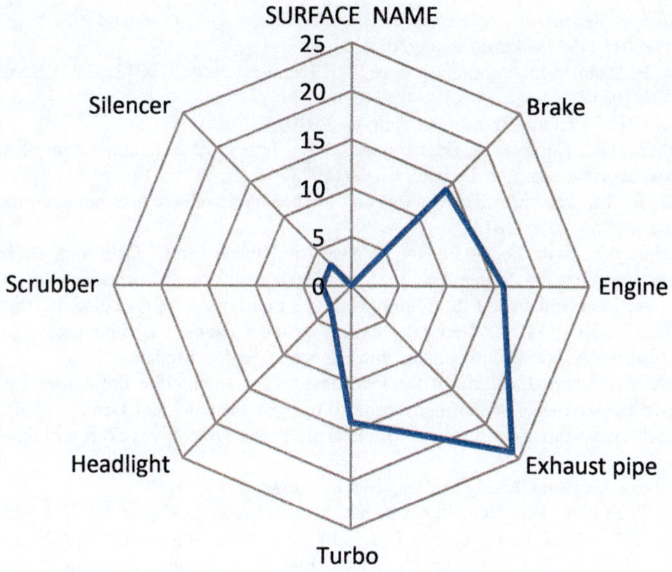

Fig. 41.8 Common surfaces involved in beginning of vehicle fire

41.4 Conclusions and Recommendations

There are some beneficial outcomes of this investigation which will definitely be helpful to the society, insurance company, automobiles industries, etc. This investigation will help the automobiles industries to detect point of the fire so that R&D Department can analyze additional reasons to overcome instances of fire such as the use of high-quality material for that part. From the last few years as we all know that fire accidents are increasing in automobiles. The client tries to claim the insured value of his vehicle according to terms and conditions of the policy and the insurer has to pay the money failing which a legal case can be started against the insurance company. These Fire accidents are causing great loss to the insurance companies also.

References

1. World Health Organization. Road Traffic Injuries. 2020. https://www.who.int/en/newsroom/fact-sheets/detail/road-traffic-injuries Accessed 15 Feb 2020
2. MORTH, Govt. of India "Road Accident in India" (2018). https://morth.nic.in/sites/default/files/Road_Accidednt.pdf
3. Sowah, R., Kwame O.: A fire-detection and control system in automobiles. IEEE Ind. Appl. Mag. (99), XX6–XX6 (2019)
4. Topical Fire Report Series, "Highway Vehicle Accident (2014–16)", vol. 19, Issue 2 (2018)

5. Guenther, D., Goodwin, L., Thaman, R.: Forensic analysis of automobile fires. SAE Technical Paper 810011 (1981). https://doi.org/10.4271/810011
6. Cooley, P.: Motor vehicle non-crash fires. SAE Technical Paper 810012 (1981). https://doi.org/10.4271/810012
7. Marty, A.: U.S. Vehicle Trends and Patterns (2010)
8. Guenther, D.A., Goodwin, L.G., Brininger, F.L.: Temperature determination of a fire from mechanical properties. Fire Technol. 61–66 (1978)
9. Hansen, R.: Investigation on fire causes and fire behavior: vehicle fires in underground mines in Sweden 1988–2010 (2013)
10. McCurnin, J.J.: State Forest Ranger, "Causes of Vehicle Fires," California Department of Forestry, Fire Prevention Notes No. 14, Sacramento, California, September 1979
11. Fire Protection Handbook, T4th Edition, National Fire Protection Association, Boston, Mass
12. Khatoon, S., Kim, M.-H.: Thermal Comfort in the Passenger Compartment Using a 3-D Numerical Analysis and Comparison with Fanger's Comfort Models (2020)
13. Tamura, Y., Suzuki, J.: Study of the investigation for automobile fire caused by electrical system. Characteristics of the melt beads formed by the trucking; Denkikeisharyokasai no hakkagen'inchosanikansurukenkyu. Trucking de keiseisaretanijiyoyukon no tokucho. Japan (1996)
14. Automotive Air Conditioning Training Manual, ariazone (2017)
15. Cooley, P.: Fire in Motor Vehicle Accidents: An HSRI Special Report, HS-014-838 (1974)
16. Chan, C.Y-K., Chi, L.L., Pagni, P.J.: Design of a Fire Proof Vehicle, US-019-085 (1975)
17. Horak, J., Schmerold, I., Wimmer, K., Schauberger, G.: Cabin air temperature of parked vehicles in summer conditions: life-threatening environment for children and pets calculated by a dynamic model. Theor. Appl. Climatol. (2017). https://doi.org/10.1007/s00704-016-1861-3
18. John, H., Kelly, V., Shew, R.: Fuel-Fed Vehicle Fires, HS-022-281 (1978)
19. Fires and fire losses classified 1975; Fire J. (1976)
20. Osei, P., Twum, B.O., Asante, B.: Vehicle Fire Outbreak Detection Communication System (2019). https://doi.org/10.9790/2834-1204041028
21. GRAMKO: Electronic Correspondence with Leif Karlsson (2011)
22. De Rosa, M.I.: Analyses of Mobile Equipment Fires for All US Surface and Underground Coal and Metal/Nonmetal Mining Categories. Industrial Hygienist, Pittsburgh Research Laboratory, National Institute for Occupational Safety and Health, Pittsburgh, PA (2004)
23. Appleby, M.: The Existence or Non-Existence of Safety Related Defects in Vehicles In-Use. Presentation November 4, 1977 at U.S. Department of Transportation Headquarters, Washington, D.C. (1977)
24. Results of the 1973 National Survey of Motor Vehicle Fires. Fire J. **69**(2) (1975)
25. Bahouth, G.: Post-crash exterior crush patterns and motor vehicle fire occurrence. SAE Paper 2006-01-0789 (2006)
26. Perette, L., Wiedemann, H.K.: CNG buses fire safety: learnings from recent accidents in France and Germany. The open archive HAL (2014)
27. Rasche, C.: Modern Composite Flaschen und die Anforderungen fur derensicherenBetriebausSicht der Speicherhersteller. Dynetek Europe GmbH Presentation, Ratingen, Germany (2003)
28. Milojević, S.: Reconstruction of existing city buses on diesel fuel for drive on Hydrogen. J. Appl. Eng. Lett. **1**(1), 16–23 (2016)
29. Jankowski, K.P., Gidlewski, M., Jemioł, L.: Comparative study of vehicle absorbed energy determination for road accident reconstruction. In: XVI EVU Annual Meeting Kraków (2007)
30. Prasad, A.K.: Energy absorbed by vehicle structures in side-impacts. SAE Paper No 910599 (1991)
31. Prasad, A.K.: Energy dissipated in vehicle crush—a study using the repeated test technique. SAE Paper No 900412 (1990)
32. Rose, N.A., Fenton, S.J., Ziernicki, R.M.: Crush and conservation of energy analysis: toward a consistent methodology. SAE Paper 2005-01-1200 (2005)

33. https://www.cnbc.com/2017/07/14/honda-recalls-1-point-2-million-accords-battery-sensors-can-catch-fire.html
34. NHTSA: 2011 chevrolet volt battery fire incident report regarding NHTSA testing and analysis OF subject vehicle HV battery fires. NHTSA Investigation Report (2012)
35. Bauouth, J., Kuznetsov, A.: Automotive and fire safety research. Fire Saf. Libr. Data Anal. 1–97 (2005)
36. Moreno, J., Ortuzar, M.E., Dixon, J.W.: Energy-management system for a hybrid electric vehicle, using ultracapacitors and neural networks. IEEE Trans. Ind. Electron. **53**(2), 614–623 (2006)
37. Zadeh, L.A.: Is there a need for fuzzy logic? Inf. Sci. **178**(13), 2751–2779 (2008)
38. Foo, S.Y.: A rule-based machine vision system for fire detection in aircraft dry bays and engine compartments. Knowl. Based Syst. **9**(8), 531–540 (1996)
39. Foo, S.Y.: A fuzzy logic approach to fire detection in aircraft dry bays and engine compartments. IEEE Trans. Ind. Electron. **47**(5), 1161–1171 (2000)
40. Lee, K.C., Lee, H.H.: Network-based fire-detection system via controller area network for smart home automation. IEEE Trans. Consum. Electron. **50**(4), 1093–1100 (2004)
41. Highway Vehicle Fires (2014–2016) TFRS **19**(2)

Chapter 42
Examination and Analysis of Thermal Steam Boiler Using Power Plants

Pradeep Kumar Singh, Rishabh Chaturvedi, and Manoj Kumar

Abstract The steam evaporator may shut holder where water or different liquids are warmed under tension and the steam discharged by the kettle is utilized for different warming applications. The fundamental contemplations in planning a kettle for a specific application are structure and warm investigation, structure for creation, physical measurements and cost. In this postulation, the progression of steam in the steam kettle (without diverters and with redirectors) is displayed utilizing the CREO parametric plan programming. The postulation will concentrate on warm and CFD examination with various info speeds (20, 30, 40 and 50 m/s). In this theory, CFD investigation decides the warmth move coefficient, the warmth move rate, the figure stream rate and the weight drop. Warm inspection to choose the temperature movement, the warmth stream for the two models of steam kettle without diverters and steam evaporator with redirectors. Discover which model is the best. 3D displaying in the parametric CREO programming and dissects acted in ANSYS.

42.1 Introduction

Boilers are pressure Vessels proposed to warm water alternately make steam, which might in this route a chance to be used will provide for space warming or water warming should a structure. Over the benefits of the business building warming applications, those warming sourball in the evaporator may be a vaporous petrol copier. It will be moreover possible to use oil burners what's more electric block warmers. Steam over warmed water will be favoring ensure applications, including

P. K. Singh · R. Chaturvedi (✉)
IET Department of Mechanical Engineering, GLA University, Mathura, Uttar Pradesh, India
e-mail: risabh.chaturvedi@gla.ac.in

P. K. Singh
e-mail: pradeep.kumar@gla.ac.in

M. Kumar (✉)
Greater Noida Institute of Technology, Noida, Uttar Pradesh, India

© The Author(s), under exclusive license to Springer Nature Singapore Pte Ltd. 2022

V. V. Rao et al. (eds.), *Computational and Experimental Methods in Mechanical Engineering*, Smart Innovation, Systems and Technologies 239,
https://doi.org/10.1007/978-981-16-2857-3_43

maintenance cooling, kitchens, laundries, sterilizers, and steam machines [1]. Boilers bring different qualities that bring them an ordinary part of structures. They have a long life, could finish efficiencies of up to 95% or more, provide for an effective techno babble to warming a structure and because from claiming steam frameworks, require practically no siphoning vitality. For any case, fuel liabilities can make impressive, standard backing, what's more assuming that upkeeps may be postponed, fixing it can be a chance to be an exorbitant.

Those support systems to the development, movement and more upkeep about boilers convey the going with assets: Tenets for the advancement about warming boilers, evaporator code. Furthermore weight vessels, segment IV-2007, Prescribed models to that attention, and also action of warming boilers, evaporator code, weight vessel, and segment VII-2007. Boilers are every now and again a standout amongst those best vitality customers done a structure. For consistently a warmer schema may be left unattended, pot costs can augment toward around 10% (1). Along these lines, evaporator movement is of a more help that will be a OK for the start stage The point when seeking to methodologies to diminish vitality usage and put aside money [2, 3].

How boilers work:

Gas and oil boilers utilize controlled fuel burning to warm water. The key parts of the kettle engaged with this procedure are the burner, the ignition chamber, the warmth exchanger and the controls.

42.2 Types of Boilers

Boilers are grouped into various sorts dependent on working weight and temperature, kind of fuel, suction strategy, size and limit and whether they consolidate water fume into the exhaust. Boilers are in some cases additionally portrayed by their key parts, for example, heat exchanger materials or cylinder structure [4]. These different capacities are talked about in the accompanying segment on key heater parts.

Two fundamental kinds of boilers incorporate fire-tube and water-tube boilers. In a fire-tube heater, hot burning gases course through a progression of funnels encompassed by water, in Water tube evaporator.

42.2.1 Key Segments of Boilers

The key components of a kettle incorporate the burner, the ignition chamber, the warmth exchanger, the release stack and the controls. Heater fittings, including the vent gas economizer, are likewise usually utilized as a powerful technique for recuperating heat from an evaporator and will be talked about quickly in the best practices area for proficient activity [5].

Petroleum gas containers using one of two sorts of flames, barometrical burners, moreover named common draft flames and inhibited draft rings known as electric burners. Since of the most severe government and state air excellence models, low NOx flames and premix flames are progressively utilized and even required in certain territories. By guaranteeing a productive blend of air and fuel when it enters the burner, this sort of burner can guarantee that NOx emanations are diminished.

42.2.2 General Majority of the Data

Boilers would use to produce steam that that point gives high temperature or force. Water may be changed over with steam in the heater. This steam goes through the warming mechanical assembly which might make any bit from claiming supplies that obliges steam to operation. Those cooled steam will be after that condensed under water returns of the heater should start the cycle once more. Supplies plan. There would three standard sorts for steam boilers: shoot hose, water pipe, furthermore cast iron by fire-tube boilers. Ignition vapors make a trip inside the cylinders to warm the encompassing water.

In water-tube boilers, be that as it may, water streams inside the channels and warms outside, as appeared previously. Cast iron reservoirs are like water-tube reservoirs, yet water is controlled in the cast iron segments instead of the channels. The accompanying outline demonstrates the segments of a fire-tube evaporator.

- **Advantages**

Water and fuel, the two main nourishments for this from the heater, are rich and modest.

Intelligent to deal with huge limits and high weight.

- **Disadvantages**

May consume throughout stoppage stages.

Water essential be blessed to receive anticipate amassing.

Here is constantly a warmth is fortune.

Finite Element Analysis of Condensation Boiler Used In Power Plants Limited component investigation of the steam kettle is utilized in 1 M control plants [6]. A heater or steam generator is a shut compartment that is utilized to produce steam by applying warm vitality to the water. Throughout the steam age process, the steam heater is exposed to huge warm and auxiliary burdens. To accomplish the productive activity of the plant, it is important to structure a structure to help these warm and basic burdens. The utilization of CAD and CAE software design is the driven plan technique of these structures before building a model. In this task, an inspection of the limited components of the steam kettle was performed to approve the undertaking for the genuine working conditions. The fundamental exercises engaged with the task are the execution of the 3D displaying of the boilers and the investigation of the completed components [7, 8]. In this task, the advancement of the structure of the kettle is likewise performed based on the outcomes acquired from the warm and auxiliary examination. The CAD NX programming is utilized for 3D planning and displaying. ANSYS programming is utilized to perform limited component investigation [9].

The steam kettle is a shut holder wherein water or different liquids are warmed under tension and the steam discharged by the evaporator is utilized for different warming applications. The fundamental contemplations in structuring a kettle for a specific application are plan and warm examination, structure for generation, physical measurements and cost. In this work, a fire-tube evaporator is examined to decide the static and warm burden. The geometric model of the kettle is made in the CATIA V5 programming as per the drawing [10]. This model is brought into HYPERMESH through the IGES position and the FEA model with merged work is created utilizing shell components. Different burden conditions apply to this FEA model, for example, plan pressure, warm loads and working conditions. One of the help legs stops every which way and different stops just in the X, Z and all revolutions bearings. This is made utilizing HYPERMESH and traded to ANSYS to acquire an answer for getting deviations, pressures [11, 12].

The thermodynamics of fare cogeneration necessitates that higher HP steam conditions produce greater power delivered, which has come out on top to arrive at increasingly elevated weights and temperatures without thinking about the economy. By and by, as conditions increment, the expense of extra capital exceeds extra power. HP's higher conditions likewise drove a subsequent pattern: the reception of single drum boilers as opposed to proceeding to utilize two drum plans. Despite the fact that this is the correct methodology in raised conditions, care ought to be taken when choosing configuration subtleties. A Study Analysis and Performance of High-Pressure Boilers with its Accessories. An examination and investigation of the presentation of high-weight boilers with their frill India Power is the fundamental and basic contribution for advancement. In this cutting-edge situation, vitality assumes a crucial job, both in modern improvement, which thus prompts flourishing age structures created in AP to satisfy the emerging need for vitality. Utilizing adornments in the evaporator. Plant proficiency increments. For instance, frill, for example, the economizer increments the stock water temperature, while the superheater builds the temperature of the steam delivered in the evaporator. The air preheater builds the temperature of the approaching air, which enters the broiler. The primary target of this undertaking work is to break down the proficiency of the economizer, superheater and air preheater by changing the diverse parameters in the evaporator area, planning and examination of the kettle model for controlling the steam pressure.

To accomplish the vitality proficient activity of the power plant, it is important to adequately control the steam weight. In this manner, an exertion is made in this record to controller such a basic parameter, in particular the fume pressure, by building up a heater unit model that uses a PLC-based PID controller that uses the IMC method to change the PID parameters. The work exhibited additionally incorporates process displaying and reproduction was performed with the proper exchange work utilizing the propelled input control system. Other pragmatic and hypothetical answers were thought about. Open circle approval was likewise performed to approve the model.

42.3 Problematic Explanation

The details of this undertaking is to variety a 3D model of the steam heater and concentrates the CFD and warm conduct of the steam kettle through limited component investigation, 3D demonstrating programming.

Master Engineer was utilized to plan and examine, programming (ANSYS) was utilized for CFD and warm investigation.

The strategy surveyed in the undertaking is the accompanying:

Create a 3D model of the steam evaporator bunch utilizing the parametric expert designer programming.

Convert the surface model into a strong record for and significance the model into ANSYS for examination.

Perform warm investigation on the steam kettle bunch for warm loads.

Carry out a CFD examination on the current surface steam heater model for speed section to realize the mass stream rate, the warmth move rate and the weight drop.

42.4 Modeling and Analysis

Models of steam boiler using CREO.

The steam evaporator is demonstrated utilizing the particulars and plan equation of the information book. The isometric perspective on the steam heater appears in the accompanying figure. The profile of the body of the external packaging of the steam heater is attracted a visual artist and afterward turned by a point of 3600 utilizing the upset choice and the funnels are structured and amassed in the steam kettle utilizing the expulsion choice (Fig. 42.1).

Steam boiler 3D model.

Steam boiler 2D model.

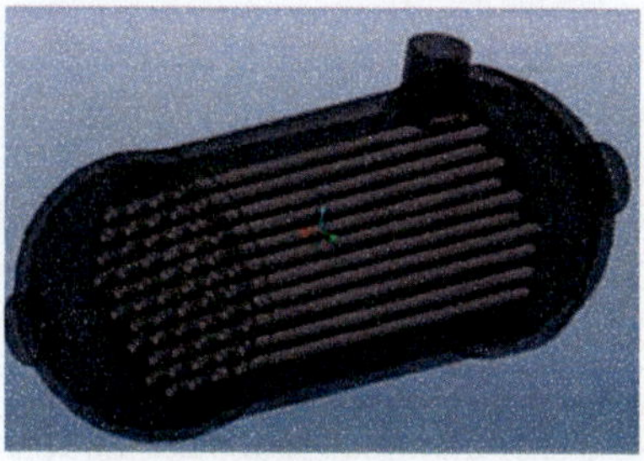

Fig. 42.1 Steam boiler surface model

Fig. 42.2 Heater exchanger coil boiler

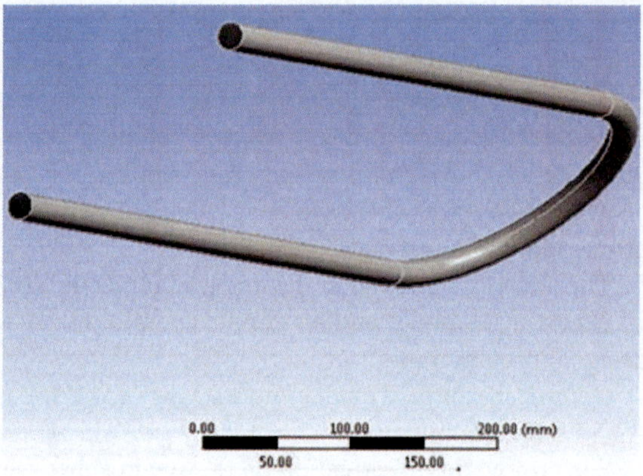

42.4.1 CFD Analysis of Heater Exchanger Coil Boiler

Mass flow rate – 0.006, 0.008 & 0.015 kg/s velocity 20, 30, 40 & 50 m/s.

FLUID—steam (Fig. 42.2)

$$\rightarrow \text{Ansys} \rightarrow \text{workbench} \rightarrow \text{select analysis system}$$
$$\rightarrow \text{fluid flow fluent} \rightarrow \text{double click}$$

$$\rightarrow \text{Select geometry} \rightarrow \text{right click} \rightarrow \text{import geometry}$$
$$\rightarrow \text{select browse} \rightarrow \text{open part} \rightarrow \text{ok}$$

$$\rightarrow \text{select mesh on work bench} \rightarrow \text{right click edit}$$
$$\rightarrow \text{select mesh on left side part tree}$$
$$\rightarrow \text{right click} \rightarrow \text{generate mesh} \rightarrow \text{MASS FLOW RATE} - 0.02\,\text{kg}/.$$

42.4.2 Thermal Analysis of Steam Boiler Heat Exchanger

Exposed workbench 14.5 > choice stable thermal in analysis systems > select geometry > right click on geometry > import geometry > select IGES file > open.

Materials used aluminum, 6061 aluminum alloy and copper Properties of the copper material.

$$\text{Thermal conductivity} = 385\ \text{w/m - k Specific heat} = 0.385\ \text{j/g}\ ^\circ\text{C}$$
$$\text{Density} = 0.00000776\ \text{kg/mm}^3$$

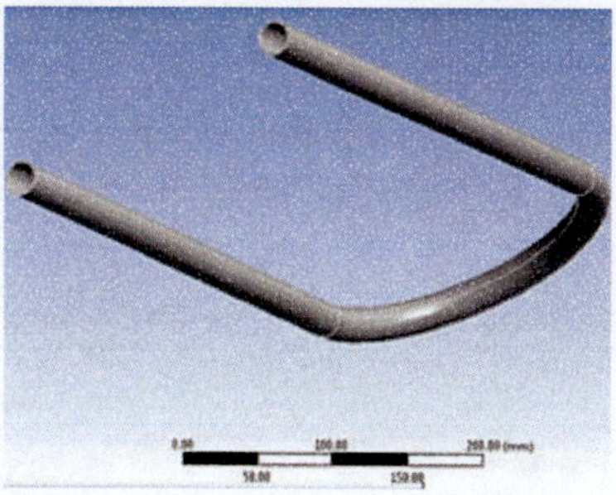

Fig. 42.3 Stream boiler

Properties of the aluminum material.

210 w/m - k Specific heat = 0.9000 j/g °C Density = 0.0000026989 kg/mm^3

Properties of the 6061-aluminum alloy material Thermal conductivity = 180 w/m-k

Specific heat = 0.896 j/g °C Density = 0.00000270 kg/mm^3.

42.4.3 *Introduced Model*

See Fig. 42.3.

42.5 Results and Analysis

See Tables 42.1 and 42.2, Fig. 42.4.

Table 42.1 CFD analysis

Mass flow rates (kg/s)	Velocity (m/s)	Pressure (Pa)	Velocity (m/s)	Mass flow rate (kg/s)	Heat transfer rate (W)
0.006	20	2.98e+00	2.43e−02	6.65e−06	2.0788574
0.008	30	4.21e+00	3.14e−02	4.69e−06	1.4709473
0.015	40	8.06e+00	5.59e−02	3.08e−06	0.9536132
0.02	50	1.13e+01	7.29e−02	1.70e−05	5.3261719

Table 42.2 Thermal analysis

Material	Temperature (°C)		Heat
	Max	Min	Flux (w/mm^2)
Aluminum	100	89.189	1.8845
Aluminum alloy 60	100	90.041	1.9072
Copper	100	95.191	2.0515

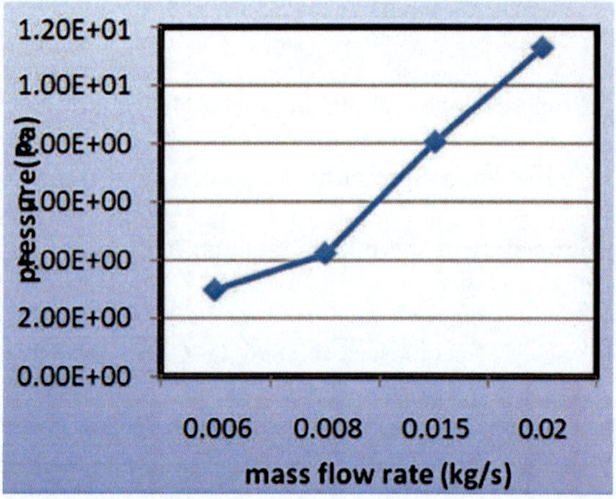

Fig. 42.4 Graphs pressure plot

42.6 Conclusion

In this project, the flow of steam in the heater pipes of the steam boiler is demonstrated by means of PRO-E design software. The proposition is to emphasis on current and CFD examination by dissimilar flow rates (0.006, 0.008, 0.015 and 0.02 m/s) and different speeds. When CFD analysis is observed, the weight drop, speed, mass flow rate and heat transfer rate increase with cumulative inlet flow rates. When thermal analysis is observed, the heat flux value is higher for copper material than for aluminum and 6061 aluminum alloy. It is achieved that the copper substantial is improved for the steam boiler.

References

1. Arai, N., Taniguchi, H., Mouri, K., Nakahara, T.: Exergy analysis on combustion and energy conversion processes. Energy **30**(2–4), 111–117 (2005)
2. Khaliq, A., Kaushik, S.C.: Thermodynamic performance evaluation of combustion gas turbine cogeneration system with reheat. Appl. Therm. Eng. **24**(13), 1785–1795 (2004)
3. Singh, P.K. et al.: Effect of sonication parameters on mechanical properties of in-situ amine functionalized multiple layer graphene/epoxy nanocomposites. J. Sci. Ind. Res. **79**, 985–989

(2020)
4. Datta, A., Ganguly, R., Sarkar, L.: Energy and exergy analyses of an externally fired gas turbine (EFGT) cycle integrated with biomass gasifier for distributed power generation. Energy **35**(1), 341–350 (2010)
5. Koroneos, C.J., Fokaides, P.A., Christoforou, E.A.: Exergy analysis of a 300 MW lignite thermoelectric power plant. Energy **75**, 304–311 (2014)
6. Sharma, A. et al.: Optimization techniques to optimize the milling operation with different parameters for composite of AA 3105. Mater. Today Proc. (2021)
7. Bhoursae, A., Shah, J., Bhatt, N.: Plan and investigation of the model kettle for the control of the steam pressure
8. Lou Roussinos, P.E.: Evaporator design and efficiency [online]
9. Shukla, M.K., Sharma, K.: Effect of functionalized graphene/CNT ratio on the synergetic enhancement of mechanical and thermal properties of epoxy hybrid composite. Mater. Res. Express **6**(8), 085318 (2019)
10. Murdock, K.L.: Book of scriptures 3ds max 9. Wiley Publishing Inc., Indianapolis, Indiana (2007)
11. Yadav, A., et al.: Investigating the effects of amine functionalized graphene on the mechanical properties of epoxy nanocomposites. Mater. Today Proc. **11**, 837–842 (2019)
12. Tiwari, A.K., Hasan, M.M., Islam, M.: Exergy analysis of combined cycle power plant: NTPC Dadri, India. Int. J. Thermodyn. **16**(1), 36–42 (2012)

Chapter 43
Study on the Development and Problems Faced in Electric Vehicles

Ankit Sharma, Dhruv Sethi, Ishu Kumar, Jatin Yadav, Siddhant Bhatia, and Amardeep

Abstract In vehicles, zero-emission production and wheels propelled with their own battery energy is possible only through electric vehicles. The long-term maintenance cost is very low as well as the lower center of gravity gives better cornering and turning stability. These advantages have influenced and have made electric vehicles as a new generation transport both in India and internationally in the automobile engineering sector. From the time EVs came into picture, there have been a lot of modifications in their design and some major components like battery pack and motor. In spite of these advancements and advantages, EVs are still not much popular in the automobile market, if we look at their growth, there is an increase but it is still less as compared to S.I and C.I engine automobiles. If we look at the reason behind that, there are some major setbacks in EVs like charging time, mileage, support of electrical components like headlights and air conditioners while driving. It's very essential to look at the issues in EVs and their causes and to find out the ways how to overcome them with few modifications keeping the pocket friendly cost in mind. A number of companies all around the globe are rigorously working on improving the performance of EVs, India is also actively participating in promoting and providing a better platform for EVs. As everyone knows those EVs are the future of automobiles and hence their development is a necessity. So, it will be very intriguing to see the future of EVs in India and around the globe.

Abbreviations

EV	Electric vehicle
U.S.	United states
S.I.	Spark ignition
DC	Direct current
C.I	Compressed ignition

A. Sharma · D. Sethi (✉) · I. Kumar · J. Yadav · S. Bhatia · Amardeep
Department of Mechanical Engineering, GL Bajaj Institute of Technology and Management, Greater Noida 201306, India

© The Author(s), under exclusive license to Springer Nature Singapore Pte Ltd. 2022

V. V. Rao et al. (eds.), *Computational and Experimental Methods in Mechanical Engineering*, Smart Innovation, Systems and Technologies 239,
https://doi.org/10.1007/978-981-16-2857-3_44

AC	Alternate current
PEV	Plug in electric vehicle
PHEV	Plug in hybrid electric vehicle
BEV	Battery electric vehicle
HEV	Hybrid electric vehicle
EREV	Extended range electric vehicle
BMW	Bavarian motor works
UK	United Kingdoms
EPCO	Electric public charge office
OEM	Original equipment manufacturer
INR	Indian national rupee
PMP	Phased produced programme

43.1 Introduction

EVs were invented during the mid-nineteenth century, at that time electricity was a popular strategy for automobile propulsion, hydrocarbon cars of that point couldn't reach the amount of comfort and easy operation. IC engines were the dominant propulsion methodology for vehicles for nearly a century; however power has remained constant in different types of vehicles. It is hard to give all the credit of the invention of the electric vehicle to a minimum of one inventor or country. Instead, it is a combination of various events from battery to electric motor which led to the success of the first electric vehicle. At the starting of the twenty-first century metal–oxide–semiconductor (MOS) technology was used in the newest electrical road vehicles. MOSFET (MOS field-effect transistor) power converters allowed operation at very high frequencies, making it easier to drive and as well as reducing power losses, resulting in reduction of cost, whereas single-chip microcontrollers were managing all aspects of the drive management and were capable of battery management too. In 2010, the electric vehicle market achieved the mark by delivering a combined one thousand units of all kinds of electric vehicles which were delivered globally. This reached 3.3 million units by Dec 2018. The global quantitative relation between annual sales of battery electrical cars and plug-in hybrids went from 56:44 in 2012 to 74:26 in 2019. As of March 2020, the Tesla Model three is the world's nonpareil best marketing plug-in electric car, with over 500,000 units [35, 36].

Approximately 2/3 of deaths from pollution on land are caused due to exhaust emissions from diesel vehicles, nearly about 385,000 people lost theirs in 2015, according to a study. On-road diesel vehicles were answerable for nearly 0.5% of health damage to people worldwide in 2015. As several as fourteen of the world's twenty most-polluted cities are in India, as mentiones in 2018 World Health Organization (WHO) report. These stats are horrific and keeping them in mind. In May 2017, India was the first to announce plans to sell solely electrical vehicles by 2030. The advantages of electric vehicles can be easily seen [1].

43.2 The Key Components in EV

There are many parts of an EV as shown in Fig. 43.1 and they are discussed below:

Traction Battery Pack: The battery's operation in an electrical automotive is as associate in nursing electricity arcade system within the type of direct-current electricity (DC). If it gets indicated by the controller, the battery can flow DC electricity to the electrical converter to then be accustomed to drive the motor. The foremost widely used is that of Li-ion batteries.

Power Inverter: The electrical converter functions to vary the DC on the battery into Associate in Nursing electricity (AC), then this electricity is employed by an electrical motor. Additionally, the electrical converter on an electrical automotive, vary the alternating current after regenerative braking to DC current then battery is recharged.

Controller: This unit is used to control the flow of electricity carried by the traction battery, dominating the velocity of the electrical traction motor and also the torsion it generates. It controls working of electric vehicle. The controller can do alternative things too—it moderates the facility and it will also act as a convertor—where it converts power from DC to AC [2].

Electric Traction Motor: It is the device that draws power from the traction battery pack and this motor propels the vehicle's wheels. Some hybrid electrical cars use a kind of generator-motor that performs the functions of propulsion and regeneration.

Charger: It is a device used to charge the battery. Chargers take electricity from external sources, like the utility grid or alternative energy plants. AC electricity is regenerated into DC electricity then held on within the battery.

Transmission: The transmission uses the mechanical power generated by the electrical traction motor to rotate the wheels.

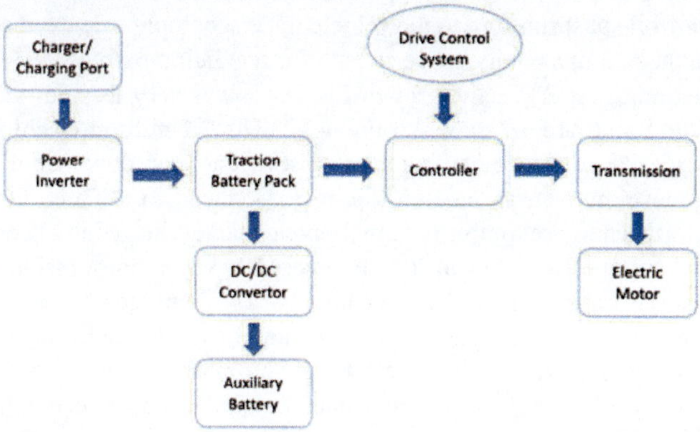

Fig. 43.1 Global electric car stock, 2010–2019 [10]

DC/DC Converter: This one amongst the electrical automotive components that convert higher-voltage DC power produced by the traction battery pack to the lower-voltage DC power required to run vehicle accessories and recharge the subsidiary battery.

Battery (auxiliary): In an electrical drive vehicle, the auxiliary battery is required to provide electricity to power vehicle accessories.

Thermal System-Cooling: This system helps to maintain a correct operation temperature of the engine, motor, power physical science and alternative elements.

Charge Port: The charge port permits the vehicle to attach to an Associate in nursing superficial power offered so as to charge the traction battery pack [3–7].

43.3 Growth in EV

In the wake of coming into modern business sectors inside the most recent 5 years, car deals have continued. In 2010 exclusively concerning seventeen electrical vehicles were on the streets. By 2019, that reach had gotten vainglorious to 7.2 million, forty seventh of that were inside the People's Republic of ("China"). Nine nations had a serious hundred electrical vehicles out and about. At least twenty nations arrived at pieces of the overall industry on top of I Chronicles. Deals of electrical vehicles lidded two. 1 million universally in 2019, outperforming in 2018—effectively a record year—to zest up the stock to seven. 2 million electrical cars. 1 electrical vehicle, that represented two. 6% of overall auto deals and concerning I Chronicles of overall car stock in 2019, enlisted a four-hundredth year-on-year increment. In 2019, indication of a strong move from direct arrangement to moral quality proceed toward that acknowledge a lot of on restrictive and different essential measures-similarly as zero-dispatch vehicles orders and effectiveness standards have set clear, since quite a while past run sign to the vehicle trade accomplice degreed clients that help the change in a financially property path for regulating party [8, 9].

In the beginning of 2010, the growth EV stock was very less but with time it increased with rapid rate, as shown in Fig. 43.2. The 2.1 million machine deals in 2019 addresses 6% advancement from the most recent year, down from numerous years' deals development, at least on top of 30% since 2016. Three fundamental reasons put forth a defense for this pattern. Vehicle markets had contracted. Absolute vehicle deals' volumes were low in 2019 for every few key nations. Inside the 2010s, intrusive business sectors like India and China for a wide range of vehicles had less deals in 2019 than in 2018. Against this backcloth of moderate deals in 2019, the 2.6% piece of the overall industry of electrical automotive in overall vehicle deals set up a record. In 2019 China (at 4.9%) and Europe (at 3.5%) accomplished new records in electrical vehicle piece of the pie. Buy sponsorships were decreased in key business sectors. China cut machine buys allocation by concerning half in 2019 (as a piece of a progressive finish of direct support started in 2016). With 90th of overall machine deals centered in China, Europe and thus the U.S., this influenced

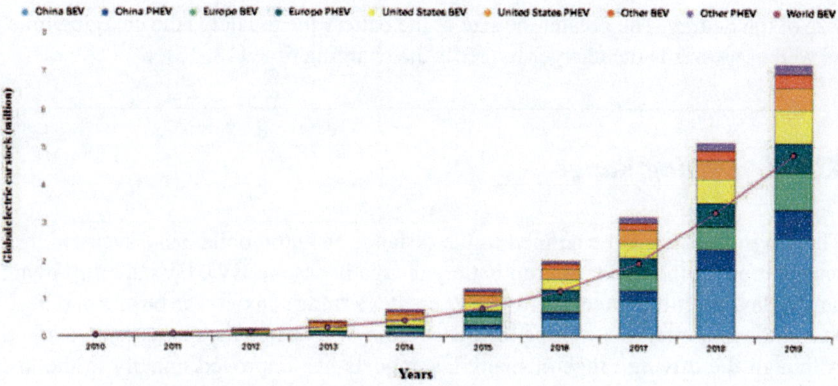

Fig. 43.2 Global electric car stock, 2010–2019 [10]

worldwide deals and over-shadowed the remarkable five hundredth deals expansion in Europe in 2019, accordingly fasting the development fever.

The 2018–19 renditions of some basic machine models show battery energy thickness that is 20–100% past their partners in 2012. Further, battery costs have contracted by more than eighty-fifth since 2010. The conveyance of most recent mass-market models like the Tesla Model three caused a spike in deals in 2018 in key business sectors like the U.S. Automakers have proclaimed a different menu of electrical vehicles, a few of that are normal in 2020 or 2021. For progressive 5 years, automakers have announced designs to release another 200 new machine models, a few of which are inside the chic machine vehicle market stage [11, 12].

43.4 Major Setbacks and Their Causes

The problems of EV and their causes which are barriers in their growth are as follows:

43.4.1 Charging Time

The total time taken by EV to fully charge an EV is called charging time. Fueling up our petrol or diesel car takes a few minutes, but charging an EV takes hours. Our car should be at rest and plugged into charging for the charging time [14]. The charging time of an electric car can be as less as half an hour or can be longer than a half-day. This relies on the size and charging speed of the battery.

A normal EV (60kWh battery) takes less than 8 h to completely charge with a 7 kW charger. For many EVs we can travel up to 161 km in approx. 35 min of charging time with a 50 kW fast charger. The charging time is proportional to the size of the battery. The greater the size of the battery the greater is the charging time, as well as slower is the charger longer is the charging time [13, 15].

43.4.2 Driving Range

The driving range can be defined as the distance an automobile can travel with full residual fuel in the fuel tank (full battery charge in case of EV). EVs are improving day by day, but still limited driving range creates a "range anxiety" in buyers making it one of the major hurdles halting the penetration of EVs into the mainstream market. Although the driving range of many EV models has improved notably in the last few years, a limited driving range does give a challenge to many drivers [16]. A cross-country ride in an EV would require some inconvenient stops to recharge them because of short driving range [17].

43.4.3 Charging Infrastructure

The basic structures and facilities needed for charging EVs are known as charging Infrastructure or charging stations. This problem varies from location to location. If you don't travel long distances and can charge your EV at home or office, then it might not be a problem for you. But in many cities or apartment residents, unfortunately, are not able to charge at home due to a lack of a garage. They must rely on public charging stations or. Even if vehicle charging stations are conveniently located, they might be occupied because of long charging time. The EV charging station varies from area to area, but this is an important point to consider before buying an EV [13].

43.4.4 High Upfront Cost

Electric vehicles are more normal in worldwide streets than any time in recent memory but, stay out of financial plan for many. The cost of a battery inside EV frames the significant lump of the final cost of the actual vehicle.

Fortunately, electric vehicles are descending in expense when contrasted with their gas-controlled counterparts. Additionally, tax breaks and state impetuous can remove a lump from the complete cost [18]. At stream rates and pack assesses, the typical battery cost for an ordinary electric vehicle is about $7,350, that has plunged a great deal, 87% over the earlier decade, as shown by Bloomberg NEF. Regardless, the typical pack cost of $156 each kilowatt-hour (from about $1,183 in 2010) is at this point more than the $100 edge at which the cost of an electric vehicle should facilitate a vehicle with an inward burning motor. That would help trigger mass gathering.

43.4.5 Limited Vehicle Choices

Recalling, 2019 was an extreme year for the electric vehicle market. Various new models were dispatched and vehicle clients have a bigger number of models to pick from than some other time. Passage is managing introducing the all-electric F-150 pickup and the Mustang Mach-E, a crossbreed SUV. General Motors needs to deliver an electric pickup in 2021 moreover. Disregarding basic progression, there are still way minimum electric models to peruse and altogether less great vehicles [13, 19].

43.5 Latest Technology Breakthroughs Solving the Problems of EV

43.5.1 Charging and Range

It is the latest technology for battery which is literally very close to the future. It's hearsay that Dyson has supplied heavily during this space which the Dyson car thanks to seem in 2020 is also the central applying a solid-state battery. Different automobile manufacturers demanding this technology are Tesla, BMW and Toyota Square. In solid-state battery analysis, the UK government has shown interest and is ready to contribute an amount of £42 million grant for car growth. Lithium-ion batteries in present electrical automobiles utilize the liquid solution. This fluid provides a path between the positive and negative terminals and generates the current that allows ions a passage between the positive and negative terminals and creates the current that moves outside of the battery to try to work. In a very compactly packed particles state battery, the solution and electrodes on the battery terminals square measure each solid. A diminished, airy and better power battery with an integrated life cycle is made [20]. Poor varies and a poor charging infrastructure wreaks a terrible combination and as long as Bharat is one among the various markets trying to ban fossil fuel-powered vehicles by 2030, there's a desire for widespread adoption of the technology. Whereas solid-state batteries don't solve the present infrastructural problems, they are doing double the gap associate degree energy unit will travel on one charge.

Solid state batteries are without doubt a stop-gap measure. However, if they're as effective as Toyota (and VW) claims they'll be, they will finally place fossil fuel-powered cars to rest. At present, most automotive brands will stop producing fossil fuel-powered cars within the next five–six years, which implies there'll be a bigger range of electrical cars on sale, in the near future [21]. The downside of a supercapacitor is that it is unable to carry upon a charge for a longer period of time. If supercapacitors were utilized in electrical cars because of the main supply of capacity, it'd take some time to get completely charged, however drivers would want to restore each mile. This can be wherever the invention by super dielectrics provides a serious advance [22]. Grabbing a charge is incredibly rational by chemical compounds prepared by super dielectrics. Along with metropolis and Surrey

Universities, this chemical compound has been utilized in supercapacitors to extend power depots.

The chemical compound will be organized into one layer and in an exceedingly polygon lattice to produce an enormous area, that interprets to a broadly hyperbolic potentiality to carry a charge. This could provide the supercapacitor the flexibility to charge quickly, in addition to the quirk to carry this charge forward. For electric four-wheelers, this could work out the prevailing issues of prolonged charging times and vary eagerness. The utilization of supercapacitors as the major supply of capacity for electrical automobiles could also be an extended means within the future however would offer rapid charging of simply a couple of minutes, and good amount of energy supply with a protracted lifetime of 3,000–5,000 charging cycles [23, 24].

43.5.2 Charging Infrastructure

With the assistance of legitimate arranging, we can have electric chargers in each territory of the country. These PCOs (Public charge workplaces) would have three segments viz. batteries to change, a bank of chargers to accuse them of and food while we backup. They could be online business bundle drop focuses, biometric offices for Aadhaar and various conveniences. Clean power coming from the network and sun-based force produced charges EV. Planning charging time and installment should be possible by cell phones. Legitimacy is that actually like PCO, ePCO will advance business. Youthful EV chiefs will be slammed in light of the fact that they will see their pictures on some ePCO dividers as PCO proprietors look at these chargers regularly [25].

WiTricity is involved with the Chinese wireless EV charging standardization procedure through its output with China Electric Power Research Institute (CEPRI), China Automotive Technology and Research Center (CATARC) and the CEC. Zhejiang VIE Science & Technology Co., Ltd. subsidiary (VIE) and Anjie Wireless Technology (Suzhou) Co. Ltd. (Anjie) have permitted WiTricity's technology and industry-leading designs which comprises all required outlying systems such as foreign object detection, position detection and communications. These companies manufacture and commercialise wireless EV charging systems of high efficiency for sale to leading automaker brands in the China market. VIE has finished manufacturing its pilot production facility to assemble wireless charging systems for its initial OEM customers. Anjie lately demonstrated a fully autonomous parking and wireless charging experience on an Xpeng Motors' EV, showing the world what is upcoming in autonomous, EV market will stand in the upcoming years [26, 27].

43.6 Future of EV in India

India is tremendously moving toward the adoption of electric vehicles, but the "National electrical quality Mission came upon (NEMMP) 2020" was really undraped in 2013 beneath the previous government. Due to improper efforts and wrong perspective electric cars are facing difficulty in India. Since then, the country is willing to adopt EVs for a better tomorrow [33]. In India major push was given to two-wheeler and three-wheeler electric vehicles rather than electric cars, as they were more popular in India. Despite the challenges and difficulty being faced, major machine manufacturers are presently observing this wide untapped sector with hope [30].

According to a survey, 95% of the total sales in the Asian country are of two-wheelers. Only about 1500 electric cars were sold between 2018 and 2019, but the sales of electric two-wheelers dropped by 94% in the first six months of 2020. In 2018 about 79% electric vehicles sold were scooters. Once people may find out that electric cars are more reliable than electric scooter that might increase the sales of electric cars. Cars are more reliable in terms of safety and space and provide a better service in this sensitive market of electric vehicles. Electric vehicle market is very fragile in terms of demand and stability. But soon, Asian countries will shift to electric cars.

As per the report, a client said they'd take under consideration shopping for an electric vehicle by 2022. This may be a couple of years before the assumptions made by experts. However, a pair of thirds (67%) of shoppers in Asian countries said they are adopting a "wait and see" approach. Consumers in land area unit seeking a less expensive worth purpose for EVs than customers in numerous countries, with the planet average tipping worth purpose for EVs being at Rs twenty-seven lakh [28, 32].

The Indian government is on a mission of making EVs popular in India. Government has planned various projects by the year 2030. The steps taken in 2019 to promote electrical vehicles inside the country include:

- Various strategies like dynamic GST are being implemented on EVs.
- INR 1.5 integer tax reduction on loans to buy for electrical vehicles.
- INR 10 K elements assigned to FAME II to push electrical quality through standardization.
- Government has projected customs exemption on various elements such as electrical drive assembly, aboard charger, e-compressor and a charging gun that will help minimizing the cost [29, 31, 34].

43.7 Conclusion

If we analyze for the alternative of IC engine automobiles then EV has the best chance to replace them. IC engines have various user-friendly merits but have a major setback that is pollution and it is a serious concern as the number of deaths due to air pollution caused from automobiles is increasing every year. While EVs are almost zero-emission vehicles resolving the major problem of pollution, along with this has some other merits also like overall maintenance cost is low and light-weighted (due to fewer components). Whereas EVs have some of their own setbacks which result in very less EVs growth and popularity in the automobile market. Gradually as EVs start becoming a necessity of the globe, modern technologies are developing which are solving EVs problems. If we analyze the growth of EVs, resolving of problems and growth of EVs go hand in hand. With every passing year, EVs are becoming much more popular and numerous new models are launching with advancements.

The future of EVs is bright not only globally but it is bright in India as well The Nations Government also started many policies promoting EV automobiles and the aim is to go full electric by 2030.

References

1. https://upload.wikimedia.org/wikipedia/commons/f/f7/Global_plug-in_car_sales_since_2011.png. Accessed on 28 Jan 2021
2. https://www.omazaki.co.id/en/electric-vehicle-components/. Accessed on 22 Jan 2021
3. https://afdc.energy.gov/vehicles/how-do-all-electric-cars-work. Accessed on 22 Jan 2021
4. https://news.hyundaimotorgroup.com/Article/Understanding-EV-Components. Accessed on 17 Jan 2021
5. https://www.energy.gov/eere/electricvehicles/electric-vehicle-basics#:~:text=There%20are%20two%20basic%20types,Cell%20Electric%20Vehicles%20(FCEVs). Accessed on 18 Jan 2021
6. https://www.edfenergy.com/for-home/energywise/how-do-electric-cars-work#:~:text=Electric%20cars%20function%20by%20plugging,they%20feel%20lighter%20to%20drive. Accessed on 19 Jan 2021
7. https://www.ergon.com.au/network/smarter-energy/electric-vehicles/types-of-electric-vehicles. Accessed on 25 Jan 2021
8. https://www.jato.com/ev-prices-have-been-growing-during-the-last-8-years/. Accessed on 28 Jan 2021
9. https://www.virta.global/global-electric-vehicle-market. Accessed on 21 Jan 2021
10. https://www.iea.org/reports/global-ev-outlook-2019. Accessed on 23 Jan 2021
11. https://www.iea.org/reports/global-ev-outlook-2020. Accessed on 26 Jan 2021
12. https://www2.deloitte.com/us/en/insights/focus/future-of-mobility/electric-vehicle-trends-2030.html. Accessed on 26 Jan 2021
13. https://earth911.com/eco-tech/6-issues-with-electric-vehicles/. Accessed on 25 Jan 2021
14. https://www.res-ev.co.uk/problems-with-ev/. Accessed on 15 Jan 2021
15. https://pod-point.com/guides/driver/how-long-to-charge-an-electric-car#:~:text=A%20typical%20electric%20car%20. Accessed on 15 Jan 2021
16. https://vehiclefreak.com/what-does-range-mean-on-a-car/. Accessed on 17 Jan 2021
17. https://www.google.com/search?q=17)+resourcesmag.org%2Fcommon-resources%2Fare-we-approaching-range-serenity-electric-. Accessed on 17 Jan 2021
18. https://auto.hindustantimes.com/auto/news/why-are-electric-cars-expensive-the-cost-of-making-and-buying-an-ev-explained-41603419957680.html. Accessed on 27 Jan 2021
19. Faizal, M., Feng, S.Y., Zureel, M.F., Sinidol, B.E., Wong, D., Jian, G.K.: A review on challenges and opportunities of electric vehicles (evs). Received 20 June 2019 Accepted 29 July 2019 Available online 06 August 2019. Accessed on 27 Jan 2021

20. https://www.moneycontrol.com/news/automobile/can-toyotas-solid-state-batteries-revolutionise-the-ev-market-6237441.html. Accessed on 28 Jan 2021
21. https://asia.nikkei.com/Spotlight/Most-read-in-2020/Toyota-s-game-changing-solid-state-battery-en-route-for-2021-debut. Accessed on 14 Jan 2021
22. https://www.carmagazine.co.uk/car-news/tech/what-is-supercapacitor-battery-ev-and-hybrid/. Accessed on 14 Jan 2021
23. https://www.intechopen.com/books/new-generation-of-electric-vehicles/batteries-and-supercapacitors-for-electric-vehicles. Accessed on 19 Jan 2021
24. https://www.information-age.com/tech-problem-electric-vehicles-123471264/. Accessed on 22 Jan 2021
25. https://www.livemint.com/opinion/columns/an-answer-to-the-problem-of-ev-charging-infrastructure-in-india/amp-11571332852406.html. Accessed on 20 Jan 2021
26. https://auto.hindustantimes.com/auto/news/wireless-charging-for-electric-vehicles-could-soon-be-a-reality-41589172528134.html. Accessed on 31 Jan 2021
27. https://witricity.com/technology/. Accessed on 01 Feb 2021
28. https://inc42.com/features/what-is-the-future-of-electric-cars-in-india/. Accessed on 01 Feb 2021
29. https://m.dailyhunt.in/news/india/english/inc42-epaper-inc/what+is+the+future+of+electric+cars+in+india-newsid-n155198052. Accessed on 02 Feb 2021
30. https://www.business-standard.com/article/economy-policy/india-s-electric-vehicle-journey-so-far-a-story-of-nudges-and-trudges-119080600339_1.html. Accessed on 29 Jan 2021
31. https://m.economictimes.com/industry/auto/auto-news/most-indian-customers-could-consider-buying-an-electric-vehicle-by-2022-study/amp_articleshow/77910102.cms. Accessed on 30 Jan 2021
32. https://commons.m.wikimedia.org/wiki/File:Global_plug-in_car_sales_since_2011.png. Accessed on 21 Jan 2021
33. https://m.timesofindia.com/home/environment/global-warming/india-ranks-5th-in-global-climate-risk-index/amp_articleshow/72367505.cms. Accessed on 03 Feb 2021
34. Un-Noor F, Padmanaban SK, Mihet-Popa L, NurunnabiMollah M, Hossain E (2021) A comprehensive study of key electric vehicle (EV) components, technologies, challenges, impacts, and future direction of development. Academic Editor: Sergio Saponara Received: 8 May 2017; Accepted: 21 July 2017; Published: 17 August 2017, accessed on 04 February 2021
35. https://www.energy.gov/articles/history-electric-car. Accessed on 10 Jan 2021
36. https://blog.evbox.com/electric-cars-history-infographic. Accessed on 11 Jan 2021

Chapter 44
Enhancing Heat Transfer Rate by Optimization of Commercial Refrigeration Condenser and Its Design Metrics

Anas Islam, Vikas Kumar Sharma, and Manish Saraswat

Abstract Optimization of heat transfer rate and size of heat money changer (condenser) by significant tests met by Refrigeration structure proposals. The heat transfer rate problem is concerned with the determination of total heat transfer rate, and the sizing issue is concerned with the determination of the aggregate high temperature exchange surface range. The key element helping for higher heat exchange rate and base high-temperature exchange zone may be those refrigerant streams out in the condenser coil, likewise those framework lies over indoor of a bureau. The objective of the available task will streamline those outlined of a commercial refrigeration condenser to enhance the heat transfer rate. CFD (Computational Fluid Dynamics) and Condenser software will be used to optimize the design of the components. Structural dissection about parts will be also will be performed to dissect the deformations and anxieties happening because of the weights and temperatures of the stream. The necessary modifications are made to improve the heat transfer rate & reduce the size of the condenser that enhances the overall heat transfer rate.

44.1 Introduction

The main objective of this project is to reduce the size of the condenser and enhance the heat transfer rate. The Condenser is a device used to exchange the heat absorbed to ambient. It rejects heat to an external cooling medium (air or water) [1, 2] be those procedure of evacuating high temperature from an encased space, or from

A. Islam · V. K. Sharma (✉)
IET Department of Mechanical Engineering, GLA University, Mathura, India
e-mail: vikash.sharma@gla.ac.in

A. Islam
e-mail: anas.islam@gla.ac.in

M. Saraswat
Department of Mechanical Engineering, Lloyd Institute of Engineering and Technology, Greater Noida, UP, India

© The Author(s), under exclusive license to Springer Nature Singapore Pte Ltd. 2022
V. V. Rao et al. (eds.), *Computational and Experimental Methods in Mechanical Engineering*, Smart Innovation, Systems and Technologies 239,
https://doi.org/10.1007/978-981-16-2857-3_45

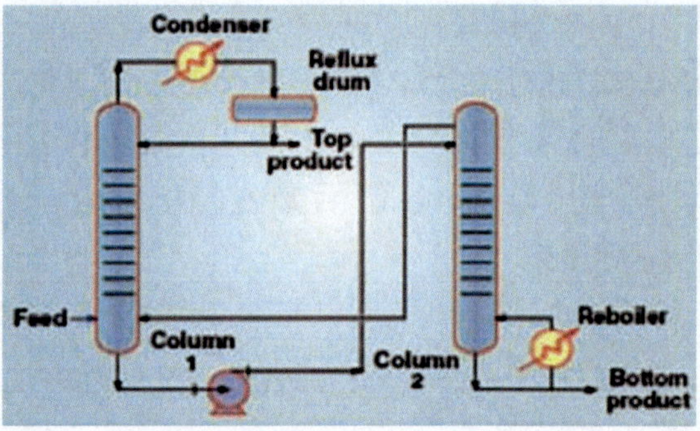

Fig. 44.1 Detailed explanation of condenser with phase change

A substance, furthermore moving it will a spot the place it cam wood a chance to be excluded in high face area, low FPI & optimum refrigerant flow through condenser coil and air flow over the condenser coil are the key for effective condenser performance [3, 4].

In the present study, the condenser coil circuit is optimized so that the refrigerant flow through the condenser flows via two tubes parallels and the heat exchange between the refrigerant and the external surface occurs very quickly [5–7] (Fig. 44.1).

44.2 Experimental Setup of Condenser Optimization

The issues that mainly affect the heat transfer rate in a required convection type condenser.

44.2.1 Air-Cooled Condenser Selection

Air-cooled condenser might be a common convection sort alternately to constrained convection sort. On practically as a relatable point, we utilize air-cooled condenser [8, 9]. In front of measuring a condenser, cautious assessment of the necessities for a particular establishment will be fundamental. The assessment ought to include, attention from claiming starting cost, operating cost, administration and aggregation, furthermore the kind about load [10–12]. A condenser that is excessively little camwood make unreasonable What's more make operating issues in easier encompassing states a under-size condenser camwood make working issues in higher encompassing states [13]. It is, therefore, paramount to think about those taking after factors in front of measuring a condenser:

- Terrible heat dismissal.
- Encompassing temperature.
- Consolidating temperature.

- Temperature distinction (TD).
- Wind stream.

Condenser ability is the capacity of the essential high-temperature exchange equation [14].

$$Qc = U \times A \times LMTD$$

here

$$\begin{aligned} \text{Qc} &= \text{Condenser capacity in Cal/h} \\ &\quad (\text{Ref.effect } + \text{Heat of Comp } + \text{Motor winding heat}) \end{aligned}$$

$$U = \text{Overall heat transfer coefficient K Cal/m}^2\text{h.}$$

$$CA = \text{Effective surface area in m}^2$$

$$\begin{aligned} \text{LMTD} &= \text{Log mean temperature difference between the} \\ &\quad \text{condensing refrigerant and the condensing medium in } ^\circ\text{C} \end{aligned}$$

Face area = Air quantity/Air velocity

The greatest speed happens between those tubes since the tubes block An and only those streams acceptably [15]. If B is those dividing among tubes in the face and c may be the tube dividing among rows, also d may be those tube breadths. The Reynolds and Nusselt number are defined as follows for this case:

$$\text{Re} = (\rho \times do \times U\infty)/\mu$$

The Grimson's correlation is as follows:

where the constants C and n are dependent upon Reynolds number (Tables 44.1 and 44.2).

Enthalpy values taken from P-h chart

$H_1 = 400$ kJ/kg $H_4 = 270$ kJ/kg $H_2 = 430$ kJ/kg $H_5 = 245$ kJ/kg $H_3 = 420$ kJ/kg $H_6 = 245$ kJ/kg

Refrigerant Effect $= H_1 - H_6$

= 155 kJ/kg

Table 44.1 Constant for Grimson's equation

Reynolds no, Re	Constant C	Constant n
0.4–4	0.978	0.44
4–40	0.920	0.358
40–4000	0.698	0.477
4000–40000	0.189	0.600
40000–400000	0.0315	0.852

Table 44.2 Observation for condenser

Design of condenser	Design consideration
Tube size	3/8th
Outside diameter of the condenser tube:	0.0080
Inside diameter of the condenser tube	0.0068
Refrigerant used	R404A
Evaporator temperature (°C)	7.2
Condenser temperature (°C)	54
Subcooling temperature (°C)	10
Superheating temperature (°C)	10.8
Compressor power (W)	678
Cooling capacity (Btu/h)	5460
Ambient temperature (°C)	43

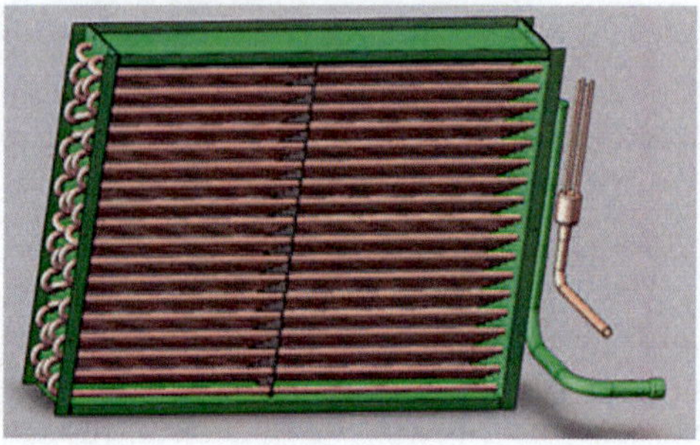

Fig. 44.2 Existing design with 12 tube single circuit

Heat Rejection Capacity (HRC) = **(Refrigeration capacity * power of compressor) * FOS**

= ((1.6) + (678/1000)) × 1.05 = 2.392 KW

Overall heat transfer coefficient

1/Uo = (At/Ai) × (1/hi condensation) + 1/ho $U_o = 47.46\ W/m^2\ K$

LMTD condensation = 7.61 °C

44.2.2 CAD Modeling

See Figs. 44.2, 44.3, 44.4, 44.5 and 44.6.

44.2.3 ANALYSIS-Structural Analysis

Boundary Conditions:

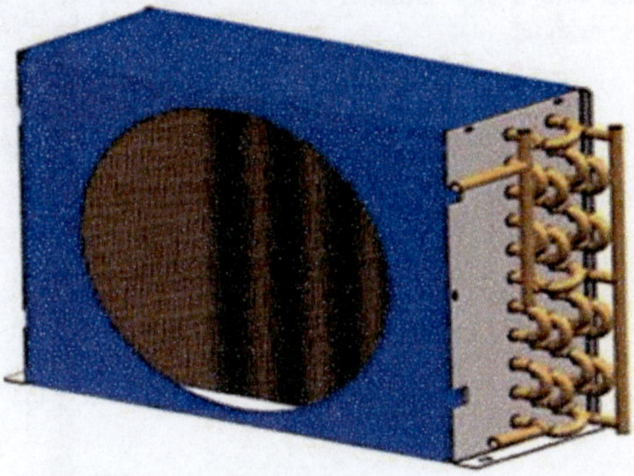

Fig. 44.3 New design with 11 tube

Fig. 44.4 Condenser coil with fins

Fig. 44.5 Condenser coil without fins

Inlet: Temperature: 85 [°C]
Outlet: Mass flow rate = 0.0137(kg/s) Ref. Pressure: 14.7 [PSI] Assumptions: Steady state single phase analysis (Figs. 44.7 and 44.8, Table 44.3).

Fig. 44.6 Physical model of the design

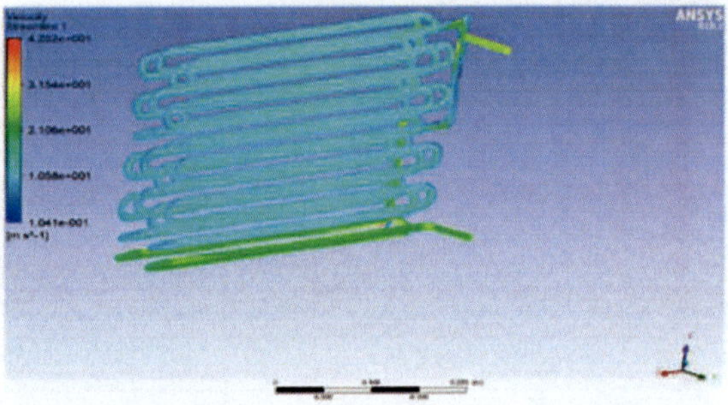

Fig. 44.7 11 *4 row condenser velocity streamlines

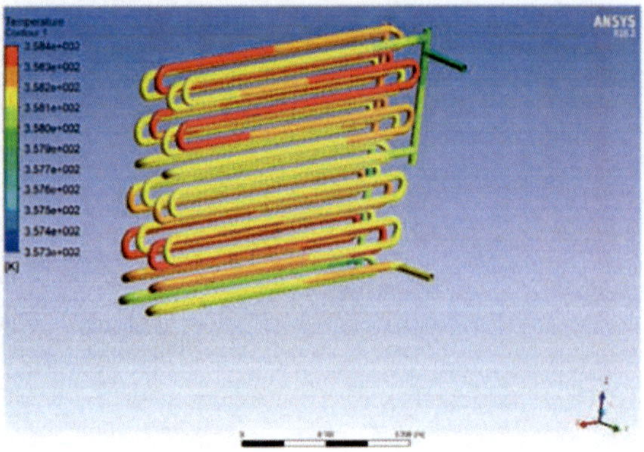

Fig. 44.8 Temperature contour

Table 44.3 Velocity and temperature

	Inlet	Outlet
MFR	1.380e02 [kg s6^−1]	−1.380e02 [kg s6^−1]
Velocity	2.691e+01 [m s^−1]	2.458e+01 [m s^−1]
Temperature	3.621e+02 [K]{85.32 °C}	3.228e+02 {50.32 °C}

44.3 Results and Discussion

By comparing the two models the outlet temperature of the new design is better. By optimizing the circuit design from one circuit to two circuit the heat transfer rate of the condenser is improved by 6% though the height of the condenser is decreased by 9% by reducing one row of the tube. The capacity increased from 3.17 to 3.37 KW. With this achievement, the higher capacity of the compressor can be used for the same refrigeration system and can be used in very compact-sized refrigeration units. The above data is calculated from the LUVATA Thest. Condenser design Software and the images of the same are shown below (Figs. 44.9 and 44.10).

Fig. 44.9 Input for 11 tubes condenser

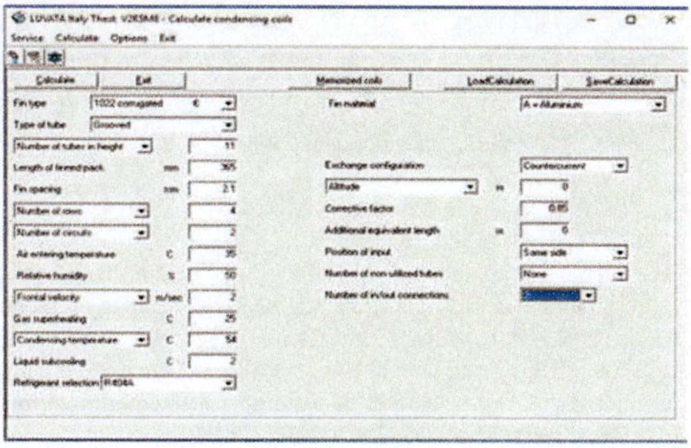

Fig. 44.10 Output for 11 tubes condenser

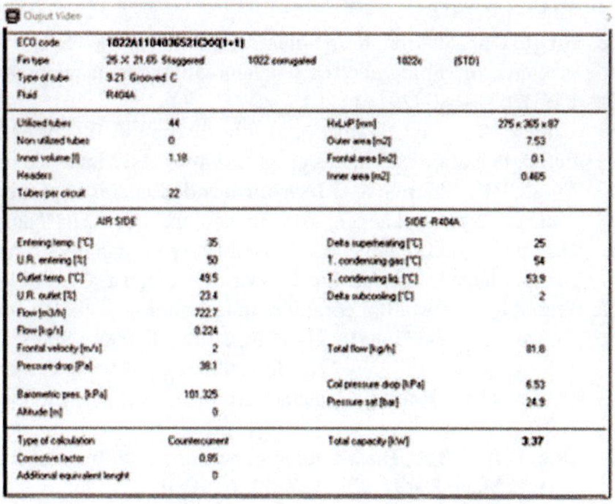

44.4 Conclusion

By optimizing the circuit designs from one circuit to two circuits the heat transfer rate of the condenser is improved by 6% though the size of the condenser is decreased by 9% by reducing one row of the tube. The capacity increased from 3.17 to 3.37 KW. With this achievement, the higher capacity of the compressor can be used for the same refrigeration system and can be used in very compact-sized refrigeration units. With this new design, the cost of the unit can be reduced by about 5–6%. In conclusion, by optimizing the circuit design the performance of the condenser can be improved.

References

1. Peng, Y., Zhang, S.J., Shen, F., Wang, X.B., Yang, X.R., Yang, L.J.: IOP Conf. Ser. Earth Environ. Sci. **93**, 012067 (2017)
2. Mallikarjun, Anandkumar, S., Malipatil, S.: Int. J. Res. Appl. Sci. Eng. Technol. (IJRASET) **2**(X) (2014). ISSN: 2321-9653
3. Kalendar, A., Galal, T., Al-Saftawi, A., Zedan, M., Karar, S.S., El-shiaty, R.: Department of Mechanical & Refrigeration Engineering, College of Technological Studies, PAAET, Kuwait
4. Firoz, R., Kulkarni, S.H.: Int. J. Emerg. Technol. Comput. Appl. Sci. (IJETCAS) 14-450 (2014)
5. Webb, R.L.: Principles of Enhanced Heat Transfer. Wiley, New York (1994)
6. Sharma, A., et al.: Effect of welding parameters on automated robotic arc welding process. Mater. Today Proc. **26**, 2363–2367 (2020)
7. Yadav, A., Kumar, A., Singh, P.K., Sharma, K.: Glass transition temperature of functionalized graphene epoxy composites using molecular dynamics simulation. Integr. Ferroelectr. **186**(1), 106–114 (2018)
8. Singh, P.K., Sharma, K., Kumar, A., Shukla, M.: Effects of functionalization on the mechanical properties of multiwalled carbon nanotubes: a molecular dynamics approach. J. Compos. Mater. **51**(5), 671–680 (2017)
9. Kumar, A., et al.: Evaluation of elastic moduli for different patterns of stone-thrower-wales defect in carbon nanotubes/epoxy composites. Mater. Today Proc. **4**(9), 9423–9428 (2017)
10. Singh, P.K., Sharma, K.: Mechanical and viscoelastic properties of in-situ amine functionalized multiple layer grpahene/epoxy nanocomposites. Curr. Nanosci. **14**(3), 252–262 (2018)
11. Han, J.C., Glicksman, L.R., Rohsenow, W.M.: An investigation of heat transfer and friction for rib-roughened surfaces. Int. J. Heat Mass Transf. **21**, 1143–1156 (1978)
12. Webb, R.L.: Toward a common understanding of the performance and selection of roughness for forced convection. In: Hartnett, J.P., et al. (eds.) Studies in Heat Transfer: A Festschrift for E.R.G. Eckert, pp. 257–272. Hemisphere Publishing Corp, Washington
13. Fenner, G.W., Ragi, E.G.: Enhanced tube inner surface heat transfer device and method, USA (1979)
14. Gee, D.L., Webb, R.L.: Forced convection heat transfer in helically rib roughened tubes. Int. J. Heat Mass Transf. **23**, 1127–1136 (1980)
15. Shukla, M.K., Sharma, K.: Improvement in mechanical and thermal properties of epoxy hybrid composites by functionalized graphene and carbon-nanotubes. Mater. Res. Express **6**(12), (2019)

Chapter 45
FPGA Implementation of Low Power Pre-processor Design for Biomedical Signal Processing Application

Kirti, Harsh Sohal, and Shruti Jain

Abstract A low-power pre-processor system is introduced to extract the informatory data from noisy biomedical signal. The proposed systems hardware is executed on ZedBoard (Zynq-7000) evaluation Field Programmable Gate Array (FPGA) board. The proposed methodology is based on the efficient pre-processor system that depletes low power & resources of FPGA: Windows and Wavelet. The comparison has been done among different windows and different wavelet-based pre-processor designs as per resource utilization and on-chip power. The inference has been drawn from the comparison that Haar wavelet consumes only 2.15% of LUTs, 12.19% of slice registers and only 25.45% of DSPs. The selected pre-processor design also consumes only 34mW of dynamic power & 110 mW of static/leakage power. In future, the introduced pre-processing system will be employed in wearable & portable biomedical equipment.

45.1 Introduction

A rapid advancement has been observed in the area of noise reduction from the biomedical signal. It plays a vital role in the accurate diagnosis of the disease by medical practitioners. A number of biomedical application's performance is directly linked with the noise removal methods [1]. The presence of low & high frequency namely Baseline Wander (BLW), Electromyography (EMG) and Power Line Interference (PLI) noises, etc., deteriorate the useful information from the Electrocardiogram (ECG) signal [2]. Popularly, digital filters are utilized to eliminate the unwanted noises from the biosignal. The microcontroller, digital signal processor (DSP) & field-programmable gate arrays (FPGA) are predominantly utilized owing to their low cost & reconfigurable property, low time to market, etc. The Programmable Logic

Kirti (✉)
GL Bajaj Institute of Information and Technology, UP Greater Noida, India

H. Sohal · S. Jain
Department of ECE, Jaypee University of Information Technology, Solan, HP, India

© The Author(s), under exclusive license to Springer Nature Singapore Pte Ltd. 2022

V. V. Rao et al. (eds.), *Computational and Experimental Methods in Mechanical Engineering*, Smart Innovation, Systems and Technologies 239,
https://doi.org/10.1007/978-981-16-2857-3_46

Devices (PLDs) devices emerged as one of the advanced technology that produces the high-end monitoring system in the hospitals. Among various processing units, FPGA is selected due to rapid prototyping design during the software simulation and synthesis. The growing number of design variations via software synthesis has coming cost-effective in comparison to the hardware prototype variations. It is distinguished that FPGAs are hugely utilized in the execution of rapid digital systems for the retrieval, processing, storage, & transmission of data.

The popular FPGA companies are Xilinx, Actel, Altera & Quick logic. Popularly, there are two designing environments for realizing digital pre-processing design filter: DSP and FPGA. The digital design based on FPGA offers re-programmability and high speed, unlike old methods namely microcontrollers, DSP devices & microprocessors. In article [3] a technique is programmed in VHDL, which themed on the optimization of the filter. In this research, a comparison has been taken into consideration with the MATLAB FDA toolbox. An adaptive filter has been suggested according to the ASIC flow with the assistance of IC design tools and MATLAB from the Mentor Graphics [4]. A transposed architecture for the FIR filter has been introduced and the programming is done on VHDL on FPGA [5]. A number of architectures have been proposed for the multiplier unit design; it provides low delay and power during the preprocessing stage for FIR filter [6]. There are various benefits of FPGA processing unit namely high speed, low power & execution time in contrast with software programmed DSPs. Popular lines as well non-linear digital filters are proposed such FIR, IIR, Wavelets, Poincare plots, etc., for the preprocessing of biomedical signals.

Previous studies illustrate that wavelet transform has emerged as an extensively utilized method for the denoising of ECG, PCG, cough signal, EEG, etc., bio-signals. In the research article [8], a threshold-based methodology employing Wavelet Transform (WT) is presented for the elimination of unwanted noise resides in the biomedical signal. The cough signal processing a multi-resolution method is given based on the basis of WT.

In this paper [8], WT centered soft-thresholding procedure is followed for bio-signal denoising. Although, for cough signal denoising soft thresholding is not an effective process. Instead, Hard thresholding utilizing Discrete Wavelet Transform (DWT) is more effective for the cough signal denoising application. An intelligent system formed on PCG has been introduced in [9]. This system employs denoising by utilizing DWT for the unwanted noisy signal mix fused with the informatory signal. The Daubechies wavelet (db6) is used to eliminate the low-frequency noise present in ECG signal [10].

In this paper, a hardware design for low-power pre-processor to remove the unwanted signal from the useful biomedical signal is introduced. FPGA is chosen as targeted hardware due to its ease of implementation and low non-reoccurring cost. To select the best pre-processor design, authors have also compared various types of linear filters using window techniques and wavelet function based on resource & on-chip power consumption. The novelty of our research is, that our proposed FPGA-based pre-processor design provides flexibility, high speed, minimum area delay and low power in comparison with other existing technologies.

The organization of this research work is drafted as: Sect. 2 constitutes the pre-processor design, Sect. 3 discusses the efficient pre-processor design for biomedical signals using FPGA. Lastly, the conclusion has been enlisted in Sect. 4 with the future scope of this application.

45.2 Pre-processor Design

In biomedical signal processing, pre-processor is termed as a system that filters useful information from the noisy biomedical signal. In this methodology, to select the best pre-processor design the authors have compared various types of linear filters using windowing techniques and non-linear wavelet functions. These FPGA-based pre-processor designs are compared based on resource utilization and on-chip power. Figure 45.1 illustrates the methodology for the selection of pre-processor design.

There are different types of analysis. The time-domain analysis is carried out by Shannon Nyquist Theorem, frequency analysis is done by Fourier Transform, Short-Time Fourier Transform is analyzed by Gabor Wavelet and the last is Wavelet Transform. The time-domain analysis constitutes of FIR filters employing various sorts of windowing methods namely Kaiser, Blackmann, Hamming, Bartlett [11]. In this research, analysis has been done by employing windowing and Wavelet transform techniques.

In data inspection, the wavelet transform emerged as an important novel time–frequency decomposition tool. It is useful for non-stationary signals analysis, like the bio-signals. The WT examines the signal concurrently in both time & frequency domains. Therefore, it enables to extract more parameters from non-stationary bio-signals. There is a difference between Fourier analysis and wavelet transform. Fourier analysis involves the splitting of any signal into sinusoidal waves of different frequencies while wavelet transform consists of the splitting of a biosignal into shift & scale version of the original wavelet that is entitled as mother wavelet. The researchers have employed different kinds of mother wavelet functions namely Haar, Daubechies, Coiflets and biorthogonal [12]. The haar wavelet's mother wavelet function can be

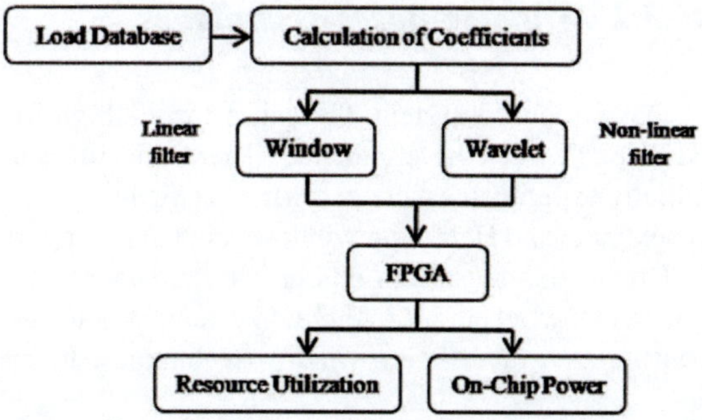

Fig. 45.1 Methodology for the selection of pre-processor design

expressed as Eq. (45.1).

$$\psi(t) = \begin{Bmatrix} 1 & 0 \le t < 1/2 \\ -1 & 1/2 \le t < 1 \\ 0 & otherwise \end{Bmatrix} \tag{45.1}$$

The haar wavelet's scaling function is shown in the Eq. (45.2).

$$\varphi(t) = \begin{Bmatrix} 1 & 0 \le t < 1 \\ 0 & otherwise \end{Bmatrix} \tag{45.2}$$

45.3 FPGA Implementation of Pre-processor Design for Biomedical Signals

This paper proposes an efficient FPGA-based pre-processor design using various kinds of windowing techniques and wavelet transform. For the implementation, MATLAB and VIVADO tools are used [13]. The entire process is executed on the Zynq-7000 evaluation board. To validate the proposed system, the authors have considered an ECG signal as a biomedical signal. The useful frequency range of informatory ECG signal lies between 0.5 and 100 Hz. In ECG prominently two types of noises are present: Base Line Wander (BLW) and Electromyography (EMG) [14]. A low-frequency noisy signal namely BLW is present below the 0.5 Hz and EMG is a high-frequency noise that occurs above 100 Hz [15].

In this paper, the authors have proposed an architecture for denoising biomedical signals. Different experiments were performed on the proposed architecture using different window techniques and different wavelet functions. Figure 45.2 illustrates the proposed architecture of pre-processor design to eliminate low-frequency and high-frequency noise.

45.3.1 Window-Based Techniques

In the introduced system, the entire low & high-frequency noises namely EMG, BLW & PLI can be eliminated. There are different types of window functions, initially experiments were performed using MATLAB out of which Kaiser, Bartlett, Blackmann and Hamming windows give the best performance [16].

Due to this reason, in this article hardware simulations were performed using four window techniques. Table 45.1 tabulated the resource utilization and power-consumption of different window techniques. It represents that all four window

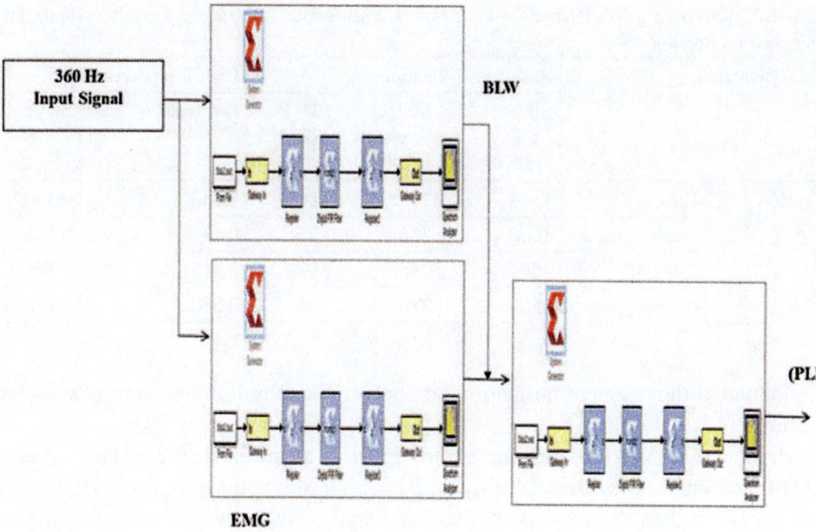

Fig. 45.2 Proposed architecture for pre-processor design

Table 45.1 Resource consumption & On-Chip power consumption of suggested pre-processor system on employing window functions

Filler architecture		Resource utilization			On-Chip power (W)	
		Slice LUT (53200)	Slice Registers (17400)	DSP (220)	Dynamic L'	Static
Windows	Kaiser	1242	2298	60	0.404	0.111
	Bartlett	1242	2298	60	0.395	0.111
	Blackmann	1242	2298	60	0.354	0.110
	Hamming	1242	2298	60	0.402	0.111

methods consume an equal amount of resources and static power. But in terms of dynamic power consumption, Blackmann utilizes only 0.354 W.

45.3.2 *Wavelet-Based Techniques*

Likewise window technology, there are different wavelet functions in literature. Therefore, simulations with different wavelet functions were carried out on MATLAB. But Haar, Daubechies, Coiflets and biorthogonal yields the best performance. For the hardware simulation, these wavelet function were considered. Table

Table 45.2 Resource utilization and On-Chip power consumption of proposed pre-processor design using wavelet functions

Filter architecture		Resource utilization			On-Chip power (W)	
		Slice LUT (53200)	Slice Registers (17400)	DSP (220)	Dynamic	Static
Wavelets	Haar	1146	2122	56	0.346	0.110
	Daubechies	1566	2862	72	0.424	0.111
	Coiflets	1566	2862	72	0.415	0.111
	Biorthogonal	1566	2862	72	0.380	0.110

45.2 tabulated the resource utilization and power consumption of different wavelet functions.

It demonstrates that an equal number of resources are utilized except Haar wavelet that utilizes only 1146 Slice LUTs, 2122 slice resisters and only 56 DSP. It also consumes very less dynamic power as compared to other wavelet functions, i.e., 0.346 W. Figure 45.3 signifies RTL schematic of proposed pre-processor design which is used to remove low-frequency and high-frequency noise. As the architecture is same, So the RTL is same for the window and wavelet technique.

This work presents the hardware implementation of pre-processor design for biomedical signal processing applications. The proposed pre-processor design removes the low-frequency and high-frequency noise present in the raw biomedical signal. After comparing various types of windowing techniques and wavelet transforms, Haar wavelet is selected due to its low power consumption and resource utilization. The Haar wavelet consumes only 21.5%, 12.19% & 25.45% of LUTs, slice registers and DSPs, respectively, it depletes 34 mW of dynamic power.

Table 45.3 tabulated the comparison of the proposed pre-processor design with the existing state-of-the-art technique. Only high-frequency noise is eliminated by the authors of [17], utilizing the Kaiser window function. It consumes 21% LUTs, 24% slice registers, 9% of input/output ports and 89mW of dynamic power targeting Spartan 3E FPGA board.

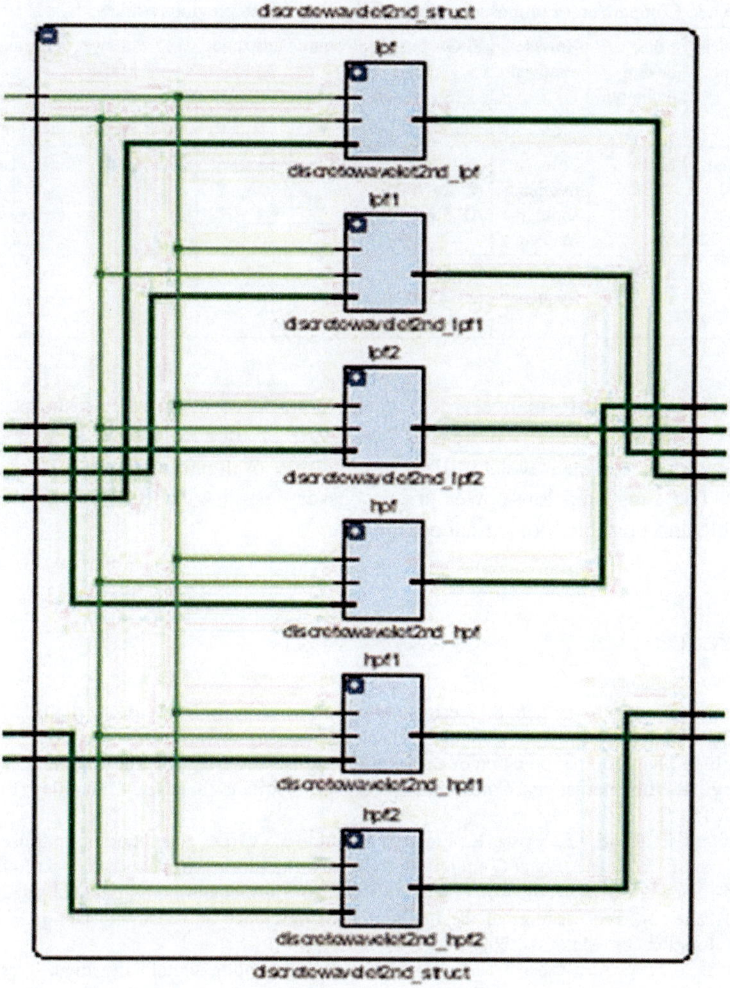

Fig. 45.3 RTL schematic of proposed pre-processing block

45.4 Conclusion and Future Scope

In this work, a low-power pre-processor architecture is designed for various biomedical signal processing applications. The proposed system is executed on the Zynq evaluation & development FPGA kit. To validate the low power and low resource consumption an exhaustive contrast has been made among different windows and different wavelet functions namely Kaiser, Bartlett, Blackmann, Hamming, Haar, Daubechies, Coiflets and Biorthogonal. Among them, Haar wavelet is selected

Table 45.3 Comparison of proposed pre-processor design with previous studies

References	Filter design technique	Noises removed	FPGA board	Resource utilization (%)			Power consumption (mW)	
				LUT	Slice register	I/O block	Dynamic power	Static power
Proposed	Haar	Low-frequency andhigh-frequency	ZedBoard (Zytiq-7000 AP SoCl	2.15	12.19	25.45	34	110
[17]	Kaiser	High-frequency	Spartan 3E (XC3S500e-4fg320)	21	24	9	89	–

because of good performance with respect to resource & power consumption. It utilizes 21.5% of Look-Up Table (LUT), 12.19% of slice Register (SR) & 25.45% DSP blocks. The Haar wavelet utilizes only 34 mW of dynamic & 110 mW of static power. The suggested low-power pre-processor design will be further utilized in wearable and portable biomedical equipment.

References

1. Prashar, N., Sood, M., Jain, S.: Design and performance analysis of cascade digital filter for ECG signal processing. Int. J. Innov. Technol. Explor. Eng. **8**(8) (2019)
2. Kirti, H., Jain, S.: Interpretation of cardio vascular diseases using electrocardiogram: a study. In: 2018 Fifth International Conference on Parallel, Distributed and Grid Computing (PDGC), vol. 159 (2018)
3. Daitx, F.F., Rosa, V.S., Costa, E., Flores, P., Bampi, S.: VHDL generation of optimized FIR filters. In: 2nd International Conference on Signals, Circuits and Systems, pp. 1–5 (2008)
4. Ab-Rahman, A.A.H., Kamisian, I., Shaameri, A.Z.: ASIC modeling and simulation of adaptive FIR filter. In: Processdings of the International Conference on Electronic Design, vol. 12, pp. 1–4. Penang, Malaysia (2008)
5. Razak, A.H.A., Abu Zaharin, M.I., Haron, N.Z.: Implementing digital finite impulse response filter using FPGA. In: Processdings of Asia-Pacific Conference on Applied Electromagnetic, vol. 12, pp. 1–5. Melaka, Malaysia (2007)
6. Tripathi, K., Narkhede, P., Kottath, R., Kumar, V., Poddar, S.: Design considerations of orientation estimation system. In: 5th International Conference on Wireless Networks and Embedded Systems (WECON), pp. 1–6 (2016)
7. Nekoei, F., Kavian, Y.S., Strobel, O.: Some schemes of realization digital FIR filters on FPGA for communication applications. In: 20th International Crimean Conference on Microwave & Telecommunication Technology. Sevastopol, Crimea, Ukraine (2010)
8. Taquee, A., Bhateja, V., Shankar, A., Srivastava, A.: Pre-processing of cough signals using discrete wavelet transform. In: Second International Conference on Computing, Communication and Control Technology, vol. 42 (2018)
9. Nabih-Ali, M., El-Dahshan, E.S.A., Yahia, A.S.: Heart diseases diagnosis using intelligent algorithm based on PCG signal analysis. Int. J. Biol. Biomed. **2** (2017)
10. Kumar, M., Pachori, R.B., Acharya, U.R.: Characterization of coronary artery disease using flexible analytic wavelet transform applied on ECG signals. Biomed. Signal Process. Control **31**, 301–308 (2017)

11. Kirti, H., Jain, S.: FPGA implementation of Power-Efficient ECG pre-processing block. Int. J. Recent. Technol. Eng. **8**, 2277–3878 (2019)
12. Oliveira, B.R.D., Duarte, M.A.Q., Abreu, C.C.E.D., Vieira Filho, J.: A wavelet-based method for power-line interference removal in ECG signals. Res. Biomed. Eng. **34**(1), 73–86 (2018)
13. Prashar, N., Sood, M., Jain, S.: Semiautomatic detection of cardiac diseases employing dual tree complex wavelet transform. Period. Eng. Nat. Sci. **6**(2), 129–140 (2018)
14. Aboutabikh, K., Aboukerdah, N.: Design and implementation of a multiband digital filter using FPGA to extract the ECG signal in the presence of different interference signals. Comput. Biol. Med. **62**, 1–13 (2015)
15. Kirti, H., Jain, S.: Comparative analysis of heart rate variability parameters for arrhythmia and atrial fibrillation using ANOVA. Biomed. Pharmacol. J. **11**, 1841–1849 (2018)
16. Venkatesan, C., Karthigaikumar, P., Varatharajan, R.: FPGA implementation of modified error normalized LMS adaptive filter for ECG noise removal. Clust. Comput. 1–9 (2018)
17. Bhaskar, M.P., Uplane, M.: FPGA based notch filter to remove PLI noise from ECG. Int. J. Recent. Innov. Trends Comput. Commun. **3**, 2246–2250 (2015)

Chapter 46
Design and Implementation of Smart Energy Meter with Real-Time Pricing

Prem Chand Vashist and Ashish Tripathi

Abstract Smart meters are an electronic device to measure energy consumption accurately, at precise time intervals. In India, Britishers introduced electricity firstly in Calcutta in 1879. After the inception of electric power, it took a very long time to become an Act named "The Electricity Act" framed in 1910. Initially, the cost of electricity consumption measured by fixed charges, after that, a measuring device as "Electric Meter" came into the market. This meter has some limitation such as precision value of meter reading, missing of date and time and kilowatt per hour etc. This was the measure concern for both the electric consumer and provider to know the exact value of the electric consumption with other essential information. To overcome these issues, several technologies have been developed and a new concept of measuring device was introduced "smart meter" introduced for an accurate result. In this paper, we discuss things that can overcome the limitations of the existing smart devices.

46.1 Introduction

After the establishment Electricity Act in India, electro-mechanical meters are being used as a measurement technique to measure the use of electric consumption for individual as well as commercial purposes. The traditional meters have some basic problems such as less accuracy (Class 2 accuracy [1], easy manipulation of reading through reversing meter [2], measurement issue (only measure reactive power at a time). In the energy measuring world, the traditional meters have played a significant role from its inception but due to technological advancement the smart energy meters are taking their place in the current scenario. In the current system, digital energy meters are fixed in the premises for calculating the electricity consumption in kilowatt

P. C. Vashist (✉) · A. Tripathi
Department of Information Technology, G. L. Bajaj Institute of Technology and Management, Greater Noida, Uttar Pradesh, India

P. C. Vashist
CST, Lucknow, Uttar Pradesh, India

© The Author(s), under exclusive license to Springer Nature Singapore Pte Ltd. 2022
V. V. Rao et al. (eds.), *Computational and Experimental Methods in Mechanical Engineering*, Smart Innovation, Systems and Technologies 239,
https://doi.org/10.1007/978-981-16-2857-3_47

per hour. The recorded data received from the energy meter of the account holder is taken by the meter reader and provided to the electricity board. On the basic of meter reading, the account holder has to pay the balance/dues to the electricity board as per the applicability of the tariff [3, 4].

This process does not end here, and a process of distribution of a hard copy of the bill against consumed electricity is being done. After that consumers pay their bills. Consequently, to make the process easy an Android Application-based system 'to track consumer electricity usage' has been proposed in this paper. In the proposed work a tool has been developed embedded with the existing digital meters to transmit the electricity consumption data to the server. Similarly, a mobile application has been developed to provide real-time information to the user about individual electricity usage. For a detailed analysis of electricity consumption, a web application has also been developed. The proposed work will help to overcome the limitations of the existing digital meters such as reducing work overload (reducing the involvement of one dedicated person for meters of a defined locality), real-time information of the electricity consumption, error-free reading/on-time bill delivery, etc.

46.2 Limitations of the Traditional Electricity Meter

(a) These meters are unreliable due to unpredictable reading of electricity consumption in the result user has to be dependent on the monthly electricity bill.
(b) Such meters require a huge number of employees to measure the reading from the respective locations.
(c) Also these meters are based on mechanical structure that supports the measurement of the electricity consumptions, which does not give accurate result as compared to the smart meters [5].
(d) Expensive in terms of maintenances cost, and labor overhead.
(e) Problem in introducing the hourly basis tariffs to consumers for the corresponding meter.
(f) Development and maintenance of software for meter and corresponding network infrastructure support are complex.

46.3 Practical Utility

Electricity Consumption analytics. Consumer can view consumption at different levels such as aggregated consumption with respect of multiple functions such as min, max, sum or average. Also, real-time usage can be seen at any instant of the day. Discovering Patterns of Consumption. In Smart meter data, the load is influenced due to various factors, such as the appliances used and outdoor weather temperature [6–8].

In pattern discovery, the system can display the raw data as is, and also the load graphs. Also, knowledge of hourly consumption of customers can be gained by utilities from the histograms provided by the system. Forecasting. For authorities, it will be helpful for them to avoid power failures by predicting the periods when the demand will be at peak. Attaining Feedback. Setting rules and sending alert messages to customers may help in getting quality service. Automated messages are transmitted with a priory set time interval after the feedback rule.

The biggest drawback of legacy system is that plenty of electricity goes waste due to power theft and human mistakes by meter readers. Since power availability is limited, it is important that energy is used in an improved and productive manner as a responsible citizen.

Many countries in every corner have clear domino effects, which have been significantly diminished Global System for Mobile communication (GSM) based smart meters [9, 10]. A further benefit of smart meter base bill generation scheme is that the mistakes created by human beings can be minimized to a large number by calculating and handling costs.

Android Application-based Smart Metering system will be mounted on residential and commercial apartments where bulk electricity is consumed [11, 12]. Advancement of technology has ensured a very quick, secure and honest exchange of information. An Android Application supported and GSM-based smart meter technologies will bring dramatic improvements in electricity consumption and its billing system [13, 14].

46.4 Proposed System

The existing electricity distribution, monitoring, consumption tracking by user and billing system are inefficient and takes long duration in processing. Thus, to overcome this issue, in the last few years, the government of India along with some private shareholders, start-ups have come up with the aim of installing smart meters in households across India. Consequently, market analysis of this field describes the challenge about the current market share and its potential. The task involved in launching this product is ensuring that the developed solution is economical and scalable for Indian markets. So, to end this, a novel system of smart electricity meter with advanced functionalities whose production and deployment is economical in nature has been proposed in this paper.

46.4.1 Actual Plan of Work

Different factors on which the proposed system will work are mentioned below.

(a) The system will be able to record the electricity consumption units from the

(b) meter.
(c) Transmit the data to a server in every 15 min.
(d) The web application will be able to analyze the patterns of electricity usage.
(e) Mobile application will show the customer the digital version of his/her meter.

46.4.2 Specification

46.4.2.1 Hardware

Power Supply Block, Microcontroller Block, Relays and Loads Block, and GSM modem.

46.4.2.2 Software

Apache Spark, R, PHP, Android-based application.

46.4.3 Implementation Layers of Smart Metering Framework

(a) **Ingestion Layer**: It is responsible for Data Acquisition and Transmission of data to server.
(b) **Processing Layer**: It is responsible for upholding Data Storage mechanism and accessing the data same time.
(c) **Analytics Layer**: Analytics Layer defines Data Analytics parameters and algorithms and Data Visualization heads.

46.4.4 Description of the Working Model

Description of the proposed working model is as follows:

(a) **Data Acquisition and transmitting the data to the server**: To make the metering circuit on even if the main power supply turns into disabled, a switched mode power supply has been provided. Similarly, GSM base system allows the meter to interact with the cloud and then with the mobile app. The smart meter is having sensitive circuitry and hence it has to support various communications protocols.
(b) **Accessing the data using Apache Spark**: Apache Spark is the big data processing framework made for promptness, classy analytics, and easy to use process. As the data will be received as streams hence we will be using stream processing technique to build our data. As the data is gathered from numerous

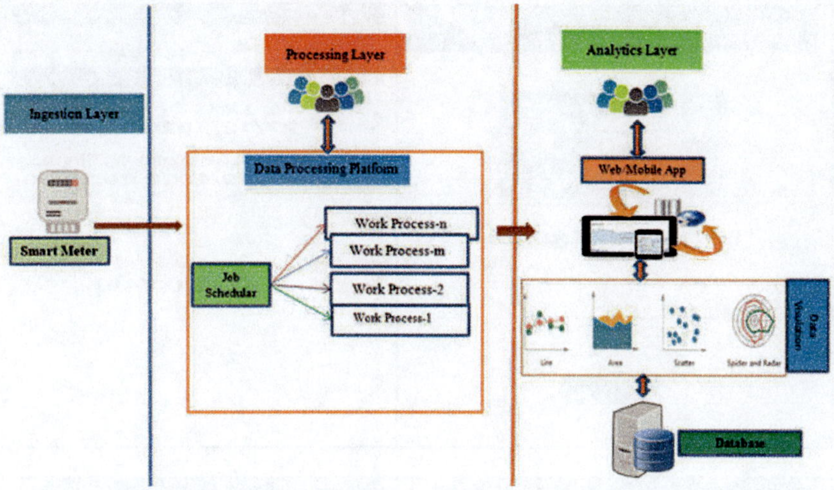

Fig. 46.1 Block diagram of the working model

sources it becomes quite significant to use Apache Spark as it can perform analytics on the data with dynamic behavior.

(c) **Data Analytics**: While working on analytic data, one of the main characterizations can be the varied electricity consumption in various climatic conditions. By using the streaming linear regression, several interesting features can be developed through these models. Another algorithm can be seen as for extracting daily consumption patterns that may occur without the effect of outside temperature.

(d) **Data Visualization**: The system works for analyzing data for supply and demand-side smart meters. Through supply-side analytics, smart grid optimization can be done. Demand-side analytics, allows us to figure electricity consumption patterns of consumers so as to provide effective services. It can also provide consumers with an insight of their own consumption, leading them to save energy.

The block diagram of the proposed working model is shown in Fig. 46.1.

46.4.5 Glimpse of the Working Model

Figures 46.2 and 46.3 show glimpses of the proposed Android App and website respectively.

Fig. 46.2 Glimpse of proposed working Android app

46.5 Conclusion

The technical development of the electrical distribution grid is an interruptive operation. In the new scheme of wireless meter readings, the measurement readings are tracked continuously and the power supply is turned off remotely when the user fails to pay the mark. It prevents human involvement, performance metering, billing error mitigation, maintenance cost control. It also reveals the corresponding device notification information on LCD. The SEM's benefits are that the amount of individuals needed is smaller, deposits are not required, power theft can be observed, a bill can be forwarded to the customer by the deadline.

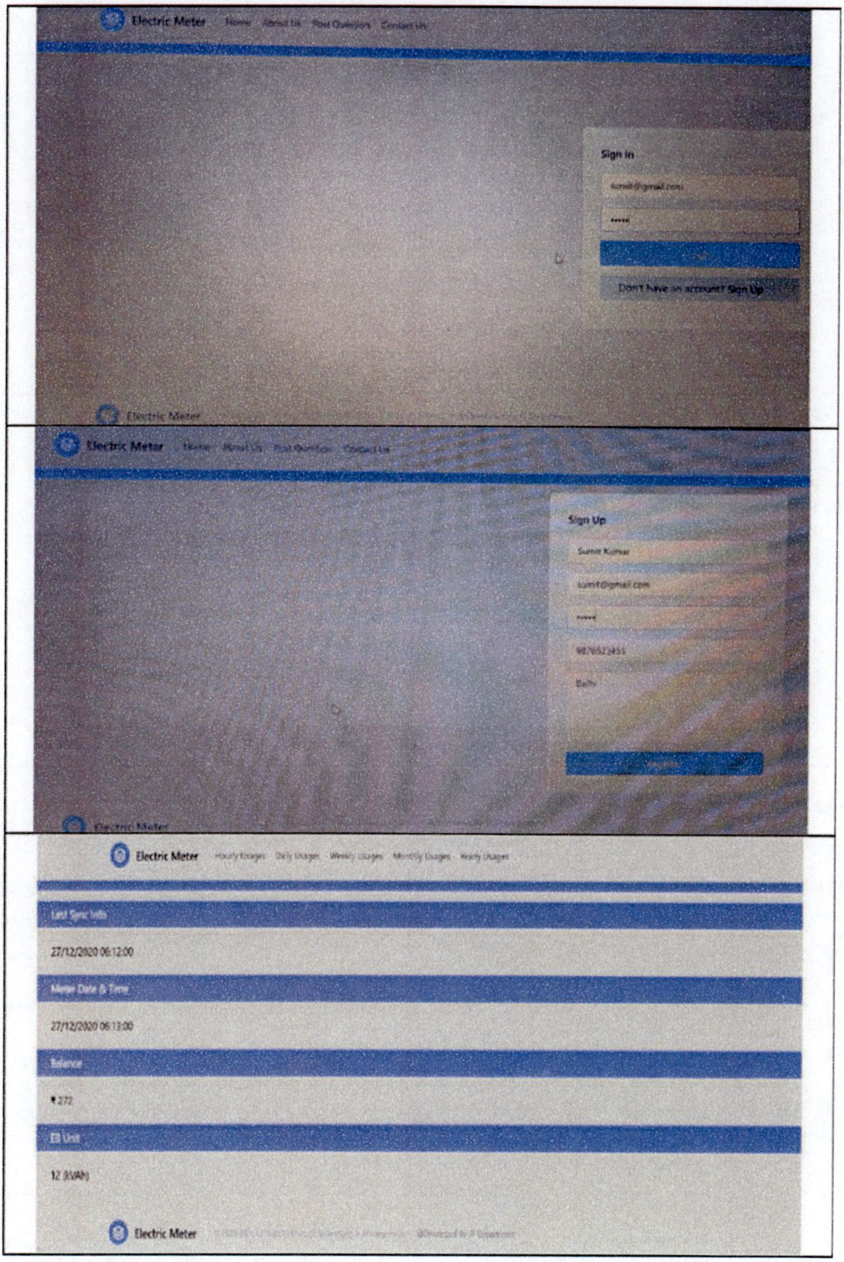

Fig. 46.3 Glimpse of proposed working website

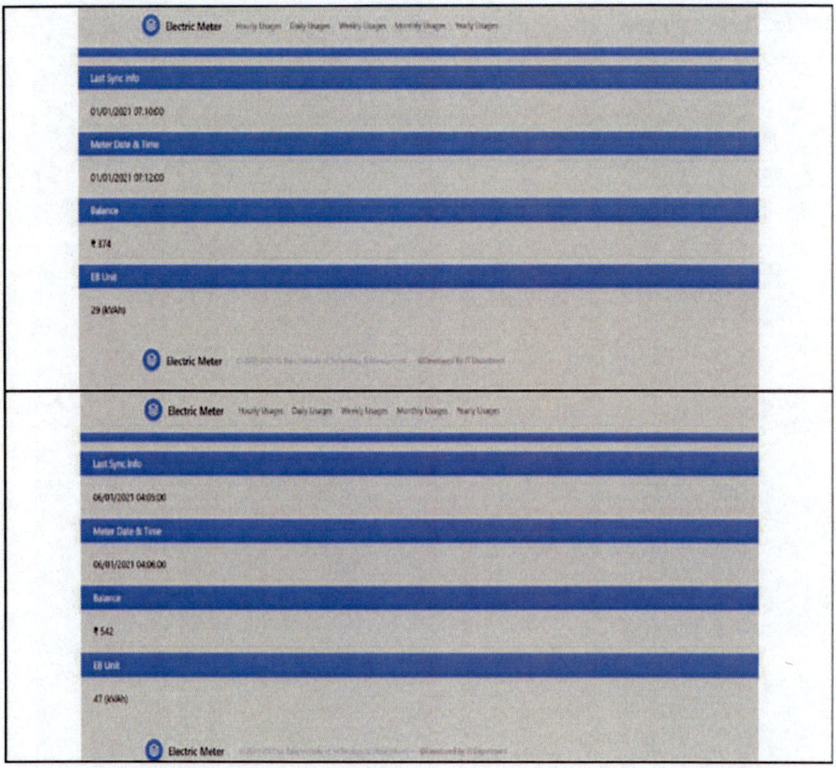

Fig. 46.3 (continued)

References

1. Special types of energy meters. Source: https://engineerexperiences.com/special-types-energy-meters.html#Types_of_Energy_meters_with_respect_to_Class_Accuracy
2. Electro-mechanical energy meters. Source: https://engineerexperiences.com/electro-mechanical-energy-meters
3. Aravind, E., Vasudevan, S.K.: Smart meter based on real time pricing. Procedia Technol. **21**, 120–124 (2015)
4. Labib, L., Billah, M., Rana, G.S.M., Sadat, M.N., Kibria, M.G., Islam, M.R.: Design and implementation of low-cost universal smart energy meter with demand side load management. IET Gener. Transm. Distrib. **11**(16), 3938–3945 (2017)
5. Qiao, L., Liu, X., Jiang, B.: Design and implementation of the smart meter in vehicle-to-grid. IEEE (2012)
6. Krishnan, P.G., Arunachaleswari, K.: Distributed generation grid infrastructure using smart meters modeled with renewable energy sources and power trading. ISSN: 2277-9655 Impact Factor: 1.852, December, 2013
7. Young, J.K., Mayne, D.: Digi International Inc., 2013. Virtual smart energy meter with integrated plug. U.S. Patent 8,378,848
8. Kuralkar, S., Mulay, P., Chaudhari, A.: Smart energy meter: applications, bibliometric reviews and future research directions. Sci. Technol. Libr. **39**(2), 165–188 (2020)

9. Fung, A.S., Mohammadi, F., Raahemifar, K., Mohassel, R.R.: A survey on advanced metering infrastructure and its application in smart grids. IEEE CCECE 2014
10. Vani, G., Reddy, V.U.: Application of smart energy meter in Indian energy context. IOSR J. JEEE **10**(3), 07–13 (2015)
11. Preethi, V., Harish, G.: Design and implementation of smart energy meter. In: 2016 International Conference on Inventive Computation Technologies (ICICT), vol. 1, pp. 1–5, August. IEEE (2016)
12. Naveenkumar, J.S.: Smart energy meter. Int. J. Tech. Res. Appl. **4**(3), 126–129 (2016)
13. Prathik, M., Anitha, K., Anitha, V.: Smart energy meter surveillance using IoT. In: 2018 International Conference on Power, Energy, Control and Transmission Systems (ICPECTS), pp. 186–189, February. IEEE (2018)
14. Singh, S., Selvan, M.P.: A smart energy meter enabling self-demand response of consumers in smart cities of Tamil Nadu. In: 2019 IEEE International Conference on Smart Cities Model (ICSCM), January, pp. 1–6. IEEE (2019)

Chapter 47
Optimization of Glass, Carbon and Graphite Fiber Mono Composite Leaf Spring Using Genetic Algorithm

Aatif Ameer, Vikas Tiwari, Vansh Pokhriyal, Alok Saxena, Divakar Tiwari, and Ranjeet Kumar Singh

Abstract Leaf springs are an essential part of automobiles because it provides directional stability, ride comfort and improves vibrational characteristics. Fatigue stresses induced in leaf springs directly influenced fatigue life. Fatigue life will increase under low fatigue stress. This is achieved by reducing unsprung weight. In automobiles, leaf springs are contributed 10 to 20% of unsprung weight. Thus, researchers tried to minimize the weight of leaf spring under the same loading conditions. Optimal design and selection of suitable materials are the ways to fulfill this objective. In this paper, double tapered composite leaf spring is selected for optimization using a genetic algorithm. Optimization has been performed for glass fiber, carbon fiber and graphite fiber composite leaf springs. The optimization result reveals that the weight of mono leaf spring made by carbon fiber composite is 91.92% lighter than multi-leaf spring. Thus carbon fiber composite is the most suitable candidate in comparison to glass and graphite fiber composite fiber mono leaf.

47.1 Introduction

Genetic Algorithm is a natural selection process to obtain the best quality results with the satisfaction of various constraints. Problem formulation is an important step in optimization. By optimizing using GA, problem formulation included objective function, design variables, design parameters, design constraints and fitness function. Further, various parameters like density of material, modulus of elasticity, fatigue strength and Poisson's ratio of the selected material as well as design constraints like factor of safety, thickness, width and length of leaf spring are other most important parameters. Rajendran et al. [1] have used genetic algorithm for the optimization of conventional and glass fiber composite leaf spring. GA includes various design constraints and selection procedures. The fittest over the all is selected for designing. Observation has showed 8% reduction of steel spring weight using genetic algorithm. Kassie et al. [2] have performed design optimization on mono composite leaf spring.

A. Ameer · V. Tiwari · V. Pokhriyal · A. Saxena · D. Tiwari · R. K. Singh (✉)
G.L. Bajaj Institute of Technology and Management, Greater Noida, UP 201306, India

© The Author(s), under exclusive license to Springer Nature Singapore Pte Ltd. 2022

V. V. Rao et al. (eds.), *Computational and Experimental Methods in Mechanical Engineering*, Smart Innovation, Systems and Technologies 239,
https://doi.org/10.1007/978-981-16-2857-3_48

Optimization was done by using genetic algorithm. During optimization weight is chosen as objective function. It is observed that 68.14% of weight reduction with same load-carrying capacity and strength. Qian et al. [3] have used genetic algorithm for the optimization of FRP leaf spring. Thickness is considered as a design variable. Observation has showed 0.58% improvement in fatigue life and stiffness of the leaf spring after optimization. Li et al. [4] have used carbon fiber, glass fiber and Aluminum Alloys for the fabrication of composite leaf spring suitable for electric aircraft. Investigation has showed optimum position of fibers w.r.t. spring axis is ±45°. The weight of composite leaf spring is reduced from 4.8 to 3.8 kg after using GA as an optimization tool. Genetic algorithm was used for optimization of conventional steel leaf spring. However, GAS was adopted by authors for optimal design of different structures [5–8].

The first paper reported on the design optimization of leaf spring through genetic algorithms was reported by Rajendran et al. [1]. In the present work, the conventional leaf spring is considered as a reference which was followed by [9]. The main objective of this research is to minimize weight of conventional leaf spring by using genetic algorithm as an optimization tool and different composites. The objective function has considered as weight reduction during optimization. The problem formulation has been considered similar as followed [1] to get optimum weight for three different composites. Optimization results have revealed that for the same material (GFRP), there is little deviation in weight. Further, carbon fiber composite leaf has the least weight in comparison to graphite and glass fiber composites. Thus carbon fiber composite is the most suitable candidate.

47.2 Problem Formulation

In this section, the similar objective function, design variable, design constraints and fitness function have considered as followed by [1] to compare results obtained from modern version GA tool available in MAT lab with older version tool followed by [1]. For better comparison, the similar material is considered and for optimum weight, CFRP and graphite fiber-reinforced composite materials are also considered.

47.2.1 Objective Function

The objective function for the minimization of weight of leaf spring is made up of different materials with given parameters. The objective function for this problem is given below:

$$f(w) = \rho l b t \tag{47.1}$$

ρ is the material density, t is the thickness of the leaf spring at the center, b is the width of the leaf spring at the center and l is the length of the spring under straight condition.

47.2.2 Design Variables

An optimization and design problems involve many design parameters, which are used for the optimization purposes and these parameters are called design variables. In the present problem, the design variables are: (1) width at the center, b and (2) thickness of leaf spring at the center, t.

The upper bound and lower bound of design variables are:

$$b_{\min} = 20\,\text{mm} \quad b_{\max} = 50\,\text{mm},$$

$$t_{\min} = 10\,\text{mm} \quad t_{\min} = 50\,\text{mm}$$

47.2.3 Design Parameters

It is defined as the parameters which are usually fixed or given in relation to the design variables. In this problem, the length of the leaf spring under straight condition (l), applied design load (W), modulus of elasticity or young's modulus (E), maximum allowable stress (σ_{max}) and maximum deflection (δ_{max}).

47.2.4 Design Constraints

Design constraint is defined as the constraint which shows some functional relationship between the design variables and design parameters. In this problem, design constraints are bending stress (σ_b) and vertical deflection (δ). Values of variables used in Eqs. (47.2) and (47.3) has taken from Table 47.1.

$$\sigma_b = \frac{1.5WL}{bt^2} \tag{47.2}$$

$$\delta = \frac{WL^3}{4Ebt^3} \tag{47.3}$$

$$\text{FOS} = \frac{\sigma_{b\max}}{\sigma_b} \tag{47.4}$$

Here, FOS is the factor of safety whose value is 2.5 and for Eqs. (47.4) and (47.5), value of σ_b and δ is taken to be maximum from the boundary conditions:

$$\sigma_{bmin} = 400\,\text{MPa} \quad \sigma_{b\,\text{max}} = 550\,\text{MPa},$$

$$\delta_{\text{min}} = 120\text{ mm} \quad \delta_{\text{max}} = 160\,\text{mm}.$$

47.2.5 Fitness Function

The fitness function $f(x)$ is derived from the objective function. For the maximization process, the fitness function is same as the objective function but for the minimization process, it is different with some modifications. So, our fitness number is:

$$objfun(x) = f(w) + A * \text{FOS} \tag{47.5}$$

Here $objfun(x)$ is the new objective function for the optimization, A is the violation parameter whose value is considered to be 1. It has been seen that GA works very ideally for unconstrained optimization problems, this problem is of constrained optimization problem. So, it is necessary to convert the problem into unconstrained form. Tables 47.1 and 47.2 have used for the calculation of obj fun (x).

47.3 Parameters of Genetic Algorithm

It is necessary to define the GA parameter for the optimization problem. One has to fix the parameters according to the particular problem.

47.3.1 Numbers of Parameters

Basically, the number of parameters is same as the design variables. In this problem, number of parameters are 2, i.e., thickness at center and width at center.

47.3.2 Population Size

In Genetic Algorithm, the population size is an important parameter which directly influences the ability to search an optimum solution in the search space. The population size in this problem is taken as 25 for all the materials.

47.3.3 Maximum Generations

It is the process of termination of the loop that has been carried out fixing the maximum number of generations, here it is 25.

47.3.4 Cross-Over Probability

Cross-over probability is defined as the probability that cross-over will occur at a particular mating; that is, not all mating must reproduce by cross-over. It recommended to have a high cross-over probability, i.e., 0.9.

47.3.5 Mutation Probability

This mutation probability determines how many chromosomes should be mutated in one generation and it is recommended to have a low mutation probability, i.e., 0.001 (Tables 47.1 and 47.2).

Table 47.1 Input parameters of glass epoxy, carbon epoxy and graphite epoxy

S. No.	Input parameters	Glass epoxy	Carbon epoxy	Graphite epoxy
1	Spring length under straight condition (mm)	1220	1220	1220
2	Density of material (kg/m^3)	2600	1600	1824
3	Young's modulus (MPa)	32.5×10^3	120×10^3	73×10^3
4	Deflection (mm)	160	160	160
5	Load (N)	4500	4500	4500
6	Maximum allowable stress (MPa)	550	550	550

Table 47.2 Input GA parameters of glass epoxy, carbon epoxy and graphite epoxy

S. No.	Input parameters	Glass epoxy	Carbon epoxy	Graphite epoxy
1	Number of parameters	2	2	2
2	Population size	25	25	25
3	Mutation probability	0.001	0.001	0.001
4	Cross-over probability	0.9	0.9	0.9
5	Maximum generations	25	25	25
6	Initial population	[20 10]	[20 10]	[20 10]
7	Minimum and maximum bound for width (mm)	20–50	20–50	20–50
8	Minimum and maximum bound for thickness (mm)	10–50	10–50	–50

47.4 Result and Discussion

Genetic algorithm is a global optimization tool that works in random fashion. Due to this reason variables are fluctuating highly in initial generations. After 5 generations, optimal values of each variable are obtained. It has been shown in Figs. 47.1, 47.2 and 47.3. Table 47.3 represents the optimal values of design variables (width and thickness) and objective function (weight) for the leaf springs made by glass, carbon and graphite fiber composites. Figure 47.4 represents a comparison between optimal values of weights of glass, graphite and carbon fiber mono composite leaf springs.

The steps followed during weight reduction process are shown in Fig. 47.5. Initially consider conventional leaf spring which has 9.28 kg weight. During optimization, multi-leaf spring was replaced by a mono leaf spring without any alteration

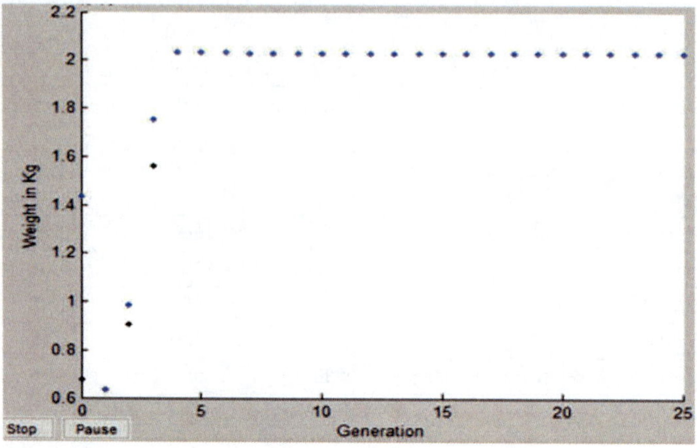

Fig. 47.1 Variation of weight of glass epoxy with number of generation

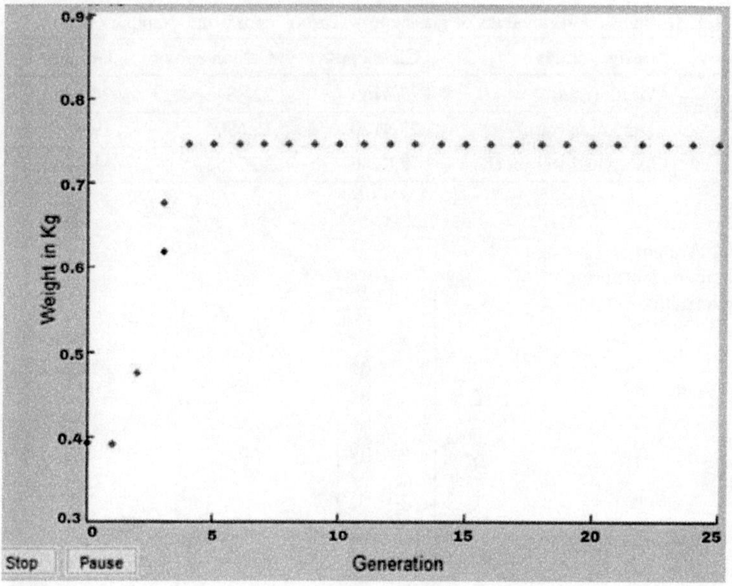

Fig. 47.2 Variation of weight of carbon epoxy with number of generation

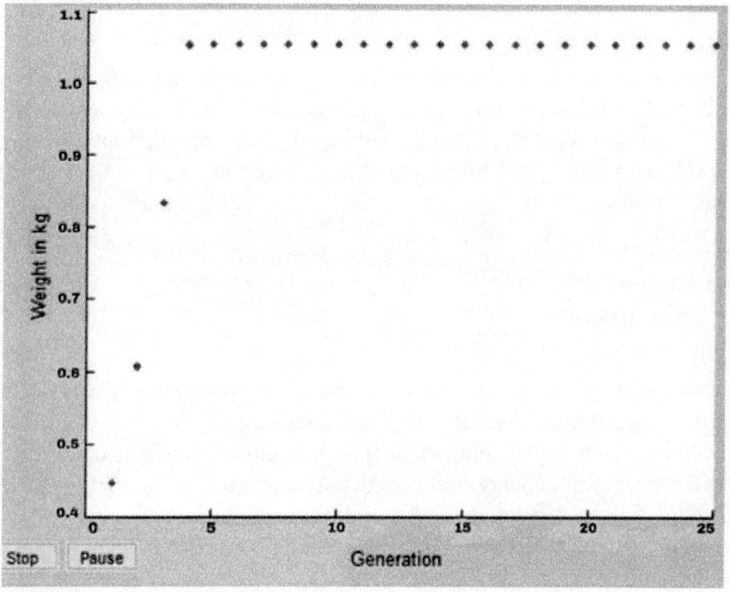

Fig. 47.3 Variation of weight of graphite epoxy with number of generation

Table 47.3 Optimum design values of glass epoxy, carbon epoxy and graphite epoxy

S. No.	Design values	Glass epoxy	Carbon epoxy	Graphite epoxy
1	Width (mm)	27.3165	22.8841	24.6455
2	Thickness (mm)	23.4120	16.6904	19.2169
3	Estimated weight (Kg)	2.0286	0.74555	1.0539

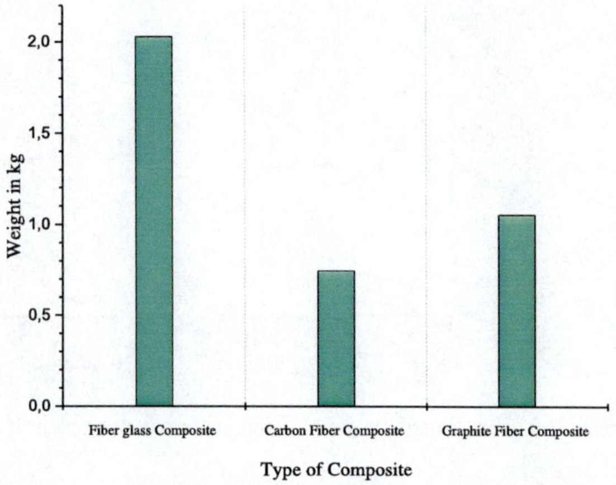

Fig. 47.4 Weight of different mono leaf spring after optimization

in load-carrying capacity and length of leaf spring. In the next step, three types of composite materials are considered as replacements of conventional leaf spring. For identical composite (GFRP) material, problem formulation, and optimization tool used by [1], the result is slightly deviated. It is found that carbon fiber mono leaf spring has the least weight.

47.5 Conclusion

1. In comparison to other optimization tools, GA is optimized globally due to this reason, it is a more powerful optimization technique.
2. In comparison to conventional leaf spring, mono composite leaf springs made by GFRP, Graphite fiber reinforcement polymer and CFRP are 79.13, 88.69 and 91.925% lighter, respectively, after optimization.
3. In comparison to GFRP leaf spring, Graphite fiber reinforcement polymer and CFRP leaf springs are 48.28 and 63.05% lighter, respectively.
4. CFRP leaf spring has minimum weight in comparison to GFRP and Graphite fiber reinforcement polymer leaf spring.

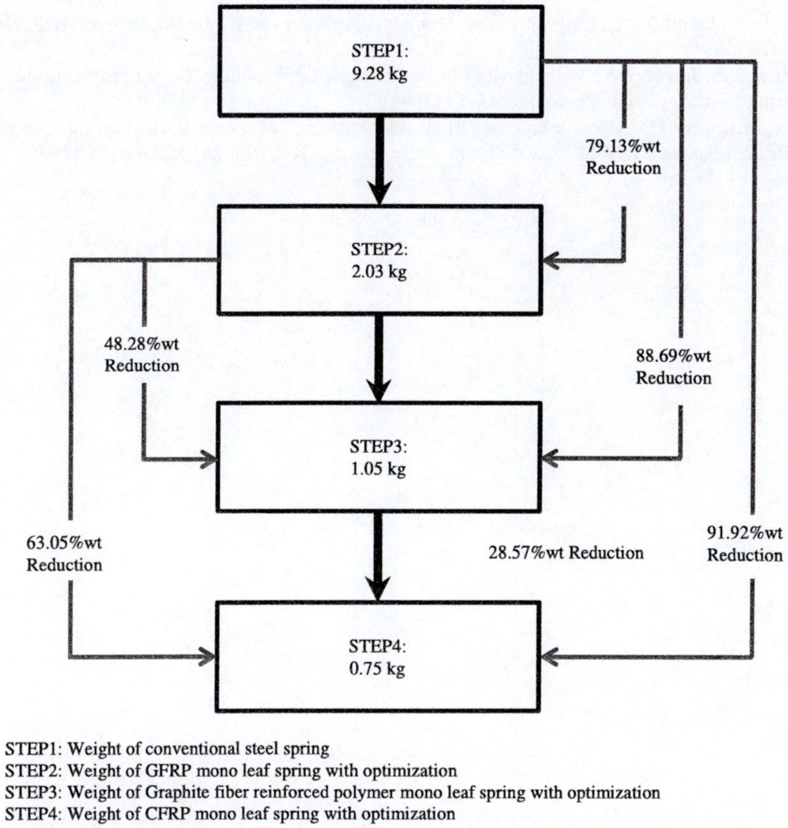

Fig. 47.5 Steps followed during optimization

References

1. Rajendran, I., Vijayarangan, S.: Optimal design of a composite leaf spring using genetic algorithms. Comput. Struct. **79**(11), 1121–1129 (2001). https://doi.org/10.1016/S0045-7949(00)00174-7
2. Kassie, A.A.: Design of single composite leaf spring for light weight vehicle. Int. J. Mech. Eng. Robot. Res. 34–41 (2014). https://doi.org/10.23883/ijrter.2017.3192.jtt1n
3. Qian, C., Shi, W., Chen, Z., Yang, S., Song, Q.: Fatigue reliability design of composite leaf springs based on ply scheme optimization. Compos. Struct. (168), 40–46 (2017). https://doi.org/10.1016/j.compstruct.2017.02.035
4. Li, D., Fan, Z., Zhang, Y., Zang, J., Yang, F.: Optimum design and experiment of composite leaf spring landing gear for electric aircraft. Chin. J. Aeronaut. **33**(10), 2649–2659 (2020). https://doi.org/10.1016/j.cja.2020.03.010
5. Yu, W.J., Kim, H.C.: Double tapered FRP beam for automobile suspension leaf spring. Compos. Struct. **9**, 279–300 (1998)
6. Goldberg, D.E.: Genetic Algorithms in Search, Optimization, and Machine Learning. Addison-Wesley, Reading, MA (1989)

7. Deb, K.: Optimization for Engineering Design: Algorithms and Examples. Prentice-Hall, New Delhi (1996)
8. Raol, J.R., Jalisatgi, A.: From genetics to genetic algorithms solution to optimization problems using natural systems. Resonance 43–54 (1996)
9. Yu, W.J., Kim, H.C.: Double tapered FRP beam for automotive suspension leaf spring. Compos. Struct. Elsevier BV (4), 279–300 (1988). https://doi.org/10.1016/0263-8223(88)90049-9

Chapter 48
Prospects of bioCNG in Modified Diesel Engine

Rahul Sharma, Amit Pal, and N. A. Ansari

Abstract As fossils fuels are depleting at a faster rate and from the standpoint of the environmental concept, it is necessary to use alternative sources of energy that could replace fossils fuels in existing engines. Dual fuel engines are economically attractive replacement for traditional diesel engines or obsolete diesel engines with slight variation in engine design. Furthermore, this engine suffers from inferior performance, slow-burning rate and higher levels of carbon monoxide and un-burnt hydrocarbons at lower loads. Various engine experts have also found an increased delay period and knocking tendency while exploring dual-fuel engines. Because of the accompanying reasons, numerous scientists have discovered a different strategy for the substitution of diesel, one of which is the utilization of biofuel as an enhancement for diesel fuel. World over there is growing trends for the generation of bioCNG with proper management of organic waste. The extraction of power from biogas (bioCNG) under dual fuel mode has risen as a potential vitality transporter to address the environmental issue associated with the compression ignition (CI) engine. This review significantly focused on the utilization of gaseous fuel (bioCNG) to CI diesel engine under dual fuel mode with diesel and diesel–biodiesel blend as pilot fuel. Different engine characteristics, such as ignition, performance and exhalations of the dual-fuel engine utilizing gaseous fuels have been gathered and assessed. Findings of various literature suggest that CI engine performance in dual fuel mode slightly deteriorates but the corresponding reduction in harmful pollutants occurs with the introduction of gaseous fuel. Overall, it can be concluded that utilizing bioCNG in pilot diesel and diesel–biodiesel blend results in improved performance with reduced exhalations at higher loads.

R. Sharma (✉)
Department of Mechanical Engineering, G.L. Bajaj Institute of Technology and Management, Greater Noida, India

A. Pal · N. A. Ansari
Department of Mechanical Engineering, Delhi Technological University, Delhi 110042, India

© The Author(s), under exclusive license to Springer Nature Singapore Pte Ltd. 2022

V. V. Rao et al. (eds.), *Computational and Experimental Methods in Mechanical Engineering*, Smart Innovation, Systems and Technologies 239,
https://doi.org/10.1007/978-981-16-2857-3_49

48.1 Introduction

Biogas can be produced utilizing an anaerobic digestion of natural assets (animal sewage, rural deposits, landfills and so on) additionally as a spotless and sustainable fuel is consistently accessible at any spot for IC engine. Thus, it very well may be a fantastic option as a vaporous fuel in diesel engine in dual fuel mode, the biogas stream rates influence the brake thermal efficiency (BTH) of the engine and brings about improved BTH. The double-fueled CI engine has displayed higher specific fuel utilization when contrasted with neat diesel [1]. For biodiesel activity at high compression ratio (CR), the diesel engine would correspond to better performance, combustion and emission than low value of CR. This may be because of increased chamber temperature and pressure bringing about better ignition [3]. Methane and biogas showed good antiknock tendency due to which it can be used in engine rather than petrol engines and thus, gains higher brake thermal efficiency and power [4]. Bio-fuel (biodiesel and biogas) conveyed from biomass presents a practical elective choice to supplant petroleum products. In this circumstance, the European Commission Directive 2009/28/EC for transportation zone intended to supersede 10% of non-environmentally friendly power source with bio-fills continually by 2020 [5]. Comparable properties of biodiesel make it the most encouraging option of traditional diesel fuel. As indicated by research, Bio-CNG is a form of biogas where all undesirable gases are eliminated to create a high measure of pure methane gas. Due to its high degree of methane and less carbon dioxide, bioCNG can be harmless to the ecosystem with low emissions level. bioCNG consists of 92–98% methane and carbon dioxide is around 2–8%. bioCNG is having calorific value around 52 MJ/kg, i.e., 167% higher than biogas. The potential for biogas in India from different sources at around 62 million tons for each annum. Due to the problem of disintegration of metal with the presence of water fumes, hydrogen sulphide and carbon dioxide make biogas more destructive and unsatisfactory. To handle this issue, biogas redesigning and sterilization are preferred. Biogas redesigning or sanitization is the cycle by which impurities in the crude biogas stream are consumed or cleaned, leaving more methane per unit volume of gas. Biogas sterilization is the cycle by which pollutants in the rough biogas stream are burned through or cleaned. Now a day's advancement for biogas is water cleaning, pressure swing adsorption, layer and chemical scrubbing. Biomethane can likewise be compressed and packaged into chambers. bioCNG is assessed to supplant 66% of India's natural gas imports, which is currently at 429 billion cubic feet. At present 17 biogas plants are operational in India which are spread in nine states and Maharashtra showed the highest share [38] (Fig. 48.1 and 48.2).

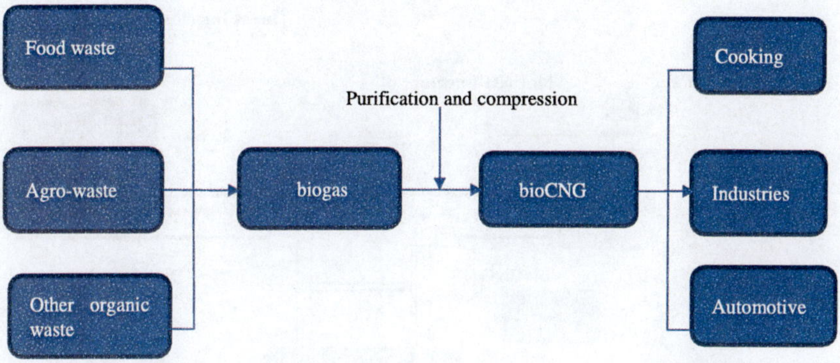

Fig. 48.1 Biogas redesigning process

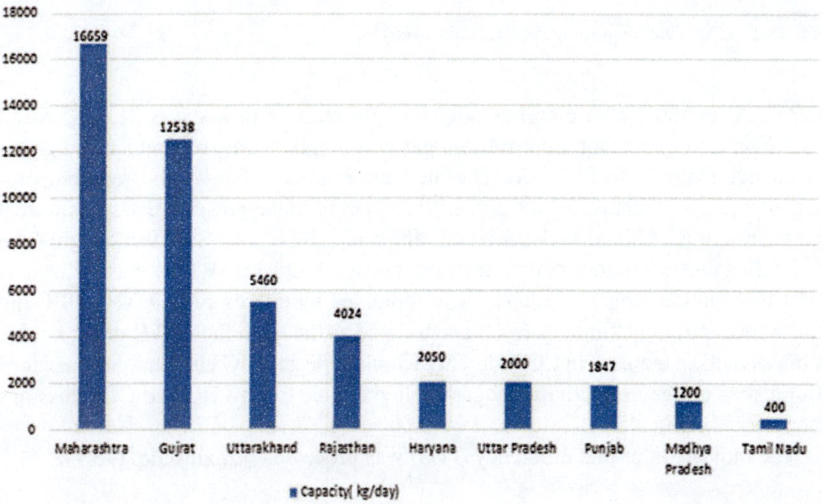

Fig. 48.2 BioCNG capacity in 9 states of India

48.2 Flammable Gas (BioCNG) as an Essential Fuel Utilized in Diesel Engine

Reza Mahmoodi demonstrated that reformed biogas (RBG) supplanting extent increases with equivalence ratio of 0.43, the mean ignition temperature decreases to 1354, 1312, 1292 K which are about 3.5, 6.6, 7.9% lower than the standard diesel. In-chamber pressure increases up to 22.63%. NOx emissions against CO are reduced significantly. RBG more than 40% showed marginal reduction in residue and UHC emissions [1]. Performance of engine with 12.65 CR was evaluated at 30°, 35° and 40° BTDC. Highest brake power developed by the engine was found at advance

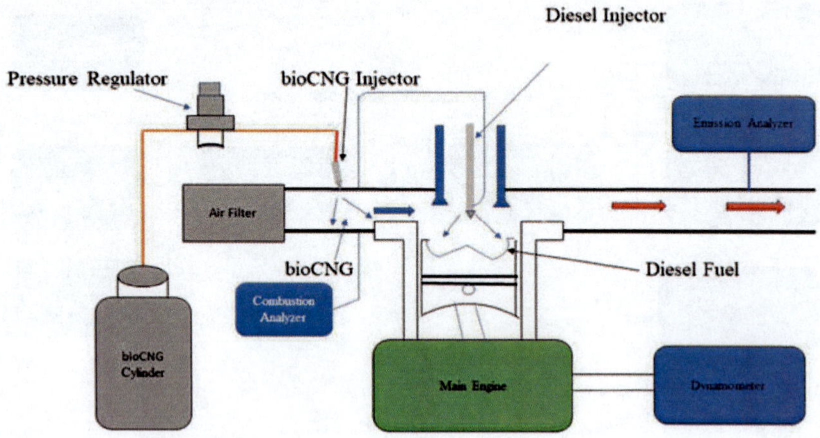

Fig. 48.3 Schematic diagram of dual fuel diesel engine

of 35°. As contrast with diesel as pilot fuel, decrease in power was 31.8, 35.6 and 46.3% on CNG, methane improved biogas and rough biogas, separately, in light of its change from CI to SI mode. The methane improved biogas showed comparative engine performance as of CNG with respect to brake power (BP), specific fuel consumption (SFC) and brake thermal efficiency (BTH) [4]. Sunmeet Singh Kalsi and colleagues presented that "pongamia pinaata biodiesel (B100) was utilized as pilot fuel though biogas structure" was simulated by mixing of CO_2 (30, 40% and half by weight) in natural gas (NG) and utilized as the main fuel. BTH of the engine reduces with an increase in CO_2 content in flammable gas. NOx and smoke emissions diminished altogether with the biogas-fuelled engine though HC and CO emissions increases [5] (Fig. 48.3).

The indicated thermal efficiency (ITH) was processed utilizing Eq. (48.1):

$$\text{ITH} = \frac{3600\text{P}}{\text{mdiesel} \times \text{H}\mu\text{diesel} + \text{mbioCNG} \times \text{H}\mu\text{bioCNG}} \tag{48.1}$$

where P represents for indicated power, m_{diesel} is the mass of petroleum diesel and m_{biocng} is the mass of bioCNG.

The m_{diesel} was processed utilizing Eq. (48.2):

$$\text{mdual} = \text{mdiesel} + \frac{\text{H}\mu\text{biocng} * \text{mcng}}{\text{H}\mu\text{diesel}}. \tag{48.2}$$

48.3 Literature Review of Dual Fuel Engine Characteristics

Xiangyu Meng et al. experimentally investigated that on a volume basis, 80% diesel and 20% n-butanol as pilot fuel leads to a decrease in both NOx and CO emissions as compared with neat diesel. At medium load condition, the blend of pilot fuel and n-butanol can improve indicated thermal efficiency (ITH) for low CNG replacement rates. There is a decrease in CO emission but an increment in NOx emissions. It was found that THC emissions are more delicate to the CNG replacement rate [7]. Abhishek Paul and colleagues experimentally investigated that diesel ethanol mix D95E5 (95% diesel 5% ethanol) with use of low proportion of CNG creates a superior emission characteristic as compared with base diesel operation and diesel-ethanol mix operation. Results depicted ethanol's potential in diminishing NOx emission, brake specific energy consumption (BSEC) and smoke opacity [8]. Arat, Hüseyin Turan and colleagues experimentally investigated that 30HCNG fuel blend with half diesel substitution rate influence the engine performance as torque, BP and BSFC value were diminished 4.3, 8, 26.8% individually as well as NOx and CO_2 emissions were diminished by 47.2, 16.7% separately while 8% increase in CO emission [9]. Olivier Bordelannea and colleagues presented that utilization of crude biogas can be conceived in a CNG engine for lean combustion. In such a case natural gas can be mixed with up to 70% volume of crude biogas [10]. Probir Kumar Bose and colleagues experimentally investigated that at 80% load condition, NOx level for hydrogen advancement without exhaust gas recirculation (EGR) is 1211 ppm while with 20% EGR NOx level is 710 ppm as well as a decrease in NOx emissions due to decrease in peak ignition temperature [11]. Kraipat Cheenkachorn and colleagues presented that the greatest part of NG in the double fuel engine operation was up to 77.90% at 1300 rpm. Double fuel activity indicated less SFC and BTH [12]. Sunmeet Singh and colleagues experimentally investigated that BTE of the engine improved up to 64% of HCNG energy share. Share of HCNG energy is less as compared with base CNG.NOx emissions increases with HCNG however it is less than base biodiesel [13]. Kyunghyun Ryu experimentally found that indicated mean effective pressure (IMEP) of biodiesel-CNG duel fuel combustion (DFC) mode is lower than that of diesel single fuel ignition (SFI) mode at higher infusion pressure. smoke pollutants were decreased at all working conditions and NOx emissions additionally displayed decreasing pattern aside from full load engine operation in biodiesel–CNG DFC. DFC yields higher CO emissions as compared with diesel SFI at all engine operation aside from 100% burden [17]. Euijoon Shim and colleagues presented that optimization strategies like throttle and hot-EGR caused HC and CO emissions reduction, improvement in combustion efficiency and enhancing combustion stability while fulfilling NOx and PM spreads under EU-VI guidelines [18]. X. Meng et al. and colleagues experimentally found that at 7.5 bar IMEP, B10CNG60 and B20CNG60 caused improvement in ITH and lower the THC emissions, while keeping same degree of NOx discharge comparative with the cases at the 80% CNG replacement rate [20].

It is critical to take note of that indigenous ability to create natural gas vehicles has empowered huge scope to gaseous fuel. As the fuel replacement system extended the business needed to react to take into account the developing business sector. The business, particularly the transport business, needed to adjust and indigenize the innovation rapidly. Industry reaction anyway shifts across vehicle portions. More industries are reacted more emphatically to the item sections that had administrative directs for transformation like transports, three-wheelers, cabs and light business vehicles. This comprehensive review gives a forthcoming assessment to the researchers examining the methods for improving the emission and engine characteristics by utilizing bioCNG in dual-fuel diesel engine (Table 48.1).

48.4 Summary [1–37]

Considering the above-written review, it was found by all the researchers that bioCNG can be utilized in diesel and diesel/biofuel engine with little or minor engine modification. The engines performance, ignition and emissions characteristics in generalized form can be summarized in the following points:

1. Proper injection strategy, injection pressure of bioCNG in a diesel and diesel/biofuel engine operation results in decrease in ignition delay period as observed higher during biodiesel blend compared to diesel fuel.
2. Low-temperature combustion and lower hydrogen to carbon ratio, high amount of energy is transferred to the exhaust ultimately reduce power output. To overcome this problem, preheat the fuel (diesel and diesel/biofuel blend) using exhaust gas.
3. Proper control of fuel injection timing, injection pressure as well as premixed burning of pilot fuel leads to peak in-chamber pressure.
4. In most of the cases the use of bioCNG with pilot fuel (diesel and diesel/biofuel blend) results in greater heat release rate.
5. BSFC is higher in case of biodiesel blends with respect to diesel fuel. It declines with the increase in engine speed and brake power. Lower heating value and higher density incremented BSFC.
6. BTE in some cases is higher and in some cases lower, which eventually depends on the biofuel used. The essential reason for lower BTE is higher viscosity, density than diesel.
7. Increasing engine load corresponds to an increase in NOx emissions but with use of bioCNG in pilot fuel (diesel and diesel/biodiesel blends) lowered it at all load conditions as compared to diesel fuel.

Table 48.1 Summary of experimental Investigation by various researchers

Author	Details about equipment's used	Summary of engine characteristics		
		Performance	Emission	Combustion
Reza Mahmoodi et al. [1]	Single cylinder CI engine	–	Reduction of 50% in NOx emission With increase in RBG above 40%, Soot and Unburned hydrocarbon emissions slightly decreased	With increase in RBG substitution ratio above 0.43, reduction in mean combustion temperature In-chamber pressure increases to 22.63%
Lijiang Wei et al. [2]	Detailed review	BTH of double fuel mode is lower at low and medium engine loads while under high burden conditions, it is comparative or somewhat higher contrasted with normal diesel mode	Diminishing the NO_X, CO_2 and particulate matter emissions, yet the HC and CO increases	In-chamber pressure is lower Ignition delay is higher
Manjunath Channappagoudra et al. [3]	Modified CI engine	Increase in brake thermal efficiency by 21.29% for 0.48 kg/hr Bio-CNG flow rate using B20+Bio-CNG5 (In maximum load condition)	Decrease in HC and CO emission to 21.66 and 10.78%, respectively	Increase in chamber pressure and heat release rate (HRR) by 18.17 and 19.22%, respectively

(continued)

Table 48.1 (continued)

Author	Details about equipment's used	Summary of engine characteristics		
		Performance	Emission	Combustion
R Chandra et al. [4]	Modified engine (5.9 kW)	Brake thermal efficiency is highest for methane enriched biogas as compared to biogas and specific fuel consumption increases as compared to CNG	–	–
Sumeet Singh Kalsi et al. [5]	Bio-diesel fueled diesel engine	Reduction in brake thermal efficiency (BTH) of the engine with an increase in proportion of CO_2 in natural gas	NOx and smoke decrease for biogas-fueled engine Slightly increase in value of HC and CO emissions	–
Devendra Pratap Singh et al. [6]	A Review	–	Mustard oil is a promising new biofuel for NOx reduction. and inophyllum is completely non-edible and trans-esterified oil shows similar engine performance and emission characteristics like other biofuels	–

(continued)

Table 48.1 (continued)

Author	Details about equipment's used	Summary of engine characteristics		
		Performance	Emission	Combustion
Xiangyu Meng et al. [7]	AVL 5402 diesel engine	Increase in thermal efficiency for 90% diesel and 10% n-butanol blend	Decrease in NOx and CO emission for 20% n-butanol in blend	–
Abhishek Paul et al. [8]	4-stroke, single cylinder CI engine	95% diesel and 5% ethanol blend showed better performance as compared to base diesel and diesel ethanol blend	Reduction in NOx Decrease in brake-specific energy consumption and smoke opacity	–
Arat,Hüseyin Turan et al. [9]	3.6 L, 4-cylinders, four-stroke, water-cooled compression ignition engine	Better performance characteristics at 30 H-CNG fuel mixture with 50% diesel	CO emission increases by 8% Reduction in NOx and CO_2 to 47.2 and 16.7%, respectively	–
Olivier Bordelannea et al. [10]	Hybrid CNG vehicle	–	Reduction in emission level by 87% for biomethane fuelled hybrid vehicle	–
Probir Kumar Bose et al. [11]	5.2 kW 4-stroke single cylinder vertical diesel engine	Increase in BTH by 12.9% without exhaust gas recirculation for a mass flow rate of 0.15 kg/h and decreases with EGR	At 80% load NOx value decreases from 1211 to 710 ppm (with 20% EGR.) HC emission decreases by 57.69% without EGR. and it became increases with EGR	–

(continued)

Table 48.1 (continued)

Author	Details about equipment's used	Summary of engine characteristics		
		Performance	Emission	Combustion
Kraipat Cheenkachorn et al. [12]	6-cylinder inline water-cooled diesel engine	–	CO and HC emissions increases while NOx and CO_2 decreases	–
Sunmeet Singh Kalsi et al. [13]	7.4 kW single cylinder diesel engine	BTH increases to 64%	CO and HC emission decreases for a blend of HCNG. and B100	–
R. Mohsin et al. [14]	6-cylinder direct injection diesel engine	Torque and power became higher for a blend of B20-CNG	Increase the CO emission by (15–32%) and NOx emission by (6.67–7.03%) Reduce UBHC (5.76–6.25%) and carbon dioxide (0.47–0.58%)	–
Cheolwoong Park et al. [15]	6-cylinder natural gas engine	–	CO and HC emission increased significantly while NOx emission decreases	–
Amin Paykani et al. [16]	Modified heavy duty CI engine	Improvement in performance	Results indicate that HC and CO emissions increases Reduction in soot and NOx emissions	–

(continued)

Table 48.1 (continued)

Author	Details about equipment's used	Summary of engine characteristics		
		Performance	Emission	Combustion
Kyunghyun Ryu [17]	Single cylinder diesel engine	–	Higher CO emission at all load except full load conditions	Combustion stability is high at higher injection pressure IMEP is lower as compared to single diesel fuel mode
Euijoon Shim et al. [18]	Heavy-duty single cylinder engine	–	Increase in CO and HC emission and decrease with optimum injection strategy	–
Euijoon Shim [19]	Single cylinder heavy duty diesel engine	Attained indicated thermal efficiency (ITH) of 45.3%	It is found that all ACTs lead to decrease in the NOx and PM emissions to achieve EU-VI	–
X. Meng et al. [20]	4-stroke single cylinder diesel engine	B10CNG60 and B20CNG60 showed improvement in ITH	Decrease in HC emission as compared to 80% substitution of CNG	–
Ibrahim Elgarhi et al. [21]	4-stroke diesel engine	With an introduction of HHO gas, BTH increases and BSFC decreases	–	–

(continued)

Table 48.1 (continued)

Author	Details about equipment's used	Summary of engine characteristics		
		Performance	Emission	Combustion
M. Mofijur et al. [22]	Detailed study	–	Biodiesel–diesel and ethanol–biodiesel–diesel blending (5 to 10% ethanol and 20–25% biodiesel) correspond to a decrease in CO, HC and PM emissions	–
Dori Yuvenda et al. [23]	CNG-diesel DF engine	Retarding IT would result decrease in CO, HC and NOx emission	Chamber pressure and heat release rate for injection timing (130° After TDC)	–
Murugan Rangasamy et al. [24]	Modified automotive diesel engine	Increase in BTH for dual fuel RCCI engine as compared to CDC and lower exergy loss in M/BD blend as compared to blend of M/D	-	–
Fatemeh Sadat Mirhashemi et al. [25]	Comprehensive review	–	Decrease in NOx emission with blending of biofuel through proper control on injection parameter and blending proportions	–

(continued)

Table 48.1 (continued)

Author	Details about equipment's used	Summary of engine characteristics		
		Performance	Emission	Combustion
M. Krishnamoorthi et al. [26]	A single cylinder, 4-stroke water cooled compression ignition engine	Higher BTH about 22% for Syngas/diesel blending mode with an injection duration of 19° bTDC	77% lower oxides of nitrogen (NOx) and PM respectively as compared to conventional diesel	–
Jinwen You et al. [27]	6 cylinder turbocharged heavy duty engine	EGR using proper intake throttling cause improvement in BTH	Decrease in emissions for diesel/natural gas DF engine at low load	–
Ganesh Duraisamy et al. [28]	3-cylinder, turbocharged, CRDI, diesel engine	At a premixed mass part of 80%, Highest brake BTH is about 31% for both Methanol/Diesel and Methanol/PODE blend which is 3.5% higher than the regular diesel burning	BSNOx and ash emissions are decreased for both double fuel RCCI ignition with an increase in methanol mass portion	–

(continued)

Table 48.1 (continued)

Author	Details about equipment's used	Summary of engine characteristics		
		Performance	Emission	Combustion
Hanyu Chen et al. [29]	A 6-cylinder prototype marine engine	It is shown from performance characteristics that brake specific fuel consumption is higher and brake thermal efficiency is lower in light hydrocarbons PCCI engine contrasted with diesel engine	Results indicate that there is decrease in CO and NOx emissions although THC emission is higher with respect to diesel Engine	In-chamber peak pressure of light hydrocarbons PCCI engine is decreased by 18.1% and HRR presents a consistent variation trend
Manjunath Channappagoudra et al. [30]	Kirloskar, 3.5 kW (TV1) diesel engine	It was found that 29. deg.bTDC IT has presented improved performance for dual fuel engine	Results showed that emission characteristics were better for 29 bTDC IT as compared to infusion timings 20, 23, 26 and 32. deg.bTDC	Ignition characteristics is better for 29.deg.bTDC IT
Mohd Hafizil Mat Yasin et al. [31]	–	Results indicate that there is an increase in BSFC as well as slight variation in BP and torque between B5 and diesel due to minute effect of biodiesel properties in B5	Decrease in CO, UHC and NOx emissions is obtained for B5	–

(continued)

Table 48.1 (continued)

Author	Details about equipment's used	Summary of engine characteristics		
		Performance	Emission	Combustion
Md. Hasan Ali et al. [32]	–	–	Lower emissions of UHC, CO though higher level of NOx is obtained	–
A.M. Liaquat et al. [33]	1-cylinder 4-stroke diesel engine	Increase in BSFC for biodiesel blended fuels for entire speed range as compared to neat diesel fuel	CB5 and CB15 would correspond to the reduction in CO and HC emissions whereas, CO_2 and NOx emissions were increased as compared with conventional diesel	
Byungchul Choia et al. [34]	Turbocharged CRDI diesel engine	–	THC and CO emissions increased as compare with D100 mode at the low engine load conditions NOx emissions increased for a blend of n-butanol/diesel	–
Gaurav Paul et al. [35]	2-cylinder DI diesel engine	It was found that the use of jatropha as a biodiesel would result in an increase in BSFC whereas BTH decreases	Emissions like NOx and Carbon dioxide increases while smoke and PM emissions decreases	An increase in maximum chamber pressure and decrease in IT with an increase in biodiesel blends share
I Wayan Surata et al. [36]	–	To enhance the performance of the engine, 20% LPG was admitted to 80% of biogas	–	–

(continued)

Table 48.1 (continued)

Author	Details about equipment's used	Summary of engine characteristics		
		Performance	Emission	Combustion
M.Shahabuddinet al. [37]	Conventional diesel engine	Blend, B20+1% results in 1.73% and 9% hike in brake power as compared with B20 and OD, respectively Blend, B20+1% show 26% and 6% decrease in SFC as compare with fuel B20 and OD	Blend, B20+1% caused reduction in CO, NOx and CO_2 emissions as compared with other fuels	–

48.5 Conclusions

Utilization of bioCNG in modified diesel engines requires several factors to be considered. Following conclusions could be drawn from the present study:

1. HC and CO emissions increase for diesel-bioCNG engine operation. From the literature survey, it is found that if oxygenated fuel is used and blend with diesel, reduction in both emissions occurs.
2. Use of biofuel blend caused slight penalty on performance in terms of BTE and BSFC due to low calorific value of biofuel. Comparable performance is achieved with use of diesel-bioCNG engine operation for low substitution ratio and high engine load condition.
3. Particulate matter and soot emissions and smoke opacity decrease up to a certain extent in biofuel blends.
4. bioCNG with lower substitution ratio of diesel with biofuel can be utilized with little or minor engine modifications.

References

1. Mahmoodi, R., Yari, M., Ghafouri, J., Poorghasemi, K.: Effect of reformed biogas as a low reactivity fuel on performance and emissions of a RCCI engine with reformed biogas/diesel dual-fuel combustion (2020)
2. Wei, L., Geng, P.: A review on natural gas/diesel dual fuel combustion, emissions, and performance. Fuel Process. Technol. **142**, 264–278 (2016)
3. Channappagoudra, M.: Comparative study of baseline and modified engine performance operated with dairy scum biodiesel and Bio-CNG. Renew. Energy (2019)
4. Chandra, R., Vijay, V.K., Subbarao, P.M.V., Khura, T.K.: Performance evaluation of a constant speed IC engine on CNG, methane enriched biogas and biogas. Appl. Energy **88**, 3969–3977 (2011)
5. Kalsi, S.S., Subramanian, K.A.: Effect of simulated biogas on performance, combustion and emissions characteristics of a bio-diesel fueled diesel engine. Renew. Energy (2017). https://doi.org/10.1016/j.renene.2017.01.006
6. Sanjid, A., Masjuki, H.H., Kalam, M.A., Rahman, S.M.A., Abedin, M.J., Palash, S.M.: Impact of palm, mustard, waste cooking oil and Calophyllum nophyllum biofuels on performance and emission of CI engine. Renew. Sustain. Energy Rev. 27, 664–682 (2013)
7. Meng, X., Zhou, Y., Yang, T., Long, W., Bi, M., Tian, J., Lee, F.F., Chia, : An experimental investigation of a dual-fuel engine by using bio-fuel as the additive. Renew. Energy **147**, 2238–2249 (2020)
8. Paul, A., Bose, P.K., Panua, R.S., Banerjee, R.: An experimental investigation of performance-emission trade off of a CI engine fueled by diesel compressed natural gas (CNG) combination and diesel ethanol blends with CNG enrichment. Energy **55**, 787–802 (2013)
9. Arat, H.T., Baltacioglu, M.K., Aydin, K., Özcanli, M.: Experimental investigation of using 30HCNG fuel mixture on a non-modified diesel engine operated with various diesel replacement rates. Int. J. Hydrog. Energy 3199–3207 (2016)
10. Bordelannea, O., Monteroa, M., Bravina, F., Prieur-Vernata, A., Selmia, O.O., Pierrea, H., Papadopoulob, M., Muller, T.: Biomethane CNG hybrid: A reduction by more than 80% of the greenhouse gases emissions compared to gasoline. J. Nat. Gas Sci. Eng. **3**, 617–624 (2011)

11. Bose, P.K., Maji, D.: An experimental investigation on engine performance and emissions of a single cylinder diesel engine using hydrogen as inducted fuel and diesel as injected fuel with exhaust gas recirculation. Int. J. Hydrog. Energy **34**, 4847–4854 (2009)
12. Cheenkachorn, K.: Poompipatpong, Chedthawut, Ho, Choi Gyeung, performance and emissions of a heavy-duty diesel engine fuelled with diesel and LNG (liquid natural gas). Energy **53**, 52–57 (2013)
13. Kalsi, S.S., Subramanian, K.A.: Experimental investigations of effects of hydrogen blended CNG on performance,combustion and emissions characteristics of a biodiesel fueled reactivity-controlled compression ignition engine (RCCI). Int. J. Hydrog. Energy 1–13 (2017)
14. Mohsin, R., Majid, Z.A., Shihnan, A.H., Nasri, N.S., Sharer, Z.: Effect of biodiesel blends on engine performance and exhaust emission for diesel dual fuel engine. Energy Convers. Manage. **88**, 821–828 (2014)
15. Park, C., Kim, C., Lee, S., Lee, S., Lee, J.: Comparative evaluation of performance and emissions of CNG engine for heavy-duty vehicles fueled with various caloric natural gases. Energy **174**, 1–9 (2019)
16. Paykani, A., Kakaee, A.H., Rahnama, P., Reitz, R.D.: Effects of diesel injection strategy on natural gas/diesel reactivity-controlled compression ignition combustion. Energy 1–13 (2015)
17. Ryu, K.: Effects of pilot injection pressure on the combustion and emissions characteristics in a diesel engine using biodiesel–CNG dual fuel. Energy Convers. Manage. **76**, 506–516 (2013)
18. Shim, E., Park, H., Bae, C.: Intake air strategy for low HC and CO emissions in dual-fuel (CNG-diesel) premixed charge compression ignition engine. Appl. Energy **225**, 1068–1077 (2018)
19. Shim, E., Park, H., Bae, C.: Comparisons of advanced combustion technologies (HCCI, PCCI, and dual-fuel PCCI) on engine performance and emission characteristics in a heavy-duty diesel engine. Fuel **262**, 116436 (2020)
20. Meng, X., Tian, H., Long, W., Zhou, Y., Bi, M., Tian, J., Lee, C.F.F.: Experimental study of using additive in the pilot fuel on the performance and emission trade-offs in the diesel/CNG (methane emulated) dual-fuel combustion mode. Appl. Therm. Eng. (2019). https://doi.org/10.1016/j.applthermaleng.2019.113718
21. Elgarhi, I., El-Kassaby, M.M., Yehia, A.: Eldrainy, enhancing compression ignition engine performance using biodiesel/diesel blends and HHO gas. Int. J. Hydrog. Energy **45**(46), 25409–25425 (2020)
22. Mofijur, M., Rasul, M.G., Hyde, J.: Recent developments on internal combustion engine performance and emissions fuelled with biodiesel-diesel-ethanol blends. Procedia Eng. **105**, 658–664 (2015)
23. Yuvenda, D., Sudarmanta, B., Wahjudi, A., Muraza, O.: Improved combustion performances and lowered emissions of CNG-diesel dual fuel engine under low load by optimizing CNG injection parameters. Fuel **269**, 117202 (2020)
24. Rangasamy, M., Duraisamy, G., Govindan, N.: A comprehensive parametric, energy and exergy analysis for oxygenated biofuels based dual-fuel combustion in an automotive light duty diesel engine. Fuel **277,** 118167 (2020)
25. Mirhashemi, F.S.: Sadrnia, Hassan, N_Ox emissions of compression ignition engines fueled with various biodiesel blends: a review. J. Energy Inst. **93**(1), 129–151 (2020)
26. Krishnamoorthi, M., Sreedhara, S., Duvvuri, P.P.: Experimental, numerical and exergy analyses of a dual fuel combustion engine fuelled with syngas and biodiesel/diesel blends. Appl. Energy **263**, 114643 (2020)
27. You, J., Liu, Z., Wang, Z., Wang, D., Xu, Y., Du, G., Fu, X.:The exhausted gas recirculation improved brake thermal efficiency and combustion characteristics under different intake throttling conditions of a diesel/natural gas dual fuel engine at low loads. Fuel **266**, 117035 (2020)
28. Duraisamy, G., Rangasamy, M., Govindan, N.: A comparative study on methanol/diesel and methanol/PODE dual fuel RCCI combustion in an automotive diesel engine. Renew. Energy **145**, 542–556 (2020)

29. Chen, H., Wang, X., Pan, Z.: A comprehensive study of fuel composition, combustion and soot nanostructure characteristics of a diesel/light hydrocarbons premixed charge compression ignition engine. Fuel **274**, 117858 (2020)
30. Channappagoudra, M.: Effect of injection timing on modified direct injection diesel engine performance operated with dairy scum biodiesel and Bio-CNG. Renew. Energy **147**, 1019–1032 (2020)
31. Yasin, M.H.M., Paruka, P., Mamat, R., Yusop, A.F., Najafi, G., Alias, A.: Effect of low proportion palm biodiesel blend on performance, combustion and emission characteristics of a diesel engine. Energy Procedia **75**, 92–98 (2015)
32. Ali, M.H., Mashud, M., Rubel, M.R., Ahmad, R.H.: Biodiesel from neem oil as an alternative fuel for diesel engine. Procedia Eng. **56**, 625–630 (2013)
33. Liaquat, A.M., Masjuki, H.H., Kalam, M.A., Fattah, I.R., Hazrat, M.A., Varman, M., Mofijur, M., Shahabuddin, M.: Effect of coconut biodiesel blended fuels on engine performance and emission characteristics. Procedia Eng. **56**, 583–590 (2013)
34. Choia, B., Kim, Y.K., Jung, G., Lee, C., Jiang, X.: Inchul, effect of diesel fuel blend with biobutanol on the emission of turbocharged CRDI diesel engine. Energy Procedia **61**, 2145–2148 (2014)
35. Paul, G., Datta, A., Mandal, B.K.: An experimental and numerical investigation of the performance, combustion and emission characteristics of a diesel engine fueled with jatropha biodiesel. Energy Procedia **52**, 626–632 (2014)
36. Surata, I.W., Nindhia, T.G.T., Atmika, I.K.A., Negara, D.N.K.P., Putra, I.W.E.P.: Simple conversion method from gasoline to biogas fueled small engine to powered electric generator. Energy Procedia **52**, 626–632 (2014)
37. Shahabuddina, M., Masjuki, H.H., Kalam, M.A., Mofijur, M., Hazrat, M.A., Liaquat, A.M., Effect of additive on performance of C.I. engine fueled with bio diesel. Energy Procedia **14**, 1624–1629 (2012)
38. Skill council for green jobs.: Future Job Roles in Operations and Maintenance of Bio-CNG Plants, Department of international development, UK government (2020)

Author Index

© The Editor(s) (if applicable) and The Author(s), under exclusive license to Springer Nature Singapore Pte Ltd. 2022

V. V. Rao et al. (eds.), *Computational and Experimental Methods in Mechanical Engineering*, Smart Innovation, Systems and Technologies 239,
https://doi.org/10.1007/978-981-16-2857-3